Reservoir Characterization

Academic Press Rapid Manuscript Reproduction

Reservoir Characterization

Edited by

Larry W. Lake
Department of Petroleum Engineering
University of Texas at Austin
Austin, Texas

Herbert B. Carroll, Jr.
National Institute for Petroleum and Energy Research
Bartlesville, Oklahoma

1986

ACADEMIC PRESS, INC.
Harcourt Brace Jovanovich, Publishers
Orlando San Diego New York Austin
London Montreal Sydney Tokyo Toronto

Artwork contained in the articles entitled Permeability Prediction From Well Logs Using Multiple Regression, by W. A. Wendt, S. Sakurai, and P. H. Nelson, Simulator Parameter Assignment and the Problem of Scale in Reservoir Engineering, by Helge H. Haldorsen, and How Heterogeneity Affects Oil Recovery, by K. J. Weber, reprinted by copyright permission of the Society of Petroleum Engineers. The authors gratefully acknowledge the use of this material.

ACADEMIC PRESS, INC.
Orlando, Florida 32887

United Kingdom Edition published by
ACADEMIC PRESS INC. (LONDON) LTD.
24–28 Oval Road, London NW1 7DX

Library of Congress Cataloging in Publication Data
Main entry under title:

Reservoir characterization.

"Proceedings of the Reservoir Characterization Technical Conference, held April 29-May 1, 1985, at the Westin Hotel–Galleria in Dallas"–Pref.
Includes index.
1. Reservoirs–Congresses. 2. Engineering geology–Congresses. I. Lake, Larry W. II. Carroll, Herbert B. III. Reservoir Characterization Technical Conference (1985 : Dallas, Tex.)
TC1167.R47 1986 622'.3382 85-48294
ISBN 0–12–434065–2 (alk. paper)

PRINTED IN THE UNITED STATES OF AMERICA

86 87 88 89 9 8 7 6 5 4 3 2 1

Contents

Contributors

Numbers in parentheses indicate the pages on which the authors' contributions begin.

N. A. Abdelmalek (393), *Gulf Research and Development, Houston, Texas 77701*

David J. Bouquet (39), *Exxon Corporation, Lafayette, Louisiana 70502*

Robert E. Bretz (585), *New Mexico Petroleum Recovery Research Center, New Mexico Institute of Mining and Technology, Socorro, New Mexico 87801*

Herbert B. Carroll, Jr. (643), *National Institute of Petroleum and Energy Research, Bartlesville, Oklahoma 74005*

D. M. Chang (373, 445), *Sohio Petroleum Company, San Francisco, California 94101*

T. R. Clifton (63), *Developmental Geosciences Department, Sohio Petroleum Company, San Francisco, California 94101*

Olivier Dubrule (223), *Sohio Petroleum Company, Dallas, Texas 75208*

Ronald A. Evenson (39), *Department of Geology and Tertiary Oil Recovery Project, University of Kansas, Lawrence, Kansas 66045*

Robert J. Finley (1), *Bureau of Economic Geology, The University of Texas at Austin, Austin, Texas 78712*

Michael L. Fowler (341), *Cities Service Oil and Gas Corporation, Tulsa, Oklahoma 74101*

G. W. Geehan (63), *Developmental Geosciences Department, Sohio Petroleum Company, San Francisco, California 94101*

J. R. Gochnour (393), *Gulf Research and Development, Houston, Texas 77701*

Allan L. Gutjahr (265), *Department of Mathematics, New Mexico Institute of Mining and Technology, Socorro, New Mexico 87801*

H. H. Haldorsen (223, 293, 373, 445), *Sohio Petroleum Company, San Francisco, California 94101*

Charles L. Hearn (341), *Cities Service Oil and Gas Corporation, Tulsa, Oklahoma 74101*

Thomas A. Hewett (83), *Chevron Oil Field Research Company, La Habra, California 90631*

John P. Hobson (341), *Cities Service Oil and Gas Corporation, Tulsa, Oklahoma 74101*

K. F. Inman (63), *Developmental Geosciences Department, Sohio Petroleum Company, San Francisco, California 94101*

Andre G. Journel (249), *Department of Applied Earth Sciences, Stanford University, Stanford, California 94305*

H. Klob (63), *Developmental Geosciences Department, Sohio Petroleum Company, San Francisco, California 94101*

Clement Kostov (249), *Department of Applied Earth Sciences, Stanford University, Stanford, California 94305*

Larry W. Lake (545, 643), *Department of Petroleum Engineering, The University of Texas at Austin, Austin, Texas 78712*

Thomas J. Lasseter (545), *Schlumberger-Doll Research, Old Quarry Road, Ridgefield, Connecticut 06877-4108*

T. F. Lawton (63), *Developmental Geosciences Department, Sohio Petroleum Company, San Francisco, California 94101*

Boris L. Litvak (561), *Texaco Inc., Houston, Texas 77701*

John R. MacMillan (265), *Geoscience Department, New Mexico Institute of Mining and Technology, Socorro, New Mexico 87801*

P. H. Nelson (181), *Sohio Petroleum Company, Dallas, Texas 75208*

K. E. Nitzberg (63), *Developmental Geosciences Department, Sohio Petroleum Company, San Francisco, California 94101*

Franklin M. Orr, Jr. (585), *Department of Petroleum Engineering, Stanford University, Stanford, California 94305*

C. Ozgen (373), *Sohio Petroleum Company, San Francisco, California 94101*

Douglas H. Rofheart (39), *Mobil Oil Corporation, New Orleans, Louisiana 70183*

J. J. Rosenzweig (393), *Gulf Research and Development, Houston, Texas 77701*

S. Sakurai (63, 181), *Developmental Geosciences Department, Sohio Petroleum Company, San Francisco, California 94101*

Robert M. Specter (585), *Department of Finance, College of Business Administration, Arizona State University, Tempe, Arizona 85281*

Fred I. Stalkup (141), *Arco Resources Technology, Atlantic Richfield Company, Plano, Texas 75086*

Noel Tyler (1), *Bureau of Economic Geology, The University of Texas at Austin, Austin, Texas 78712*

John R. Waggoner (545), *Department of Petroleum Engineering, University of Texas at Austin, Austin, Texas 78712*

Anthony W. Walton (39), *Department of Geology and Tertiary Oil Recovery Project, University of Kansas, Lawrence, Kansas 66045*

K. J. Weber (487), *Shell International Petroleum Company, The Hague, The Netherlands 2501AN*

W. A. Wendt (181), *Sohio Alaska Petroleum Company, Anchorage, Alaska 99506*

Marvin D. Woody (39), *Marathon Oil Corporation, Casper, Wyoming 82602*

Preface

This book contains the proceedings of the Reservoir Characterization Technical Conference held April 29–May 1, 1985, at the Westin Hotel—Galleria in Dallas. The conference was sponsored by the National Institute for Petroleum and Energy Research (NIPER), Bartlesville, Oklahoma. This preface details some of the motivations for the conference.

At the first meeting of the organizing committee for this conference, we tried to establish what exactly "reservoir characterization" was. This committee, composed of several knowledgeable individuals with diverse backgrounds, surprisingly agreed upon essentially the same definition. Reservoir characterization is a process for quantitatively assigning reservoir properties, recognizing geologic information and uncertainties in spatial variability. The immediate application of the process is to numerical simulation, although the committee felt that the information so obtained has intrinsic merit. Since reservoir characterization remains very broad, the committee endorsed the idea of a conference to bring the variety of disciplines together.

The conference was organized into four technical sessions each with an invited keynote speaker: Geological and Petrophysical Basis: Dr. Robert J. Finley, University of Texas Bureau of Economic Geology; Spatial Variability: Dr. John Wilson, New Mexico Institute of Mining and Technology; Simulator Parameter Assignment: Dr. Helge H. Haldorsen, Sohio Petroleum Company; and Fluid Flow in Heterogeneous Media: Dr. K. J. Weber, Shell Internationale Petroleum.

Each session contained presentations based on prepared manuscripts which, in edited form, appear here. The conference contained a workshop session whose proceedings also appear here. There were more than 130 attendees from all over the world.

No endeavor of this magnitude succeeds without the help of an industrious committee. Our thanks to the organizing committee for their labors and insights: Robert J. Finley, Bureau of Economic Geology, University of Texas; Helge H. Haldorsen, Sohio Petroleum Company; Thomas A. Hewett, Chevron Oilfield Research Company; Thomas J. Lasseter, Schlumberger–Doll Research; Larry E. Murray, Union Oil Company; A. C. (Sandy) Riese, Atlantic Richfield

Company; Bijon Sharma, NIPER; Fred I. Stalkup, ARCO Resources Technology; Anthony W. Walton, University of Kansas; and J. Keith Westhusing, U.S. Department of Energy.

In addition, we are most indebted to Ms. Debra Kerr Fassnacht and Ms. Virginia Foreman, NIPER, for their organizational, editorial, and management skills. The efficiency with which the conference proceeded is a tribute to their energy and good cheer.

Larry W. Lake
Austin, Texas

Herbert B. Carroll, Jr.
Bartlesville, Oklahoma

GEOLOGICAL CHARACTERIZATION OF SANDSTONE RESERVOIRS[1]

Robert J. Finley
Noel Tyler

Bureau of Economic Geology
The University of Texas at Austin
Austin, Texas, U.S.A.

I. INTRODUCTION

Current efforts to maximize oil production from mature hydrocarbon provinces in the lower 48 States are becoming increasingly dependent on geological reservoir characterization. Experience in Texas for the period 1973-1982 underscores the need for greater insight into reservoir characterization in existing fields. These are the reservoirs for which we have the greatest volume of data. During the 1973-1982 period, new-field wildcat oil discoveries accounted for 11 percent, or 650 million barrels, of reserve additions, whereas reserve growth from existing fields accounted for 89 percent, or 5.4 billion barrels, of reserve additions (Fisher, 1985). Extension and infill drilling accounted for more than four-fifths of this in-field reserve growth. However, the largest part of the in-field reserve growth potential is yet to be realized in Texas, which, as a major oil province, has historically accounted for 40 percent of the production of crude oil in the United States.

[1] *Publication authorized by the Director, Bureau of Economic Geology, The University of Texas at Austin.*

Of the 156 billion barrels of oil discovered in Texas, some 54 billion barrels are currently estimated to be ultimately recoverable (Fisher and Galloway, 1983). With the addition of an estimated 20 billion barrels of in-place oil remaining to be discovered and a typical recovery rate of a little less than 35 percent, at least 116 billion barrels of in-place oil is now classed as unrecoverable by conventional means. The volume of this oil that may ultimately be recovered is not known, but estimates range from 5 to 40 percent (Fisher and Galloway, 1983). The Texas situation is comparable to the national oil recovery outlook in that nationwide 323 billion barrels of oil, or two-thirds of the discovered oil, will remain in known reservoirs, according to data on current recovery technology (National Petroleum Council, 1984).

Although recovery efficiency for primary recovery and secondary recovery processes (principally waterflooding) is typically about one-third of the original oil in place (OOIP), recoveries from specific reservoirs can range from less than 5 percent up to 80 percent (National Petroleum Council, 1984). In Texas, average recovery efficiencies for major terrigenous clastic reservoirs vary from 15 percent (range of 6 to 21 percent) for slope/basin depositional systems to 53 percent (range of 38 to 69 percent) for barrier strandplain depositional systems (Tyler and others, 1984). Further variation occurs with drive mechanism in that the greatest recovery efficiency is associated with water drive (up to 80 percent for one deltaic sandstone), and the lowest efficiency (6 percent) is related to solution gas drive, as well as to low permeability, in the slope/basin depositional system (Tyler and others, 1984).

A. Facies and Depositional System Defined

In Texas, oil reserve growth in existing fields has been primarily due to extension and infill drilling, which often precede secondary recovery efforts. Lesser reserve growth has occurred through new-pool discovery, tertiary recovery processes, and delayed abandonment (Fisher, 1985). To realize the maximum hydrocarbon potential of these reservoirs, the geological reservoir framework must be thoroughly understood in terms of the facies present and their overall depositional system. A facies is a three-dimensional body of rock having an environmental origin that can be inferred from a set of characteristics including external geometry, internal geometry, sedimentary structures, lithology, organic content, stratigraphic relations, and associated sedimentary facies. An assemblage of facies, often termed lithogenetic facies in reference to the importance of rock type origin, that is linked by depositional environment and associated processes forms a depositional system (Fisher and McGowen, 1967). Associated occurrence of several depositional systems forms a systems tract (Brown and Fisher, 1977). Such a tract may include, for example, fluvial, deltaic, shelf, and slope depositional systems. In this example, coeval systems reflect a paleoslope from source area through basin margin to deep marine environments. Thus, expected mutual relationships between depositional systems can be defined to provide a regional setting within which studies of individual systems and their contained facies provide a guide to reservoir delineation within trends. This in turn provides a basis to extrapolate between trends of similar depositional origin.

B. Application of Facies Concepts to Reservoir Fluid Distribution

As a result of basic facies variations resulting from different modes of deposition and subsequent histories of burial, physical compaction, deformation, and cementation, all reservoirs, whether gas or liquid filled, show varying degrees of inhomogeneity in internal petrophysical properties that affect fluid distribution. Even with high-quality reservoirs of average to good permeability, sufficient heterogeneity exists to preclude complete drainage of fluids with conventional well spacings and recovery techniques. In reservoirs with a high degree of internal compartmentalization and geometric complexity, as much as half of the potentially recoverable oil commonly is not drained by primary and secondary recovery programs. Such geologic variations, which complicate the efficient exploitation of good reservoirs, are magnified in tight formations, in formations with less efficient drive mechanisms, or in formations with high-viscosity liquid hydrocarbons.

The study of modern depositional systems and their ancient counterparts has led to the development of models for major clastic depositional systems (Brown and Fisher, 1977; Selley, 1978; Fisher and Brown, 1984; Walker, 1984). Much of this work has been specifically aimed at aiding the exploration process. The importance of extending the process of geologic depositional modeling to include the smaller scale features of known reservoirs should not be overlooked, however, because of the increasing recognition of in-field reserve growth potential. Assignment of quantitative geological predictors to interwell areas is a relatively new endeavor for many reservoir geologists but is one that is critical to the reservoir engineer's analysis of fluid flow and hydrocarbon productivity. The following sections of this paper include descriptions of two major clastic depositional systems with illustrations of how facies within these systems have controlled hydrocarbon distribution. References will be made to important elements of interwell reservoir geology rather than to interfield exploration considerations.

II. AN APPROACH TO RESERVOIR ANALYSIS

The genetic approach to reservoir facies analysis involves an understanding of facies as a rock unit having distinctive characteristics and of depositional systems as an assemblage of facies having a related origin. The term facies relates to the aspect, or motif, of a rock body that encompasses lithological, structural, and organic elements that may be recognized in the field, or, less completely, be seen in core (Walker and Harms, 1971; Middleton, 1978; Walker, 1984). An environmental interpretation may be added to the pure description of the rock character itself to become part of the facies analysis process (Middleton, 1978). The distribution of facies having a related origin can be used in a predictive sense without detailed consideration of formal nomenclature and stratigraphic codes (Fisher and Brown, 1984). Thus, the genetic approach offers the reservoir geologist and engineer the best opportunity to characterize the interwell area using fundamental rock units that are likely to have some similarity from area to area. Without this expectation of similarity, rooted in the origin of the rock body, the reservoir geologist cannot make full use of studies of modern depositional facies and associated depositional processes, and of outcrop and subsurface studies aimed at key questions of reservoir geometry and distribution of petrophysical properties. More and more of these studies focus on the within-field, or interwell, scale of reservoir description and are an important extension of studies that have been heavily utilized in guiding hydrocarbon exploration.

The geological characterization of reservoirs requires full utilization of all types of data from the subsurface. These data may include geophysical well logs, cores (full diameter and sidewall), core analyses, mud logs, production histories, drill-stem and other test data, pressure data, and injection profiles. A physical framework for reservoir analysis is defined by the most abundant data (invariably well logs), and cores are carefully compared to the distribution of lithologies, porosity, and hydrocarbon-bearing reservoir rock indicated by the well logs. Interpretations made from slabbed, full-diameter core are carefully tested against inferred vertical and horizontal distributions of the same rock body seen on well logs, and all data taken together are compared to facies models. The model itself is defined by a general summary of a specific sedimentary environment (Walker, 1984) that allows satisfactory prediction from limited data. Use of the model in turn leads to its refinement; for reservoir characterization studies, better understanding of reservoir heterogeneity at the scale of typical well spacing is particularly needed.

III. GEOLOGICAL CHARACTERIZATION OF SHORE-ZONE RESERVOIRS

Shore-zone system tracts are composed of two principal components: strandplain systems and barrier island/lagoon systems. In the rock record, these deposits are an important source of oil. In Texas, for example, they contain approximately 12 percent of Texas' in-place oil resource in sandstones and display an above-average recovery efficiency of 50 percent (Tyler and others, 1984). Strandplains are marine-process-dominated depositional features welded onto coastal mainlands. In contrast, barrier islands are separated from the adjacent coastal plains by extensive lagoons or bays. Strandplains are further classed into two broad groups: sand-rich beach-ridge plains and mud-rich chenier plains.

Barrier islands, beach-ridge plains, and chenier plains are dominantly progradational features, shaped by the relations among sediment texture and rate of supply, coastal physiography (including slope), and wave and tidal energy. These systems can occur as a continuum, as exemplified by the microtidal regressive coastline of the Gulf Coastal

Plain west of the Mississippi River. From the spatial arrangement of mud- and sand-rich strandplains and barrier islands, their vertical superposition in the stratigraphic column can be predicted. Indeed, this is true for the Oligocene upper Frio Formation along the central Texas Gulf Coast, where a major component of this formation is composed of stacked barrier-lagoon, sand-rich strandplain, and mud-rich strandplain depositional systems. Forty-six major oil reservoirs in this play collectively contain over 4.2 billion barrels of oil, making this trend one of the richest of the Gulf Coast. Detailed studies of two reservoirs in the North Markham - North Bay City field in the Frio barrier island/strandplain play (fig. 1) point to considerable macroscopic heterogeneity even within sand-rich depositional systems (Tyler and Ambrose, in press).

A. Strandplain Reservoirs in the North Markham - North Bay City Field

The North Markham - North Bay City (NM-NBC) field, Matagorda County, Texas, produces oil and gas from multiple stacked barrier and strandplain sandstones of the upper Frio Formation (fig. 2). The trap is an elongate rollover anticline downdip of a growth fault. Two of the three principal oil reservoirs in the field display considerable macroscopic heterogeneity. The Cayce reservoir (fig. 2), which had produced 10 million barrels of oil by 1984, was deposited in a composite, sand-rich, fluviodeltaic/strandplain system, whereas the Cornelius reservoir, having a cumulative production of 17 million barrels of oil, was deposited in a mud-rich strandplain system (Tyler and Ambrose, in press). Both reservoirs are characterized by high porosity; the mud-rich Cornelius displays considerably lower average permeabilities than the sand-rich Cayce. Other pertinent properties are presented in table I.

B. Cayce Reservoir--A Sand-Rich Strandplain

1. Reservoir stratigraphy

Conventional sand-thickness and sand-percent mapping defined three strike-parallel sand-rich belts in the Cayce, as well as considerable variation in sand content over the crest of the anticline. A more useful tool for determining

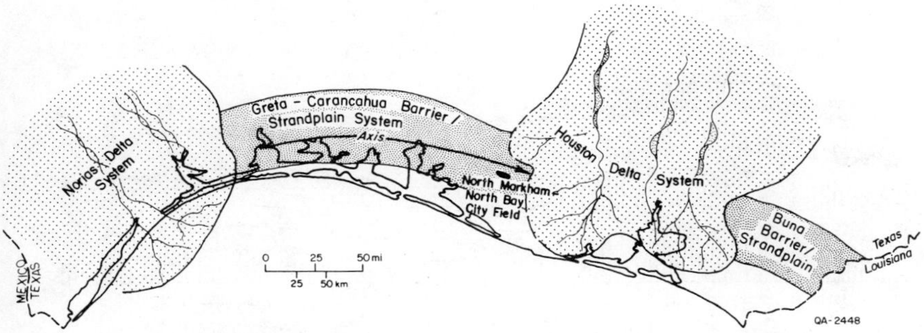

Figure 1. Frio depositional systems and location of the North Markham - North Bay City field.

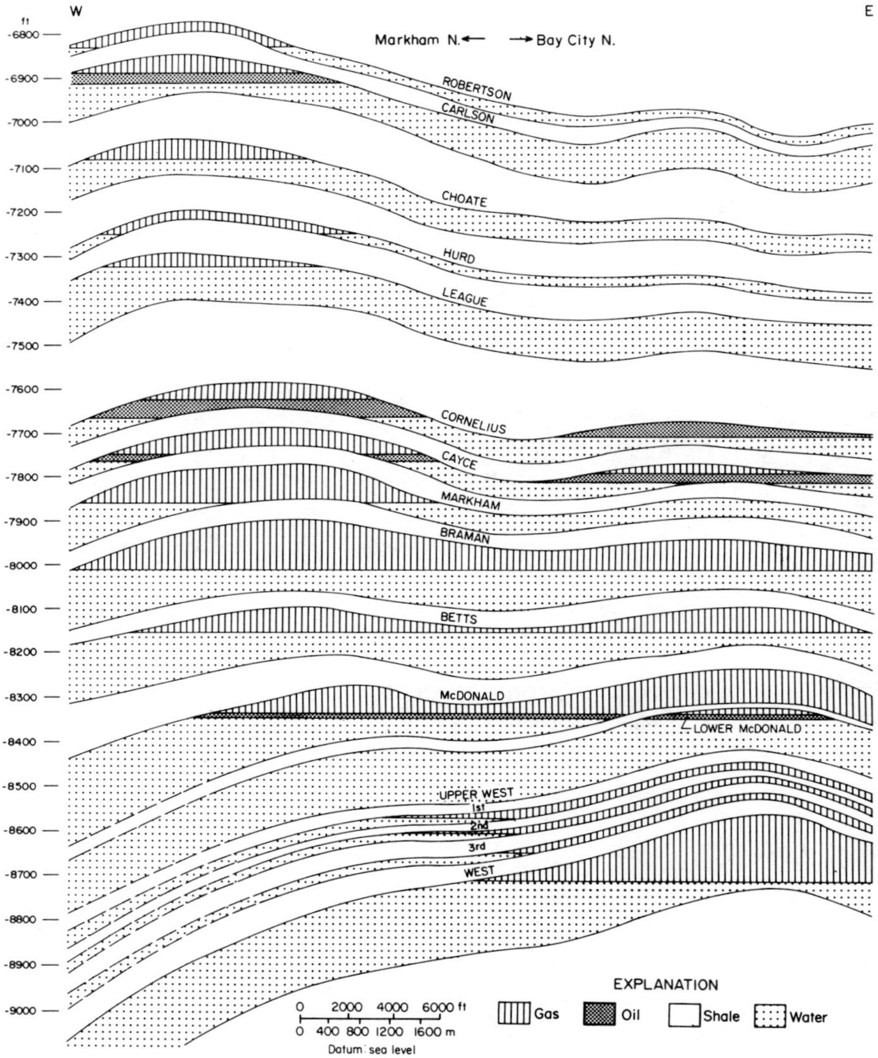

Figure 2. Generalized strike section across the North Markham - North Bay City field. Principal oil reservoirs are the Cayce, Cornelius, and Carlson. Modified from Marathon Oil Co., Railroad Commission of Texas Docket No. 3-23116, exhibit 2.

Table I. Geologic, fluid property, engineering, and production characteristics of the principal oil reservoirs of the North Markham - North Bay City field[a].

	Reservoir	
	Cornelius	Cayce
Drive	Water drive and gas-cap expansion	Water drive and gas-cap expansion
Depth (ft/m)	7,700/2,347	7,800/2,377
Oil column (ft/m)	40/12	20/6
Porosity (%)	26	28.5
Permeability (average md)	293	1,737
Permeability (log range)	2-3	2-3
Initial water saturation (%)	28	30
API gravity (°)	36	35
Initial gas-to-oil ratio (scf/STB)	640	628
Initial pressure (Psig)	3,572	3,623
Temperature (°F/°C)	193/89	196/91
Pressure maintenance	gas injection	gas injection
Unitization date	52	52
Well spacing (acres)	40	40
Original oil in place (million barrels)	36	15.7
Cumulative production (million barrels)	17.4	9.8
Ultimate recovery (million barrels)	22.0	10.4
Recovery efficiency (%)	61	66

[a] Data from hearing files at the Railroad Commission of Texas.

reservoir complexity was the lateral variation in spontaneous potential (SP) log character across the field area. In the western half of the reservoir, log patterns are commonly serrate; locally thin sandstones define upward-coarsening cycles (fig. 3). Upward-fining and blocky patterns, rarely present in the western half of the field, dominate the eastern half. Recognition of progradational and aggradational facies, as well as areas in which these two primary building blocks are mixed in varying proportions, from SP profiles (fig. 3) facilitated mapping of the architecture of component facies (fig. 4) of this composite sandstone body.

Progradational, aggradational, and mixed facies are, in turn, subdivided more specifically, as illustrated in the explanation in figure 4. Progradational and "blocky" SP facies are the framework of the Cayce sandstone, which is characterized over much of the area by variations of a blocky SP response; upward-coarsening log patterns are concentrated basinward. Transecting the suite of mixed and progradational facies is a north-south-oriented aggradational system consisting of simple and serrate upward-fining sandstones. Both sandstone-rich and mudstone-rich serrate log patterns are closely associated with the crosscutting system. Downdip, this aggradational facies becomes strike oriented, and the sands, although retaining their dominantly upward-fining character, become increasingly massive.

a. Interpretation. The massive ("blocky") sandstones that compose the bulk of the Cayce were deposited in a prograding sand-rich beach-ridge plain system. The Cayce beach-ridge plain was transected by a river that eroded and redeposited strandplain sediments as upward-fining fluvial sandstones. According to log facies data in areas of dense well control, the fluvial deposits are flanked by areally restricted sandstone-rich and sandstone-poor zones that display serrate SP responses and probably represent levee and floodplain deposits. The upper part of the river system consists of three discrete channels merging seaward into a single tract; these separate channels probably represent avulsed courses of the river. Remnant beach-ridge plain deposits separate the channels. Adjacent to the convergence zone of the channels is a local area of upward-coarsening sandstones. The restricted distribution and lobate geometry of these sandstones suggest deposition as a crevasse splay.

Downdip, the orientation of the aggradational system changes to strike-parallel (fig. 4) and the SP response becomes more blocky. Adjacent seaward deposits are

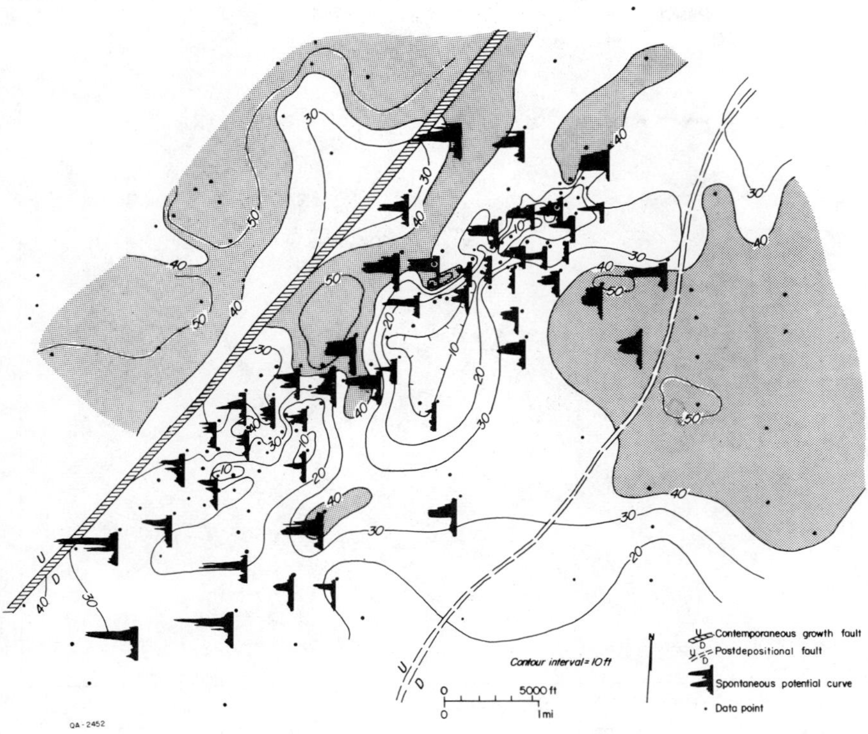

Figure 3. Log-character and net-sandstone map, Cayce reservoir.

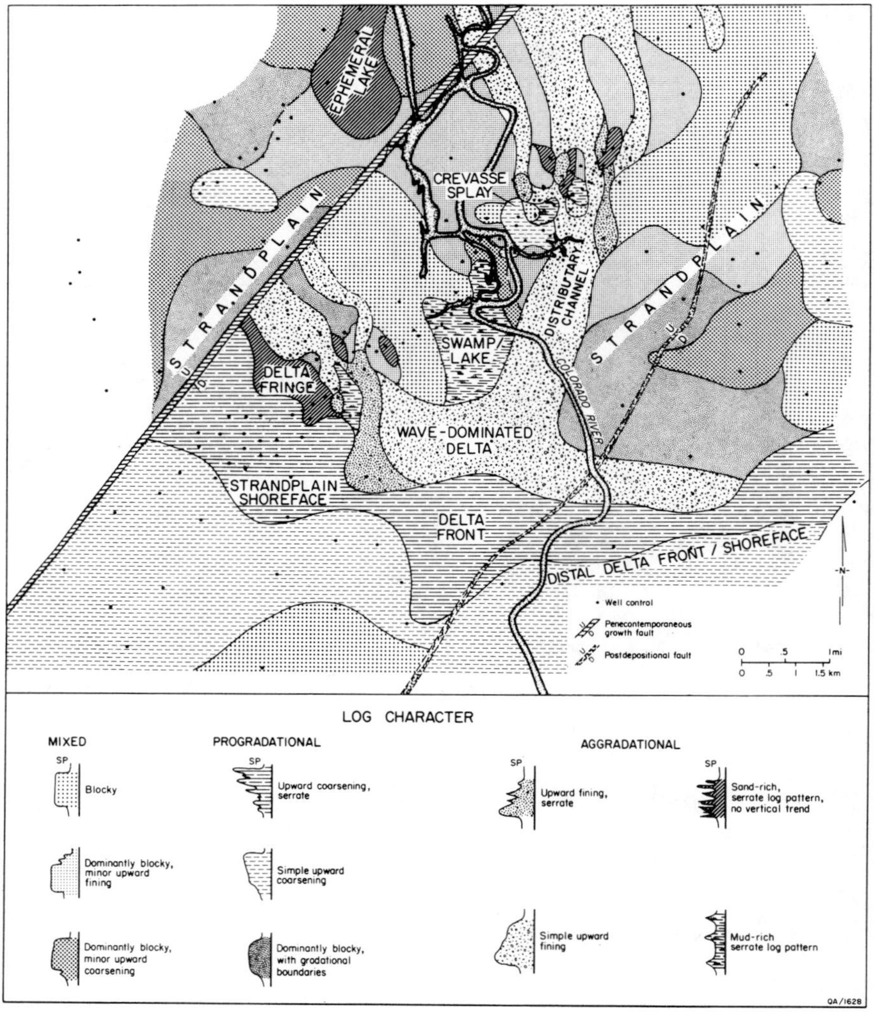

Figure 4. Facies anatomy of the Cayce sandstone adapted from vertical SP profiles shown on figure 3.

typically upward-coarsening, progradational shoreface or delta-front sandstones, or both. This system of strike-parallel aggradational-to-blocky sandstones merging seaward with progradational sandstones is interpreted to be a wave-reworked delta.

2. Reservoir continuity

The component facies of the Cayce sandstone directly control reservoir continuity, ultimate recovery efficiency, and fluid migration paths. The lateral continuity of the Cayce is well illustrated by cross sections on which the magnitude of the resistivity deflection is contoured. When considered in conjunction with log-facies maps, the resistivity sections also show the relation between sedimentary facies and the areal distribution of hydrocarbons. Predictably, reservoir continuity is greatest in beach-ridge deposits (fig. 5). The crosscutting fluvial complex disrupts the continuity of the reservoir and imparts considerable lateral heterogeneity. Even though the beach-ridge plain and channel deposits are both sandstone-rich, the interface between these facies, possibly composed partly of fine-grained levee deposits, acts as a flow boundary. Well-completion data show that remnant strandplain deposits isolated within the channel complex exhibit lower initial production than do the adjacent channel sandstones. For example, wells 71 and 74, which were both completed in remnant beach-ridge plain deposits, initially produced an average of 182 barrels per day, whereas well 72, completed in the channel sandstone separating the two beach-ridge remnants, yielded 225 barrels per day (fig. 5). The two facies clearly have different production characteristics. Without whole-core data, it is difficult to determine whether the differences are a result of permeability variation or of disparity in distribution of original oil in place. However, fluid flow between the two facies is clearly restricted.

The lower production from the sandstones of the remnant beach-ridge plain isolated in the fluvial complex is anomalous when compared to production from beach-ridge plain sandstones over the entire east Cayce reservoir. Average initial production from beach-ridge deposits of the east Cayce exceeds that for fluvial-channel sandstones by 50 barrels per day. (Production from beach-ridge plain sandstones averages 248 barrels per day; production from fluvial-channel sandstones averages 194 barrels per day.)

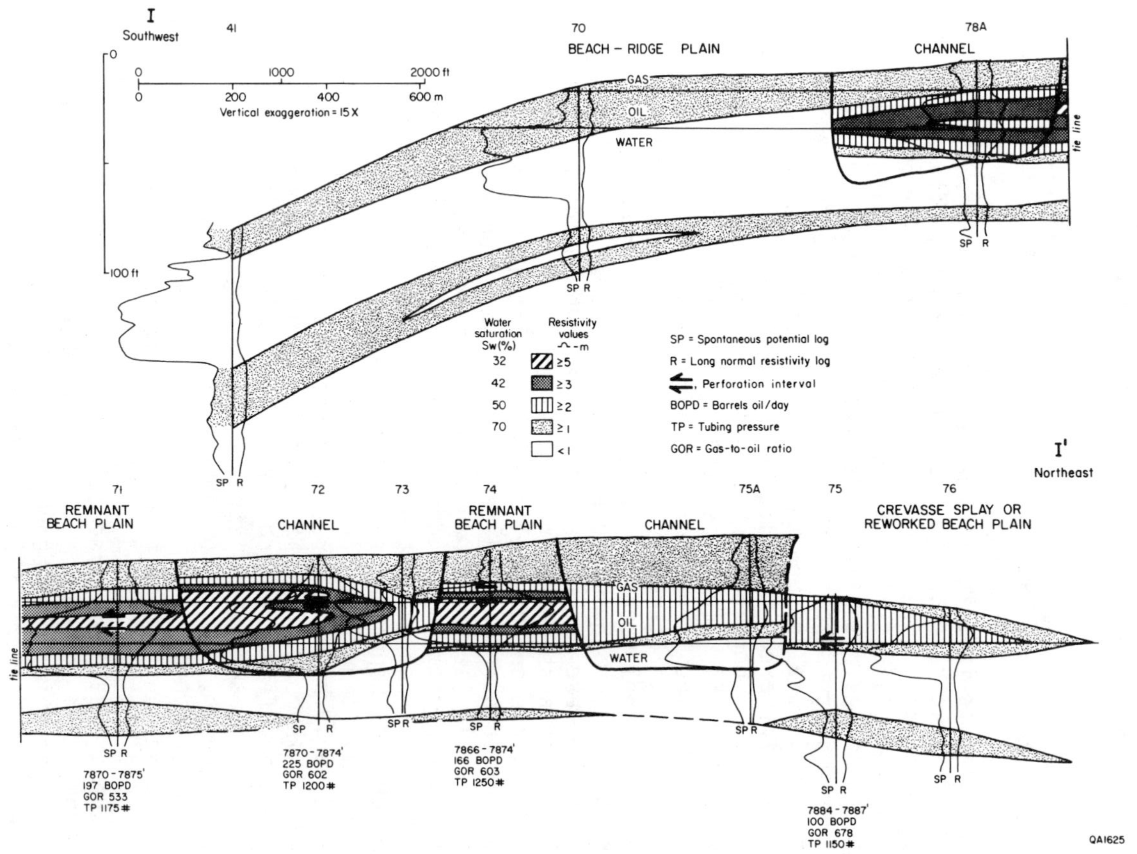

Figure 5. Structural resistivity cross section, Cayce reservoir, showing hydrocarbon continuity, or lack thereof, in component facies of this reservoir.

Production from the associated crevasse splay sandstones is the lowest of all Cayce sandstone facies, averaging 90 barrels per day.

3. Fluid migration paths

Reservoirs in the Frio Formation, particularly along the Central and East Texas Gulf Coast, are characterized by strong water drives. Natural water invasion in these reservoirs is frequently erratic and irregular, being controlled by the genetic architecture of the sediment. Defining preferential water-influx paths using water-cut maps is a critical step before undertaking secondary injection because the injected fluids will behave similarly to the primary reservoir fluids.

In the Cayce reservoir, water invasion followed two distinct modes. The fluvial sandstones focused early water invasion into the pool. As the channel sandstones cut directly across the reservoir, the resultant pattern was one of narrow fingers of high water cut reaching toward the crest of the structure. In contrast, the laterally continuous beach-ridge plain deposits displayed broad zones of uniform edgewater migration into the pool. Certain facies retard water invasion, most notably the interbedded sands and muds of the delta fringe, which are characterized by serrate SP patterns. Other facies displaying low water productions are levee and sand-poor abandoned-channel-fill and marsh deposits. In general, these argillaceous interbedded sand and mud facies are characterized by production with higher gas-to-oil ratios than those of the associated fluvial-channel and beach-ridge plain sandstones.

C. Cornelius Reservoir--A Mud-Rich Strandplain Complex

1. Reservoir stratigraphy

The sedimentary architecture of the Cornelius reservoir contrasts strongly with that of the directly underlying sand-rich Cayce (fig. 2). In the Cornelius, narrow strike-parallel bands of high sand content alternate with axes of much lower sand content (fig. 6). These alternating axes are on the order of 1,000 to 1,500 ft (304 to 456 m) wide and suggest a beach-ridge-and-swale paleotopography. Integration of sand mapping with log-facies mapping shows that the ridges and swales lie immediately landward of an upward-

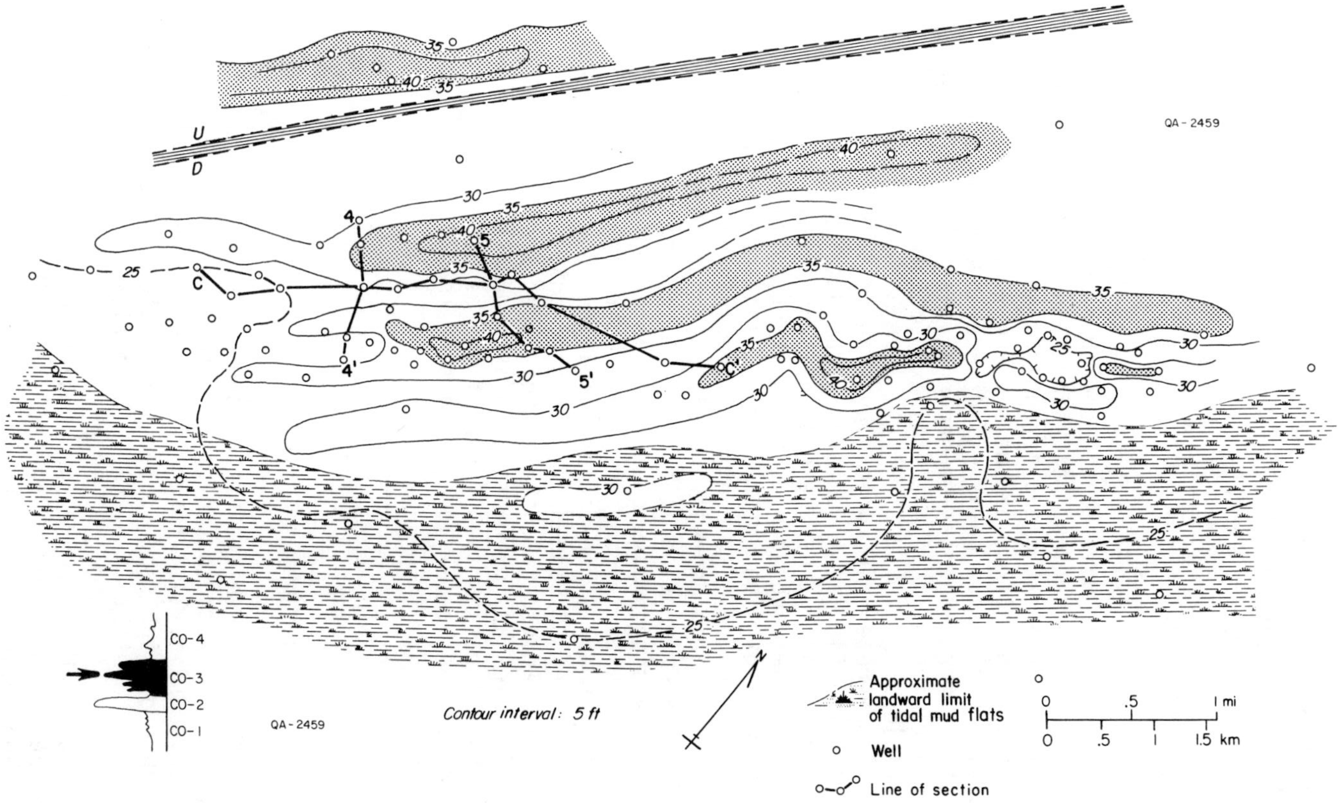

Figure 6. Detailed net-sandstone map of the third Cornelius sandstone (CO-3, see inset), illustrating an area of alternating sandstone thicks and thins that lie immediately landward of tidal mud-flat facies. Cross section C-C' is shown on figure 7.

fining tidal mud flat much like those forming today along the western Louisiana coastline. The Cornelius sandstone is therefore interpreted as a mud-rich strandplain system consisting of a beach-ridge and swale complex flanked seaward and landward by coastal mud flats.

2. Reservoir continuity

The continuity of mudstone-rich beach-ridge plain reservoirs varies greatly. Beach-ridge deposits exhibit great lateral continuity parallel to the shore zone. Normal to the shore zone, however, continuity is variable to poor. Resistivity cross sections (fig. 7) illustrate the impersistent and nonuniform distribution of hydrocarbons in ridge and interridge areas. Where strong ridge-and-swale paleotopography exists, the continuity between chenier and adjacent upward-fining interridge deposits is poor. Initial potential tests of wells perforated in ridge sandstones displayed high initial productions and pressures and low gas-to-oil production ratios, whereas wells in the muddier swale deposits were characterized by high gas-to-oil production ratios and low initial yields and pressures, regardless of location on the structure. Thus, mudstone-rich beach-ridge plain reservoirs can be characterized as composite bodies composed of a series of stringerlike subreservoirs that may or may not be isolated from adjacent bodies by intervening swales. Interbedded sandstones of washover origin in the swales contain hydrocarbons and furthermore may act as conduits between subreservoirs.

The well-defined thick and thin sandstone trends of the ridge-and-swale facies grade marginally into broad areas of similar sandstone content, indicating a widespread tidal sand flat on the flanks of the beach-ridge complex. Here, hydrocarbon distribution is uniform in the reservoir.

3. Fluid migration paths

As in the Cayce reservoir, the depositional fabric of the Cornelius strongly influenced reservoir drainage patterns, particularly during the first 30 years of production. Sequential water production maps illustrate ridges of high water cut separated by troughs of lower water production (fig. 8, 1950, 1960). The strike-parallel linear zones of high water influx conform to the axes of the beach ridges in the Cornelius sandstone, and troughs separating high

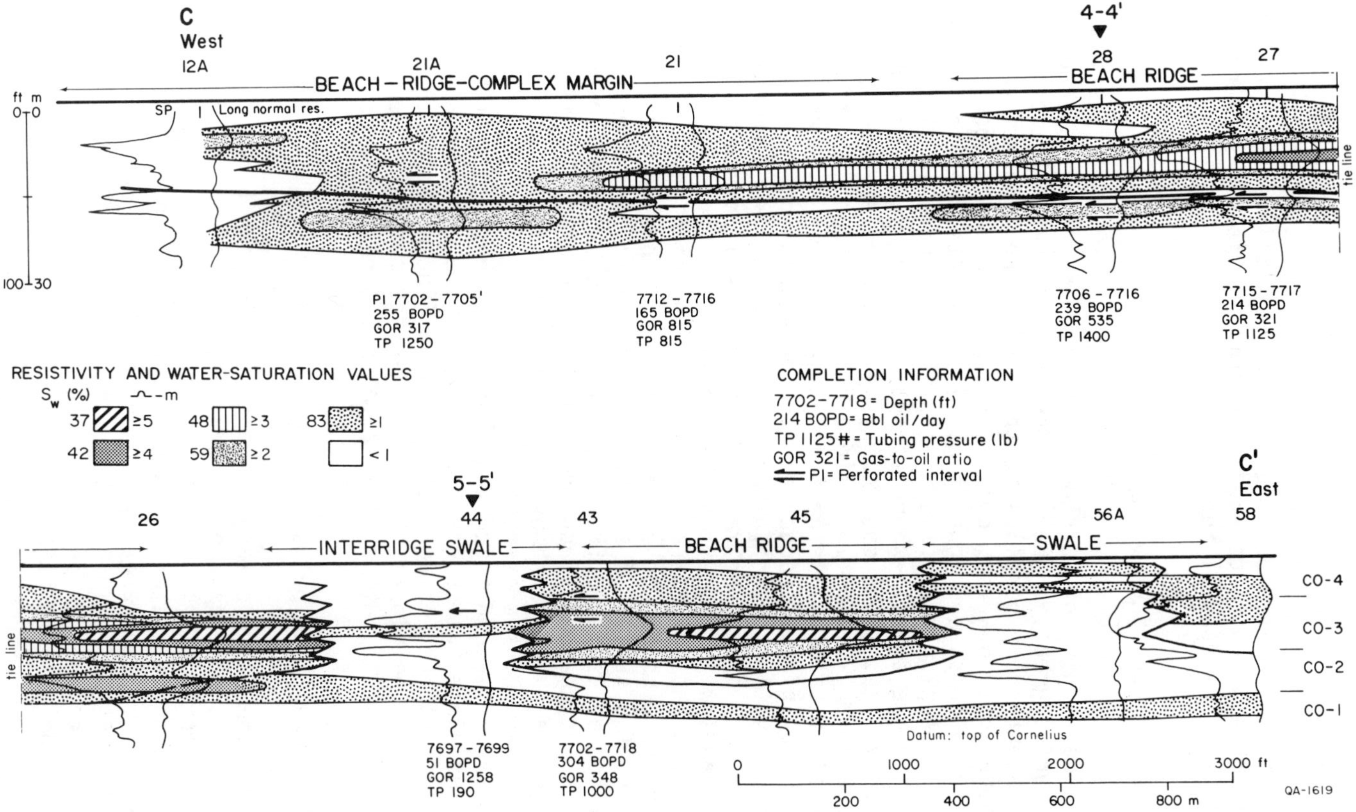

Figure 7. Resistivity section aligned obliquely to the strike of the Cornelius sandstone, showing that the zone of hydrocarbon impregnation in beach-ridge sandstones is interrupted by the intervening swales. Line of section shown in figure 6.

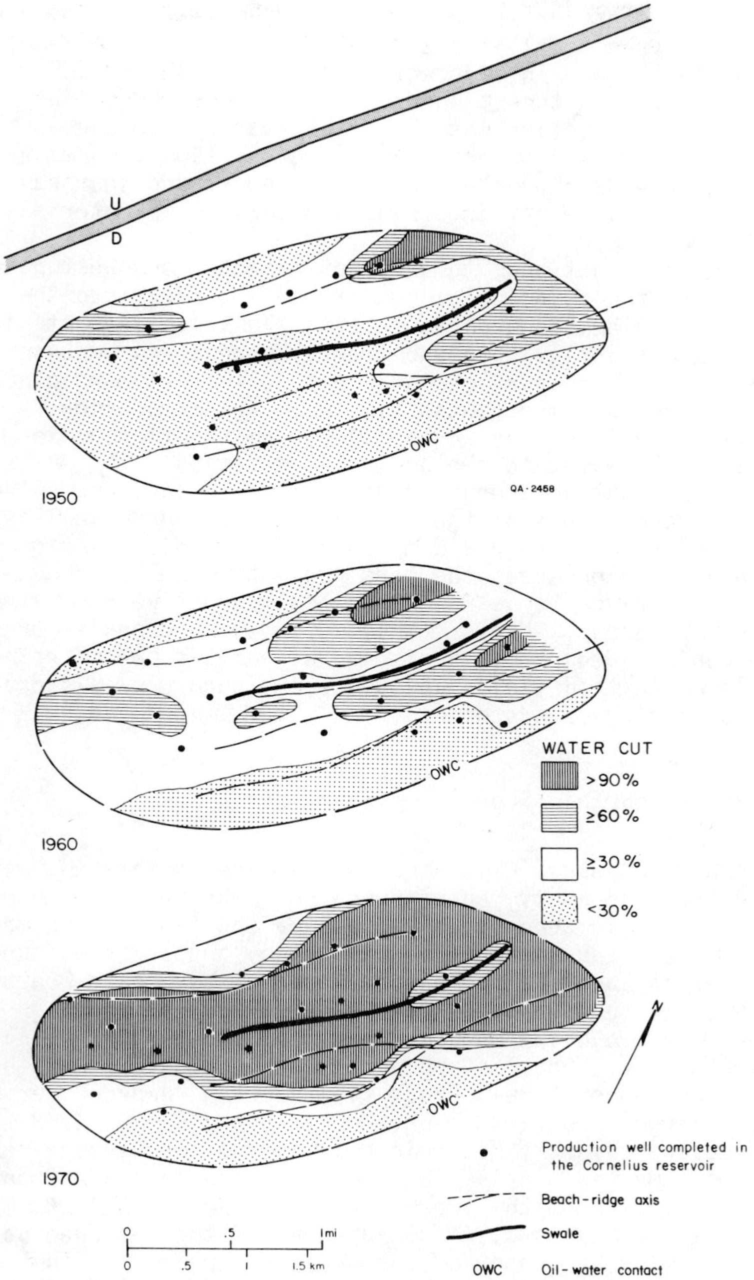

Figure 8. Sequential water-cut maps of west Cornelius showing the preferential influx of formation water in beach-ridge facies.

water-cut zones correspond to interridge swales. The production (hydrologic) gradient between producing wells can be extremely steep. In adjacent wells (only 660 ft [200 m] apart) in the northeast quadrant of the west Cornelius reservoir, production was 28 percent water from one and 95 percent water from the other (fig. 8, 1960). Similar steep gradients characterize the northwest and southeast margins of the reservoir during all three time intervals that were mapped.

The northeast and, to a lesser extent, southwest quadrants were the principal zones of water influx into the reservoir. The seaward (southeast) flank was characterized by low water influx throughout early production because that area of the reservoir lies adjacent to the landward pinch-out of the tidal mud flats (fig. 6). These mud-rich deposits functioned as an aquiclude, restricting updip migration of water to the beach-ridge facies.

Oil production trends in the Cornelius conform to the facies architecture of the reservoir. Contoured well-by-well yields are linear and match sandstone distribution patterns. In general, the more productive wells produce from beach-ridge facies and the less productive wells from interridge areas. Wells completed in swale deposits are more rapidly depleted and display higher gas-to-oil ratios and lower initial yields and pressures than do wells draining beach-ridge sandstones.

D. Rocky Mountain Shore-Zone Systems

Lower and Upper Cretaceous shore-zone systems are well known from the Rocky Mountain region. Many of these barrier and strandplain deposits originated along the western margin of the Cretaceous epicontinental seaway and are now found in a series of Laramide-age basins from New Mexico to Montana. The barrier sandstones of the upper Almond Formation (Mesaverde Group) at Patrick Draw field in the Green River Basin (Weimer, 1966; McCubbin and Brady, 1969) and of the Muddy Sandstone at Bell Creek field in the Powder River Basin (Berg and Davies, 1968; McGregor and Biggs, 1968) are well-described Rocky Mountain oil reservoirs in shore-zone systems. Gas is widely distributed in barrier and strandplain systems from the Upper Cretaceous Point Lookout, Cliff House, and Pictured Cliffs Sandstones of the San Juan Basin to the Almond Sandstone of the Green River Basin. The Corcoran and Cozzette Sandstones (Mesaverde Group) of the

Piceance Creek Basin, Colorado, are representative of Upper Cretaceous regressive shoreline sequences in the Rocky Mountain region and form low-permeability gas reservoirs.

E. Corcoran and Cozzette Sandstones

The Late Cretaceous Corcoran and Cozzette Sandstones, members of the Price River Formation, are fine to very fine grained blanket-geometry barrier and strandplain sandstones that trend northeast across the southern Piceance Creek Basin of Colorado. These sandstones form low-permeability gas reservoirs primarily within Garfield and Mesa Counties, Colorado. These sandstones are part of the widespread, highly gas productive Mesaverde Group although the gas resource within the Corcoran-Cozzette is not large (approximately 4 Tcf). The Mesaverde Group is a major regressive wedge of terrigenous clastics containing continental and shoreline sandstones and is present in several Rocky Mountain Basins. Within the Piceance Creek Basin the Corcoran and Cozzette are time-transgressive marginal marine sands that separate the marine Mancos Shale from the continental deposits that make up the bulk of the Mesaverde Group above the Rollins Sandstone.

1. Depositional systems

The Corcoran, Cozzette, and overlying Rollins Sandstones are generally considered marginal marine and of "beach and bar origin" (Dunn, 1974). Lorenz (1982) studied outcrops of these sandstones along the Grand Hogback at Rifle Gap, Colorado, and described these sandstones as blanket, wave-dominated shoreline deposits. An oyster bed overlying the Cozzette Sandstone at Rifle Gap indicates that the Cozzette may have built upward to form a barrier island backed by a brackish lagoon (Lorenz, 1982). Coal immediately overlying siltstones and mudstones in part of the Cozzette at Rifle Gap suggests that bay or lagoon environments filled with fine sediment to the extent that sufficient vegetative matter accumulated and coal ultimately formed. The Corcoran and Cozzette each consists of a lower, generally upward-coarsening sequence, the cleanest and thickest sandstone development being at the top of the interval, overlain by a more variable upper sequence that contains sandstones and intervening siltstones and mudstones (fig. 9). Sandstones in the upper sequence of each member show gamma-ray log

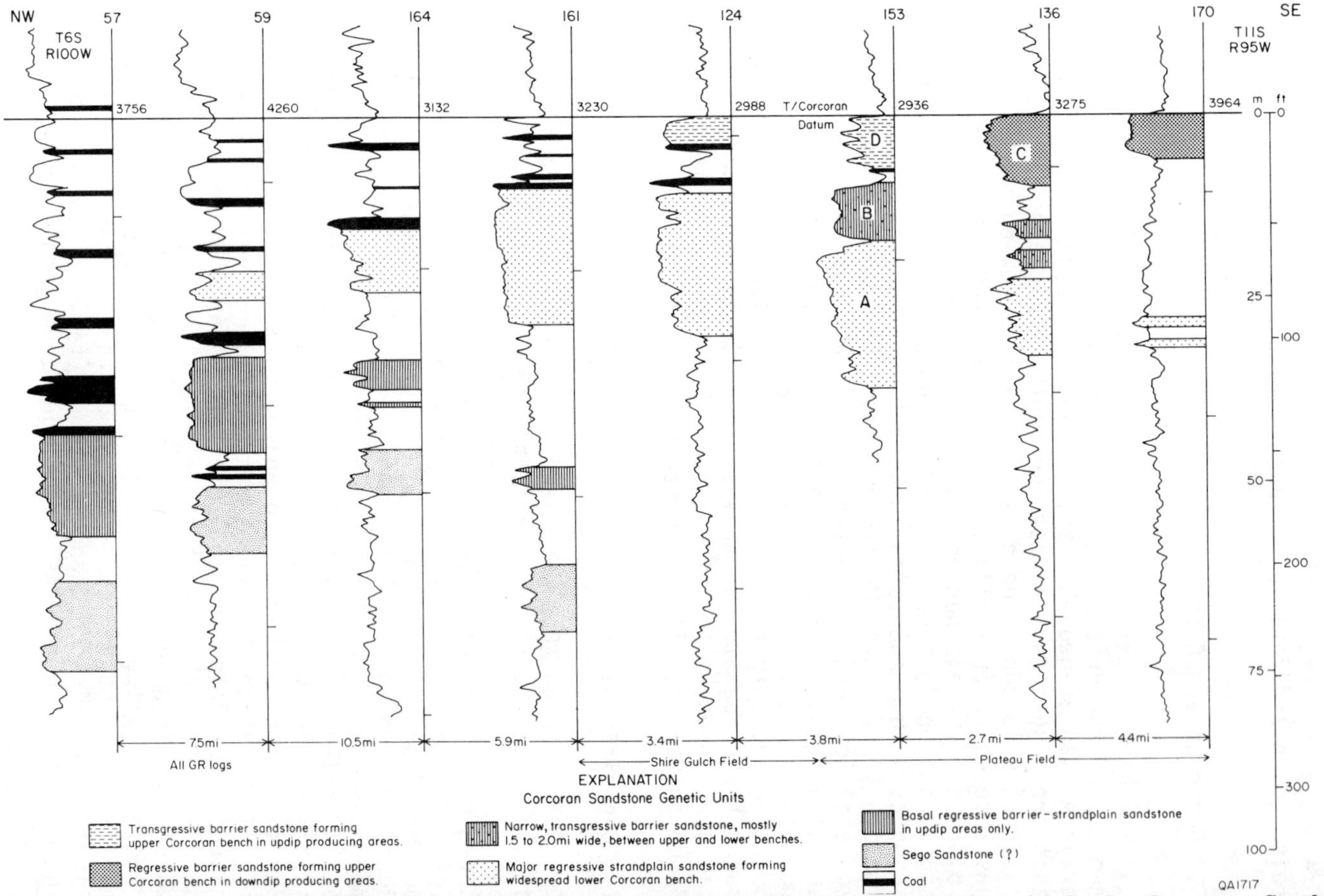

Figure 9. Cross section illustrating four genetic depositional barrier/strandplain units in the Corcoran Sandstone.

character that ranges from upward-coarsening to blocky and can be used to define genetic depositional units. The Corcoran and Cozzette therefore cannot each be regarded as only a sandstone unit; each member also includes mudstone, siltstone, and coal that form stratigraphic boundaries between discrete sandstone bodies, each of which represents a different depositional episode (Finley, 1985).

The Corcoran and Cozzette Sandstones actually contain seven discrete barrier and strandplain depositional units in Plateau and Shire Gulch fields, the major producing areas in the south-central Piceance Creek Basin. In the lower Corcoran, gamma-ray log facies show a mostly well defined upward-coarsening sequence having a variable thickness of blocky sandstone at the top (A, fig. 9). This laterally continous depositional unit is interpreted as a regressive strandplain. No lagoon fill is distinctly associated with this sandstone, and much of the preserved sand body is interpreted as lower and upper shoreface facies. The upper Corcoran consists of three vertically and areally distinct sandstones. Lowermost in the sequence is a transgressive barrier sandstone that is present as only a narrow belt on the updip side of Plateau field (B, fig. 9). A subsequent regressive barrier sandstone (C, fig. 9) is confined by marine shale both above and below in downdip (southeastern) depositional areas. The uppermost stratigraphic unit in the upper Corcoran (D, fig. 9) is in part defined by its abrupt termination depositionally updip (northwestward) and by its position between lagoon fill or distal coastal plain sediments of the upper Corcoran and marine shale of the lowest Cozzette.

The Cozzette Sandstone may be subdivided into three genetic sandstone units in the Shire Gulch - Plateau field area, the lowest of which is a basal regressive strandplain similar to that of the lower Corcoran (A, fig. 10). An upward-coarsening gamma ray log character is characteristic of this unit, and the preserved sand body is interpreted as being primarily lower shoreface. A regressive barrier or marine bar sandstone (B, fig. 10) merges with the top of the underlying strandplain updip but is separated from the latter downdip by marine shale. This unit is distinguishable only in limited areas. The uppermost unit in the Cozzette Sandstone (C, fig. 10) represents a regressive strandplain, according to the assumption that the thin shale interval below the sandstone body is marine rather than lagoonal. This assumption seems reasonable because this shale interval has fewer associated thin (lagoonal?) coals

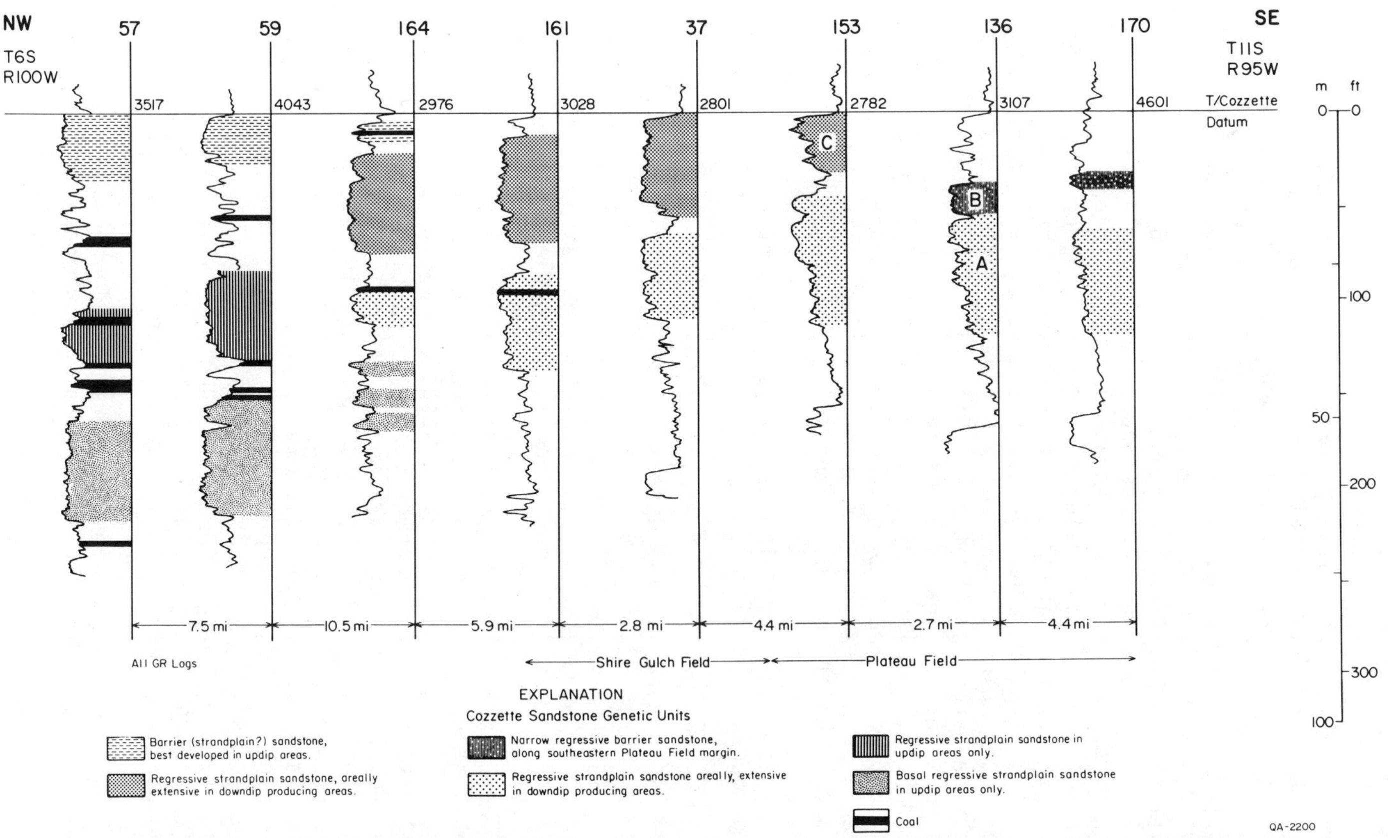

Figure 10. Cross section illustrating three genetic depositional barrier/strandplain units in the Cozzette Sandstone.

and sand splays than have probable lagoonal deposits in the Corcoran Sandstone.

Interpretations of depositional systems where adequate well control is available show, therefore, that individual stratigraphic units in the producing area, generally described as beach or bar sandstones, can be further subdivided into genetic depositional units. The correlation of hydrocarbon productivity is complicated, however, by the vertical stacking of as many as seven such units and the commingling of gas by operators from multiple units of the Corcoran and Cozzette, and in some cases also from the Rollins Sandstone. As a consequence of commingled production, the gas yield and production decline rates of a single genetic depositional unit are masked. Furthermore, in the case of tight gas reservoirs such as these sandstones, variations in completion practices, including the magnitude of the hydraulic fracture treatment, complicates the assessment of depositional heterogeneities from production histories. Production tests are rarely performed individually on reservoirs in the closely stacked vertical sequence, and fracture treatments are usually not isolated by zone when completing in the Corcoran and Cozzette. Initial potential flow (IPF) shows no consistent pattern by stratigraphic unit or combinations of units. A plot of total volume of proppant used in hydraulic fracture treatments compared to IPF gives little indication that greater gas productivity will be achieved by increasing the magnitude of the treatment, even though plotted data were segregated by sandstone member or combinations of members.

In view of these difficulties in reservoir evaluation, another method was required to relate hydrocarbon productivity, or potential productivity, to the genetic depositional unit within the Shire Gulch - Plateau field area. Although calculation of net pay is common practice in evaluating a given well for completion, field-wide calculation of net pay as a proportion of net sandstone has not been applied on a large scale. For tight gas sandstones, summation of net pay, despite the acknowledged difficulties in well log analysis of these tight sandstones (Kukal and others, 1983) offers a means of areally defining potential gas productivity; however, because of uncertainties in log analysis, the computed net pay is best considered potential net pay.

2. Net pay in Corcoran and Cozzette Sandstones

Potential net pay in each of the seven genetic depositional units in the Corcoran and Cozzette Sandstones was defined using limits of 7 percent or more for porosity and 70 percent or less for water saturation. Computations were made at 2-ft intervals for 55 wells, but not all wells contain all depositional units. Individual genetic depositional units show varying distributions of potential net pay related to their origin and hence offer different expectations of how the gas resource is distributed between different parts of a field.

The basal strandplains of the Corcoran and Cozzette are the most areally extensive genetic depositional units in each sandstone member. In the Corcoran, strandplain A is an upward-coarsening progradational unit for which potential net pay averages 63 percent of net sandstone (fig. 11). This strandplain has a gradational downdip boundary, and only 33 percent of net sandstone forms net pay below the 20-ft net pay contour. Basinward edges of such widespread basal strandplain units are therefore likely to be poor hydrocarbon producers because of greater silt and clay content.

Barrier sandstones in the Corcoran-Cozzette have a range of 72 to 87 percent of net sandstone as potential net pay. The highest value, 87 percent, occurs for the transgressive barrier unit B in the Corcoran, which shows sharp, nongradational upper and lower boundaries (fig. 12). This unit appears to have developed after a minor transgression of the distal edge of the underlying strandplain and a later relative sea-level stillstand and period of aggradation. Its restricted, linear distribution parallel to the inferred shoreline direction is typical of barrier sandstones. Reservoir quality is typically good to the margins of the unit because the barriers in both the Corcoran and Cozzette have developed on sandy strandplain platforms without extensive interfingering with offshore shales. An exception may be the depositionally downdip margin of a regressive barrier.

3. Resource distribution

Conventional approaches to understanding resource distribution within discrete depositional units of the Corcoran-Cozzette are not applicable because of commingling of production and variations in completion procedures.

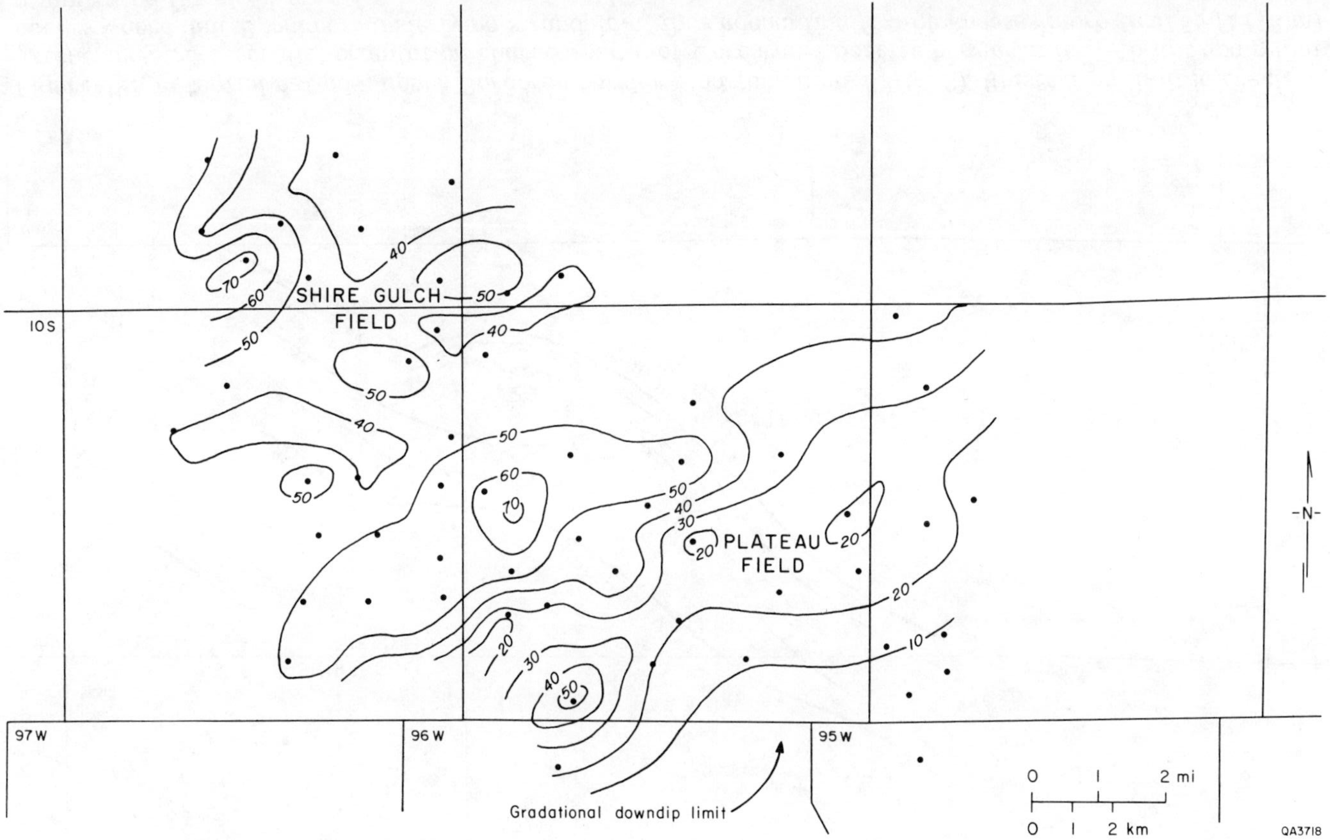

Figure 11. Potential net pay, lower Corcoran strandplain sandstone (unit A, fig. 9), illustrating relatively widespread sandstone distribution and gradational downdip limit. Contour interval is 10 ft (3 m).

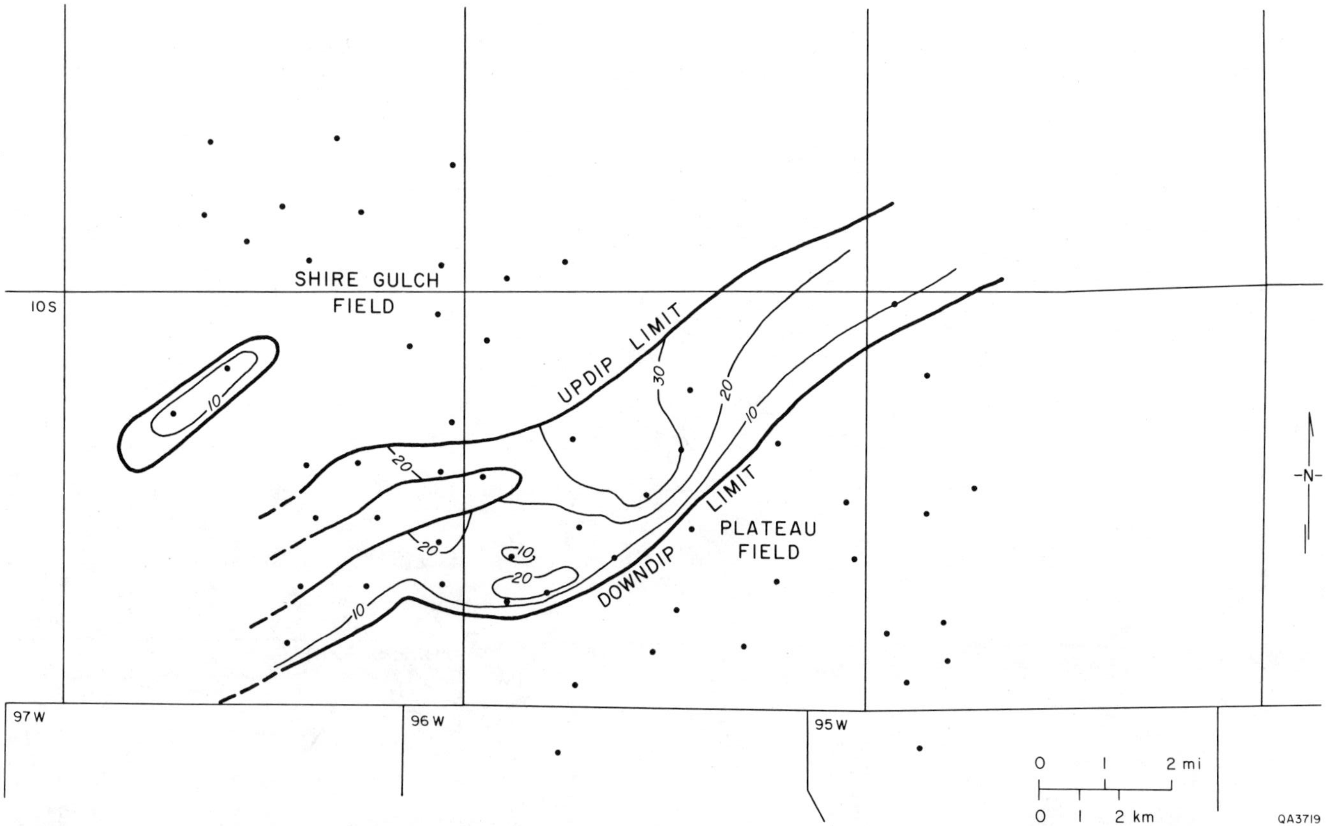

Figure 12. Potential net pay, upper Corcoran barrier sandstone (unit B, fig. 9), illustrating limited areal extent and strike-parallel orientation characteristic of Corcoran-Cozzette barrier units. Abrupt updip limit occurs where unit B welds to underlying strandplain, thus accounting for thicknesses more than 50 ft (15 m) of unit A on figure 11.

Instead, defining potential net pay in relation to net sandstone offers an alternative for understanding resource distribution. Although porosity and water saturation limits may be changed for net pay, patterns of distribution of potential net pay remain tied to the net-sandstone distribution. Potential net pay, when defined for each genetic depositional unit in a sandstone reservoir, provides a way of characterizing the areal distribution of the tight gas resource that may be extended to other reservoirs. This appears especially useful where production data cannot be definitively related to a single sandstone body because wells are completed in multiple depositional units.

IV. GEOLOGICAL CHARACTERIZATION OF DELTAIC RESERVOIRS

Sandstones of deltaic origin compose a most important class of oil reservoir. In Texas, for example, production from deltaic deposits will account for between 60 and 70 percent of all the oil recovered from sandstones at an ultimate recovery efficiency of 68 percent (Tyler and others, 1984). This far exceeds the average recovery from sandstones.

Deltaic reservoirs display a considerable but predictable range of recovery efficiencies. The three principal and related causes for the wide range of reservoir yields are reservoir genesis, diagenetic overprint, and drive mechanism. Dip-elongate, fluvial-dominated deltaic reservoirs historically display low to average recoveries, whereas laterally extensive wave-dominated deltas exhibit above-average recovery efficiencies. Large wave-modified deltas of the upper Texas Gulf Coast that produce by strong water drives are also highly productive. By contrast, one major deltaic play, the San Miguel - Olmos play of South Texas, in which reservoir sandstones were deposited in wave-modified to wave-dominated deltaic systems, displays anomalously low recovery efficiencies. The influence of the interrelated variables of diagenesis and drive mechanism is illustrated in the San Miguel - Olmos play, where intense diagenesis coupled with biogenetic modification of deltaic sandstones results in relatively tight reservoirs and weak solution-gas reservoir-drive mechanisms. One of the larger reservoirs in this play, the Big Wells (San Miguel), provides an example of the relation between megascopic reservoir heterogeneity and oil recovery.

A. The Big Wells (San Miguel)--A Wave-Dominated Deltaic Reservoir

The Big Wells field produces from one of a series of Late Cretaceous wave-dominated deltaic sandstones in South Texas. The trap at Big Wells results from updip porosity pinch-out in a deltaic sandstone, which grades landward from porous beach-ridge plain deposits into low-porosity coastal-plain sandstones and mudstones. Broad and gentle anticlinal warping of the sandstone about a dip-parallel axis imparts partial structural closure to the trap.

As is the Cornelius reservoir in the North Markham - North Bay City field, the Big Wells sandstone is a composite body composed of five genetic reservoir units. To understand the origin of the reservoir, it was necessary to subdivide the sandstone into genetic components and then to map sand distribution patterns for each genetic unit. The four lower sandstone units (lower stringers LS1-4) display an offlapping (progradational) geometry; in contrast, the fifth and uppermost sandstone thickens landward and was deposited during marine transgression. This fine-grained, argillaceous, and calcareous sandstone seals the reservoir.

Conventional sand mapping delineated the architecture of each genetic reservoir unit. For example, in the uppermost of the progradational units (LS1) a crosscutting distributary channel is flanked by a ridge-and-swale complex, which merges further downdrift with a second distributary/beach-ridge plain complex (fig. 13). Sand content decreases seaward into lower shoreface and shelf siltstones and mudstones. Detailed investigation of sedimentary facies displayed in whole cores showed that areas of thickly developed sandstone in the southern half of the field corresponded to upper shoreface and foreshore (beach) deposits. These sandstones are better sorted and more mature than the thinner and muddier sediments found closer to the northern distributary channel. The megascopic gradation from thick and mature beach and upper shoreface sediments in the south to thinner, less mature sediments in the north imparts a fundamental control on the engineering and recovery characteristics of the reservoir (Tyler and others, in press).

Two key parameters that control recovery from the pool are effective permeability to oil and induced-fracture half lengths. These values were estimated from pressure buildup analysis. The reservoir is characterized by uniformly low effective permeabilities in the range from less than 1 to 27 md. The southern half of the reservoir is characterized

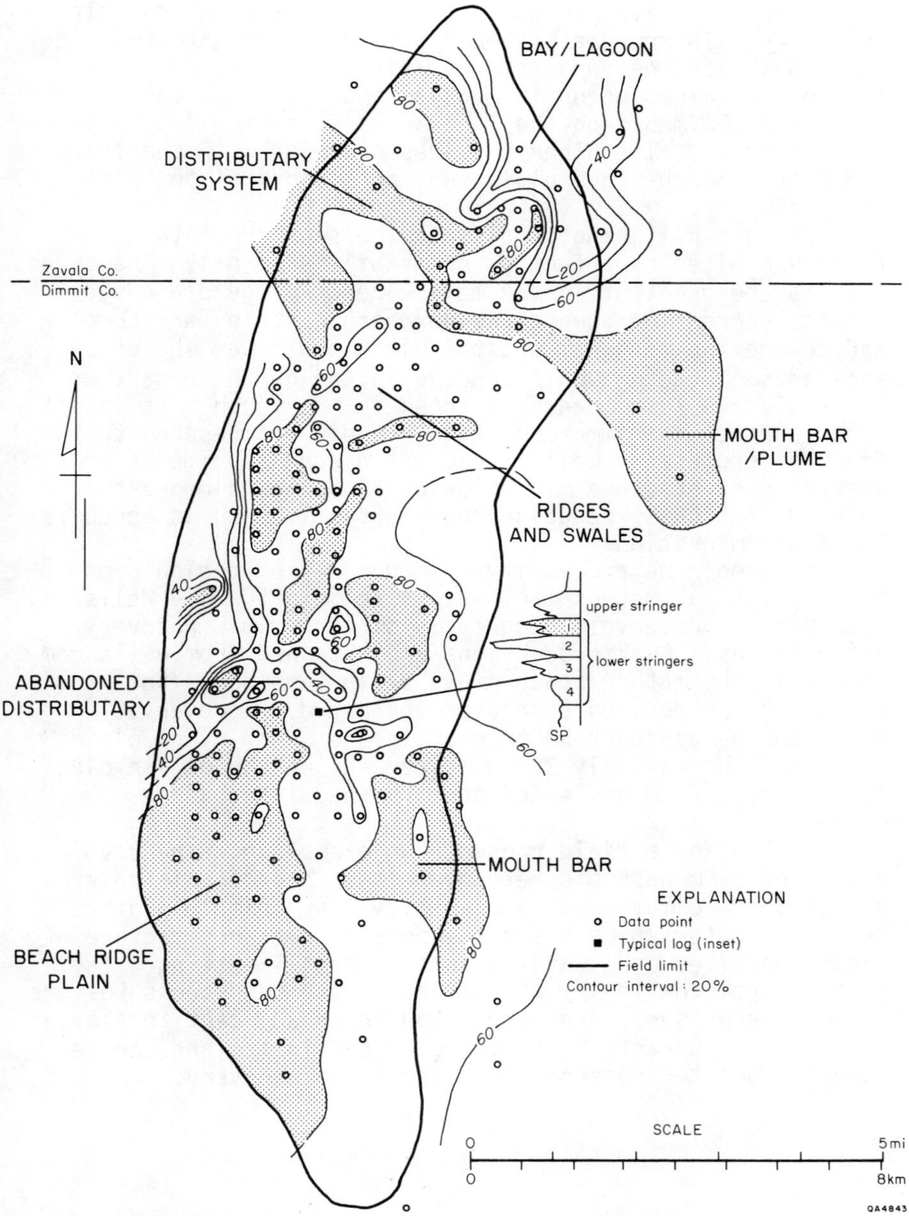

Figure 13. Facies architecture of lower stringer 1, Big Wells (San Miguel) reservoir, based on net-sand mapping.

by higher effective permeabilities than the northern half (fig. 14). Intermediate values define an elongate belt that coincides with the depositional axis of the reservoir, and the lowest values occur in muddier coastal plain and lower shoreface sediments on the landward and seaward flanks of the reservoir. The highest values coincide with the thick and mature beach-ridge plain sediments south of the distributary complex.

Fracture half lengths display the same areal trends as those exhibited by effective permeability to oil. Fracture lengths are greatest in the mature beach-ridge plain sediments, decreasing toward the northern distributary complex and toward the marginal coastal plain and lower shoreface sediments. Fracture half lengths in the northern part of the field are less than half those in the south. It is not known whether the immature sediments did not respond to the fracture process as well as did the cleaner sediments or whether post-fracture migration of fines has clogged the induced fractures, reducing their effectiveness as conduits for fluid migration.

Megascopic heterogeneity as deduced from geologic and engineering parameters manifests itself in the Big Wells (San Miguel) reservoir by greatly differing oil recovery efficiencies. Peak productions are obtained from wells completed in the mature beach-ridge plain sandstones to the south, and production decreases northward. Recovery efficiencies approach 50 percent in the beach-ridge plain deposits, whereas only 25 to 30 percent of the oil-in-place is recovered from wells adjacent to the distributary complex.

The Big Wells field presents an example of a reservoir dominated by megascopic heterogeneity. Macroscopic heterogeneity, or well-to-well variability, is minimal. Both of these characteristics are a response to the architecture of sands deposited in a wave-dominated coastal setting. The good lateral continuity of the shore-zone sandstones fosters oil recovery; sandstones deposited in or adjacent to fluvial systems are characterized by poorer continuity and, consequently, greater heterogeneity and lower recovery.

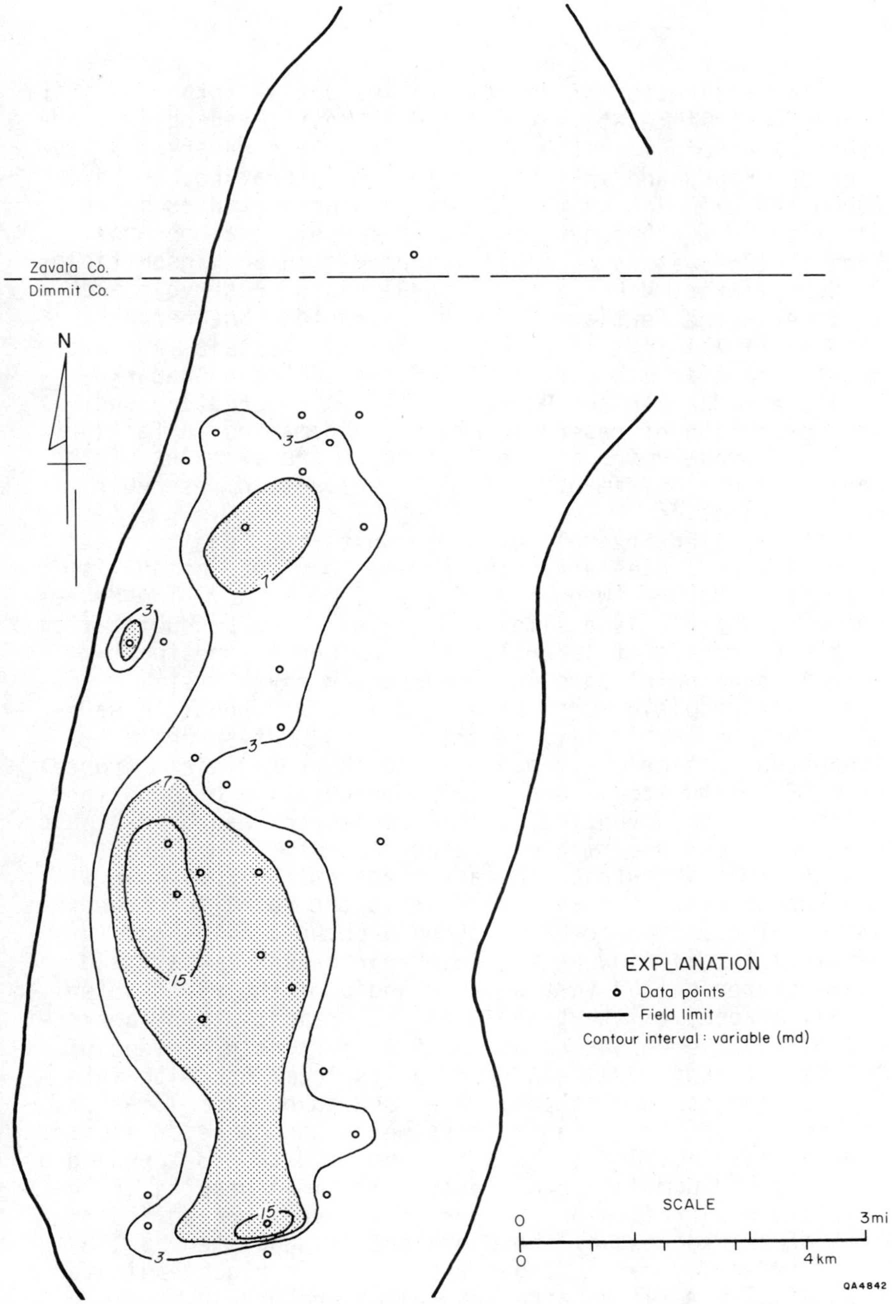

Figure 14. Calculated effective permeability to oil showing higher values in clean beach-ridge plain sediments.

V. APPLICATIONS

The application of genetic facies analysis to reservoir development is relatively new and is most advanced in sandstone reservoirs. Although many fields and reservoirs have been described and interpreted in the literature, we have found few examples of the results of these studies being utilized in further development of the hydrocarbon pool. Nanz's (1954) study of fluvial channels in Seeligson field in South Texas, Weber's (1971) analysis of reservoir architecture in the Tertiary Niger delta system, Sneider and others' (1977) investigation of fluvial, deltaic, and shorezone systems in Elk City field of the southern Anadarko Basin, and Hartman and Paynter's (1979) description and interpretation of reservoir drainage anomalies in Tertiary sandstone reservoirs of the Gulf Coast are examples of the application of sedimentological techniques in reservoir characterization.

An excellent example of the results that can be obtained by utilizing facies analysis in reservoir development is present in Neches (Woodbine) field, East Texas. The Neches (Woodbine) field is a simple anticlinal trap producing from a stacked series of laterally discontinuous sandstones deposited as point bars in a meandering river system. Continuous floodplain mudstone and shale units separate sandstone bodies vertically, imparting local but strongly expressed vertical heterogeneity to the reservoir. Truncation of the mudstones and local superposition of sandstone units results in vertically interconnected reservoirs, which originally had a common oil-water contact.

Of great importance to management of the reservoir was the recognition of clay plugs within the point-bar sandstone units. These impermeable abandoned channel fills act as barriers to oil flow as the reservoir drains. The field operator recognized that areas downdip of the plugs potentially trapped oil that would not be drained at the conventional 40-acre well spacing. Detailed structural maps of the top of individual sandstone units, combined with interpretive facies information, were used to outline locations for infill wells. Because these wells had to be drilled off regular spacing, locations were submitted to and approved by the Railroad Commission of Texas. Specific results of the in-field exploration program are not given, but an indication of the success of the operators is suggested by the estimated recovery of 63 percent of oil in place indicated for this 210-million-barrel (in place) reservoir.

VI. CONCLUSIONS

Presented herein are examples of the geological characterization of selected shore-zone and deltaic reservoir systems. Delineation of detailed reservoir character in these and other clastic depositional systems is dependent on recognition of genetic depositional units. The definition of these units is enhanced by understanding of facies, depositional systems, and systems tracts based on study of modern and ancient occurrences.

One product of reservoir characterization research will certainly be a set of advanced geological reservoir models that focus on the within-field scale of reservoir character. These models contrast with the broader scale of models that have been incorporated into modern exploration strategies. A critical interface between geology and reservoir engineering will be bridged by these new second-generation geological models, which will serve to reduce the uncertainty that engineering, well log analysis, and other disciplines must deal with in reservoir evaluation and well completion. Furthermore, geological quantification of reservoir geometry through newly applied analytical technologies will bridge the gap between purely descriptive models and entirely numerical approaches. By providing better definition of the geologic host rock, the geologist and engineer can form a much more effective team for the recovery of hydrocarbons and avoid assumptions not based on the most accurate and complete description of the reservoir.

REFERENCES

Berg, R. R., and Davies, D. K., 1968, Origin of Lower Cretaceous muddy sandstone at Bell Creek field, Montana: American Association of Petroleum Geologists Bulletin, v. 52, no. 10, p. 1888-1898.

Brown, L. F., Jr., and Fisher, W. L., 1977, Seismic stratigraphic interpretation of depositional systems: examples from Brazilian rift and pull-apart basins, *in* Payton, C. E., ed., Seismic stratigraphy--applications to hydrocarbon exploration: American Association of Petroleum Geologists Memoir 26, p. 213-248.

Dunn, H. L., 1974, Geology of petroleum in the Piceance Creek Basin, *in* Murray, D. K., ed., Energy resources of the Piceance Creek Basin, Colorado: Rocky Mountain Association of Geologists Guidebook, 25th Field Conference, p. 217-223.

Finley, R. J., 1985, Reservoir properties and gas productivity of the Corcoran and Cozzette tight sandstones, Colorado: Proceedings, 1985 Society of Petroleum Engineers/Department of Energy Joint Symposium on Low Permeability Gas Reservoirs, Denver, Colorado, SPE/DOE Paper No. 13852, p. 33-45.

Fisher, W. L., 1985, Texas crude oil--some positive trends: Interstate Oil Compact Commission, midyear meeting, Austin, Texas (preprint), 12 p.

Fisher, W. L., and Brown, L. F., Jr., 1984, Clastic depositional systems--a genetic approach to facies analysis; annotated outline and bibliography: The University of Texas at Austin, Bureau of Economic Geology, 105 p.

Fisher, W. L., and Galloway, W. E., 1983, Potential for additional oil recovery in Texas: The University of Texas at Austin, Bureau of Economic Geology Geological Circular 83-2, 20 p.

Fisher, W. L., and McGowen, J. H., 1967, Depositional systems in the Wilcox Group of Texas and their relationship to occurrence of oil and gas: Gulf Coast Association of Geological Societies Transactions, v. 17, p. 105-125.

Hartman, J. A., and Paynter, D. D., 1979, Drainage anomalies in Gulf Coast Tertiary sandstones: Journal of Petroleum Technology, v. 31, p. 1313-1322.

Kukal, G. C., Biddison, C. L., Hill, R. E., Monson, E. R., and Simons, K. E., 1983, Critical problems hindering accurate log interpretation of tight gas sand reservoirs: Proceedings, Society of Petroleum Engineers/Department of Energy Joint Symposium on Low Permeability Gas Reservoirs, Denver, Colorado, SPE/DOE Paper No. 11620, p. 181-190.

Lorenz, J. C., 1982, Sedimentology of the Mesaverde Formation of Rifle Gap, Colorado and implication for gas-bearing intervals in the subsurface: Sandia National Laboratories, Report No. SAND 82-0604, 46 p.

McGreagor, A. A., and Biggs, C. A., 1968, Bell Creek field, Montana--a rich stratigraphic trap: American Association of Petroleum Geologists Bulletin, v. 52, no. 10, p. 1869-1887.

McCubbin, D. G., and Brady, M. J., 1969, Depositional environment of the Almond reservoirs, Patrick Draw field, Wyoming: The Mountain Geologist, v. 6, no. 1, p. 3-26.
Middleton, G. V., 1978, Facies, *in* Fairbridge, R. W., and Bourgeois, J., eds., Encyclopedia of sedimentology: Stroudsburg, Pennsylvania, Dowden, Hutchinson and Ross, p. 323-325.
Nanz, R. H., Jr., 1954, Genesis of Oligocene sandstone reservoir, Seeligson field, Jim Wells and Kleberg Counties, Texas: American Association of Petroleum Geologists Bulletin, v. 38, p. 96-118.
National Petroleum Council, 1984, Enhanced oil recovery: Washington, D.C., 96 p. plus appendices.
Polasek, T. L., and Hutchinson, C. A., Jr., 1967, Characterization of nonuniformities within a sandstone reservoir from a fluid mechanics standpoint: Proceedings, Seventh World Petroleum Congress, v. 2, p. 397-407.
Selley, R. C., 1978, Concepts and methods of subsurface facies analysis: American Association of Petroleum Geologists, Continuing Education Course Notes Series No. 9, 82 p.
Sneider, R. M., Richardson, F. H., Paynter, D. D., Eddy, R. E., and Wyant, I. A., 1977, Predicting reservoir rock geometry and continuity in Pennsylvanian reservoirs, Elk City field, Oklahoma: Journal of Petroleum Technology, v. 29, p. 851-866.
Tyler, N., and Ambrose, W. A., in press, Facies architecture and production characteristics of strandplain reservoirs in the Frio Formation, Texas: The University of Texas at Austin, Bureau of Economic Geology Report of Investigations No. 146.
Tyler, N., Ambrose, W. A., and Gholston, J. C., in preparation, Genetic stratigraphy and oil recovery from a wave-dominated deltaic reservoir, Big Wells (San Miguel) field, South Texas: The University of Texas at Austin, Bureau of Economic Geology Report of Investigations.
Tyler, Noel, Galloway, W. E., Garrett, C. M., Jr., and Ewing, T. E., 1984, Oil accumulation, production characteristics, and targets for additional recovery in major oil reservoirs of Texas: The University of Texas at Austin, Bureau of Economic Geology Geological Circular 84-2, 31 p.
Walker, R. G., ed., 1984, Facies models (2d ed.): Geological Association of Canada, Reprint Series 1, 317 p.

Walker, R. G., and Harms, J. C., 1971, The "Catskill Delta"--a prograding muddy shoreline in central Pennsylvania: Journal of Geology, v. 79, no. 4, p. 381-399.

Weber, K. J., 1971, Sedimentological aspects of oil fields of the Niger delta: Geologie en Mijnbouw, v. 50, p. 569-576.

Weimer, R. J., 1966, Time stratigraphic analysis and petroleum accumulations, Patrick Draw field, Sweetwater County, Wyoming: American Association of Petroleum Geologists Bulletin, v. 50, p. 2150-2175.

CHARACTERIZATION OF SANDSTONE RESERVOIRS IN THE CHEROKEE GROUP (PENNSYLVANIAN, DESMOINESIAN) OF SOUTHEASTERN KANSAS

Anthony W. Walton
David J. Bouquet[2]
Ronald A. Evenson
Douglas H. Rofheart[3]
Marvin D. Woody[4]

Department of Geology
and Tertiary Oil Recovery Project
University of Kansas
Lawrence, Kansas, USA

ABSTRACT

Fluvial shoestring sandstones of the Cherokee Group comprise five facies: conglomerate; cross-bedded sandstone; laminated sandstone; ripple-bedded sandstone; and interbedded sandstone, siltstone, and clayshale. Average porosity of the cross-bedded and ripple-bedded sandstone lithofacies are approximately the same, that of the conglomerate is slightly lower. Although the range of permeability values of the cross-bedded and ripple-bedded sandstone and the conglomerate overlap, cross-bedded sandstone is the most permeable and the ripple-bedded sandstone is the least. The permeability has been reduced by about three orders of magnitude from initial values by compaction. Cementation and growth of clays have further reduced the permeability of certain rocks, even

---

[1]Supported by the Tertiary Oil Recovery Project and the Kansas Geological Survey.
[2]Present address: Exxon Corporation, Lafayette, LA
[3]Present address: Mobil Oil Corporation, New Orleans, LA
[4]Present address: Marathon Oil Corporation, Casper, WY

rendering some impermeable. This has created a strongly skewed frequency distribution for permeability.

Deposition of these shoestrings was in a series of episodes. Scour surfaces were formed at the beginning of each episode. These surfaces commonly either transect an underlying low-permeability rippled sand body or are overlain by low-permeability conglomerate. The reservoirs thus consist of a series of depositional units. Within each unit, the cross-bedded sandstone facies is most permeable. Cross-bedded sandstone has greatest permeability parallel to the flow of the currents that deposited it, along the long axis of cross-bed troughs. Examination of the petrophysical properties of the sediment in each facies reveals the heterogeneity of the reservoir and leads to a more realistic understanding of the natural architecture of the sand body.

I. INTRODUCTION

The problem of characterizing hydrocarbon reservoirs is basically one of finding simple, quantitative means of describing extremely variable entities. A description must indicate the flow paths and permeability barriers as well as the distribution of pores and interstitial fluids. All of these features might seem initially to be random. If they were truly random, and thus in a sense homogeneous, the problem of reservoir description would actually be fairly easy to solve. Unfortunately, reservoirs are heterogeneous and anisotropic. As a first step in understanding heterogeneity of reservoirs, it is best to determine how their properties are related to the processes that formed the rocks

Over the past 25 years, geologists have developed a number of conceptual models that aid in determining environments of deposition of sedimentary rocks (Scholle and Spearing, 1982; Galloway and Hobday, 1983). In addition to their heuristic contribution to the purely academic study of stratigraphy, these models have had vast practical application to exploration both in predicting trends in areas of unknown lithology and as aids in interpreting seismic data from frontier and offshore areas. An essential aspect of these models is their recognition that certain sets of depositional processes characterize each environment.

Processes of deposition exert close control on initial values of the fundamental properties of the sedimentary rocks: mineral composition, grain size, grain shape, grain orientation and particle packing. Variations of these properties within different parts of the depositional environment permit recognition of sediments from various parts of the depositional environment, where different processes predominated during accumulation. Recognizably different sediments are assigned to different "facies" because they have different characteristics. In addition to having characteristic small-scale properties, each facies within a reservoir should have a preferred extent and degree of continuity that depend on the particular facies and the properties of the local system.

For several years, graduate students in geology working with the Tertiary Oil Recovery Project at the University of Kansas have been studying oil reservoirs in the Cherokee Group in southeastern Kansas. The major emphasis in these studies has been describing the internal subdivisions of the reservoirs as depositional facies, mapping of the extent of these facies, interpreting their origin, determining the post-depositional changes in the rock, and establishing their petrophysical properties. The results of these studies provide one means of describing heterogeneity of reservoirs and understanding its causes.

II. GEOLOGIC BACKGROUND

The Cherokee Group is of Pennsylvanian (Desmoinesian) age and is the host for over a billion barrels of oil in southeastern Kansas (Ebanks, 1975). Most Cherokee production in Kansas is from the Cherokee Basin, really a shallow northward extension of the Arkoma Basin of Oklahoma and Arkansas (Fig. 1). Some Cherokee oil fields lie in the southern part of the Forest City Basin in east central Kansas and west central Missouri. These basins formed in response to the stresses induced in the Midcontinent during the Ouachita orogeny.

Cherokee Group consists mostly of clayshale and siltstone, with a smaller amount of sandstone, and some coal, and limestone (Fig. 2; Harris, 1985). The Cherokee Group crops out in a belt extending SSW across the southeastern corner of Kansas and dips at angles averaging less than one degree to the WNW into the Cherokee Basin. In the Kansas portion of the basin, the Cherokee ranges from 100 to 150m and is thickest in Labette County (Harris, 1985). This area has

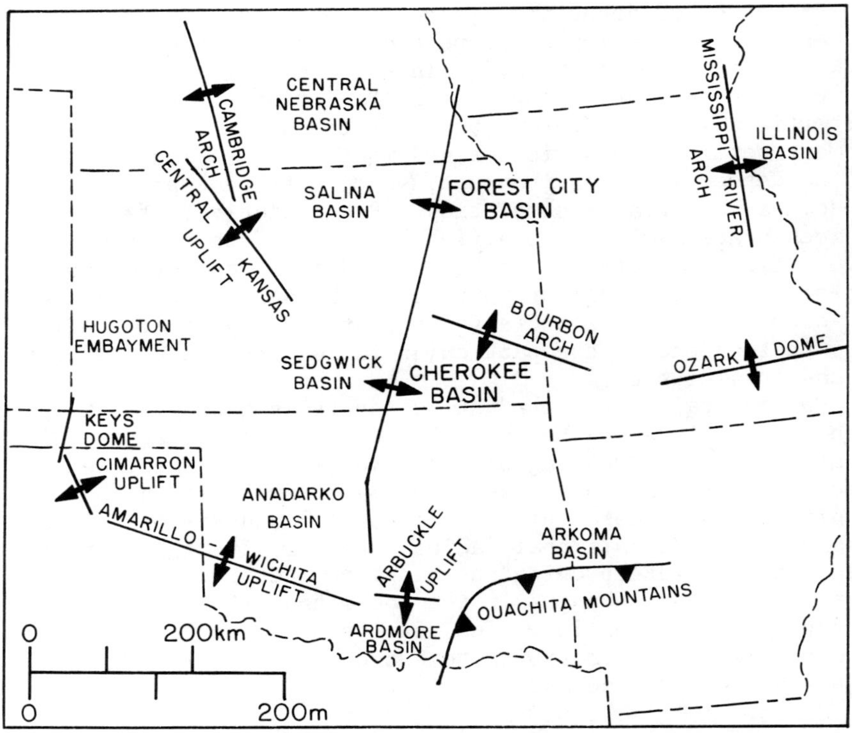

Figure 1. The Midcontinent of the US, showing major tectonic elements. Fluvial shoestrings are characteristic of the Cherokee and Forest City Basins.

experienced little deformation, other than minor normal faulting, since the middle Pennsylvanian, and Cherokee Group rocks are not extensively fractured. The cores upon which this study is based are not reported to have vertical fractures in them.

The group accumulated in a complex of deltaic, shoreline, and shallow marine environments associated with a sea that cyclically transgressed and regressed across the Midcontinent in response to glacial advances and retreats elsewhere in the world. Individual layers in this sequence, especially radioactive marine black shales, can be traced laterally for hundreds of kilometers. Sandstones, in contrast, are predominantly either elongate, lenticular shoestrings or are smaller and more equant in plan, but still lenticular in cross section. This discussion will concentrate on the shoestring sandstones because they are such characteristic oil

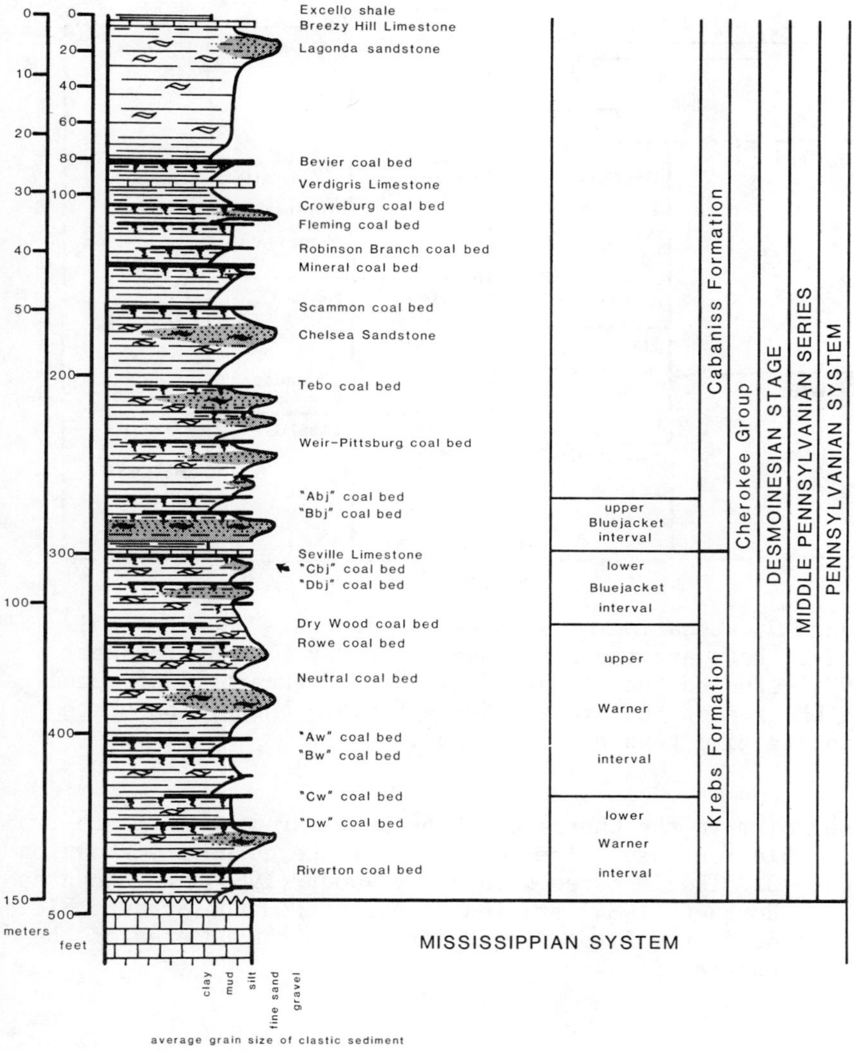

Figure 2. Composite section of the Cherokee Group in Crawford and Cherokee counties, Kansas. Sandstones are highlighted, shales dashed, and coals are heavy lines with underclays shown as root marks beneath. From Harris (1985).

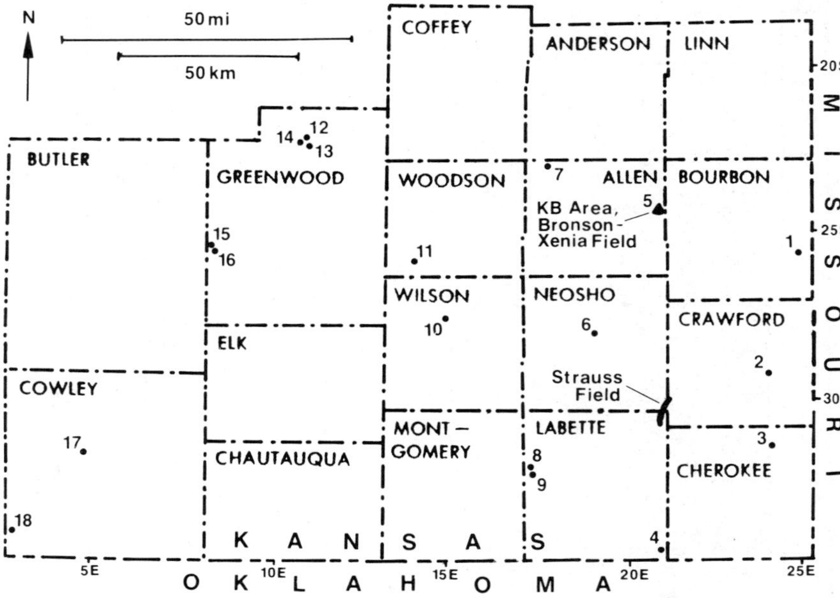

Figure 3. Location of sources of materials used for this study. Dots are the cores used by Woody (1984), Bouquet (1984) studied the Strauss Field, and Rofheart (1985) studied the KB area of the Bronson-Xenia Field. Numbers on state boundary give township and range.

reservoirs in the Cherokee and because far more data are available for them. The locations of the cores are shown on Figure 3. The 18 cores studied by Woody (1984) are shown as dots. Bouquet (1984) studied 24 cores from the Strauss Field and Rofheart (1985) studied 9 cores from the KB area of the Bronson-Xenia Field.

III. SHOESTRING SANDSTONES

Shoestring sandstone reservoirs were first recognized in the Cherokee Basin in the 1920's. Early workers believed they originated as offshore bars in a marine environment (Bass, 1936), and several other origins have been proposed over the years. Observation of sedimentary structures in cores establishes a fluvial origin for all of the shoestring reservoirs we have examined. The sand bodies are a few hundred meters wide, a few meters to a few tens of meters thick and extend distances measured in kilometers or tens of kilometers.

Oil fields in these shoestring sands may be quite shallow; depths to productive horizons are as little as 50m. Reservoir temperatures and pressures do not favor good recovery of oil, but drilling costs are cheap. Common well spacings (330 feet, approximately 100 m) are smaller than the scale of the sedimentary facies in the reservoir beds. Wells are evaluated by analysis of a core of the productive horizon rather than by drill-stem test or log analysis. Close spacing and abundant cores therefore provide a wealth of subsurface information. Consequently, these sandstones are ideal for use as natural laboratories for developing methods of reservoir characterization.

A. Component Facies

Clear differences of sedimentary structures and grain size permit division of Cherokee reservoirs into five separate facies, each of which represents deposition in a different part of the fluvial sedimentary environment (Fig. 2). One of the facies is conglomerate, three are sandstone, and one consists of interbedded sandstone, siltstone, and clayshale. The framework composition of the sandstone ranges from phyllarenites to subarkose with some quartzarenites (nomenclature of Folk, 1980).

1. Conglomerate

Beds of conglomerate up to a meter thick are common at the bottom of shoestring sandstones and overlying scour surfaces within them. Clasts within the conglomerate beds are mostly shale and carbonaceous material of local origin, or concretions eroded from slightly older beds nearby. Conglomerates probably represent lag deposits of coarse material at the bottom of fluvial channels.

2. Cross-bedded sandstone

The most productive facies is cross-bedded, very fine-grained to medium-grained sandstone. It is difficult to determine in cores whether these are trough or planar cross beds, but similar sediments in outcrop are normally trough cross-bedded. Such sandstone forms on sand bars in fluvial

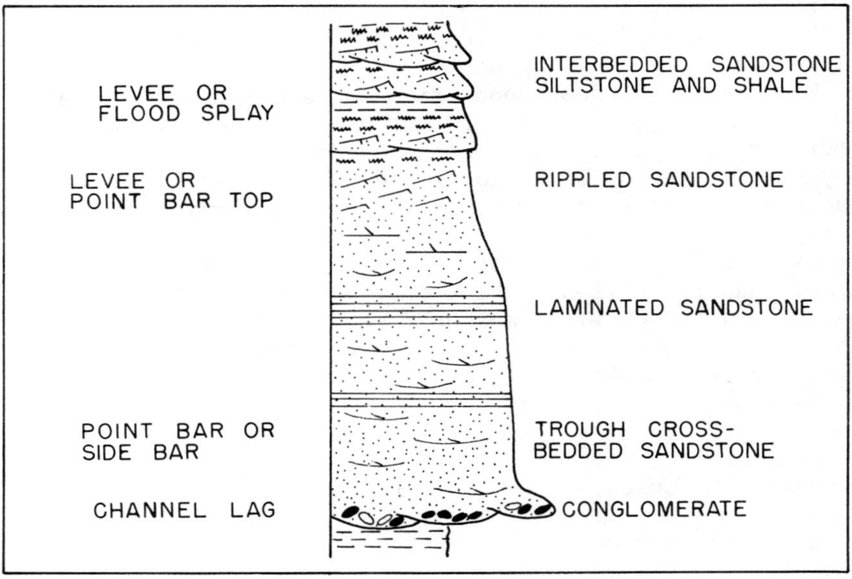

Figure 4. Idealized vertical sequence of lithofacies in a Cherokee shoestring sandstone deposited by a meandering river. The sand body would be about 10 m thick. Compare with Figure 12.

channels, either as point bars in meandering rivers or as side bars in straight channels, such as distributaries in some deltas. Cross-bedded intervals are up to a few meters thick.

3. Plane-laminated sandstone

An uncommon facies consists of very fine to fine grained, plane-laminated sandstone. This facies is most commonly interbedded with cross-bedded sandstone but forms from rapidly-flowing or shallower water on fluvial sand bars. Like cross-bedded sandstone, plane-laminated sandstone is very good reservoir rock.

4. Rippled sandstone

Rippled sandstone is very fine sandstone with a substantial admixture of silt or interlaminated with siltstone. Layers of rippled sandstone may be up to a meter thick without interruption, but they commonly contain interstratifications of shale or siltstone spaced as closely as a few centimeters apart. Ripples form in the upper parts of fluvial sand bars and in levees and flood plains adjacent to the channel. Ripple-bedded sandstones make productive reservoirs where clay layers are rare, but have lower permeability than cross-bedded or plane-laminated sandstones and conglomerates.

5. Interbedded sandstone, siltstone, and shale

Interbeds of three rock types on a scale of a few decimeters are common at the top of sandstone sequences. These consist of rippled sandstone, which has a sharp, sometimes scoured base, and gradationally overlying siltstone and shale. These beds represent deposits of waning currents, probably during floods, on the levee or flood plain of rivers that formed the sandstones. They have poor permeability and porosity and are not generally favorable as reservoirs.

6. Structureless sandstones

Because of the heavy oil stain on many Cherokee rocks and their lack of contrasts of grain size, many intervals appear to be structureless. In a few instances, transmitted X-rays indicate the existence of sedimentary structures in these rocks. Petrophysical properties of these intervals are similar to those of associated rocks that show structures, either ripples or large-scale cross beds (Hulse, 1978). Structureless beds probably represent deposition in the same environment as associated beds with sedimentary structures. Consequently, they are not treated as a separate facies.

B. Petrophysical Properties

Summary data on petrophysical properties for samples that have been assigned to different facies are given in Figure 5. Too few data are available on plane-laminated sandstone, structureless sandstones and interbedded sandstone, siltstone, and shale to place any quantitative significance on their properties. These data are from three different studies. Data in Woody (1984) and Rofheart (1985) were taken in the laboratory of the Tertiary Oil Recovery Program at the University of Kansas. Porosity was determined by brine saturation, using a brine similar to oil field waters recovered from the Cherokee in southeastern Kansas. Permeability is permeability to brine for both of these studies, except that Woody measured the permeability to nitrogen for nine samples of very low permeability. Bouquet (1984) used data provided by Inco Energy for the Strauss Field. These values were measured by Oil Field Research Co. in Chanute, Kansas. Permeability measurements on these samples were made with air. Fracture permeability is presumed to be small in these reservoirs because few fractures were noted in cores. In all studies the grain size was measured by determining the long axis of individual grains in thin section. Composition of rocks was determined by counting 400 points on thin sections.

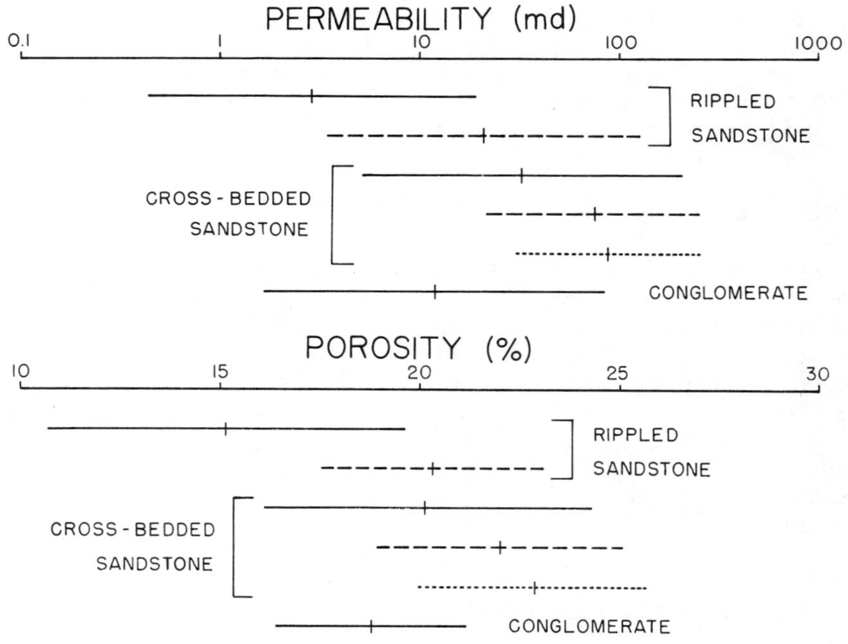

Figure 5. Average values and variability of porosity and permeability of rocks from different lithofacies in the Cherokee. Length of bar is one standard deviation, tick mark gives average. Data are from Woody (1984, solid line, K to brine except for 9 samples of low permeability), Bouquet (1984, long dashes, K to air), and Rofheart (1985, short dashes, K to brine).

1. Porosity and permeability

Ranges of porosity values (expressed as one standard deviation) and their averages do not differ substantially from facies to facies, except porosity of the the conglomerate facies is lower. Maximum permeability values in the cross-bedded sandstone facies are greater than those in the rippled sandstone facies, although the range of the two groups of samples overlaps. This produces scatter plots such as Figure 6, where the distributions of values overlap.

It is likely, however, that two features cause the extent of overlap on plots like Figure 6 to be artificially large. One is that the samples of the cross-bedded sand facies may be taken at any angle to the strike of the cross beds, while the maximum permeability is in the plane of the cross beds (Dodge and others, 1971). Rofheart (1985) avoided this problem by cutting his plugs only parallel to strike of cross beds. The other feature that increases the amount of overlap is that rippled sequences commonly have occasional beds of sandstone a few cm thick without any mudstone drapes. In many cores, these sand-rich parts of ripple-bedded intervals have been selectively sampled, resulting in higher permeability averages for this facies.

2. Influence of diagenesis

Differences in permeability between facies without differences of porosity are consistent with the grain size differences between them (Pettijohn, Potter, and Siever, 1973, p. 94). The cross-bedded sandstones are coarser than the ripple-bedded sandstones, very fine- to medium-grained sand as opposed to very-fine sand. However, several facts argue that other influences have changed the permeability of these rocks considerably since they were deposited. First is that although some sets of data show a statistically significant positive correlation of grain size and permeability (Bouquet, 1984), other sets show no such pattern. In fact, some sets of data show a slight decrease of permeability with increasing grain size!

A second indicator of the importance of post-depositional changes is that permeability of modern fluvial sands is considerably greater than the values measured for the Cherokee sandstones. Pryor (1973) reports values of 50 to 100 darcys for sands from some point bar sandstones, compared to the averages of about 100 millidarcys in Cherokee sandstones. Pryor's sands were slightly coarser.

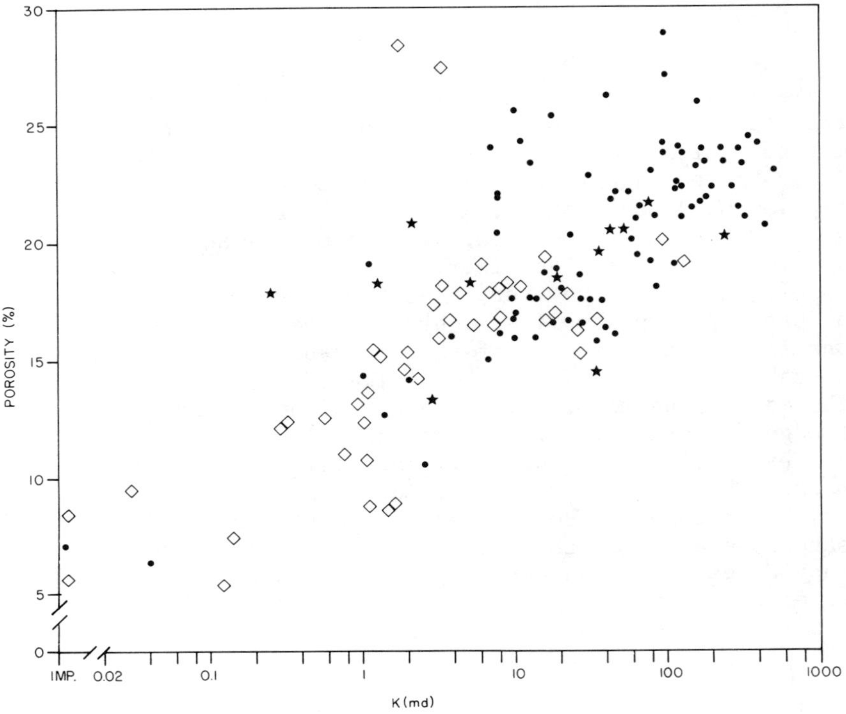

Figure 6. Porosity and permeability of samples from various facies in several different sandstone bodies in the Cherokee Group (Woody, 1984). The 9 lowest values are permeability to nitrogen, the others are to brine. Dots are samples from the cross-bedded sandstone facies; diamonds are from the rippled sandstone facies; and stars are from the conglomerate facies.

This substantial decline of permeability is partly related to compaction (Hsu, 1977), but Figure 7 suggests that cementation is also a major factor. The histogram at the base of the figure and the concave-upward shape of the curve both show that the permeability is skewed toward the lower values. Similar skewness characterizes rippled sandstones and conglomerates in both Woody's data and that of Bouquet (1984; Fig. 8). Skewness is quantified in these diagrams by using calculations of the third moment about the mean of the distribution, after grouping the data as shown in the histogram (Griffiths, 1958). This skewness is the third

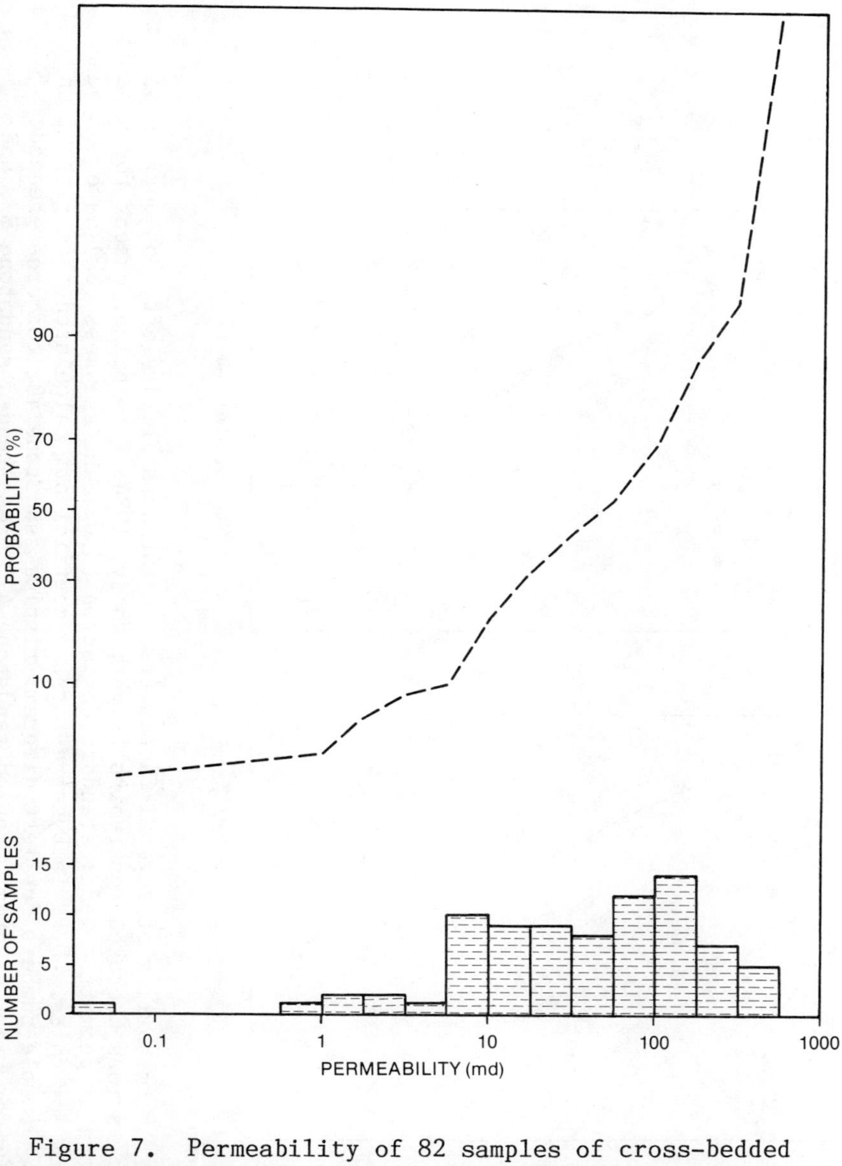

Figure 7. Permeability of 82 samples of cross-bedded sandstone from several different Cherokee reservoirs (Woody, 1984). Histogram shows the distribution of grouped data. Probability curve constructed from the grouped data shows skewed distribution by its concavity.

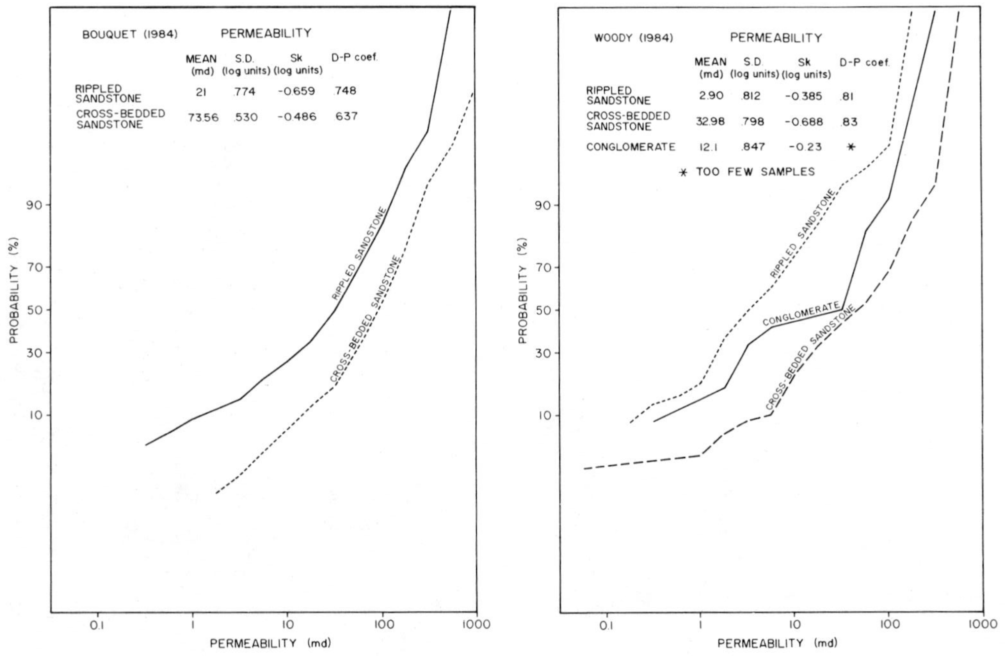

Figure 8. Permeability of several facies from the Strauss Field (left; Bouquet, 1984; K to air) and from various Cherokee reservoirs (right; Woody, 1984; K to brine, except for nine low K samples). Mean is geometric mean; standard deviation (S.D.) and skewness (Sk) have not been converted from log units. D.P. is Dykstra-Parsons coefficient, calculated from plot of ungrouped data. Sample numbers for left diagram: rippled sandstone 43, for cross-bedded sandstone, 80; for right diagram: rippled sandstone 47, cross-bedded sandstone 82, and conglomerate and conglomeratic sandstone, 12.

indication of substantial influence of post-depositional changes of the sediment on permeability.

Clearly post-depositional changes of the sediment, diagenesis, have influenced permeability, as suggested in Figure 9. Compaction has reduced the average permeability of cross-bedded sandstones is reduced from initial values of 10 to 100 darcys to about 100 millidarcys. In certain samples, cementation and growth of authigenic clays have reduced the permeability still farther, perhaps rendering the rocks impermeable, and leaving the skewed distribution. The existence of this skewness of permeability distribution is extremely important in characterizing the reservoir.

C. Diagenetic Sequence

The diagenetic sequence in sandstones of the Cherokee Group is similar to that in other sandstones of similar initial composition and burial history (Woody, 1984; Fig. 10). Although the sequence of changes did not differ from facies to facies, the extent to which each change took place did vary. The initial phase was precipitation of early carbonate, either siderite or calcite. These minerals are concentrated in a few rocks, not widespread in reservoirs. Compaction followed, affecting especially the rocks in which neither early carbonate formed. A phase of chlorite formation was the next step; this was favored in rocks that had substantial concentration of clay-rich intraclasts, but is rare in Cherokee sandstones. Syntaxial overgrowths grew on quartz grains after the chlorite formed. An episode of dissolution followed, affecting feldspar and the early carbonates. Scattered but ubiquitous crystals of a Ca, Fe, Mg carbonate, called Ca-ankerite for ease of reference, grew after the episode of dissolution both in intergranular pores and in some of the spaces formed by dissolving grains and cements. The final diagenetic event, introduction of pore filling kaolinite cement into pores and dissolved feldspar grains, may have been preceded by another cycle of dissolution that affected both feldspar and all of the carbonates formed earlier. Introduction of oil terminated the diagenetic process.

Bouquet (1984) showed how the pore filling materials differ from facies to facies. His data, summarized in Figure 9, are in percentage of the space between framework grains that is occupied by each diagenetic or matrix phase, or in percentage of "potential pore space". Results from Rofheart (1985) and Woody (1984) generally agree. Siderite is most common in conglomerates, averaging 64% of available space,

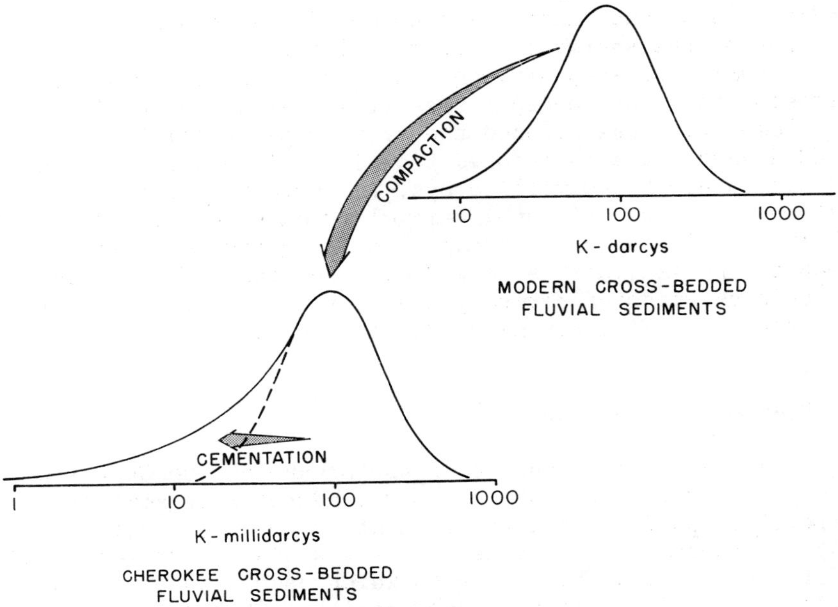

Figure 9. Post-depositional reduction of permeability includes a drop of 2–3 orders of magnitude by compaction and a further reduction by cementation. Selective cementation of certain parts of the reservoir produces the skewed distribution.

but it is present in both cross-bedded and rippled sandstones. Quartz overgrowths occupy 13% of the potential pore space in cross-bedded sandstones and 11% in rippled sandstones, but are absent in conglomerates, probably because, as Rofheart reports, grain-coating chlorite is common in rocks with abundant intraclasts, and this inhibits growth of quartz (Pittman and Lumsden, 1968). Calcium-ankerite cement is most common in cross-bedded sandstones (16% of potential pore space), and present in both rippled sandstone (11%) and conglomerate (12%). Kaolinite is abundant in both cross-bedded sandstone (37%) and rippled sandstone (33%) and present in conglomerate (10%). Clay deposited with the sample, or matrix, and intergranular fine material that results from crushing rock fragments, called pseudomatrix, are absent from conglomerates where early siderite cement prevents crushing, but total 5% in cross-bedded sandstones and 15% in rippled sandstones.

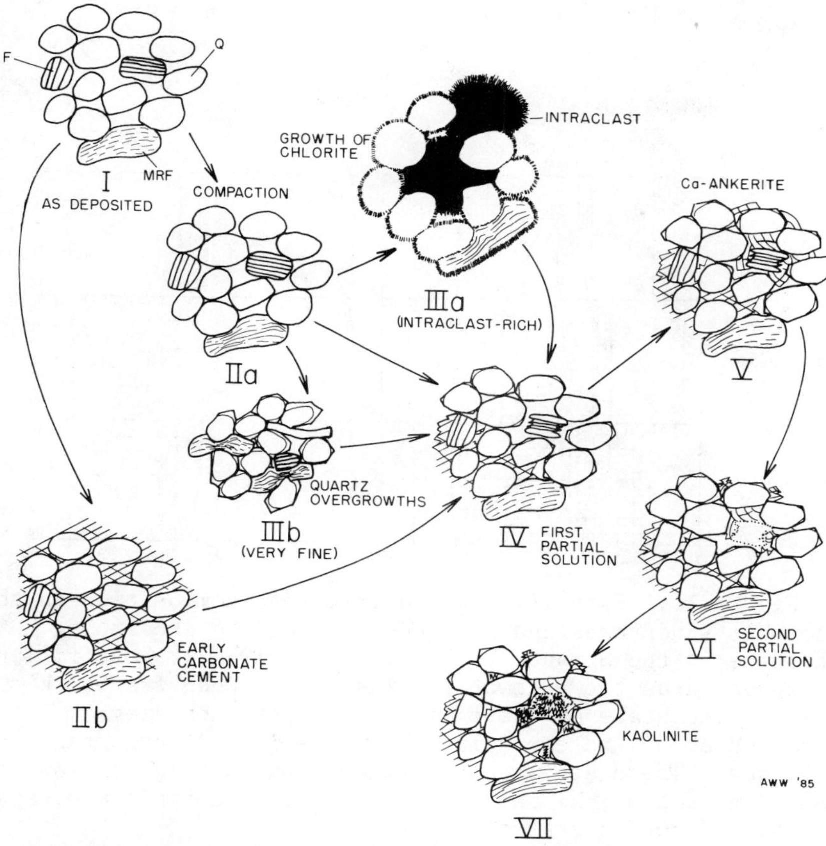

Figure 10. Diagenetic sequence in Cherokee sandstone reservoirs. Evidence for the second period of dissolution and the relative order of formation of kaolinite and Ca-ankerite is equivocal.

These data show clearly that the effects of diagenesis are different from facies to facies. Bouquet (1984) states that original detrital constitution may contribute to this (Fig. 11). Micas and rock fragments, mostly metamorphic rock fragments and intraclasts, are more common in the rippled sandstones and conglomerates than in the cross-bedded sandstones. Rock fragments are easily crushed during compaction, reducing porosity and permeability, but compaction is prevented in conglomerates by early siderite cement. Rofheart (1985) has shown that rock fragments favor growth of chlorite, and samples rich in rock fragments have more micropores. All of these reduce permeability. The higher

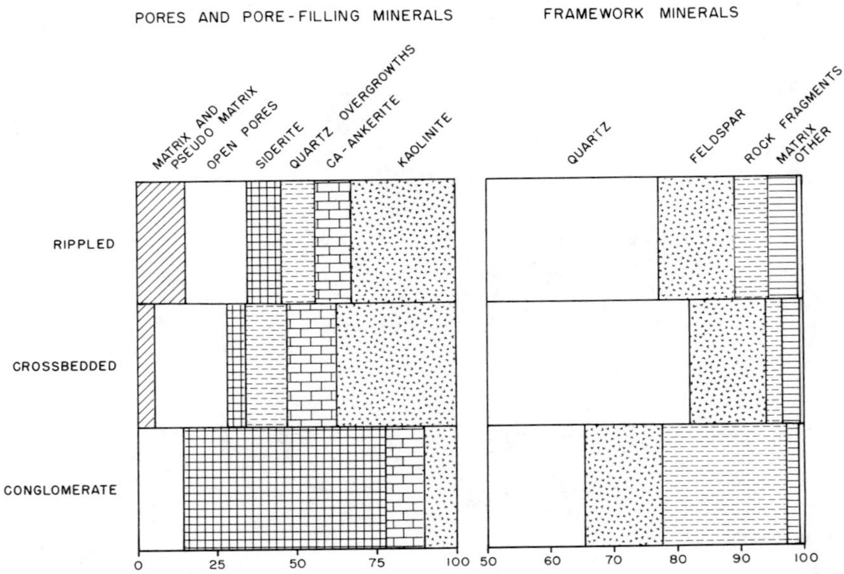

Figure 11. Pore-filling and framework composition in the reservoir sandstones and conglomerates of the shoestring sandstone in the Strauss Field (Bouquet, 1984). Pore filling is expressed as the percentage of space between framework grains. The scale of the framework-composition diagram is truncated at 50% to emphasize the variation of rock fragments and micas. These are average values from 14 thin sections of rippled sandstone, 21 samples of cross-bedded sandstone, and 1 sample of conglomeratic sandstone.

permeability of the cross-bedded sandstones are related to the higher percentage of quartz and the relatively few rock fragments and mica flakes (Bouquet, 1984). Feldspar is present in similar quantity in all three facies, consequently kaolinite, which arises from the dissolution of feldspar in ground water, is widely distributed. Its lower abundance in conglomerate reflects early cementation and abundance of siderite.

D. Distribution of Facies

The data presented above describe the extent to which depositional facies control the permeability of Cherokee reservoirs. They indicate that permeability is reduced by two or three orders of magnitude during compaction.

Precipitation of mineral cements, such as carbonates, quartz, and clays, reduces the permeability of certain samples to very low levels and introduces the characteristic skewed frequency distribution for each facies. Within individual facies, grain size plays only an equivocal role. Diagenetic processes are factors in the relationship between depositional facies and petrophysical properties. Consequently, the distribution of depositional facies will control the permeability variation of reservoirs.

1. Vertical sequencing

The normal sequence encountered in Cherokee shoestring sandstones begins with the conglomerate overlying a scour surface and passing upwards successively into a cross-bedded sandstone then into rippled sandstone and finally into the interbedded sandstone, siltstone, and shale (Fig. 12). Structureless sandstone is found with both the rippled and cross-bedded sandstones. Plane-laminated sandstone is usually found with cross-bedded sandstone.

Fining-upward sequences are rarely complete, but are interrupted by scour surfaces, with conglomerate or coarser sandstone above them, and ripple-bedded or cross-bedded sandstone beneath. Scour surfaces may thus mark horizons with lower permeability that interrupt the continuity of the reservoir sand body. Scour surfaces record episodes of channel migration that were accompanied by substantial erosion. These may have been individual floods, but more likely correspond to longer-term migrations, perhaps related to avulsion to different river courses or cutting off of meanders. The interval between scour surfaces ranges up to about ten meters.

2. Depositional units

Sand bodies defined by successive scour surfaces can be mapped by correlating kicks on gamma-ray logs or by careful examination of cores (Fig. 13; Rofheart, 1985). The sand bodies are hundreds of meters wide and have convex-downward bases and tops that may be modified by later channel migration and erosion (Fig. 14). Although the sand bodies can be traced in closely spaced wells, the degree of hydraulic interconnection is not known. Rippled-bedded sandstones and conglomerates average an order of magnitude lower in permeability than cross-bedded sandstones, but are in thin beds and may not be continuous.

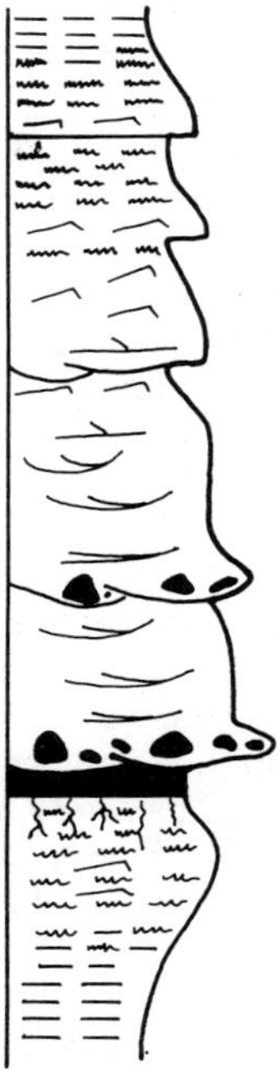

Figure 12. Vertical succession of facies in Cherokee fluvial shoestring sandstones. Intervals between scour surfaces range up to 15m thick, commonly they are 10 or less.

Within the depositional units marked by scour surfaces, the permeability should be at a maximum in the direction of the long axis of the flow. The greatest permeability in individual cross beds is in the plane of the bed. In gently dipping cross beds, the maximum permeability is then in the direction of flow because of favorable grain orientation with the long axis in the downstream direction. In sets of steeply dipping beds, the maximum permeability is perpendicular to the flow direction, along the strike of the cross bed, or across the axis of the trough. Pryor's (1973) examination of permeability of recent fluvial sand bodies, however, shows that masses of trough cross bed sets are more permeable parallel to the long axis of the trough than across it. He finds that set boundaries are zones of low permeability and that low permeability clay drapes are common features of trough cross-bedded fluvial sand bodies. These force the flow to follow the axis of the trough cross bed sets. Average orientation of these trough sets parallels the

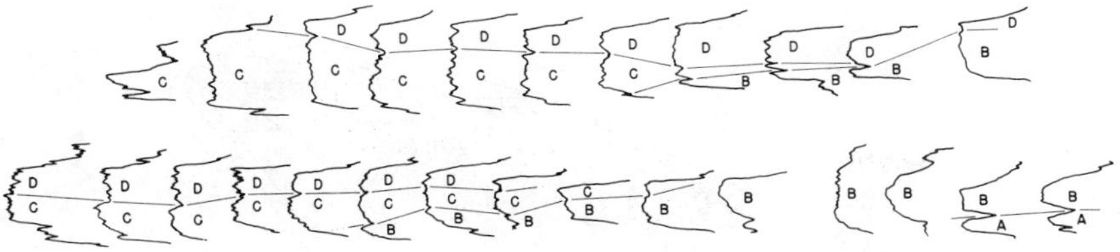

Figure 13. Two gamma-ray-log cross-sections of the Bronson-Xenia Field, showing correlation of scour surfaces bounding depositional units. Modified from Rofheart (1985).

long axis of the fluvial channel that formed them. In ancient sandstones cementation zones may follow individual cross beds, producing zones of almost no permeability at a high angle to the trough sets. Flow will divert around them into other trough sets, and permeability will generally be higher parallel to the axis of the troughs.

IV. SUMMARY

In fluvial shoestring-sandstone reservoirs of the Cherokee Group, the average permeability differs in different depositional facies. It is highest in the cross-bedded and is lower in ripple-bedded sandstones. Conglomerates have intermediate values. These differences are not simply a result of the different grain sizes of the facies, because diagenesis has greatly influenced the permeability. Compaction generally has reduced the permeability two to three orders of magnitude. Cementation has selectively reduced permeability, rendering some samples essentially impermeable and producing a markedly skewed frequency distribution in the rest. Facies-to-facies variations of permeability and diagenetic mineralogy also follow compositional changes in the framework grains as originally deposited.

The effect of permeability differences between facies is to define flow units that correspond to depositional events. Each flow unit contains a more-or-less complete vertical sequence of fluvial deposits--conglomerate, cross-bedded sandstone, rippled sandstone, and interbedded sandstone, siltstone and shale--and is separated from other depositional units by boundaries that have lower permeability. Cross-bedded sandstone, which is the most favorable reservoir facies, has greatest permeability along its long axis,

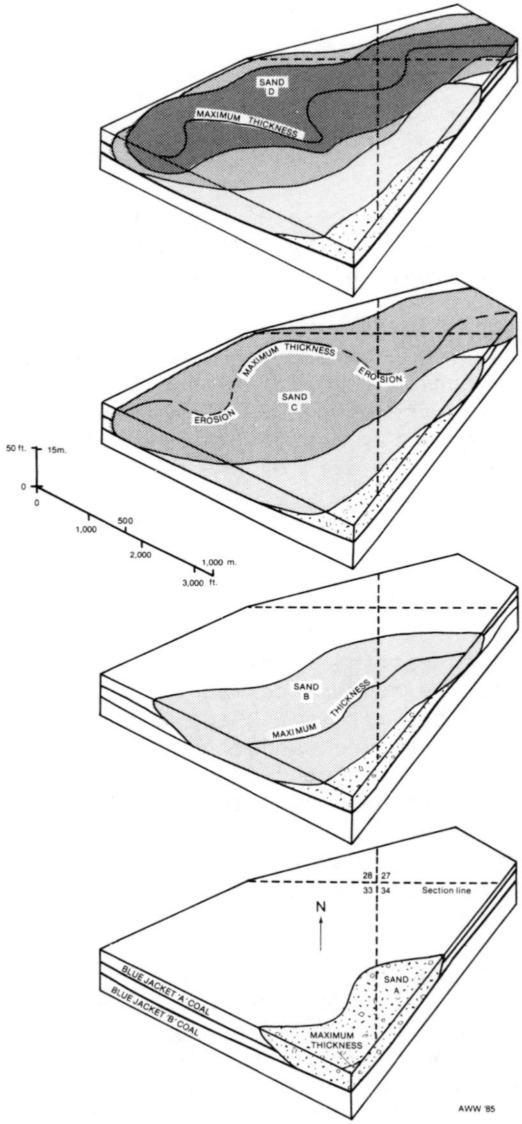

Figure 14. Development of the reservoir sandstones in a segment of the Bronson-Xenia Field (Rofheart, 1985). Each of the four sand bodies (A, B, C, and D) represents a separate episode of channel development in this area.

parallel to the direction of flow of the river that deposited it and to the troughs of its cross bedding.

These relationships present the opportunity to deal with heterogeneity of reservoirs in a rational manner. At the small scale the mineralogy, grain size, grain packing, grain shape, and grain orientation are characteristically distributed within depositional facies. Porosity, permeability, and other petrophysical properties then vary over certain ranges within each depositional facies. At the large scale, the extent and continuity of the facies determine the favorable channels for flow of fluids through an unfractured reservoir.

ACKNOWLEDGMENTS

Mack Colt and Inco Energy provided cores and subsurface information and petrophysical data. Aaron Phillips drafted the illustrations. Don W. Green and Mark McKibben read early versions of the manuscript.

REFERENCES

Bass, N.W. (1936). Kansas Geol. Surv. Bull. 23.
Bouquet, D.J. (1984). M.S. Thesis, Univ. of Kansas.
Dodge, C., Holler, D., and Meyer, R. (1971). Amer. Assoc. Petrol. Geol. Bull. 55, 1814.
Ebanks, J. (1975). Univ. of Kansas, Tertiary Oil Recov. Proj. Contribution 1.
Folk, R.L., 1980, "Petrology of Sedimentary Rocks." Hemphill Publishing Co., Austin, Tex.
Galloway, W.E. and Hobday, D.K. (1983). "Terrigeneous Clastic Depositional Systems." Springer-Verlag, New York, Berlin, Heidelburg, Tokyo.
Griffiths, J.C. (1967) "Scientific Methods in Analysis of Sediments." McGraw-Hill, New York.
Harris, J. (1985). M.S. Thesis, Univ. of Kansas.
Hsu, K.J. (1977). Amer. Assoc. Petrol. Geol. Bull. 61, 169.
Pettijohn F., Potter, P., and Siever, R. (1973). "Sand and Sandstone." Springer-Verlag, New York, Berlin, Heidelburg.
Pittman, E.D. and Lumsden, D.N. (1968). J. Sedim. Petrol. 38, 668.
Pryor, W. (1973). Amer. Assoc. Petrol. Geol. Bull. 57, 162.

Rofheart, D. (1985). M.S. Thesis, Univ. of Kansas.
Scholle, P. and Spearing, D. (1982). "Sandstone Depositional Environments," AAPG Mem. 31.
Woody, M.D. (1984). M.S. Thesis, Univ. of Kansas.

GEOLOGIC PREDICTION OF SHALE CONTINUITY PRUDHOE BAY FIELD

G.W. Geehan
T.F. Lawton
S. Sakurai
H. Klob
T.R. Clifton
K.F. Inman
K.E. Nitzberg

Development Geosciences Department
Sohio Petroleum Company
San Francisco, California

ABSTRACT

Shale distribution is a dominant geologic control on reservoir performance within the Prudhoe Bay field. The interpreted depositional environment, mostly braided-fluvial, suggests that many stratigraphically equivalent shales may not be continuous between wells spaced roughly 1900 to 2600 ft. apart. Thus, shale continuity interpretations and size estimates of discontinuous shales must rely on depositional facies information.

To be useful at Prudhoe Bay, shale facies interpretations made from core must be extended to uncored wells. A log-data based discriminant analysis procedure was developed for this purpose. Shale facies predictions obtained by this procedure, combined with discontinuous shale size-frequency data obtained from outcrops of analogous facies, form the basis of a complete shale description.

I. INTRODUCTION

Shale distribution and continuity are dominant geologic controls on reservoir performance in the Prudhoe Bay field. Where present, the continuous shales, defined as those shale beds continuous between two or more wells, divide the oil column into production intervals isolated from one another. They act as effective vertical permeability barriers and thus help prevent both gas coning and water influx. However, production under such shales has caused significant pressure anomalies in the field because direct pressure support from the gas cap is not available. When a continuous shale intersects the gas cap at Prudhoe Bay, the production-induced pressure sink below the shale causes gas underrunning. A generalized N-S cross-section through the Eastern Operating Area at Prudhoe (Fig. 1) illustrates the underrunning condition. Production beneath continuous

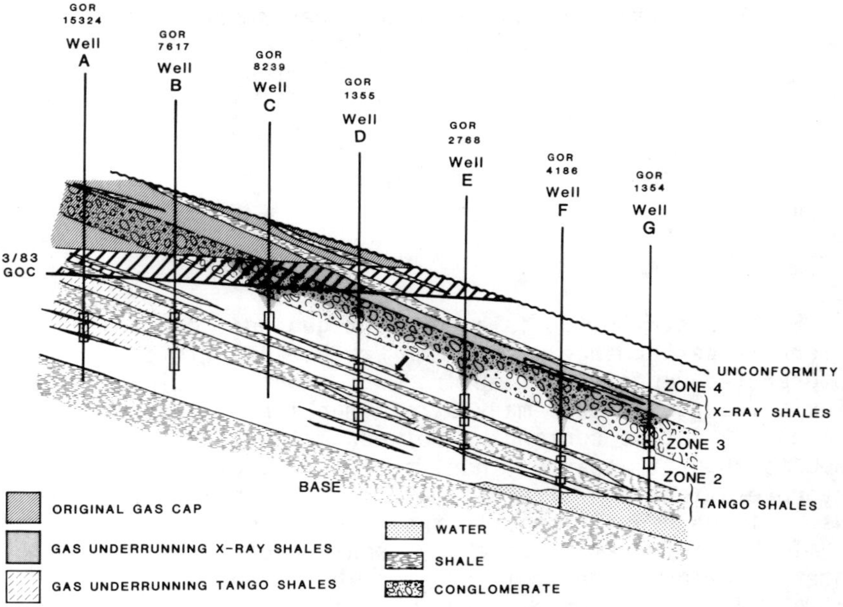

Figure 1. North-south cross-section through the eastern part of the Prudhoe Bay field showing gas underruning shales (modified from Chang et al., 1984). The arrow indicates a continuous shale which prevents coning but which does not intersect the gas cap and induce underrunning.

shales which do not intersect the gas cap (an arrow indicates one such shale in Figure 1) will not induce gas underrunning.

Discontinuous shales, defined as those shales smaller than the interwell distance occur scattered throughout the sandstone facies in the reservoir. These shales increase the tortuosity of vertical fluid flow, resulting in lowered effective vertical permeabilities. Where gas cap expansion is the primary production mechanism, small lenses of oil, which drain slowly over time, are left behind on top of the discontinuous shales. In enhanced recovery operations, specifically a water or gas injection situation, the presence of discontinuous shales should increase sweep efficiency by inhibiting gravity segregation (Haldorsen, personal communication).

Thus an accurate description of shale distribution is important to accurate predictions of reservoir performance for both primary and secondary recovery processes. Furthermore, the description must go beyond a simple correlation of stratigraphically equivalent shales. The interpretation of depositional environments (see below) suggests that many correlated time-equivalent shales may not be continuous between wells spaced approximately 1900 to 2600 ft. apart (80 and 160 acre spacing). Continuity prediction must be linked to depositional environment interpretation. In addition to continuity, the size of discontinuous shales, which controls the degree of tortuosity, must also be estimated. Interpreted depositional environment also provides the basis for prediction of shale size or lateral dimension. At Prudhoe Bay, shale continuity is also commonly interrupted by faulting; we will restrict this discussion to depositional controls on continuity.

II. INTERPRETED DEPOSITIONAL ENVIRONMENTS

The Ivishak Formation forms a largely progradational deltaic-fluvial complex 300 to 650 feet thick within the Prudhoe Bay unit (Fig. 2; Jones and Speers, 1976). The formation is subdivided into four zones based on characteristics of the gamma-ray, sonic and density logs (Fig. 2). The correspondence indicated in Figure 2 between depositional environments and petrophysical zones is reasonably consistent throughout the field.

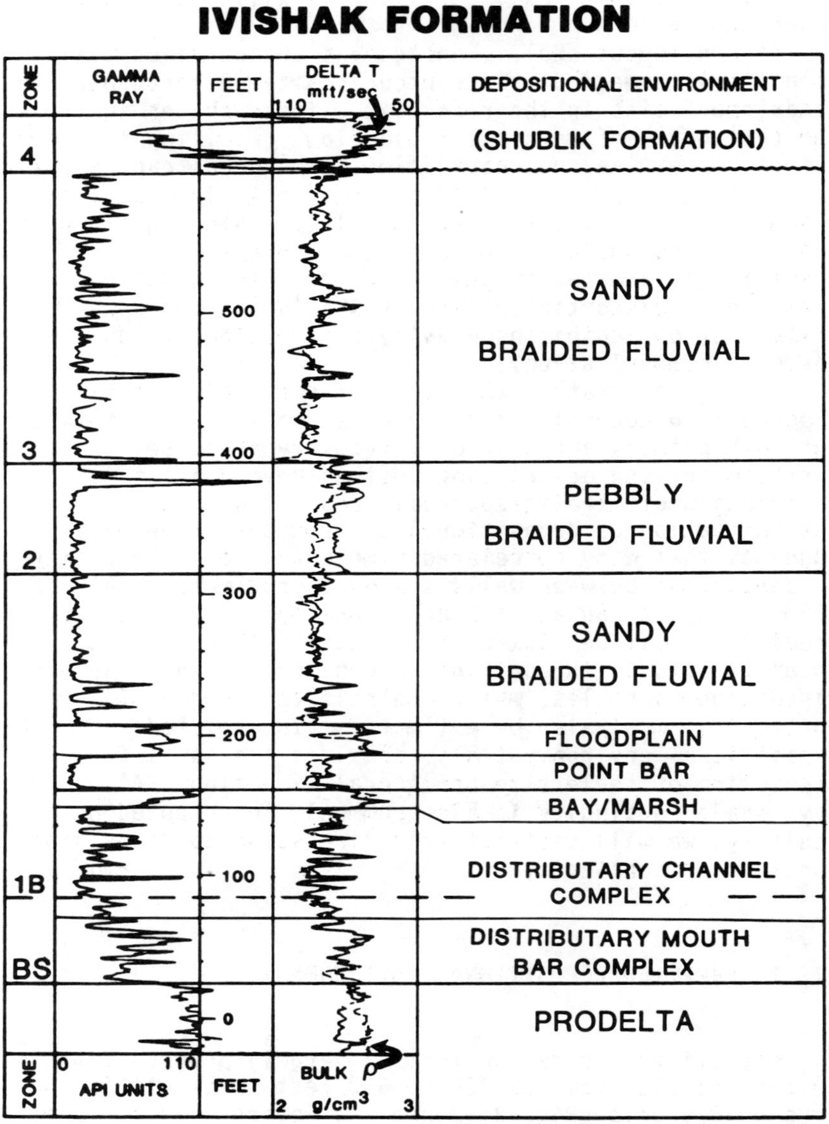

Figure 2. Petrophysical zonation, electric-log response, and interpreted depositional environments for the Ivishak Formation in a typical well at Prudhoe Bay.

The lower deltaic part of the sequence reaches 100 feet thick and rests gradationally upon prodelta deposits of the Kavik Shale. This sequence consists of two major depositional settings, a lower distributary mouth-bar complex and an overlying (and in part equivalent) distributary channel system. The mouth-bar complex is a sequence of upward-thickening and upward-coarsening very fine grained sandstone beds interbedded with thin shale beds. The distributary channel complex is composed of upward-fining cycles 6 to 10 ft. thick of medium to fine-grained sandstone with mudchip lags.

The upper fluvial part of the Ivishak Formation, ranging from 200 to 550 feet thick, was deposited by rivers of highly variable character. The lower 75 feet of the section consists of fine to medium-grained sandstone and pebbly sandstone interbedded with shales ranging from a few inches to 40 feet thick. The sandstones and pebbly sandstones were deposited by a combination of braided and meandering rivers. Thicknesses of fining-upward cycles indicates that the meandering channels reached depths of 25 to 30 feet. Above the lower 75 feet of fluvial deposits, grain size and both pebble content and diameter increase upsection from medium sandstone with few pebbles to clast-supported cobble conglomerate at the top of Zone 3. Corresponding average channel depths estimated from thickness of fining-upward cycles decrease from 10 to 4 feet. Deposition was almost exclusively by braided streams. Crossbedded fine to medium-grained sandstone of Zone 4 was deposited in braided stream channels which averaged 8 feet deep as determined from the average thickness of completely preserved fining-upward cycles. Mudchips are common as channel lags in Zone 4 sandstone.

Shale beds within the reservoir were deposited in several different depositional environments. Six genetic shale facies categories were identified. Each facies category possesses a combination of lithologic characteristics which permit its discrimination. The genetic shale facies include:

1) Thick-bedded prodelta siltstone and shale, with common slump structures, occasional thin beds of very fine sandstone, and abundant carbonaceous detritus;

2) Thin-bedded bar-fringe siltstone and shale with slump structures and abundant carbonaceous detritus;

3) Thick marsh or bay shale, thinly laminated. with very fine sandstone interbeds;

4) Abandoned channel shale, 2 to 10 feet thick, often slumped and mixed with sandstone and pebbles;

5) Thick mottled floodplain siltstone and shale with occasional mudcracks and tightly cemented sandstone interbeds;

6) Thin eddy, drape and slough shales, often rippled or laminated, and less than 2 feet thick.

The prodelta and bar-fringe categories are restricted to the deltaic sequence within the reservoir (Fig. 3). The prodelta category is extremely continuous laterally and generally occurs below the reservoir; however, thick prodelta shales have been described within Zone 1. Bar-fringe shale interfingers with sandstone of the distributary mouth-bar complex. Despite its thin-bedded character (less than 2 feet), the bar-fringe facies has high expected continuity, and may be expected to thicken laterally into adjacent prodelta shales (Fig. 3).

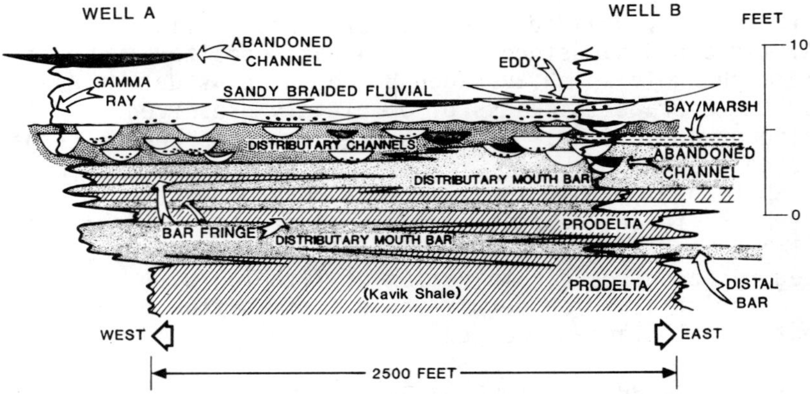

Figure 3. Interpreted depositional environments and associated genetic shale facies for Zone 1 and lower part of Zone 2 in two wells in the Eastern Operating Area, Prudhoe Bay field. Well B is a cored well for which core was described. Depositional facies picks in Well A are based on the predictive procedure described here; sandstone depositional environments are based on correlation and log response.

The marsh/bay facies is found interbedded with the distributary channel deposits and in many areas it separates the deltaic and overlying fluvial complexes. Continuity of the marsh/bay facies is variable but generally high. Correlation indicates that a combined marsh/bay and floodplain complex is continuous throughout much of the western part of the field.

The abandoned channel facies occurs in the distributary channel complex and within both braided and meandering river complexes (Fig. 4). Within these different environments it is expected to approximately conform to the shape of the original channel and so vary in cross-sectional shape as shown in Figure 4. Abandoned channel deposits can be expected to be longer in the direction of the original channel axis but restricted in width.

The floodplain facies constitutes the most extensive of the shale facies. Specific floodplains may be correlated for several miles within the field. Still other correlative

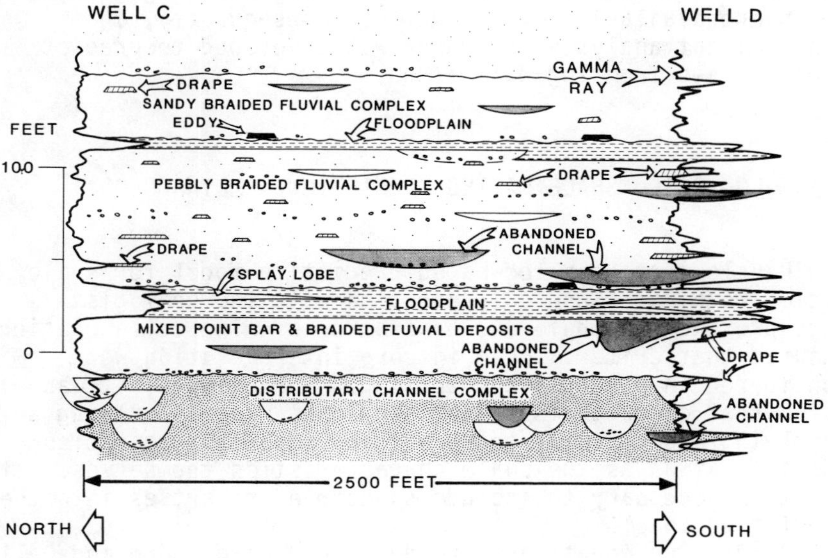

Figure 4. Interpreted depositional environments and associated genetic shale facies for Zone 2 in two wells in the Eastern Operating Area, Prudhoe Bay field. Well C is a cored well for which core was described. Genetic shale facies in Well D are based on the predictive model described here; sandstone depositional environments are based on correlation and log character.

floodplain deposits were dissected into smaller units by subsequent channel erosion.

The final facies category, eddy/drape/slough, forms a suite of small discontinuous shale types deposited in different subenvironments. Eddy deposits occur at the bottom of channels subsequently filled by sand or gravel. Drapes were deposited on sandy bars as flood stage flows waned. Slough deposits were deposited in the lee of midstream bars in sandy or pebbly braided rivers, as described by Cant and Walker (1978). Identification of the shale types within this facies depends upon their position within a fining-upward cycle, which makes petrophysical discrimination difficult. Therefore, we have combined them into a single category. Lateral extent of this facies can be expected to range from a few feet to several tens of feet.

The facies interpretations outlined derive from characteristics observable in drill core. At Prudhoe Bay, however, wells available for examination compose only 2 percent of the total wells drilled. A complete assessment of shale facies throughout the reservoir thus requires that genetic shale facies be identified using well log data, which are available for all wells. Consequently, a discriminant analysis procedure was developed to predict shale facies from log data alone.

III. LOG-BASED SHALE FACIES PREDICTION - METHODOLOGY

Development of a log-based predictive model for shale facies required close collaboration with the geologist interpreting the shale facies from core. The collaboration insured that criteria used in core interpretation were considered in predictive modelling. For example, it was noted that characteristics of sandstone beds underlying and overlying a particular shale were as important to facies identification as the shale characteristics themselves. It was thus necessary to include sandstone properties into the predictive model.

Predictive equations were developed using core and well log data from nine wells in the Eastern Operating Area. A tenth well with core observations held out was treated as an uncored well. The predicted facies in this "test well" were then compared with observations as a test of the validity of the predictive model.

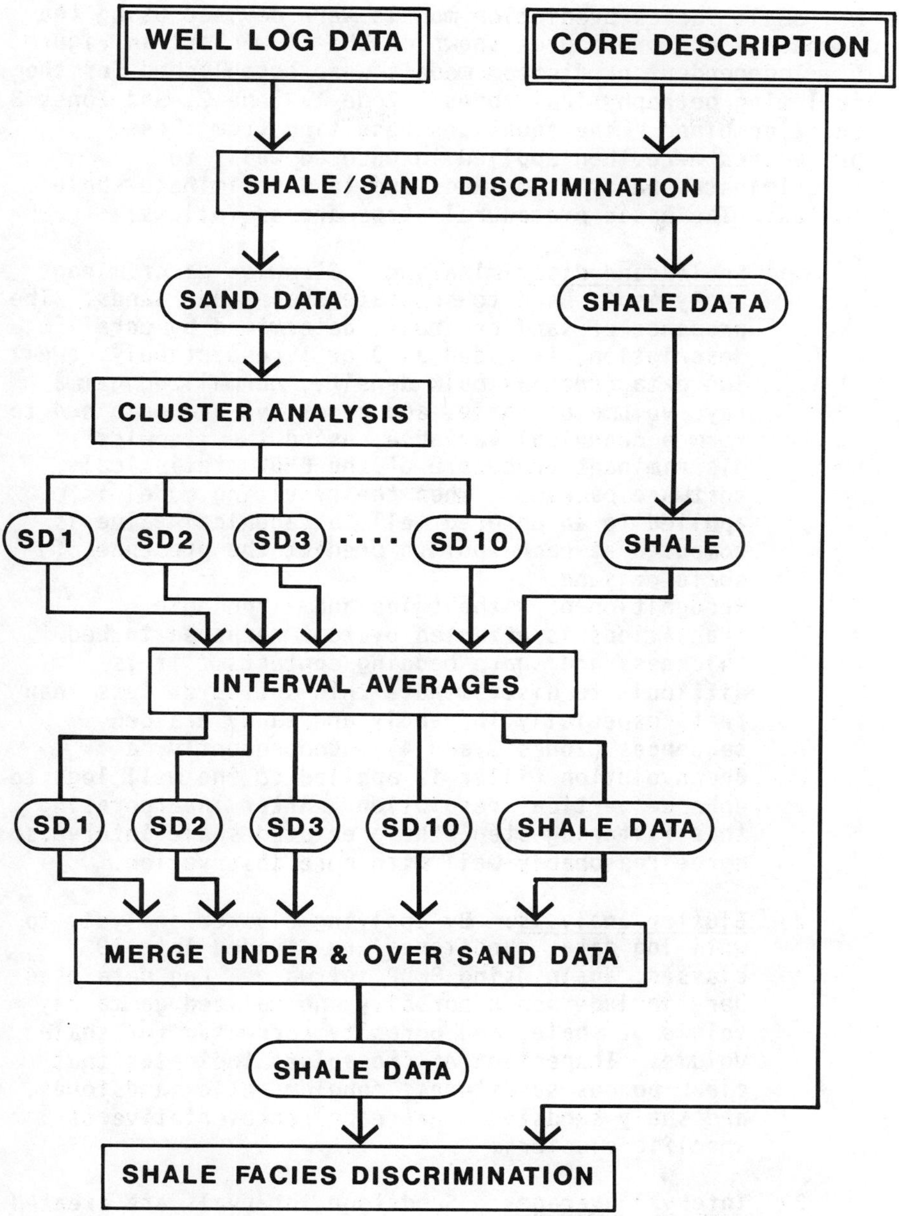

Figure 5. Flow chart showing computational procedures to develop a shale facies predictive model. SD1, SD2, etc. indicate specific log variable clusters for sandstone data.

Shale facies prediction models were devised using the computational procedures shown on the flow chart in Figure 5. Independent prediction models were established for the following petrophysical zones: Zone 1, Zone 2, and Zones 3 and 4 combined. The equations resulting from these procedures were then applied in uncored wells to discriminate shales from sands and to discriminate shale facies. The basic procedural steps are as follows:

1) Shale/sand discrimination: Stepwise discriminant analysis is used to separate shales and sands. The presence of sand or shale, determined by core description, is coded as 0 or 1 respectively. Well log data (such as bulk density, normalized gamma ray, volume of shale, and porosity) are combined to form a canonical variable, using the stepwise discriminant procedure of the BMDP statistical software package. When the resulting model is applied to an uncored well, a canonical value is computed at each foot to predict the presence of shale or sand.
   Recognition of lithologies and lithologic transitions is affected by tool response to bed thickness and sharp bedding contacts. It is difficult to discriminate thin shales of less than 2 feet, especially in pebbly and sandy braided sequences (Zones 3 and 4). Consequently, a deconvolution filter is applied to the well logs to enhance vertical resolution. After incorporating this filtering step, the predicted shale intervals agree reasonably well with core observation.

2) Cluster analysis: By applying cluster analysis to well log data, sandstones are divided into 10 classes, again using BMDP software. Log data used here include sonic porosity, normalized gamma ray, volume of shale, and porosity corrected for shale volume. Inspection of crossplots indicates that clean porous sandstones, conglomeratic sandstones, and shaly sandstones are each representative of specific clusters.

3) Interval averages: Sandstone intervals are created by grouping contiguous footages of sand having the same cluster number. Shale intervals are created wherever contiguous footages of shale occurred. The thickness of each interval is computed at this step and the well log data are then averaged over each

interval. From this point on, all computations are done on an interval basis, rather than on a foot-by-foot basis.
A sharp lithologic change, easy to recognize in core, is recorded as a gradual change on well logs because of the smoothing effect of tool resolution. Thus a transitional zone is created between sand and shale lithologies. In the model these are usually manifested as one-foot thick sandstone clusters adjacent to shales. The transitional cluster would be interpreted as shaly sandstone in almost all cases. To eliminate these transitional effects, the one-foot sand clusters are placed in the adjacent sand cluster by recalculating weighted averages.

4) <u>Merging underlying and overlying sand data</u>: This step associates the properties of adjacent sands with each shale interval, thereby creating new variables. For example, the "porosity" <u>difference</u> between a shale and underlying sand is computed and added to the list of variables for that particular shale.

5) <u>Shale facies discrimination</u>: Shale facies discrimination is handled in a manner similar to sand/shale discrimination: a model is built using the core description and then applied to uncored wells. But facies discrimination is more complicated because 5 or 6 facies are possible instead of just a choice between sand or shale. The independent variable which best separates the shale facies is entered first, followed by the second variable providing for the best combined discrimination, and so on. Classification functions are then computed for each shale facies. Finally, each shale interval is classified as the shale facies with the highest classification score.

Figure 6, an example from Zone 1, shows how well log variables can discriminate different shale facies. NGR*H (Normalized gamma ray * thickness) is the most important variable for classifying facies within delta-front to delta-plain sequences. Prodelta shales can easily be separated from other shales on the basis of their high NGR*H values. The next most important variable, the difference of Vsh (volume of shale) between the shale and underlying sand, is the ordinate of the crossplot. Most of the abandoned

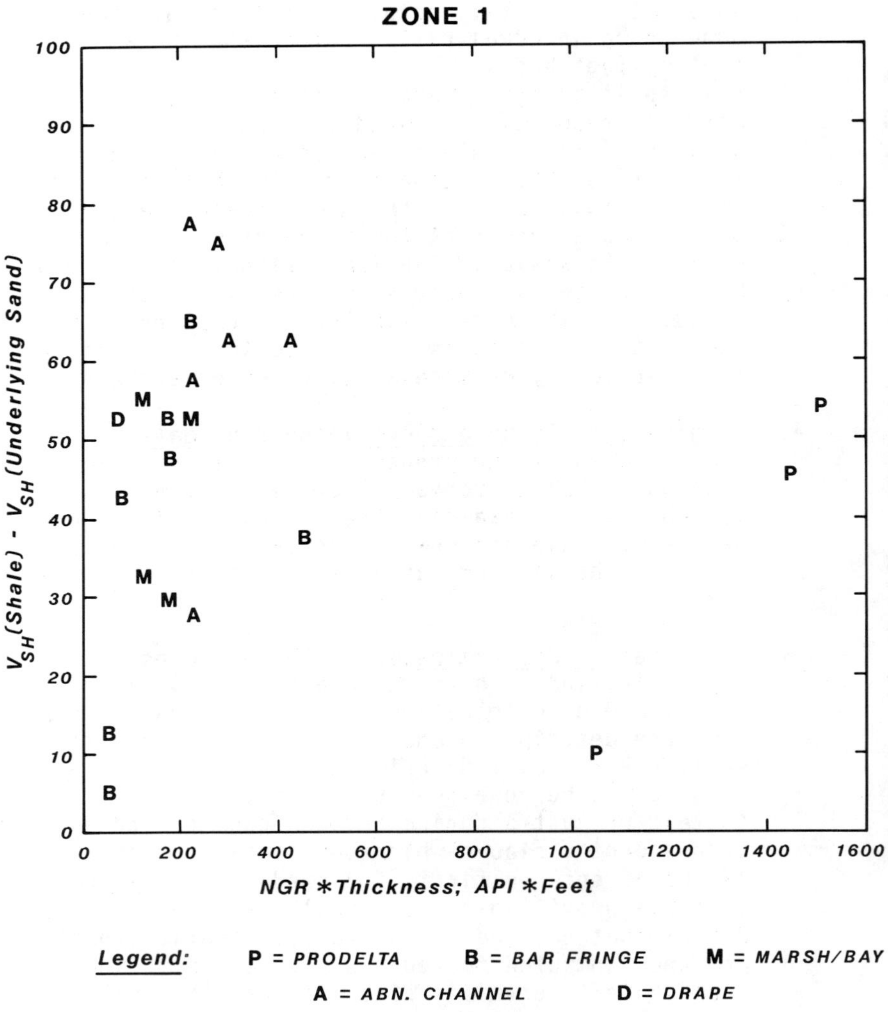

Figure 6. Crossplot showing how the first and second discriminating variables separate various shale facies within Zone 1. NGR is natural gamma ray. Vsh is volume of shale.

channel shale facies can be distinguished from other low NGR*H shale facies by their larger Vsh differences. This happens because abandoned channel fill shales often abruptly overlie cleaner sandstone sequences, while other shales have gradational lower contacts, and thus smaller Vsh differences.

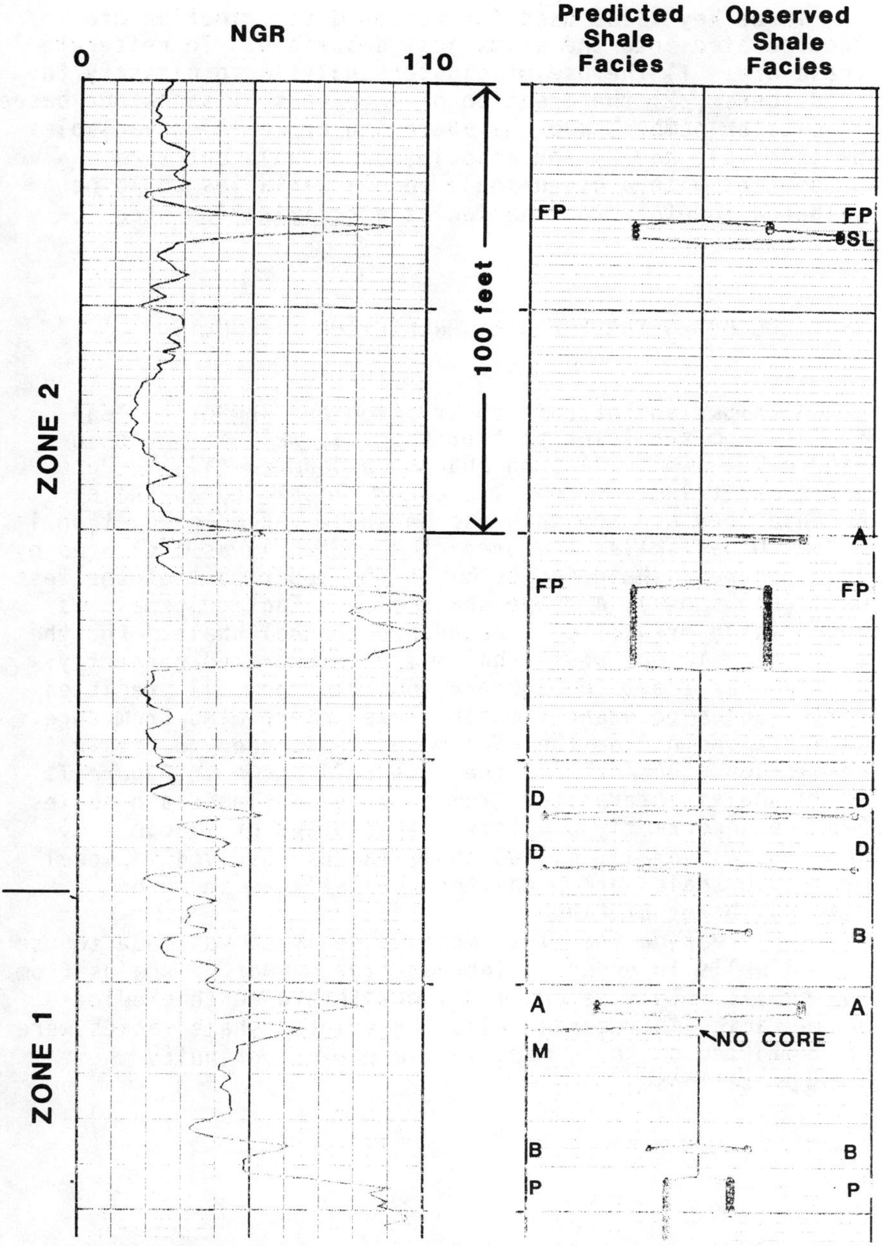

Figure 7. Comparison of core-observed shale facies and log-predicted shale facies, from the lower part of the test well. FP = floodplain; SL = slough; A = abandoned channel; D = drape; B = bar fringe; M = marsh/bay; P = prodelta.

Three key ideas used for facies discrimination are incorporated into the steps just described. To reiterate, these are: 1) the use of cluster analysis to classify the sandstones , 2) the creation of intervals of sandstone based on cluster number, and the averaging of well log variables by interval, and 3) the association of adjacent sand properties with a given shale so all variables could be gathered together for the facies discrimination step.

IV. LOG-BASED SHALE FACIES PREDICTION - RESULTS

A comparison of core observation and log predicted facies from the lower part of the test well is shown in Figure 7. Two shales, an abandoned channel fill in Zone 2 and a bar fringe in Zone 1 are not classified as shale because they are too thin, so no shale facies prediction is made. Other shales are predicted perfectly as indicated by the "observed shale facies". In the upper part of the test well (not shown), a drape shale of one-foot thickness is incorrectly predicted as abandoned channel shale. For the test well, 10 out of 11 shales are interpreted correctly.

The bar chart in Figure 8 indicates how all predicted shale facies occurences match those interpreted from core. Data are shown from the nine model wells used to create prediction equations and the test well shown in Figure 7. Of 29 shales interpreted from core to be floodplain shales, two are incorrectly predicted to be drape or slough deposits. Overall, of 100 shale facies observed in model wells, 90 shales are predicted correctly as shown on diagonal boxes of Figure 8.

For creating the final model a decision was made to use all 10 wells in order to increase the number of shales from 100 to 111. This second model was then used to predict shale facies in uncored wells. Predicted shale facies were then plotted on shale maps to aid shale continuity interpretations.

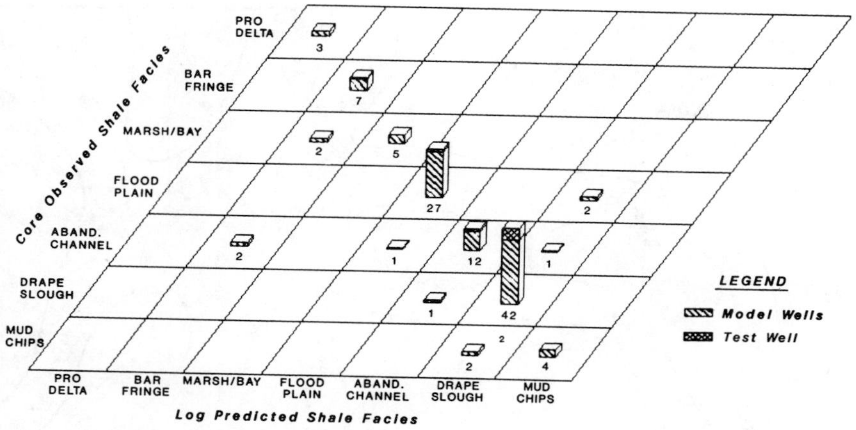

Figure 8. Bar chart showing the overall match between observed and predicted shale facies for the test well and all model wells. The numbers of individual cases are indicated below each "bar".

V. CONTINUITY PREDICTION

Figure 9 is a preliminary shale map of a portion of the reservoir. Shale thicknesses are contoured without geologic bias but with the assumption that all the stratigraphically equivalent shales are continuous between adjacent wells. The resulting shale distribution is a maximum continuity case, similar to maps generated by commercial mapping programs. Figure 10 shows the same data contoured to conform to geometry/continuity interpretations guided by predicted shale facies types. Predicted floodplain shales are shown mostly as continuous, while abandoned channel drape, and slough facies shales are shown as discontinuous. The approximate sizes of the discontinuous shales are taken from shale length data collected from outcrops of the Castlegate Sandstone in Utah (Fig. 11). The outcropping formation is believed to consist of similar depositional facies, and thus lithologic geometry and continuity, to that of the fluvial portion of the Ivishak at Prudhoe Bay.

As the discontinuous shales are believed to have dimensions generally much smaller than well spacings, it is reasonable to assume the presence of discontinuous shales between well locations. Given the frequency of discontinuous shales encountered in the well bore and

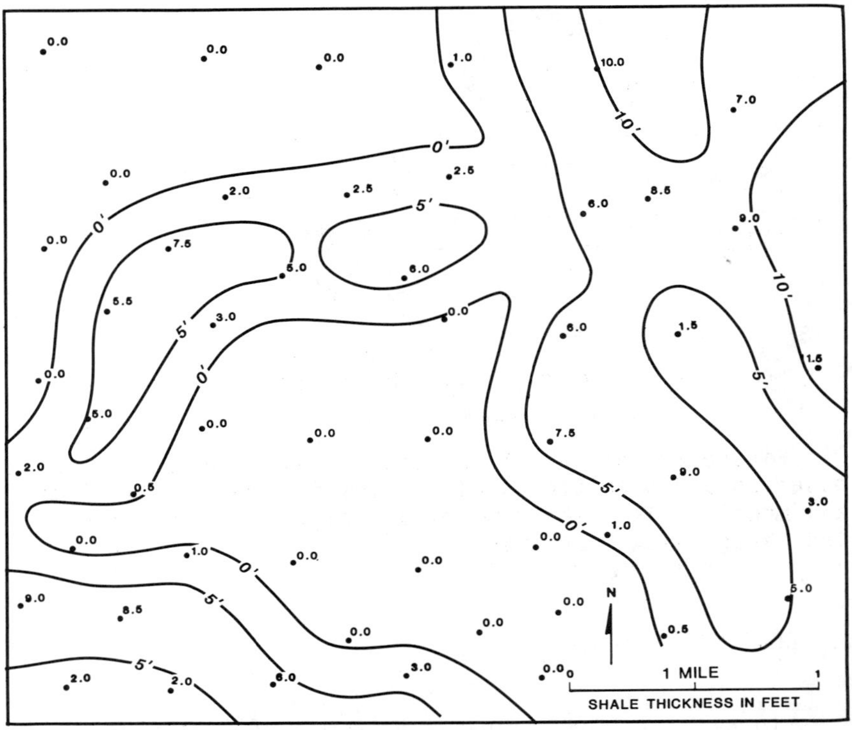

Figure 9. Contoured shale thickness data (net shale within a correlated interval 10 to 12 ft. thick): no geologic bias.

size-frequency data such as that shown in Figure 11, a stochastic model of discontinuous shale distribution can be produced (Haldorsen and Lake, 1982). Within the inset of Figure 10, the frequency of discontinuous shale types roughly equals the probability, calculated from well-bore shale occurrences, of encountering each type in the mapped interval.

Comparison of Figures 9 and 10 reveals the improved shale description that results through inclusion of predicted shale facies in the mapping. The utility of the interpreted map is increased in several ways:

1. A more realistic assessment of shale continuity will allow more accurate prediction of reservoir performance.

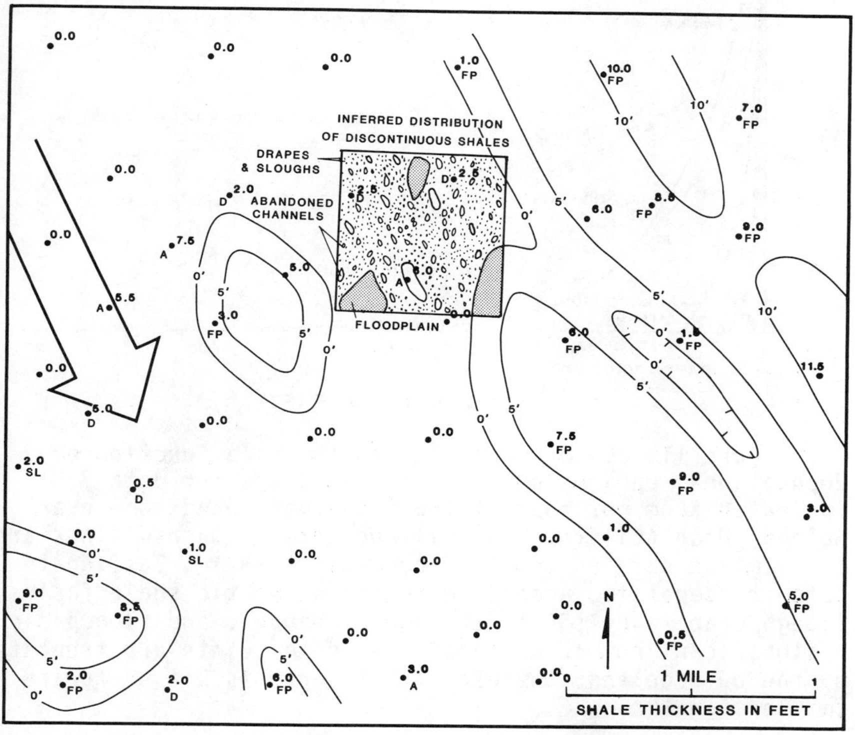

Figure 10. Contoured shale thickness data (net shale within a correlated interval 10 to 12 ft. thick): biased to conform to geometry/continuity interpretations. Shale facies codes generally follow those predicted by log-based discrimination. FP = floodplain; A = abandoned channel; D = drape; and SL = slough. The inset square diagramatically shows a distribution of discontinuous shales in the interwell spaces that might be inferred from the frequency of encounters with various discontinuous shale facies by existing well bores. The large arrow indicates the local channel-complex stream-flow direction inferred from the orientation of floodplain shale boundaries.

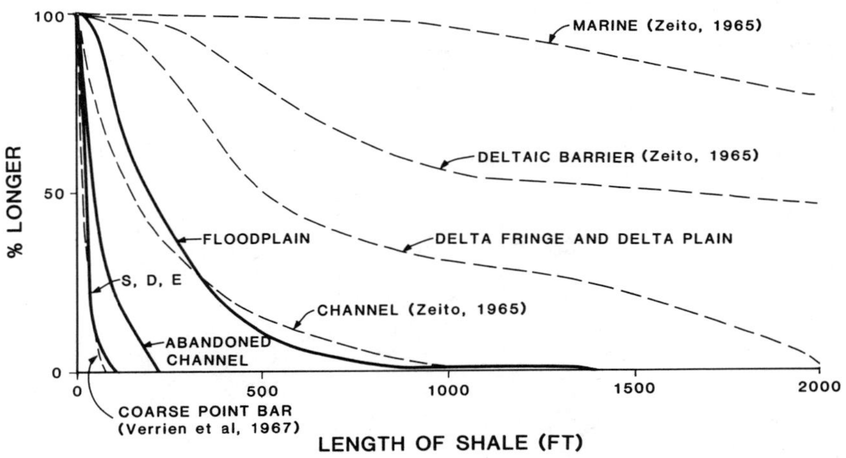

Figure 11. Observed shale lengths as a function of depositional environment. Solid lines are for data collected from outcrops of the Castlegate Sandstone near Helper, Utah (Clifton, unpublished data). Dashed lines are for similar data as compiled by Weber (1982). Castlegate data are separated according to three genetic shale facies: slough/drape/eddy (SDE), abandoned channel, and floodplain. A significant number of floodplain data points are truncated by the outcrop end; therefore that curve is biased toward smaller lengths.

2. As performance data for the map area become available, the shale continuity illustrated can be modified using geologically reasonable guidelines instead of arbitrary ones.

3. Paleocurrent trends (and their possible influence on permeability trends) can be estimated from the distribution and thickness of continuous shale bodies, as shown in Figure 10. Paleocurrent estimation is more difficult based on the information in Figure 9.

VI. CONCLUSIONS

1. Important reservoir lithologies should not be simply correlated and mapped without some consideration of the likely geometry and continuity of those lithologies.

2. Often the spacing of well data points in the subsurface is greater than the extent of lithologic units; consequently, geometry and continuity must be inferred from careful depositional facies studies and with reference to appropriate analogs.

3. Using the example of shale bodies within the Prudhoe Bay reservoir, we have shown that detailed core examinations and the application of facies analogs can significantly modify interpreted shale continuity.

4. Furthermore, we have shown that depositional facies interpretations based on core observation can be successfully extended to uncored wells with discriminant analysis of log-based variables.

ACKNOWLEDGMENTS

The application of log-based statistical discriminant procedure to shale facies prediction was conceived by Doug Lorenz and implemented with significant help from Phil Nelson. The immense job of managing and merging correlation facies, and well log data was handled by Tim Atkins, Susan Haymaker and Lynn Skinner. The authors benefitted from discussions with numerous Sohio colleagues besides those mentioned above, including Helge Haldorsen, Roger Plimmer and Mark Bradley.

REFERENCES

Cant, D.J. and Walker, R.G., 1978, Fluvial processes and facies sequences in the sandy braided South Saskatchewan River, Canada: Sedimentology, v. 25, p. 625-648.

Chang, D.M., Haldorsen, H.H., and Kirwan, P.A., 1984, History matching and predicting GOR behavior with analytical functions: Society of Petroleum Engineers of AIME paper no. 12763, p. 337-343.

Haldorsen, H.H. and Lake, L.W., 1982, "A new approach to shale management in field scale simulation models", SPE Paper No. 10976.

Jones, H.P. and Speers, R.G., 1976, Permo-Triassic reservoirs of Prudhoe Bay Field, North Slope, Alaska, in Braustein, J. (ed.) North American Oil and Gas Fields, AAPG Mem. 24, p. 23-50.

Verrien, J.P., Cournad, G., Montadert, L. 1967, Application of production geology methods to reservoir characteristics analysis from outcrop observations, Proc. Seventh World Pet. Cong., Mexico, p. 425-446.

Weber, K.J., 1982, Influence of common sedimentary structures on fluid flow in reservoir models, J. Pet. Tech., p. 665-672.

Zeito, G.A., 1965, Interbedding of shale breaks and reservoir heterogeneities, J. Pet. Tech., p. 1223-1228.

POROSITY AND MINERAL ALTERATION BY FLUID FLOW THROUGH A TEMPERATURE FIELD

Thomas A. Hewett

Chevron Oil Field Research Company
La Habra, California

I. ABSTRACT

The pattern of porosity alteration produced by an aqueous solution flowing slowly through a porous medium is considered. On the scale of geologic time, it is assumed that the fluid and rock maintain local equilibrium. As the fluid moves through gradients of temperature and pressure, the equilibrium solubilities of the different mineral phases change and cause material to be leached from or precipitated onto the solid matrix, changing the porosity and mineral composition. The pattern of this diagenetic alteration by the moving fluid and its evolution in time are determined by the temperature and flow fields and the equilibrium relations for the mineral dissolution/precipitation reactions. The theory presented permits a quantitative prediction of these patterns from a knowledge of the flow and temperature fields and thermodynamic data for the minerals involved.

When conditions are quasi-steady and the primary driving potential for mass transfer is flow through temperature gradients, it is shown that a complete description of the sequence of mineral compositions can be derived independent of the flow field. It is also shown that the position of each composition moves in a wave-like manner and that, as long as all of the mineral phases initially present remain as solid phases, the pattern of porosity and mineral alteration can be described as a continuous wave field of reversible changes. When one of the original mineral phases is depleted in some region of the flow, discontinuities in mineralogy and porosity are introduced which propagate downstream and irreversibly alter the rock.

The particular case of a fluid in a thin, gently-folded, porous layer bounded by impermeable shales is considered and solutions for the temperature and flow fields are obtained. The relative contributions of natural convection driven by lateral temperature gradients and forced flow driven by hydraulic gradients is shown to depend on a single dimensionless parameter, referred to as the Convective Flow Index, or CFI. The patterns of rock alteration predicted for a range of the CFI in a cosine-shaped fold are presented. The importance of formation structure on the pattern of diagenetic alteration is emphasized.

II. INTRODUCTION

The primary control on the spatial distributions of porosity and permeability in sedimentary rocks is the distribution of sediments at the time of burial. Significant alterations of these primary patterns can occur subsequent to burial by compaction, cementation, and mineral dissolution. There is generally a decrease in porosity and permeability with increasing depth of burial due to compaction. At depth, however, differences in reservoir quality are generally a function of the amount of cementation and dissolution which has altered the initial grain framework (Loucks et al., 1984).

The role of moving pore fluids in altering sediments subsequent to burial has received increased attention in recent years (Schmidt and McDonald, 1979; Pittman, 1979; Cathles, 1981, 1983; Wood and Hewett, 1982, 1984). The movement of fluids through the pore space of sedimentary rocks can be caused by a variety of driving potentials. These have been used to classify the different hydrologic regimes through which a sedimentary basin passes (Galloway, 1984). Although the boundaries of these regimes are not always clearly defined, they may nonetheless be useful in characterizing the types of flow dynamics to which a basin may be subjected. The meteoric regime is dominated by the infiltration of surface waters and flow is along the hydraulic gradient. The compactional regime is characterized by upward and outward expulsion of pore waters trapped within compacting sediments. The thermobaric regime lies in the deepest portions of the basin where the temperature and pressure are greatest and significant amounts of water may be released by the dehydration reactions of clay minerals and other hydrous mineral phases.

In addition to these potentials for forced convection, thermal conductivity contrasts in dipping sedimentary sequences lead to horizontal temperature and fluid density gradients which drive natural convection flows (Wood and Hewett, 1984; Davis et al., 1985). These natural convection flows may be regarded as background flows which are present throughout the evolution of a basin and upon which the others are superimposed. The importance of natural convection in diagenetic processes depends on the relative magnitudes of the driving potentials for forced and natural convection. When strata are isolated from surface hydraulic gradients, or these are weak, natural convection will predominate and diagenesis will proceed in what may be termed the thermoconvective regime.

On the scale of geologic time, most geochemical processes can be considered to maintain a state of local equilibrium between aqueous solutions and the mineral phases they contact (Helgeson,1979). If solutions flow slowly through gradients in equilibrium conditions, then mass transfer between the fluid and solid phases will occur. When a solution moves to a region where the equilibrium solubility of a mineral phase increases, it will dissolve some of that phase to maintain equilibrium. Conversely, when a solution moves to a region where a mineral has a lower equilibrium solubility, that mineral will be precipitated out of solution onto the solid phase to maintain equilibrium. The solubility of each component is determined by the equilibrium relations for all of the reactions in which the component participates.

This paper examines the nature of rock alteration by flowing aqueous solutions. First, the relations which must be satisfied for mass to be conserved in reactions among aqueous species and between solid mineral phases and aqueous solutions are reviewed. The assumption of partial local equilibrium is introduced and it is shown how mass transfer between the solid phases and aqueous solution is driven by flow through gradients of local equilibrium conditions.

The particular case of mass transfer due to flow through a temperature field is considered and it is shown that the sequence of compositions which are produced by this mechanism can be predicted independent of the details of the flow and temperature fields and that a simple mapping of this sequence onto the physical domain is possible when the flow and temperature fields are known. The patterns of porosity and mineral alteration produced by this mechanism can be described as a field of continuous waves of reverible chemical alteration which radiate from local sources of new compositions. These patterns persist until one or more of the original mineral phases is depleted in some region, at which point discontinuous fronts of change in porosity and mineralogy are introduced which irreversibly alter the rock.

Fluid flow in a thin, gently-folded porous layer bounded by impermeable shales is then considered and solutions for the flow and temperature fields are obtained. The patterns of porosity alteration in a cosine-shaped fold are then presented as an example of the influence of layer structure on the patterns of diagenetic alteration by moving pore fluids.

III. MASS CONSERVATION IN FLUID-ROCK INTERACTIONS

Mass transfer between minerals and aqueous solutions in geochemical processes is described by reactions of the form

$$\sum_{k=1}^{K} \upsilon_{kr} S_k + \sum_{\ell=1}^{L} \upsilon_{\ell r} S_\ell = 0 \qquad r=1,\ldots,R \qquad (1)$$

where S_k is the chemical formula for mineral k (k=1,...,K), S_ℓ is the formula for aqueous species ℓ (ℓ=1,..,L), and υ_{kr} and $\upsilon_{\ell r}$ are the stoichiometric coefficients for the subscripted species in reaction r (r=1,..,R). The stoichiometric coefficients are taken to be negative for reactants and positive for products. When a reaction involves only aqueous species, $\upsilon_{kr}=0$. In dissolution/precipitation reactions, $\upsilon_{kr}=-1$.

Equation 1 specifies the relative rates of mass transfer between different species. If $R_{\ell r}$ is the rate at which species ℓ is added to the aqueous phase per unit volume of aqueous solution by reaction r, then we can write

$$R_{\ell r} = \upsilon_{\ell r} \frac{\partial \xi_r}{\partial t} = \upsilon_{\ell r} \dot{\xi}_r \qquad (2)$$

where ξ_r is the progress variable for.reaction r (Prigogine and Defay, 1954; Helgeson, 1979) and $\dot{\xi}_r$ is the reaction rate expressed as moles per unit time per unit volume of solution. Similarly, if R_{kr} is the rate at which mineral k is added to the solid phases by reaction r per unit pore volume, then

$$R_{kr} = \upsilon_{kr}\frac{\partial \xi_r}{\partial t} = \upsilon_{kr}\dot{\xi}_r \tag{3}$$

Either of these expressions can be used to evaluate $\dot{\xi}_r$ when information about the rate of change of a species is known.

When chemical species can participate in more than one reaction, the total rate at which moles of ℓ are added to the solution per unit pore volume is

$$R_\ell = \sum_{r=1}^{R} R_{\ell r} = \sum_{r=1}^{R} \upsilon_{\ell r}\dot{\xi}_r \tag{4}$$

and the total rate of addition of mineral k to the solid phases is

$$R_k = \sum_{r=1}^{R} R_{kr} = \sum_{r=1}^{R} \upsilon_{kr}\dot{\xi}_r = -\dot{\xi}_k \tag{5}$$

where r=k is the dissolution reaction for mineral phase k.

When there is no migration of solid mineral species, a mass balance on a solid phase k requires the molar concentration, C_k, of that phase to change in accordance with

$$\frac{\partial C_k}{\partial t} = \phi R_k = -\phi\dot{\xi}_k \tag{6}$$

The porosity, ϕ, appears in Equation 6 to account for the fact that the reaction rate is written per unit volume of solution while the molar concentration, C_k, is for a unit bulk volume of rock. If we restrict our attention to pure mineral phases, the molar concentration can be related to other properties of interest by the relation

$$C_k = \frac{\rho_k \chi_k}{M_k} \tag{7}$$

where ρ_k is the density of mineral phase k, χ_k is the fraction of the bulk volume occupied by phase k, and M_k is its molecular weight.

When the molar concentrations of all of the mineral phases are known, the porosity can be calculated as

$$\phi = 1 - \sum_{k=1}^{K} \chi_k = 1 - \sum_{k=1}^{K} \left(\frac{M_k C_k}{\rho_k}\right) \tag{8}$$

Changes in porosity as a consequence of chemical reactions of the mineral phases with the solution phase may then be calculated as

$$\frac{\partial \phi}{\partial t} = \phi \sum_{k=1}^{K} \left(\frac{M_k}{\rho_k} \dot{\xi}_k\right) \tag{9}$$

A mass balance on a species ℓ in the solution phase requires

$$\frac{\partial}{\partial t}(\phi c_\ell) + \underset{\sim}{V} \cdot \underset{\sim}{\nabla} c_\ell = \underset{\sim}{\nabla}(\phi D_{e,\ell} \underset{\sim}{\nabla} c_\ell) + \phi R_\ell \tag{10}$$

where c_ℓ is the molarity of aqueous species ℓ, $\underset{\sim}{V}$ is the superficial or Darcy velocity, $D_{e,\ell}$ is the effective dispersion coefficient for species ℓ in the aqueous phase, and R_ℓ is the rate of addition of species ℓ to the aqueous phase by chemical reactions given by Equation 4. For flow velocities less than about 50 m/yr, the effects of mechanical dispersion are negligible and the effective dispersion coefficient can be related to the molecular diffusion coefficient of species ℓ in the aqueous phase, D_ℓ, by

$$D_{e,\ell} = \frac{D_\ell}{\tau} \tag{11}$$

where τ is the tortuosity of the pore space (Blackwell, 1962). The relative rates of convective and diffusive transport are determined by the value of the Peclet number, Pe_ℓ, defined as

$$Pe_{\ell} = \frac{|\underset{\sim}{V}|L}{D_{e,\ell}\phi} \tag{12}$$

where L is the distance over which transport occurs. In the geologic context, this number will be large when fluid flow velocities are on the order of a meter a year or more and the distances over which material is transported are greater than about a meter (Wood and Hewett, 1982). For these conditions, Equation 10 simplifies to

$$\frac{\partial(\phi c_{\ell})}{\partial t} + \underset{\sim}{V} \cdot \underset{\sim}{\nabla} c_{\ell} = \phi R_{\ell} \tag{13}$$

When the set of reactions includes redox reactions in which species change oxidation states, it is necessary to include expressions which insure electrical neutrality and conservation of electrons in the flowing system (Walsh, et al., 1984). If, however, we restrict our attention to systems which do not involve redox reactions, then conservation of charge is implicitly maintained by conservation of mass (Helgeson et al., 1970).

One further simplification is possible if we limit our attention to solutions dilute in the products of the dissolution reactions considered. For most mineral species of interest, $m_{\ell} = c_{\ell}M_{\ell}/\rho_s < 10^{-3}$ (1000 ppm). When this is true, Equation 13 can be combined with Equations 9 and 4 to show that

$$\phi \frac{\partial c_{\ell}}{\partial t} + \underset{\sim}{V} \cdot \underset{\sim}{\nabla} c_{\ell} = \phi \sum_{r=1}^{R} \upsilon_{\ell r} \dot{\xi}_r \tag{14}$$

IV. LOCAL PARTIAL EQUILIBRIUM WITH CONNATE AQUEOUS SOLUTIONS

Geologic systems are commonly in a state of partial equilibrium during geochemical processes (Helgeson et al., 1970). By this we mean that the systems are in equilibrium with respect to at least one chemical reaction, but may be out of equilibrium with respect to others. In the development that follows it will be assumed that equilibrium is maintained among species in the aqueous solution and between the aqueous solution and mineral phases dissolved or precipitated in reversible reactions. The only departures from equilibrium will be due to rate-limiting irreversible reactions which add species to the aqueous phase. Typically these reactions involve the slow dissolution of one of the mineral phases or the slow evolution of reaction products from the thermal decomposition of organic matter in neighboring strata (Franks and Forester, 1984). The determination of these irreversible reaction rates is beyond the scope of this treatment and they will be assumed to be characterized by a single kinetic progress variable, ξ_t, which is uniform throughout the region considered. This irreversible progress variable corresponds to the overall reaction progress variable of Helgeson (1979) for isothermal, isobaric conditions.

If we consider a system of R chemical reactions involving L aqueous species, then the solution composition will be described by L equations of the form

$$c_\ell = c_{\ell,i} + \sum_{r=1}^{R} \upsilon_{\ell r}\xi_{e,r} \qquad \ell=1,\ldots L \qquad (15)$$

where $c_{\ell,i}$ is the concentration of species ℓ in the fluid prior to contacting the mineral assemblage and $\xi_{e,r}$ is the progress in reaction r required to bring the solution into partial equilibrium with the mineral assemblage at the local conditions. If a subset E of the reactions are fast enough to establish equilibrium then the values of $\xi_{e,r}$ for those reactions will be constrained by E expressions for the law of mass action applied to each of the reactions.

For reactions between aqueous species, the law of mass action is of the form

$$\prod_{\ell=1}^{L} a_\ell^{\upsilon_{\ell r}} = K_r \qquad r=1,\ldots,E \qquad (16)$$

where K_r is the thermodynamic equilibrium constant for reaction r and a_ℓ is the chemical activity of aqueous species ℓ. The chemical activities are related to molar concentrations through activity coefficients, γ_ℓ, where

$$a_\ell = \gamma_\ell c_\ell \qquad \ell=1,\ldots L \qquad (17)$$

We will restrict our attention to dilute solutions for which $\gamma_\ell = 1$. For reactions between solid mineral phases and aqueous solutions, the same form of the mass action equation applies, with the activities of stoichiometric solid phases taken as unity (Helgeson, et al., 1970) and the equilibrium constant replaced by the solubility product for the mineral in question. When the product in Equation 16 is less than the solubility product, the solution is undersaturated and further progress in the reaction is arrested.

When a solution and a mineral assemblage are equilibrated statically, the solution of Equations 6 and 14 with V=0 shows that each mineral phase will dissolve until the concentrations of the aqueous species satisfy Equation 16 or until the initial amount of a minor mineral phase is depleted. The progress in the dissolution of such a minor mineral, m, will stop when

$$\xi_{e,m} = \frac{C_{m,i}}{\phi} \tag{18}$$

i.e., when all of the $C_{m,i}$ initial moles per unit bulk volume have reacted. For these minerals, the product in Equation 16 will be less than the solubility product and the reaction progress for reaction m will be a constant determined by Equation 18.

When a solution and a mineral assemblage are equilibrated by flowing the unequilibrated solution through a porous medium containing the minerals, the solution of Equations 6 and 14 with Equation 16 shows that a series of reaction fronts are initiated where the fluid enters the rock (Hoffman, 1972; Lund and Fogler, 1976; Walsh, et al., 1984). These fronts travel in the direction of fluid flow but at velocities which are greatly retarded relative to that of the fluid. Downstream of these fronts the mineral assemblage will be in partial equilibrium with a solution saturated in all of the E minerals which were initially present and participate in reversible reactions (Walsh, et al., 1984).

In all of the subsequent discussion of mass transfer due to flowing solutions it will be assumed that the progress of kinetically controlled reactions and the dissolution reactions for any undersaturated minerals are known. The solution composition and progress in the reversible reactions at partial equilibrium with the mineral assemblage can then be determined from Equation 16. Mathematically, we must solve for the L unknown solution concentrations and the R progress variables. For this we have E equilibrium Equations 16, R-E specified kinetic and minor mineral progress variables, and L Equations 15 for a total of R+L equations. The R+L equations described above are non-linear. When the number of equations are small, their solution may be obtained by algebraic elimination of variables and back substitution. For most problems, however, a numerical solution will be required. A standard Newton-Raphson technique (Carnahan, et al., 1969; Walsh, et al., 1984) is adequate for solving the systems of equations commonly encountered in chemical equilibrium calculations.

V. MASS TRANSFER IN FLOWING EQUILIBRIUM SYSTEMS

The relation between the rate of progress in fluid-rock reactions and the composition of the flowing solution required for mass conservation in the aqueous phase was given by Equation 14. When the aqueous solution and a mineral assemblage are in partial equilibrium, the solution composition is given by Equation 15, where the values of the $\xi_{e,r}$ are constrained by the equilibrium conditions of Equation 16 for reversible reactions, or are specified for undersaturated minerals and minerals dissolving in irreversible reactions. Combining these, we see that for dilute solutions

$$\frac{\partial \xi_r}{\partial t} = \frac{\partial \xi_{e,r}}{\partial t} + \frac{1}{\phi} \underset{\sim}{V} \cdot \underset{\sim}{\nabla} \xi_{e,r} = \frac{D\xi_{e,r}}{Dt} \qquad (19)$$

where

$$\frac{D()}{Dt} \equiv \frac{\partial ()}{\partial t} + \frac{1}{\phi} \underset{\sim}{V} \cdot \underset{\sim}{\nabla}(\ )$$

is the convective derivative. The convective derivative gives the rate of change experienced by a moving fluid particle due to local changes in a field and by motion through it at an interstitial velocity V/ϕ. Equation 19 simply states that the reaction proceeds at the rate required to equilibrate the moving fluid as it experiences changes in equilibrium conditions.

Since the only changes which can occur in equilibrium conditions are those due to changes in temperature, pressure, or the addition of species by irreversible reactions, this may also be written as

$$\frac{\partial \xi_r}{\partial t} = \left(\frac{\partial \xi_{e,r}}{\partial T}\right)_{p,\xi_t} \frac{DT}{Dt} + \left(\frac{\partial \xi_{e,r}}{\partial p}\right)_{T,\xi_t} \frac{Dp}{Dt} + \left(\frac{\partial \xi_{e,r}}{\partial \xi_t}\right)_{p,T} \frac{\partial \xi_t}{\partial t} \qquad (20)$$

where it has been assumed that the irreversible reactions proceed uniformly in space. For coupled reactions, the change in any one reaction in response to changes in equilibrium will be reflected in all of the reactions. This means that the derivatives of the $\xi_{e,r}$ with respect to T, p, and ξ_t must be found simultaneously. The equations available to solve for these derivatives are the 3L equations obtained by differentiating Equation 15 with respect to the three equilibrium variables (T,p, and ξ_t). These are of the form

$$\left(\frac{\partial c_\ell}{\partial T}\right)_{p,\xi_t} = \sum_{r=1}^{R} \upsilon_{\ell r} \left(\frac{\partial \xi_{e,r}}{\partial T}\right)_{p,\xi_t} \qquad \ell=1,\ldots L \qquad (21)$$

$$\left(\frac{\partial c_\ell}{\partial p}\right)_{T,\xi_t} = \sum_{r=1}^{R} \upsilon_{\ell r} \left(\frac{\partial \xi_{e,r}}{\partial P}\right)_{T,\xi_t} \qquad \ell=1,\ldots L \qquad (22)$$

$$\left(\frac{\partial c_\ell}{\partial \xi_t}\right)_{p,T} = \sum_{r=1}^{R} \upsilon_{\ell r} \left(\frac{\partial \xi_{e,r}}{\partial \xi_t}\right)_{p,T} \qquad \ell=1,\ldots L \qquad (23)$$

The derivatives of the progress variables for the dissolution reactions of undersaturated minerals are zero. This gives us 3L equations in 3L+3R unknown derivatives of c_ℓ and $\xi_{e,r}$. The remaining 3R equations required may be obtained by differentiating the equilibrium relations. Since they are expressed as products of concentrations it is easiest to first rewrite them in logarithmic form as

$$\sum_{\ell=1}^{L} \upsilon_{\ell r} \ln c_\ell = \ln K_r \qquad r=1,\ldots,R \qquad (24)$$

Then,

$$\sum_{\ell=1}^{L} \frac{\upsilon_{\ell r}}{c_\ell} \left(\frac{\partial c_\ell}{\partial T}\right)_{p,\xi_t} = \frac{1}{K_r} \left(\frac{\partial K_r}{\partial T}\right)_{p,\xi_t} \qquad r=1,\ldots,R \tag{25}$$

$$\sum_{\ell=1}^{L} \frac{\upsilon_{\ell r}}{c_\ell} \left(\frac{\partial c_\ell}{\partial p}\right)_{T,\xi_t} = \frac{1}{K_r} \left(\frac{\partial K_r}{\partial p}\right)_{T,\xi_t} \qquad r=1,\ldots,R \tag{26}$$

$$\sum_{\ell=1}^{L} \frac{\upsilon_{\ell r}}{c_\ell} \left(\frac{\partial c_\ell}{\partial \xi_t}\right)_{T,p} = 0 \qquad r=1,\ldots,R \tag{27}$$

The derivatives of the equilibrium constant can be related to the standard state heat of reaction and molar expansion as

$$\left(\frac{\partial \ln K_r}{\partial T}\right)_{p,\xi_t} = \frac{1}{K_r} \left(\frac{\partial K_r}{\partial T}\right)_{p,\xi_t} = \frac{\Delta H^\circ_r}{RT^2} \qquad r=1,\ldots,R \tag{28}$$

$$\left(\frac{\partial \ln K_r}{\partial p}\right)_{T,\xi_t} = \frac{1}{K_r} \left(\frac{\partial K_r}{\partial p}\right)_{T,\xi_t} = \frac{-\Delta V^\circ_r}{RT} \qquad r=1,\ldots,R \tag{29}$$

where ΔH°_r is the standard state heat of reaction for reaction r, and ΔV°_r is the standard state molar expansion (Prigogine and Defay, 1954). Equations 21-27 are linear in the derivatives of c_ℓ and $\xi_{e,r}$ and may be solved by matrix inversion. Since these equations also depend on the values of the c_ℓ, these must first be determined by the method described in the previous section.

The chemical coefficients in Equation 20 can be determined for any chemical system as a function of temperature, pressure, and irreversible reaction progress independent of any specification of the flow field or pressure and temperature distributions. Similarly, the convective derivatives in Equation 20 do not depend on the particular chemical system involved (except for an independent specification of ξ_t) and may be calculated separately. Equation 20 then provides the critical link which permits the mapping of chemical changes onto physical space and time. Since the distribution of equilibrium solution composition is fixed by the equilibrium relations and the distributions of temperature and pressure, the only chemical changes which remain to be calculated are the changes in the solid mineral phases due to the effects of mass transfer with the flowing solution. Combining Equation 20 with Equation 6, the rate of change of a mineral phase k is

$$\frac{\partial C_k}{\partial t} = -\phi \frac{\partial \xi_k}{\partial t} = -\phi \left[\left(\frac{\partial \xi_{e,k}}{\partial T}\right)_{p,\xi_t} \frac{DT}{Dt} + \left(\frac{\partial \xi_{e,k}}{\partial p}\right)_{T,\xi_t} \frac{Dp}{Dt} + \left(\frac{\partial \xi_{e,k}}{\partial \xi_t}\right)_{p,T} \frac{\partial \xi_t}{\partial t} \right] \quad (30)$$

The change in the volume fraction of mineral phase k can be calculated as

$$\frac{\partial \chi_k}{\partial t} = \frac{M_k}{\rho_k} \frac{\partial C_k}{\partial t} = -\phi \frac{\rho_s}{\rho_k} \left[\alpha_{T,k} \frac{DT}{Dt} + \alpha_{p,k} \frac{Dp}{Dt} + \alpha_{\xi_t,k} \frac{\partial \xi_t}{\partial t} \right] \quad (31)$$

where

$$\alpha_{T,k} \equiv \frac{M_k}{\rho_s} \left(\frac{\partial \xi_{e,k}}{\partial T}\right)_{p,\xi_t}$$

$$\alpha_{p,k} \equiv \frac{M_k}{\rho_s} \left(\frac{\partial \xi_{e,k}}{\partial p}\right)_{T,\xi_t}$$

$$\alpha_{\xi_t,k} \equiv \frac{M_k}{\rho_s} \left(\frac{\partial \xi_{e,k}}{\partial \xi_t}\right)_{T,p}$$

give the rate of change of the effective solubility of k (on a mass of mineral per unit mass of solution basis) with respect to the variables shown. ρ_s is the solution density. The rate of change of porosity is then

$$\frac{\partial\phi}{\partial t} = \frac{\rho_s}{\rho_a}\,\phi\,\left[\alpha_T \frac{DT}{Dt} + \alpha_p \frac{DP}{Dt} + \alpha_{\xi_t} \frac{\partial\xi_t}{\partial t}\right] \tag{32}$$

where

$$\alpha_T \equiv \rho_a \sum_{k=1}^{K} \frac{\alpha_{T,k}}{\rho_k}$$

$$\alpha_p \equiv \rho_a \sum_{k=1}^{K} \frac{\alpha_{p,k}}{\rho_k}$$

$$\alpha_{\xi_t} \equiv \rho_a \sum_{k=1}^{K} \frac{\alpha_{\xi_t,k}}{\rho_k}$$

and ρ_a is the density of the aggregate mineral assemblage. Equation 32 is a generalization of earlier descriptions of diagenetic changes due to flowing solutions (Wood and Hewett, 1982; Wood and Hewett, 1984) to include multiple mineral phases and the influence of irreversible reactions on the state of partial equilibrium. It should be noted that the individual values of $\alpha_{T,k}$, $\alpha_{p,k}$, and $\alpha_{\xi t,k}$ may be positive or negative so the values of α_T, α_p, and α_{ξ_t} may be small or zero even when considerable mass transfer is occurring, This corresponds to isomorphic replacement of mineral phases.

The convective derivatives in the previous equations account for changes in equilibrium conditions encountered by a moving fluid due both to changes in conditions locally and to changes encountered along the flow path. In considering the entire burial history of a porous medium, the changes in solution equilibrium due to changes in the temperature and pressure fields and by progress in irreversible reactions will be important. In many cases, however, the local rate of change in solution equilibrium due to changes in these variables will be small compared to the rate at which a moving fluid encounters changes in equilibrium conditions due to its motion. For these cases, Equations 31 and 32 may be put into a "quasi-steady state" form as

$$\frac{\partial \chi_k}{\partial t} = - \frac{\rho_s}{\rho_k} [\alpha_{T,k} \underset{\sim}{V} \cdot \underset{\sim}{\nabla} T + \alpha_{p,k} \underset{\sim}{V} \cdot \underset{\sim}{\nabla} P] \tag{33}$$

and

$$\frac{\partial \phi}{\partial t} = \frac{\rho_s}{\rho_a} [\alpha_T \underset{\sim}{V} \cdot \underset{\sim}{\nabla} T + \alpha_p \underset{\sim}{V} \cdot \underset{\sim}{\nabla} P] \tag{34}$$

Equation 34 has been presented previously (Wood and Hewett, 1984). The right hand sides of these equations may still be time dependent. The quasi-steady approximation simply assumes that the rates of mass transfer depend on the current solution composition, flow field, and pressure and temperature distributions, and that the rate of reaction due to local changes in equilibrium solution composition is small. The influence of changes in local solution composition is incorporated into time dependent values for $\alpha_{T,k}$ and $\alpha_{p,k}$ which depend on the solution composition. Changes in the flow field are accommodated in time dependent expressions for the convective derivatives. Thus, the influence of burial history on diagenetic processes can be examined with the quasi-steady approximation. The behavior of $\alpha_{T,k}$ and $\alpha_{p,k}$ as functions of the progress of irreversible reactions will be particularly important in assessing the influence on inorganic diagenesis of organic acids and other reaction products evolved during organic diagenesis (Franks and Forester, 1984).

In most groundwater flows and flows by natural convection, the velocities are low enough that the pressure distribution is essentially hydrostatic and the contributions of the pressure derivatives are small. Under these conditions, Equations 33 and 34 simplify to

$$\frac{\partial \chi_k}{\partial t} = -\left(\frac{\rho_s \alpha_{T,k}}{\rho_k}\right) \underset{\sim}{V} \cdot \underset{\sim}{\nabla} T \tag{35}$$

$$\frac{\partial \phi}{\partial t} = \left(\frac{\rho_s \alpha_T}{\rho_a}\right) \underset{\sim}{V} \cdot \underset{\sim}{\nabla} T \tag{36}$$

As indicated by these equations, the primary driving force for changes in porosity and mineralogy by flowing solutions is flow through temperature gradients.

VI. MASS TRANSFER DUE TO FLOW THROUGH TEMPERATURE GRADIENTS

In deriving Equations 35 and 36, the following assumptions were made:

(1) local partial equilibrium between the solution and mineral phases is maintained,
(2) the aqueous solution is dilute in the products of mineral dissolution reactions,
(3) the effects on solution composition of flow through pressure gradients are negligible,
(4) irreversible reaction products are distributed uniformly in space,
(5) advective transport is dominant over diffusion (Pe>>1), and
(6) the local rate of change of solution composition due to unsteady equilibrium conditions is small compared to the changes experienced by the moving fluid (quasi- steady approximation).

The integration of these equations requires a knowledge of $\underset{\sim}{V}(\underset{\sim}{x},t)$, $T(\underset{\sim}{x},t)$, and the set of $\alpha_{T,k}(\underset{\sim}{x},t)$. The $\alpha_{T,k}(\underset{\sim}{x},t)$ can be calculated from the equilibrium relations given earlier if we know $p(\underset{\sim}{x},t)$, $\xi_t(t)$, $c_{\ell,i}(\underset{\sim}{x})$, and the set of $K_r(p,T)$. Given these and the initial mineral distribution, $C^r_{k,i}(\underset{\sim}{x})$, the entire history of porosity and mineral alteration by a flowing solution can be calculated by integration of Equations 35 and 36.

Some rather general conclusions about the nature of chemical changes driven by flow through temperature gradients can be drawn if we further restrict our attention to systems which satisfy the following assumptions:

(7) all initial concentrations $C_{k,i}(\underset{\sim}{x}),c_{\ell,i}(\underset{\sim}{x})$ are uniformly distributed,
(8) none of the original mineral phases are depleted during the interval considered, and
(9) the $\alpha_{T,k}$'s may be treated as constants over the region of interest during the interval considered.

The last assumption is the most restrictive. In order for it to be satisfied, we require

$$\frac{1}{\alpha_{T,k}} \left(\frac{\partial \alpha_{T,k}}{\partial T}\right)_{p,\xi_t} \Delta T << 1 \tag{37}$$

where ΔT is the maximum temperature change. As an example, the value of $\alpha_{T,k}$ for quartz in water varies in a nearly linear fashion from a value of 1 ppm/K at 70°C to 4 ppm/K at 170°C (Wood, 1984). If the maximum variation in α_T is to be less than 10 percent, the maximum temperature variation for systems at the low end of the range is ΔT=3K. At the upper end of the range the maximum allowable temperature variation will be ΔT=13K. If the temperature variations are due to flow through the normal geothermal gradient (.025K/m.), then the maximum vertical relief which can be considered consistent with this assumption is about 100 m. at the low end of the temperature range up to about 400 m. at the high end of the temperature range.

Recognizing these restrictions, Equation 35 can be integrated to yield

$$\chi_k(\underset{\sim}{x},t) = \chi_k(\underset{\sim}{x},o) - \frac{M_k}{\rho_k} \Xi_k(\underset{\sim}{x},t) \tag{38}$$

where

$$\Xi_k(\underset{\sim}{x},t) = \int_o^t \phi\dot{\xi}_k(\underset{\sim}{x},t')dt' = \left(\frac{\rho_s \alpha_{T,k}}{M_k}\right)\int_o^t \underset{\sim}{V}(\underset{\sim}{x},t') \cdot \underset{\sim}{\nabla}T(\underset{\sim}{x},t')dt'$$

is the reaction progress per unit volume of bulk rock.

Since each reaction is driven by the same integrated value of the convective derivative, the progress of each reaction will be proportional to all of the others and the progress of all of them can be described by the progress of any one. Denoting the reaction chosen as the reference by r=1, we can define an overall thermal progress variable, Ξ_T, as

$$\Xi_T(\underset{\sim}{x},t) = \Xi_1(\underset{\sim}{x},t) = \left(\frac{\rho_s \alpha_{T,1}}{M_1}\right)\int_o^t \underset{\sim}{V}(\underset{\sim}{x},t') \cdot \underset{\sim}{\nabla}T(\underset{\sim}{x},t')dt' \tag{39}$$

By convention, we will choose the reference reaction to be the reaction for a prograde mineral so that Ξ_T is positive when $\underset{\sim}{V} \cdot \underset{\sim}{\nabla}T$ is positive.

The progress of all of the other reactions can then be written as

$$\Xi_k(\underset{\sim}{x},t) = \lambda_k \Xi_T(\underset{\sim}{x},t) \tag{40}$$

where

$$\lambda_k = \frac{(\alpha_{T,k}/M_k)}{(\alpha_{T,1}/M_1)}$$

and

$$\chi_k(\underset{\sim}{x},t) = \chi_k(\underset{\sim}{x},o) - \left(\frac{M_k\lambda_k}{\rho_k}\right) \Xi_T(\underset{\sim}{x},t)$$

$$\phi(\underset{\sim}{x},t) = \phi(\underset{\sim}{x},o) + \left(\frac{\alpha_T}{\rho_a}\right)\left(\frac{M_1}{\alpha_{T,1}}\right) \Xi_T(\underset{\sim}{x},t) \tag{41}$$

One important consequence of the result in Equation 40 is that the changes in composition which can occur by this mechanism are restricted. Denoting the change in the volume fraction of phase k by $\Delta\chi_k$, Equation 40 restricts the changes to obey the relation

$$\frac{\Delta\chi_k}{\Delta\chi_j} = \frac{(M_k\lambda_k/\rho_k)}{(M_j\lambda_j/\rho_j)} = \frac{(\alpha_{T,k}/\rho_k)}{(\alpha_{T,j}/\rho_j)} \tag{42}$$

Since we have restricted our attention to systems with constant values for the $\alpha_{T,k}$'s, this ratio will be constant throughout the system. Thus, the entire sequence of mineral compositions which can evolve by flow through temperature gradients from a uniform starting composition can be calculated independent of the particular flow and temperature field considered. The particular composition within the sequence which will reside at a particular location at a particular time depends only on the value of $\Xi_T(\underset{\sim}{x},t)$ for that location and time. Equation 40 describes the path through "composition space", i.e. the suite of all possible compositions, and Equation 39 provides the template for mapping those compositions onto physical space and time.

When the distribution of $\dot{\Xi}_T$ has spatial variations, the composition which is produced at one location at a particular time will be produced at a different location at a different time. This gives rise to an apparent motion of compositions through space which resembles the motion of continuous waves.

VII. CONTINUOUS WAVES OF REVERSIBLE ROCK ALTERATION

As long as the mineral phases are continuously distributed, $\Xi_T(x,t)$ may be treated as a continuous field of chemical change or diagenetic alteration. Starting from an initially uniform composition, this field evolves in time. At any instant we can identify contours of constant Ξ_T along which, according to Equation 40, the composition is constant. As the field evolves, the contours for particular values of Ξ_T will move in a direction normal to themselves. Thus, contours of Ξ_T may be thought of as moving composition fronts.

If we denote the direction normal to a contour of constant Ξ_T by the unit vector $\hat{n}(\underset{\sim}{x},t)$, we can calculate it as

$$\hat{n}(\underset{\sim}{x},t) = -\frac{\underset{\sim}{\nabla}\Xi_T(\underset{\sim}{x},t)}{|\underset{\sim}{\nabla}\Xi_T(\underset{\sim}{x},t)|} \tag{43}$$

where $\hat{n}$ faces in the direction of decreasing Ξ_T. Since the concentration of all of the mineral phases in a given composition must move together, we can determine the apparent velocity of a composition by the velocity of the concentration of any phase k. Denoting this velocity by

$$\underset{\sim}{V}_{C_k} = V_{C_k}\hat{n} \quad , \tag{44}$$

the speed of concentration C_k, V_{C_k}, can be computed using the chain rule for partial derivatives of a continuous function as

$$V_{C_k} \equiv \left(\frac{\partial n}{\partial t}\right)_{C_k} = -\frac{\left(\dfrac{\partial C_k}{\partial t}\right)_n}{\left(\dfrac{\partial C_k}{\partial n}\right)_t} \tag{45}$$

where n is distance in the $\hat{n}$ direction. The time rate of change of C_k can be substituted from Equation 35 to obtain

$$V_{C_k} = \frac{\dot{\Xi}_T(\underset{\sim}{x},t)}{|\underset{\sim}{\nabla}\Xi_T(\underset{\sim}{x},t)|} \tag{46}$$

This expression shows that in regions where $\dot{\Xi}_T$ increases with time, compositions move from high values of Ξ_T to lower values and that the velocities are highest where the variations in Ξ_T are the least. Similarly, in regions where Ξ_T decreases with time, compositions move from low values of Ξ_T to higher values. Thus, the highest positive values of $\dot{\Xi}_T$ and the lowest negative values of $\dot{\Xi}_T$ act as sources of new compositions and new composition waves. These waves move away from the sources toward contours of $\dot{\Xi}_T=0$ where their velocity goes to zero. The direction of motion of a composition does not in general coincide with the fluid flow direction and may even be counter to it.

It is interesting to note that in all of the discussion of mass transfer due to flow through a temperature field, the mass transfer process has been completely reversible. By this, we mean that if the sign of $\dot{\Xi}_T$ were reversed and the mass transfer process were allowed to continue, we would eventually return to the starting composition everywhere. The sequence of compositions which evolved originally would be retraced in reverse order at every location. This is the situation which would pertain if the temperature field remained constant and the fluid flow direction were reversed.

In the discussion to this point, it has been assumed that all of the minerals initially present as solid phases in equilibrium with the aqueous solution are present in sufficient quantity that they are not depleted by reaction with the flowing solution during the interval considered (see assumption 8). However, when the initial molar concentrations of some minerals are small, reactions with the fluid can deplete a region of one or more of them. When this occurs, new mechanisms for rock alteration can arise.

VIII. DISCONTINUOUS WAVES OF IRREVERSIBLE ROCK ALTERATION

A minor mineral phase m will first be depleted at a point of maximum positive $\dot{\Xi}_T$ or minimum negative $\dot{\Xi}_T$, depending on whether it has prograde or retrograde solubility. As time progresses, the point where the flow encounters the region depleted in m will move upstream to lower absolute values of $\dot{\Xi}_T$. Downstream of this location, no further progress in the dissolution reaction for m can occur and the progress of reactions with the remaining minerals must be determined with the value of Ξ_m fixed at $C_{m,i}$. This can change the rate at which the other reactions proceed, and the values of the λ_k's may change to new constant values. This represents a change in the "direction" the sequence of compositions moves through composition space.

When the mineral phases are initially uniformly distributed, the fluid will generally encounter more solid mineral m at some location downstream of the depleted region. This occurs because the point where the phase first disappears is a point of maximum alteration. Downstream of this point there is less alteration and the likelihood of excess mineral phase m. When a fluid encounters such a phase downstream of a region where it has become undersaturated, it will dissolve a finite amount of it to come back into equilibrium. In so doing, a discontinuity is introduced into the solid which propagates in the direction of flow. This discontinuity is similar to the reaction fronts which propagate away from the entry region of a reactive fluid into a porous solid (Hoffman, 1972; Lund and Fogler, 1976; Walsh, et al., 1984).

The nature of this mechanism can be understood more clearly by considering the example shown in Figure 1a. Here, a steady parallel flow in a porous layer goes through a change in elevation due to a fold in the layer. Since the geothermal gradient points vertically downward, the fluid is heated as it travels down the sloping fold. The distribution of $\dot{\Xi}_T$ with distance s measured along a streamline is shown in Figure 1b. Away from the fold, the flow is orthogonal to the temperature gradient and $\dot{\Xi}_T=0$. In the region of the fold, where the flow has a component in the direction of the temperature gradient, $\dot{\Xi}_T$ is positive and goes through a maximum at the point of maximum slope.

If we consider a minor mineral phase, m, which has a prograde solubility (solubility increases with temperature), the progression of the spatial distribution will be like that shown in Figure 1c. As time progresses, the initially uniform value of $C_{m,i}$ is dissolved away, with the amount of dissolution being proportional to the local value of $\dot{\Xi}_T$. The new values of C_m introduced at the point of maximum change move to locations upstream or downstream of it at later times. This pattern continues until the phase first disappears at the point of maximum $\dot{\Xi}_T$ at the time corresponding to

$$\Xi_{m,max} = \lambda_m \Xi_{T,max} = C_{m,i} \qquad (47)$$

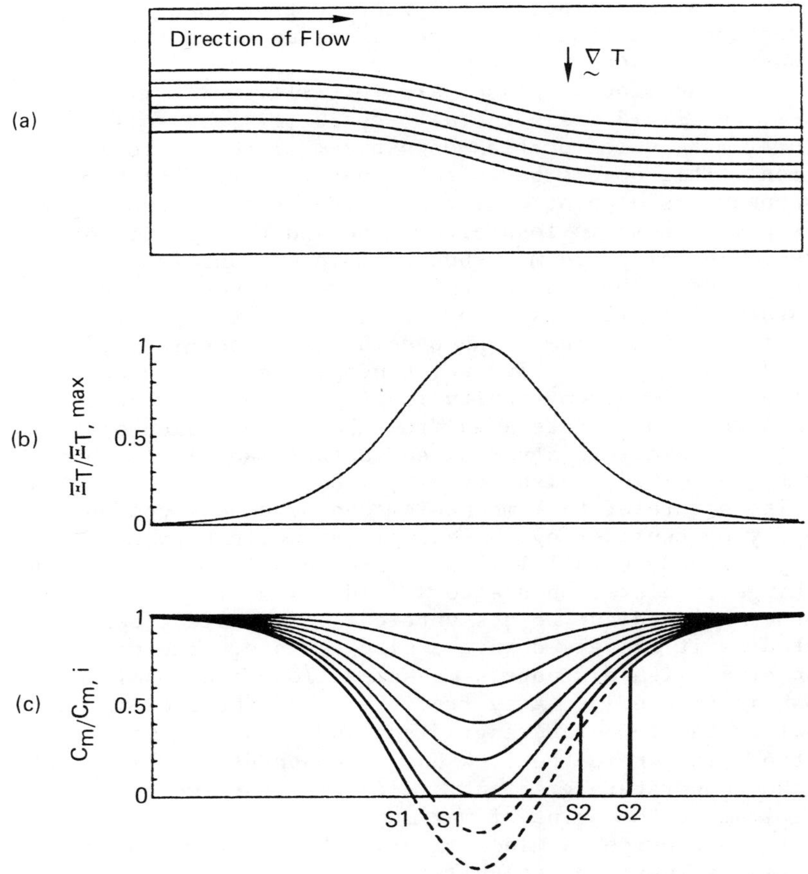

Figure 1. (a) Streamline pattern in a gently folded porous layer with a change in elevation. (b) Distribution of $\dot{\Xi}_T$ along a streamline. (c) Dissolution pattern of mineral m along a streamline.

If $\dot{\Xi}_T$ is steady, we can define a characteristic time, τ_m, for the first disappearance of phase m as

$$\tau_m = \frac{C_{m,i}}{\lambda_m \dot{\Xi}_{T,max}} \qquad (48)$$

Following the first disappearance of phase m, the point where the fluid first encounters an absence of m, S_1, moves upstream. The position of S_1 corresponds to the location of $\Xi_m = C_{m,i}$. As time progresses, this corresponds to lower and lower values of $\dot{\Xi}_T$ upstream of $\dot{\Xi}_{T,max}$. The fluid then crosses a region depleted in phase m until it again encounters excess phase m at a location S_2. At this point, the fluid dissolves enough of the excess phase to equilibrate with it. In so doing it dissolves a finite amount of the phase and moves a discontinuous dissolution front downstream. The location of this front, S_2, will be such that the integral between S_1 and S_2 of the mineral distribution which would have resulted if it were possible to dissolve to negative values of C_m (shown as a dotted line in Figure 1c) is zero. By propagating a discontinuous dissolution front into the solid phase downstream of a depleted region, the fluid dissolves an amount of material equal to that it could not dissolve in the depleted region.

As shown in Appendix A, mass conservation in a more general flow requires the motion of discontinuous fronts to be such that the following relation is satisfied.

$$\int_{S_1}^{S_2} [C_{m,i} - (\frac{\alpha_{T,m}\rho_s}{M_m}) \int_o^t (\underset{\sim}{V} \cdot \underset{\sim}{\nabla} T) dt] A ds = 0 \qquad (49)$$

The bracket inside the integral in this expression is just the value C_m would take if mass transfer to negative values were physically possible. This is shown by the dotted line in Figure 1c. Equation 49 shows that the integrated effect of propagating a discontinuous dissolution front is to supply to the fluid an amount of phase m equal to that which would have been transferred if continuous removal in the depleted region were possible.

This can be seen more clearly by differentiating Equation 49 with respect to time to obtain

$$C_{m,2} A_2 \frac{dS_2}{dt} = (\frac{\alpha_{T,m}\rho_s}{M_m}) \int_{S_1}^{S_2} (\underset{\sim}{V} \cdot \underset{\sim}{\nabla} T) A ds \qquad (50)$$

where the subscript 2 indicates conditions at S_2. This result shows that the rate at which mineral phase m is removed from the solid by the moving front equals the rate at which the fluid becomes undersaturated in m as it crosses the depleted interval. This can also be written as

$$\frac{dS_2}{dt} = V_2 \left(\frac{\alpha_{T,m}\rho_s}{M_m}\right) \frac{(T_{S_2} - T_{S_1})}{C_{m,2}} \tag{51}$$

There are several important differences in the nature of the mass transfer effected by continuous and discontinuous waves. As shown previously, the velocity of a continuous wave is proportional to the local value of $\dot{\Xi}_T$. This means that a wave introduced at a source in a region of positive $\dot{\Xi}_T$ can never cross into a region of negative $\dot{\Xi}_T$, and vice-versa. The region of influence of sources of continuous waves are bounded by contours of $\dot{\Xi}_T=0$.

In contrast, the velocity of a discontinuous wave depends on the integrated value of $\dot{\Xi}_T$ as shown in Equation 50. This means that a discontinuity introduced at a point of maximum dissolution can propagate from that point through the region of dissolution downstream of it and into a region where the flow was previously precipitating a mineral. A discontinuous dissolution front can affect the entire field downstream of the point where it is introduced.

This can be seen by considering the steady parallel flow through the doubly-folded layer shown in Figure 2a with the distribution of $\dot{\Xi}_T$ shown in Figure 2b. For times less than τ_m, the initially uniform distribution of mineral m will be altered in a pattern geometrically similar to the Ξ_T distribution as shown in Figure 2c. For short times after τ_m, the discontinuity introduced at the point of maximum dissolution will remain in the region where m is being dissolved by the flow, as shown in Figure 2d. Denoting the time for the discontinuity to reach the synclinal trough by t_1, for $t > t_1$ the discontinuity moves into the region where the pattern of continuous waves was previously precipitating phase m onto the solid. As the discontinuity moves into this region its velocity will slow as it encounters larger and larger values of $C_{m,2}$.

A second difference in the nature of the mass transfer process by discontinuous waves is that the rock alteration caused by them is irreversible. Reversing the sign of $\dot{\Xi}_T$ everywhere in the flow, as occurs when the flow direction is reversed, will stop the motion of a discontinuity, but the discontinuity itself will remain. The only exception to

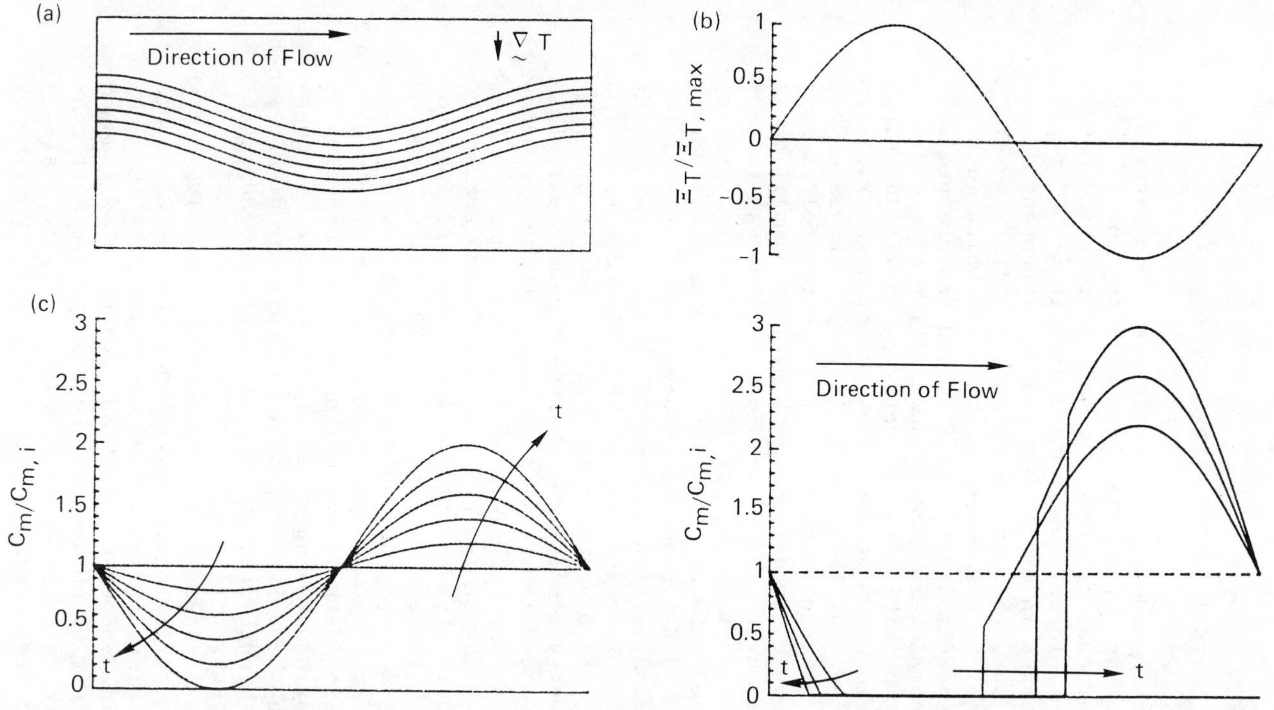

Figure 2. (a) Streamline pattern in a gently folded porous layer with two changes in elevation. (b) Distribution of $\dot{\Xi}_T$ along a streamline. (c) Alteration pattern of mineral m along a streamline before discontinuities are introduced. (d) Alteration pattern after discontinuous waves are introduced.

this is when the amount of solid phase on both sides of the discontinuity is reduced to zero by subsequent mass transfer.

IX. FLUID FLOW IN A GENTLY FOLDED POROUS LAYER

The equations developed in the preceding sections show that the sequence of changes in mineralogy due to flow through a temperature field can be predicted independent of the flow field and temperature distribution. Mapping of those changes in mineralogy and the attendant changes in porosity onto the physical plane as a function of time, however, requires a knowledge of $\Xi_T(x,t)$ which in turn depends on the details of the flow and temperature fields. Although a completely general description of flow in porous media is not possible, one geometry of particular significance in sedimentary geology does permit a general solution to be obtained. As shown by Davis et al. (Davis et al., 1985), when a fluid is confined in a gently folded layer with impermeable boundaries, mathematical simplifications in the equations describing fluid flow and heat transfer allow exact solutions for the flow and temperature fields to be obtained. Solutions for the case of natural convection in folded two and three dimensional sheet-like porous layers have been published. The results presented here extend those results to include flow due to hydraulic gradients and combined free and forced convection.

The geometry considered is shown schematically in Figure 3a and consists of a sequence of three sedimentary units: (1) an upper shale, (2) a porous sand, and (3) a lower shale. The shales are impermeable and homogeneous in all of their physical properties. The sand is also homogeneous. There is a constant heat flux, Q, imposed from below and a constant surface temperature, θ_o. In the absence of any regional hydraulic gradients we expect that a recirculating natural convection flow will develop in the sand body and transport dissolved material through the pores. When there is a regional hydraulic gradient, a net volumetric throughflow, q_o, will be superimposed on the natural convection flow.

As shown in Appendix B, when the maximum slope of the layer, S, is sufficiently small, the equation describing the temperature distribution may be uncoupled from the equations for fluid flow and exact solutions for the flow and temperature fields may be obtained. The results are most conveniently presented in terms of characteristic scales of velocity and temperature change multiplied by dimensionless

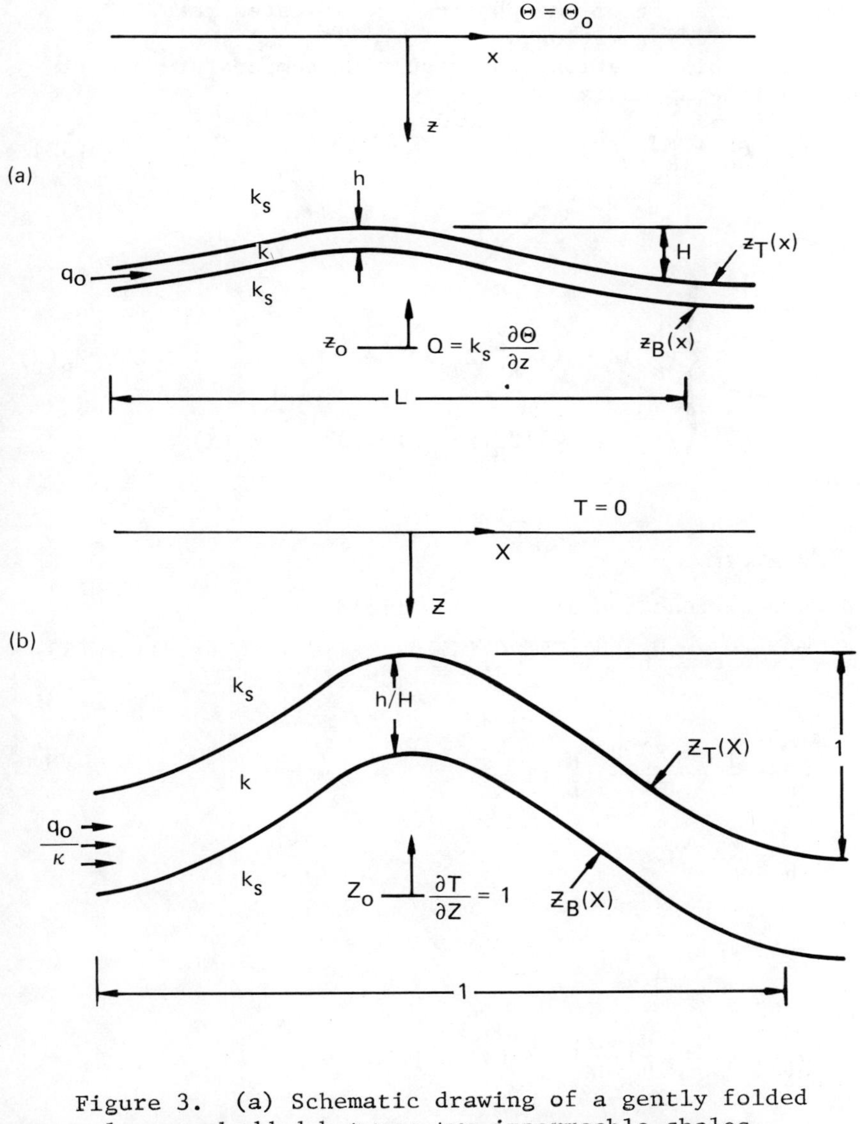

Figure 3. (a) Schematic drawing of a gently folded porous layer embedded between two impermeable shales. (b) Schematic drawing of the same layer in dimensionless coordinates.

geometric functions which give the spatial distributions for a particular layer geometry. The notation is summarized in Appndix B and Figure 3b where it may be noted that the scaling of the physical coordinates results in a vertical exaggeration.

With this notation, the resulting temperature distribution (see Appendix B) is

$$\theta = \theta_o + \frac{QH}{k_s} F(Z) \quad , \tag{52}$$

where

$$F(Z) = \begin{cases} Z & 0 \leqq Z < Z_T(X) \\ (\frac{k_s}{k})Z + (1-\frac{k_s}{k})Z_T(X) & Z_T(X) \leqq Z \leqq Z_B(X) \\ Z - (1-\frac{k_s}{k})[Z_B(X) - Z_T(X)] & Z_B(X) \leqq Z \leqq Z_o \end{cases}$$

and

$$Z = z/H$$

$$X = x/L$$

The stream function of the flow field is

$$\psi = (U_c + U_h)\ H\ [CFI\ G_1(X,Z) + (1-CFI)\ G_2(X,Z)] \tag{53}$$

where

$$U_c = \frac{1}{2}\ (\frac{k_p g\alpha Qh}{\upsilon K})(\frac{H}{L})(1-\frac{k_s}{k})$$

$$U_h = \frac{q_o}{h}$$

$$CFI = \frac{U_c}{U_c+U_h}$$

and

$$G_1(X,Z) = (\frac{H}{h})\ \frac{dZ_T}{dX}\ [Z_B(X)-Z][Z-Z_T(X)]$$

$$G_2(X,Z) = (\frac{h}{H})\ \frac{[Z_B(X)-Z]}{[Z_B(X)-Z_T(X)]}$$

Here U_c and U_h are the velocity scales associated with natural convection and hydraulic throughflow, respectively. The convective flow index, CFI, gives the fraction of the maximum velocity contributed by natural convection. The velocity components corresponding to this stream function are

$$u = -U_c G_{1,Z}(X,Z) - U_h G_{2,Z}(X,Z)$$

$$w = U_c S\, G_{1,X}(X,Z) + U_h S\, G_{2,X}(X,Z) \tag{54}$$

where

$$G_{1,Z}(X,Z) = -2\left(\frac{H}{h}\right)\left[Z - \frac{1}{2}(Z_B + Z_T)\right]$$

$$G_{2,Z}(X,Z) = -\left(\frac{h}{H}\right)\frac{1}{(Z_B - Z_T)}$$

$$G_{1,X}(X,Z) = \left(\frac{H}{h}\right)\left\{\frac{d^2Z_T}{dX^2}(Z_B - Z)(Z - Z_T) + \frac{dZ_T}{dX}\left[(Z - Z_B)\frac{dZ_T}{dX} + (Z - Z_T)\frac{dZ_B}{dX}\right]\right\}$$

$$G_{2,X}(X,Z) = \left(\frac{h}{H}\right)\left\{\frac{1}{(Z_B - Z_T)}\frac{dZ_B}{dX} - \frac{(Z_B - Z)}{(Z_B - Z_T)^2}\left[\frac{dZ_B}{dX} - \frac{dZ_T}{dX}\right]\right\}$$

$G_1(X,Z)$ and $G_2(X,Z)$ are defined such that the maximum contribution to the velocity by natural convection is U_c and the maximum contribution by hydraulic flow is U_h.

As an illustration of the use of these solutions, consider the following example in which the porous layer is a sheet of constant thickness folded to form an anticline with the shape of a cosine function. In scaled coordinates, this may be represented as

$$Z_T(X) = D - \mathrm{Cos}(X)$$

$$Z_B(X) = Z_T(X) + 1 \tag{55}$$

The isotherms of the resulting temperature distribution coincide with constant values of F(Z), shown in Figure 4 for two different values of the thermal conductivity ratio, k_s/k. When k_s/k is greater than unity, the sand is less conductive than the shales and the heat flux vector is

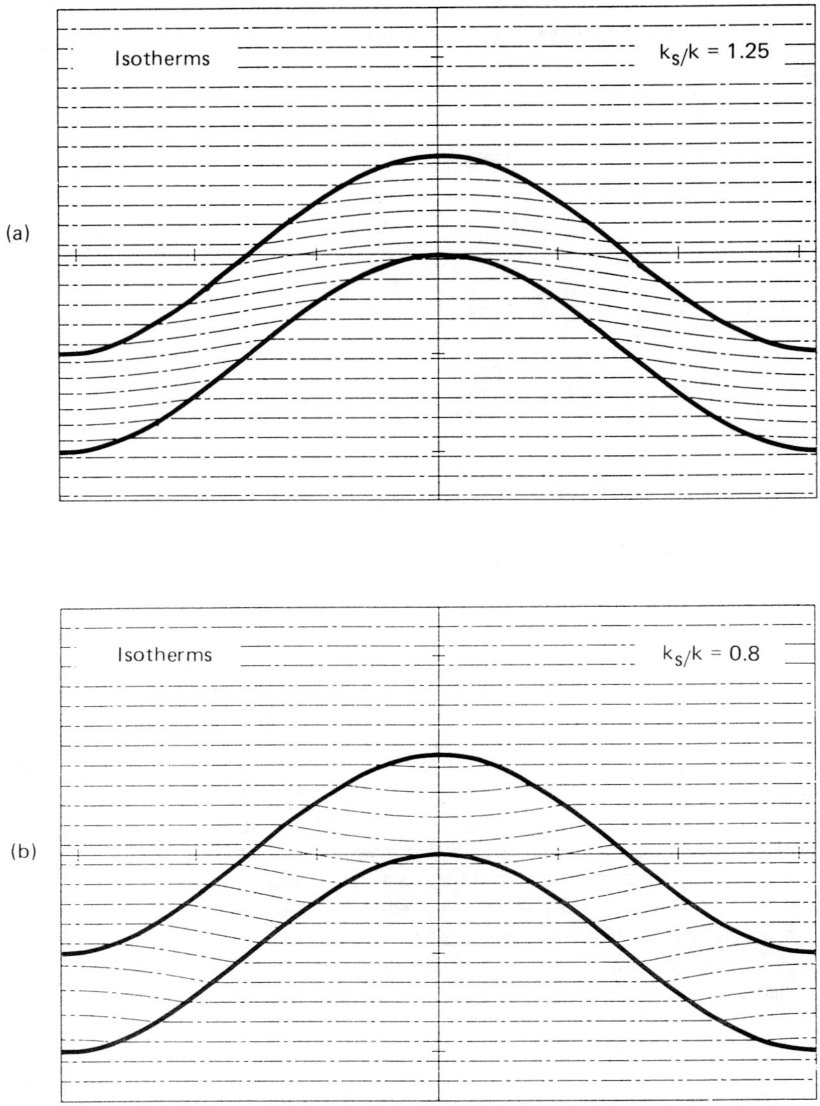

Figure 4. (a) Isotherm contours for a cosine-shaped layer with $k_s/k = 1.25$. (b) Isotherm contours for a cosine-shaped layer with $k_s/k = 0.8$.

"refracted" from the vertical to take a shorter path across the sand. This results in the isotherms being bent from the horizontal to more closely conform to the boundaries of the layer as shown in Figure 4a. When k_s/k is less then unity, the sand is more conductive than the shales and the heat flux vector takes a longer path across the sand bending the isotherms in the opposite direction as shown in Figure 4b. In both cases, the isotherms are not horizontal within the sand, so natural convection currents must arise to compensate for the resulting non-hydrostatic distribution of pressure.

The flow field depends on the convective flow index in addition to the layer geometry and thermal conductivity contrast. When CFI=1, the flow is solely due to natural convection and streamlines coincide with constant values of $G_1(X,Z)$, shown in Figure 5. The sense of rotation in these recirculating loops depends on the value of the thermal conductivity ratio. As shown in Figure 4, when k_s/k is less than unity, the temperature along any horizontal plane is higher at the top boundary of the layer than it is at the lower boundary and the warmer fluid will rise along the top boundary while the cooler fluid descends along the lower boundary. This gives a clockwise sense of rotation in the loop on the left and a counter-clockwise sense of rotation in the loop on the right. When the k_s/k is greater than unity the sense of rotation reverses. When k_s/k is one, there is no natural convection, as indicated by the expression for U_c. When CFI=0, the flow is solely due to hydraulic throughflow and the streamlines coincide with constant values of $G_2(X,Z)$, shown in Figure 6. For this constant thickness layer, the streamlines simply run parallel to the layer boundaries. In layers with a varying thickness they diverge or converge along with the changing layer thickness.

Other values of the CFI correspond to combined free and forced convection. When CFI<=0.5, the hydraulic throughflow dominates. The convective component slows the flow and spreads the streamlines in regions where it runs counter to the throughflow and speeds the flow, compressing the streamlines, where it runs with the throughflow. An example of this is shown in Figure 7 where the streamlines for CFI=0.5 are shown. When CFI>0.5, the convective velocity scale is larger than that of the throughflow and regions of recirculating flow are established. An example of this type of streamline pattern is shown in Figure 8 for CFI=0.7. At even larger values of the CFI the recirculating flow completely dominates and the throughflow is diverted around flow loops similar to those in purely

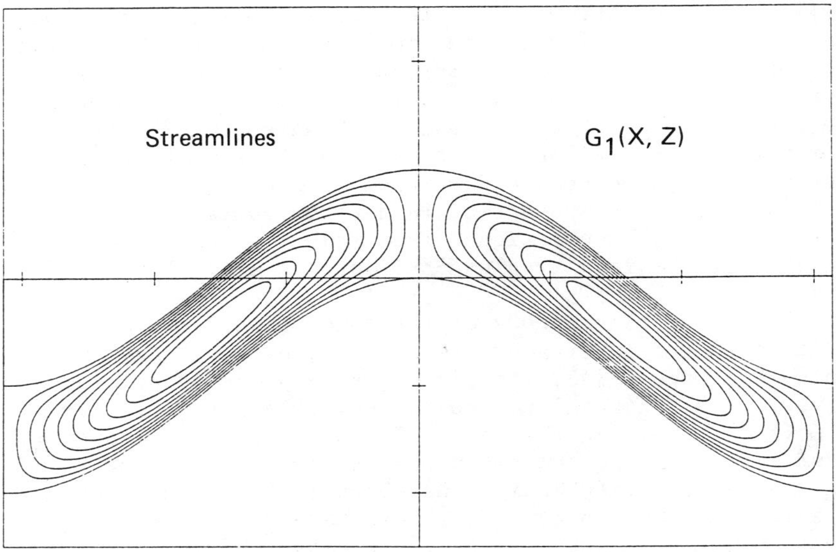

Figure 5. Geometric distribution of streamlines in natural convection flow in a cosine-shaped porous layer.

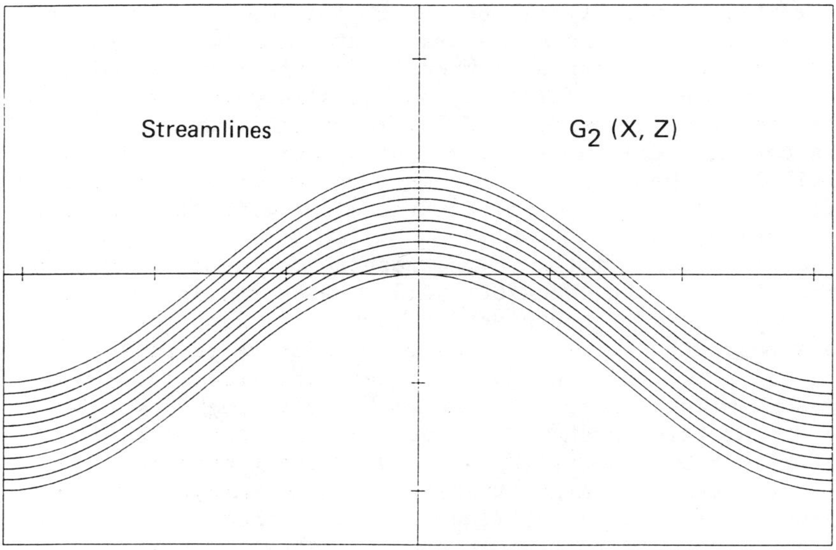

Figure 6. Geometric distribution of streamlines in flow due to hydraulic potential gradients in a cosine-shaped porous layer.

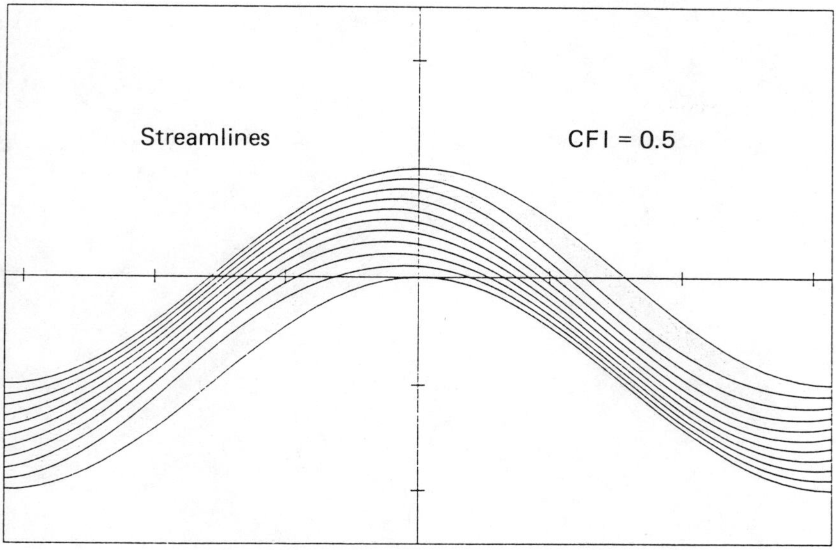

Figure 7. Streamline patterns for combined free and forced convection in a cosine-shaped layer with CFI = 0.5.

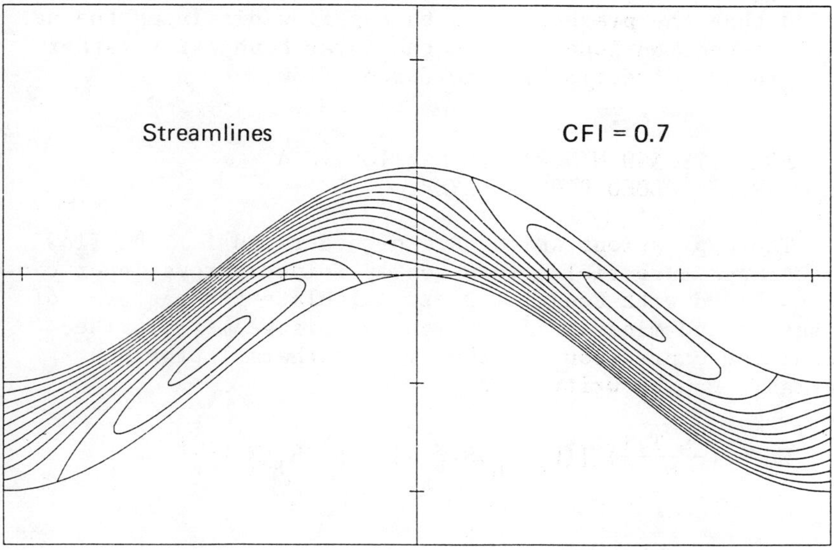

Figure 8. Streamline patterns for combined free and forced convection in a cosine-shaped layer with CFI = 0.7.

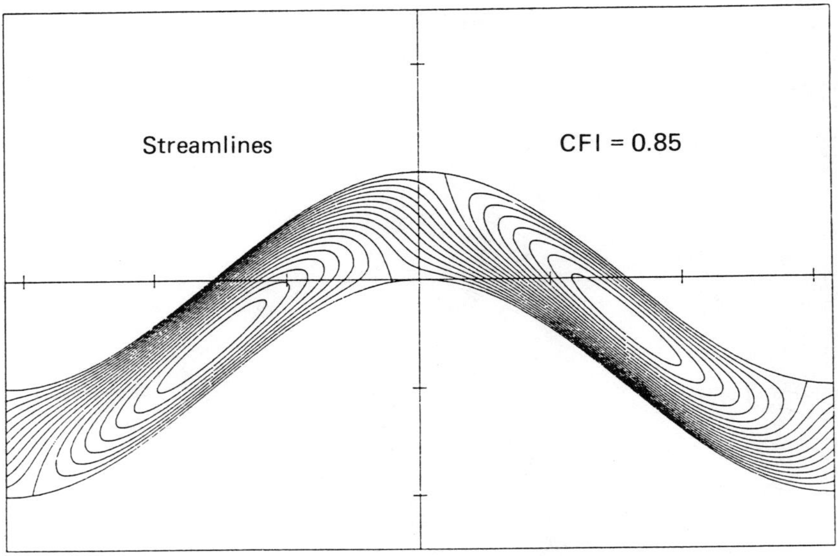

Figure 9. Streamline patterns for combined free and forced convection in a cosine-shaped layer with CFI = 0.85.

convective flow, as shown is Figure 9. It should be noted that the presence of a throughflow displaces the natural convection loops toward the layer boundaries rather than in the direction of the forced flow.

X. POROSITY AND MINERAL ALTERATION IN A GENTLY FOLDED POROUS LAYER

The expressions which have been derived for the flow and temperature fields in a gently folded porous layer can be combined with Equation 39 to calculate the patterns of porosity and mineral alteration in this geometry. The resulting expression for the overall thermal progress variable can be written as

$$\dot{\Xi}_T = \left(\frac{\rho_s \alpha_{T,1}}{M_1}\right) \left[(U_c+U_h) S\left(\frac{Q}{k_s}\right)\right] \left[\mathrm{CFI}\ H_1\left(X,Z,\frac{k_s}{k}\right) + (1-\mathrm{CFI})\ H_2\left(X,Z,\frac{k_s}{k}\right)\right] \qquad (56)$$

where

$$H_1(X,Z,\frac{k_s}{k}) = (\frac{k_s}{k})G_{1,X}(X,Z) - (1-\frac{k_s}{k}) \frac{dZ_T}{dX} G_{1,Z}(X,Z)$$

$$H_2(X,Z,\frac{k_s}{k}) = (\frac{k_s}{k})G_{2,X}(X,Z) - (1-\frac{k_s}{k}) \frac{dZ_T}{dX} G_{2,Z}(X,Z)$$

For the steady flow considered here,

$$\Xi_T = \dot{\Xi}_T t \qquad (57)$$

When the flow is purely convective (CFI=1) the contours of constant alteration or wavefronts of moving compositions will coincide with constant values of $H_1(X,Z,k_s/k)$, shown in Figure 10 for a value of $k_s/k=0.8$. Although this pattern appears very complicated, it can be understood by considering the path of a fluid particle through the temperature field. Consider flow along the outermost streamline in the left convection loop. For this value of the thermal conductivity ratio, the flow in the left loop is clockwise. Starting from the top of the anticline, the flow heats as it moves down across the layer and heats further as it moves down along the lower layer boundary, with $\dot{\Xi}_T$ reaching a maximum value halfway down the lower boundary. After the flow turns at the syncline, it cools as crosses the layer to the upper boundary and further cools as it flows up along the upper flank of the anticline, with $\dot{\Xi}_T$ going through a minimum halfway up the flank. The maxima and minima midway along the boundaries act as sources of new compositions which move out from them on wavefronts parallel to the contours shown.

When CFI=0, the flow is purely due to the hydraulic throughflow, and contours of constant alteration correspond to constant values of $H_2(X,Z,k_s/k)$, shown in Figure 11 for $k_s/k=0.8$. The pattern of alteration for this case is much simpler since the magnitude of $H_2(X,Z,k_s/k)$ is proportional to the slopes of the boundaries. Thus, as the flow proceeds from left to right, it is cooled at a maximum rate midway up the left slope, reverses from cooling to heating at the anticlinal crest, and is heated at a maximum rate midway down the right slope. New compositions are introduced and the maximum porosity alteration occurs midway down the flanks of the anticline.

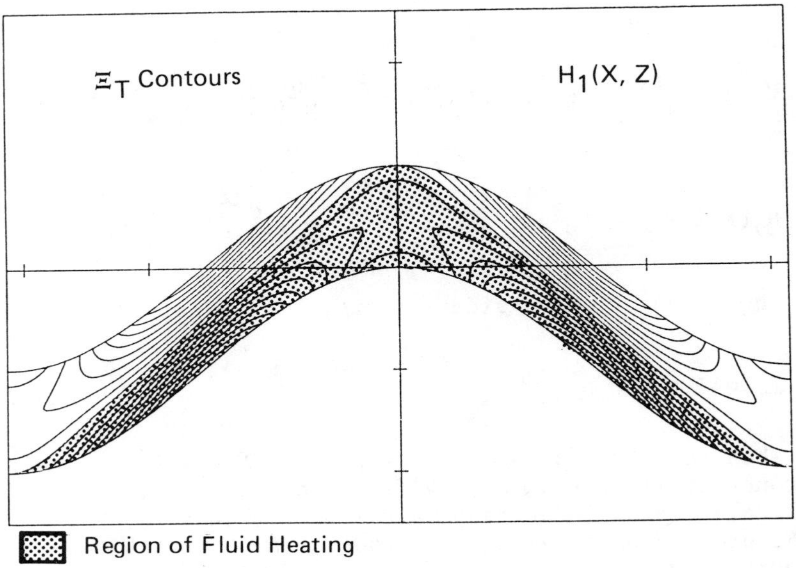

Figure 10. Contours of constant porosity and mineral alteration by natural convection flow in a cosine-shaped porous layer with $k_s/k = 0.8$.

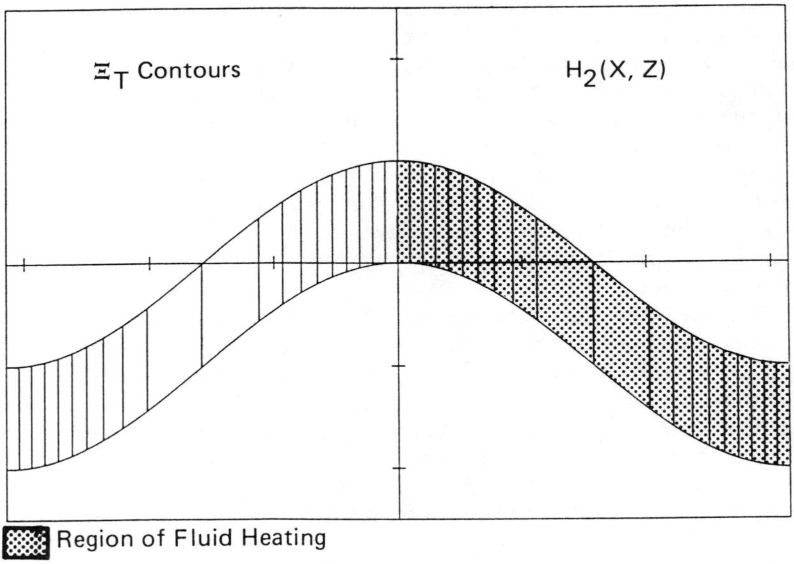

Figure 11. Contours of constant porosity and mineral alteration by hydraulic through flow in a cosine-shaped porous layer with $k_s/k = 0.8$.

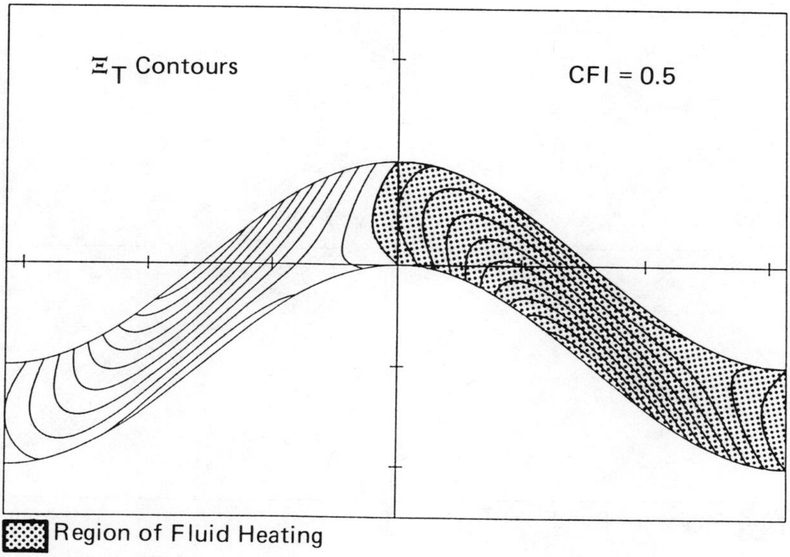

Figure 12. Contours of constant porosity and mineral alteration by combined free and forced convection in a cosine-shaped porous layer with $k_s/k = 0.8$ and CFI = 0.5.

Other values of the CFI correspond to combined free and forced convection. The pattern of contours for CFI=0.5 is shown in Figure 12. Here, the maximum cooling occurs along the upper boundary of the left slope, while the maximum heating occurs along the lower boundary of the right slope. The pattern of contours for CFI=0.7 and CFI=0.85 are shown in Figures 13 and 14 and show the same complicated features observed in the purely convective flow case. These patterns can also be understood by considering the path of a fluid particle along a streamline.

All of the patterns of alteration shown will change when one or more of the minor mineral components initially present is depleted. A characteristic time scale for alteration can be defined as

$$\tau = \left(\frac{\rho_1}{M_1}\right) \frac{1}{\dot{\Xi}_{T,max}} \tag{58}$$

This is the amount of time it takes to fill a unit volume with reference mineral 1. For flow in a gently folded porous layer

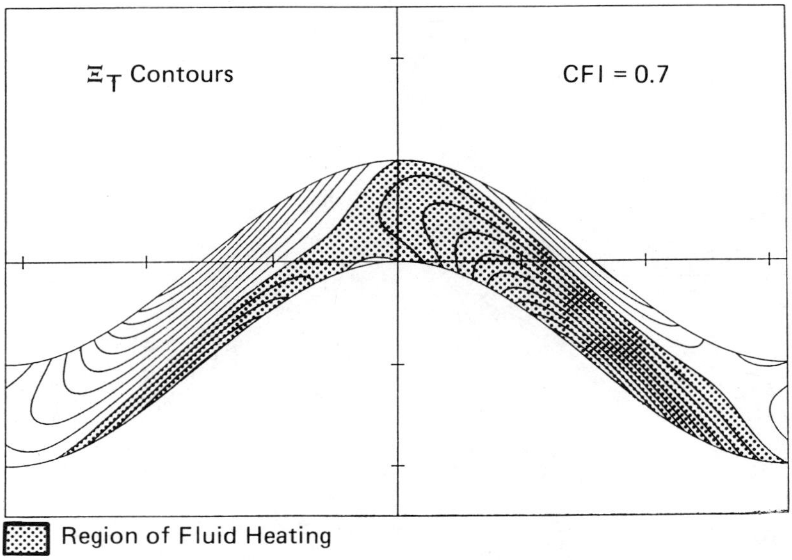

Figure 13. Contours of constant porosity and mineral alteration by combined free and forced convection in a cosine-shaped porous layer with $k_s/k = 0.8$ and CFI = 0.7.

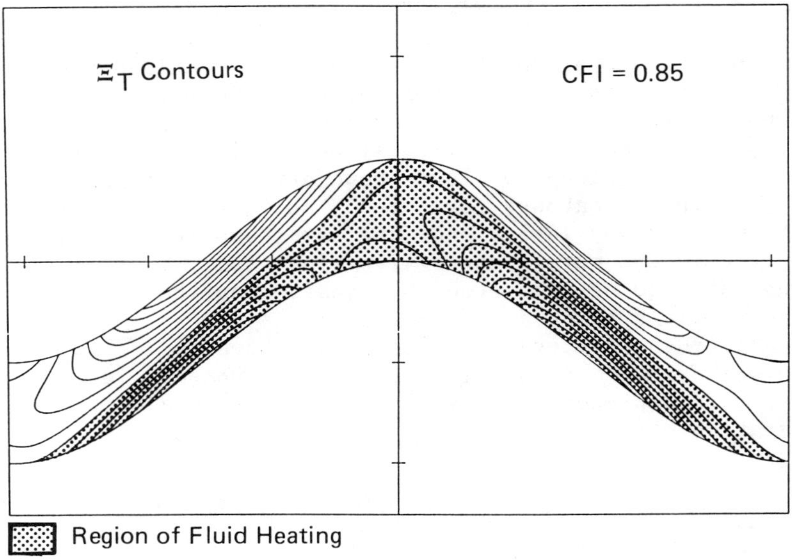

Figure 14. Contours of constant porosity and mineral alteration by combined free and forced convection in a cosine-shaped porous layer with $k_s/k = 0.8$ and CFI = 0.85.

$$\tau = \{\alpha_{T,1}(\frac{Q}{ks})(U_c+U_h)S\ [CFI\ H_1(X,Z,\frac{ks}{k}) + (1-CFI)H_2(X,Z,\frac{ks}{k})]_{max}\}^{-1} \quad (59)$$

The time for first depletion of any mineral k can then be calculated from Equation 48 as

$$\tau_k = \chi_{k,i}\ (\frac{\rho_k}{\rho_1})\ (\frac{\alpha_{T,1}}{\alpha_{T,k}})\ \tau \quad (60)$$

As an example of the kind of patterns which can evolve by irreversible mass transfer subsequent to depletion of a mineral, consider the distribution of a minor prograde ($\alpha_{T,M}>1$) mineral species in purely convective flow (CFI=1) in this geometry. To aid in visualizing the pattern of alteration predicted, we will add a third dimension to the figure which is proportional to the amount of species m present at a given time. Initially, this mineral is uniformly distributed, so the surface representing its spatial distribution is a uniform height as shown in Figure 15a. After a time $t=0.5\ \tau_m$, the original amount of the mineral will be reduced by half at the point of maximum heating along the lower boundary and will be increased by 50% at the point of maximum cooling along the upper boundary. The amount of alteration at other locations is proportional to the local values of $\dot{\Xi}_T$ as shown in Figure 15b. This pattern continues to develop until $t=\tau_m$, shown in Figure 15c. At this point, all of the original mineral at the point of maximum dissolution has been depleted and a discontinuous front will be introduced as the flow continues.

Once a discontinuous front is introduced it moves in the direction of flow leaving behind a depleted mineral zone (DMZ). The motion of the front along each streamline is governed by the integral relation of Equation 49. The shape of the DMZ produced by the moving front at several subsequent times is shown in Figure 16. At $t=2\ \tau_m$, shown in Figure 16a, the front has nearly reached the synclinal trough, leaving a strip depleted in the mineral along the lower layer boundary. By $t=4\ \tau_m$, the front has swept around the bottom of the streamline loop and slows as it moves into the region where the mineral is being precipitated by the continuous mass transfer process. At $t=6\ \tau_m$, the front continues to sweep toward the anticlinal crest at an ever decreasing velocity. The net result of this irreversible

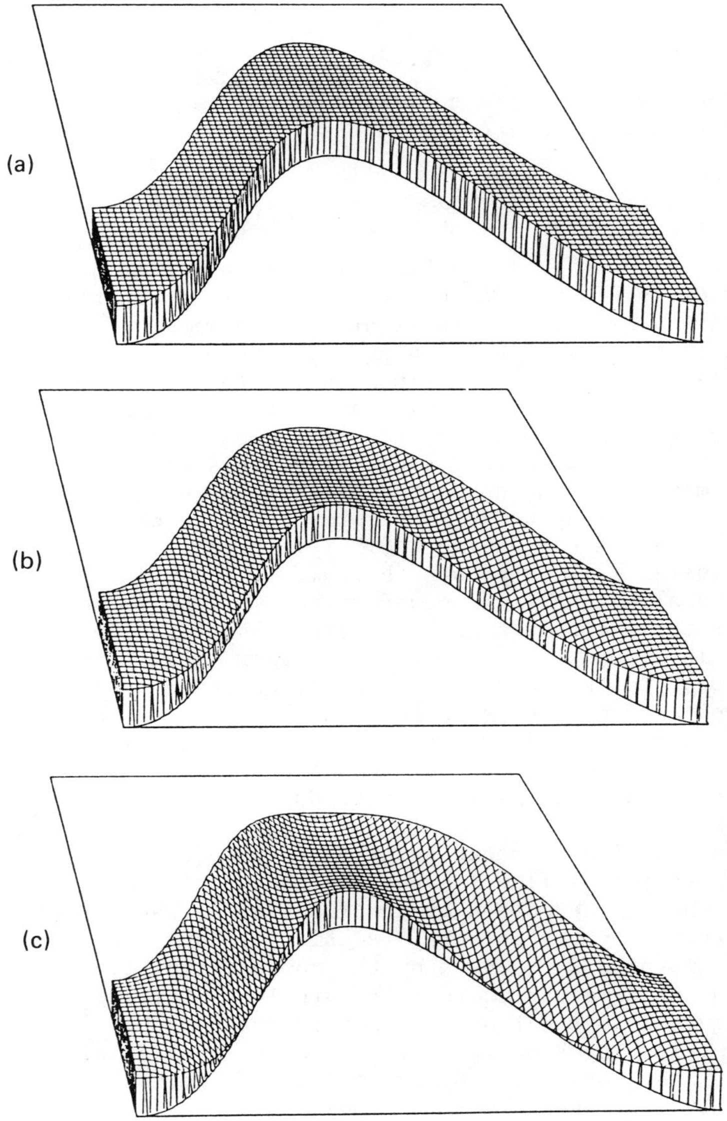

Figure 15. (a) Uniform initial distribution of minor prograde mineral m (the height of the plateau is proportional to χ_m). (b) Distribution of χ_m at $t/\tau_m = 0.5$. (c) Distribution of χ_m at $t/\tau_m = 1$.

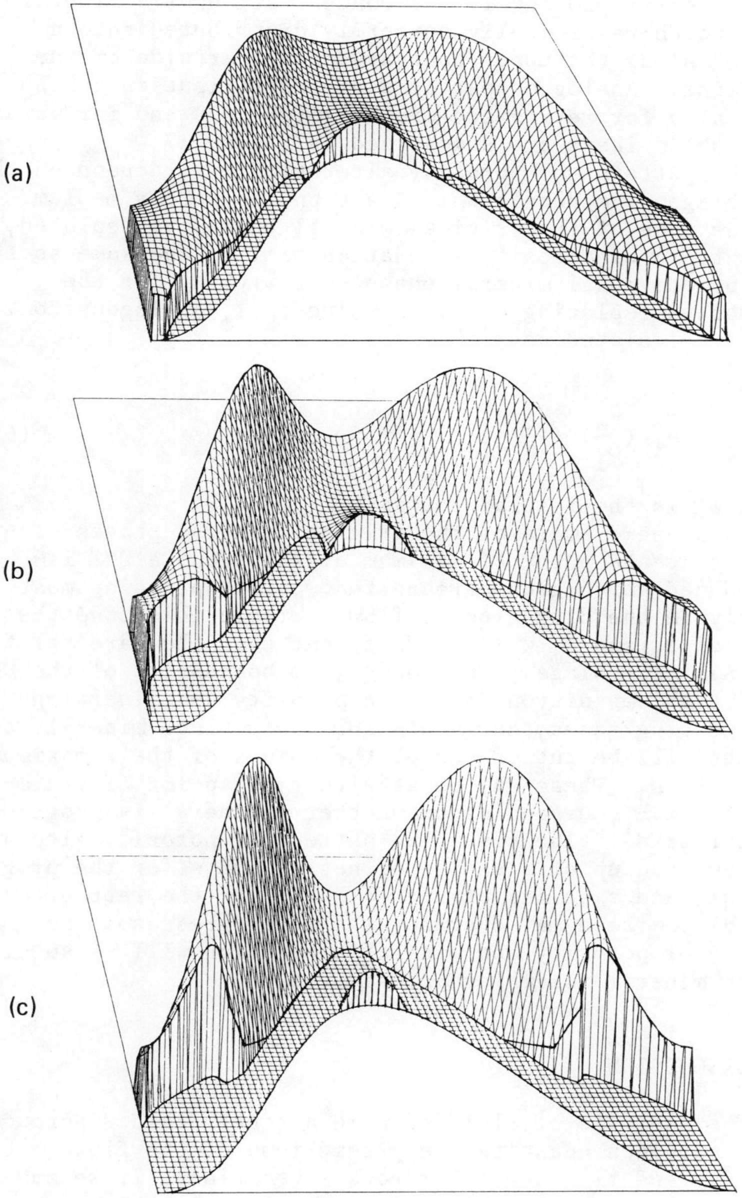

Figure 16. (a) Distribution of χ_m at $t/\tau_m = 2$ showing the growth of a depleted mineral zone (DMZ) around the point of maximum alteration. (b) Distribution of χ_m at $t/\tau_m = 4$. (c) Distribution of χ_m at $t/\tau_m = 6$.

mass transfer process is the concentration of a mineral phase which was initally uniformly distributed into a "drape" along the upper boundary on either side of the anticline. Analogous patterns of redistribution can be calculated for retrograde minerals ($\alpha_{T,m}<1$) and for values of $k_s/k>1$. These are summarized in Figure 17.

The patterns of porosity alteration will depend on the assemblage of minerals which make up the porous medium. So long as none of the original mineral phases is depleted, the pattern of porosity alteration will be the same as that for an individual mineral phase, but with α_T for the assemblage replacing $\alpha_{T,k}$. A value of τ_ϕ analogous to τ_k can be calculated as

$$\tau_\phi = \phi_i \left(\frac{\rho_a}{\rho_1}\right) \left(\frac{\alpha_{T,1}}{\alpha_T}\right) \tau \qquad (61)$$

where ϕ_i is the initial porosity

When one or more of the original mineral phases disappears at the point of maximum dissolution, a DMZ is introduced which grows around that point spreading most quickly in the direction of flow. Within this zone the value of α_T and τ_ϕ will be different than they are for the original assemblage. The downstream boundaries of the DMZ exhibit a jump discontinuity in porosity and mineralogy. As time progresses, new DMZ's for other minor mineral components will be introduced at the points of their maximum dissolution. These may be at either the point of maximum or minimum $\dot{\Xi}_T$, depending on whether a mineral is prograde or retrograde. The resulting pattern of porosity alteration is then made up of a series of nested DMZ's for the prograde minerals and a series of nested DMZ's for the retrograde minerals. These may overlap as time progresses. The pattern of porosity and mineral alteration will be specific to the mineral assemblage considered.

XI. DISCUSSION

The analysis of fluid flow in a gently folded porous layer provides quantitative predictions of the flow velocity and time scale for rock alteration. If we substitute the property values for a relatively thick, unconsolidated quartz sand shown in Table I into Equation 53, a convection velocity scale of U_c=1.5 m/yr can be calculated. This leads to a calculated alteration time scale of τ=46 MMYr, according to Equation 59. The depletion

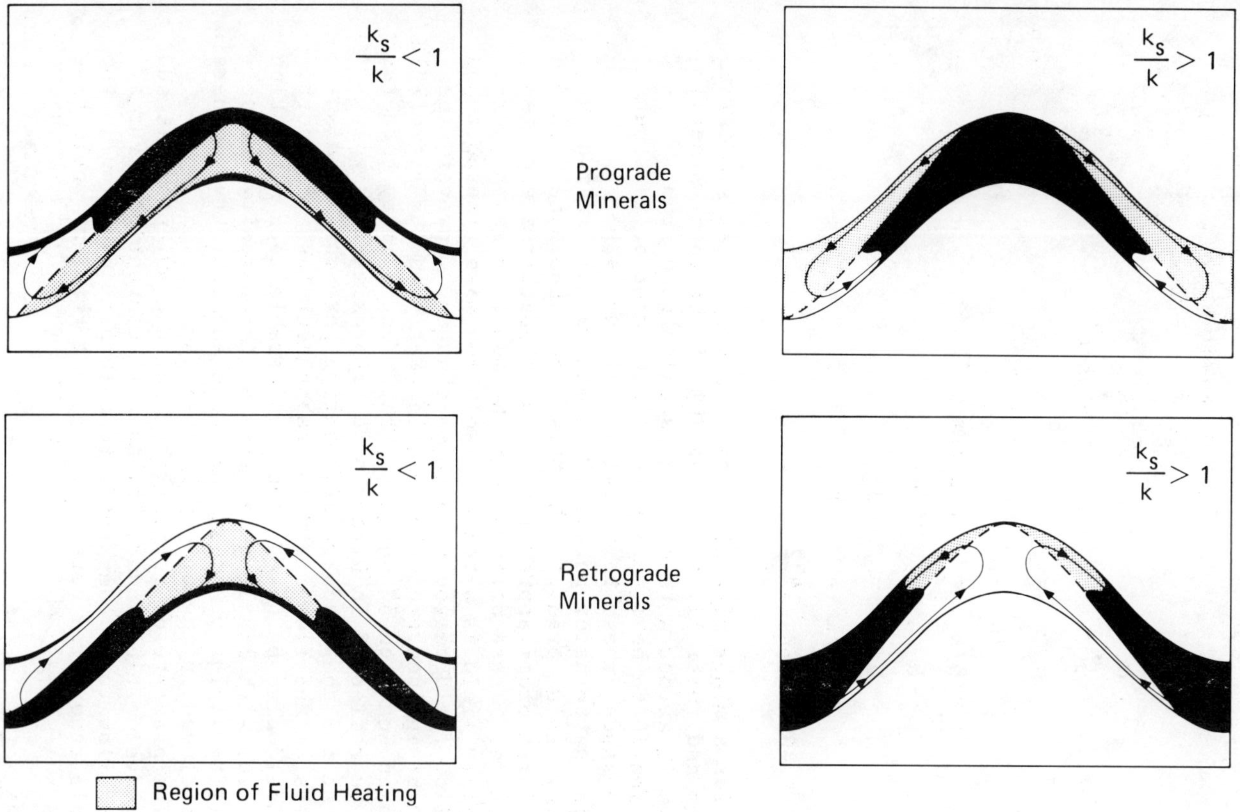

Figure 17. Patterns of redistribution of minor mineral species by discontinuous fronts.

TABLE I. Example Property Values

k_p	=	10 Darcies = $10^{-11}m^2$
g	=	9.8 m/sec^2
α	=	$2 \times 10^{-3}/K^1$
υ	=	$3 \times 10^{-7} m^2/sec^1$
$\alpha_{T,1}$	=	$5 \times 10^{-6}/K$
ρ_1/ρ_s	=	2.6
Q/k_s	=	.025 K/m
h	=	100m
H/L	=	0.3
k_s/k	=	0.8

[1]Eckert and Drake, 1972

time for a minor mineral phase with properties similar to quartz and initially occupying 10% of the total mineral volume in a rock with ϕ_i=0.25 is calculated from Equation 60 to be τ_m=3.5 MMYr. In thinner or less permeable sands, these time scales would be larger.

The patterns of porosity and mineral alteration by flowing fluids predicted by this model should be regarded as a first order approximation to the patterns expected in nature. These patterns will be modified as the permeability field is altered by solution mass transport and the flow field responds to these changes. However, the dominant influence of the layer geometry on the flow field shown in Equation 53 suggests that these primary patterns will be reflected in higher order approximations which account for the feedback of changing permeability distributions on the flow field and subsequent mass transport. It should also be noted that considerable rock alteration can occur with little alteration of the porosity and permeability fields when both prograde and retrograde minerals are included in the original assemblage.

The presence of regional hydraulic gradients will add to the velocities induced by natural convection. For a forced flow to be comparable to natural convection in the example considered here, the regional gradient would have to be

$$\frac{\partial h}{\partial x} = 1.5 \times 10^{-4} \tag{62}$$

where h is the fluid head, $h = p/\rho g - z$. When a fluid-saturated layer is in hydraulic contact with potential gradients, the effects of forced throughflow will often swamp the natural convection motions and dominate the transport process.

A procedure was outlined for determining the required chemical coefficients from the equilibrium relations for the reactions involved. It consists of first determining the solution composition from a set of non-linear algebraic equations (Equations 15 and 16) and then determining the temperature and pressure derivatives of the species concentrations from a set of linear algebraic equations derived by differentiating the same equations. An alternative to this procedure is to calculate the temperature and pressure derivatives of the species concentrations numerically by evaluating the solution composition at neighboring temperatures and pressures. This may prove to be the most straightforward approach to obtaining these coefficients when batch equilibrium geochemical computer models such as EQ3/EQ6 (Wolery, 1979) are available.

XII. CONCLUSIONS

1. When local partial equilibrium is maintained between a porous rock and an aqueous solution flowing through it, the rate of mass transfer between the rock and fluid can be calculated from a knowledge of the flow and temperature fields and the progress of any irreversible reactions in the chemical system considered.
2. All of the chemical coefficients required for computing the mass transfer by a flowing solution can be calculated from the equilibrium relations for the reactions between species and thermodynamic data for the species themselves.
3. For quasi-steady conditions in which the primary driving potential for mass transfer is flow through temperature gradients, the complete sequence of compositions which can evolve from a given starting composition can be calculated independent of the flow and temperature distributions.
4. For quasi-steady conditions in which the primary driving potential for mass transfer is flow through temperature gradients, the pattern of porosity and mineral alteration can be described as a continuous wave field of reversible rock alteration so long as all of the mineral phases originally present remain as solid phases.

5. When one of the original mineral phases is depleted in some region of the flow, discontinuities in porosity and mineralogy are introduced which propagate downstream and irreversibly alter the rock.
6. When the flow is confined to a porous layer with gently sloping boundaries, approximate analytical solutions for the flow and temperature fields can be obtained and the patterns of diagenetic alteration can be calculated.
7. The patterns of rock alteration by solution mass transport can be quite complex even in flows with a simple geometry.

NOMENCLATURE

a_ℓ - activity of aqueous species ℓ
c_ℓ - molarity of aqueous species ℓ
C_k - molar concentration of mineral phase k
CFI - convective flow index
D - molecular diffusion coefficient
D_e - effective dispersion coefficient
g - acceleration of gravity
h - thickness of permeable layer
H - total change in elevation of permeable layer
k, k_s - thermal conductivity of permeable layer and shale, respectively
k_p - permeability
K - total number of mineral phases
K_r - equilibrium constant for reaction r
L - total number of aqueous species, length of permeable layer
M_k - molecular weight of species k
p - pressure
Pe - Peclet number
q_o - volumetric flow rate
Q - heat flux
R - total number of reactions, gas constant
R_k, R_ℓ - reaction rates per unit per volume for species k and ℓ
Ra - Rayleigh member
S - Slope parameter, H/L
S_1, S_2 - upstream and downstream boundaries of depleted mineral zones
S_k, S_ℓ - chemical formulae for species k and ℓ
T - temperature
u - horizontal velocity component
U_c - natural convection velocity scale

U_h - hydraulic throughflow velocity scale
$\underset{\sim}{V}$ - fluid velocity vector
w - vertical velocity component
x - horizontal distance
X - dimensionless horizontal distance, x/L
z - vertical distance
Z - dimensionless vertical distance, Z/H
α - thermal expansion coefficient of water
$\alpha_T, \alpha_p, \alpha_\xi$ - solubility coefficients, see Equation 31
γ_ℓ - activity coefficient for species ℓ
Θ - temperature
κ - thermal diffusivity
λ_k - proportionality constant for reaction progress of species k, Equation 40
μ - fluid viscosity
$\upsilon_{kr}, \upsilon_{\ell r}$ - stoichiometric coefficients for species ℓ and k in reaction r
ξ_r - progress of reaction r per unit pore volume
ξ_e - reaction progress required for initial equilibration of minerals and aqueous solution
ξ_t - reaction progress of irreversible reactions
Ξ_k - progress in reaction r per unit total volume
Ξ_T - overall thermal progress variable, Equation 39
ρ - density
τ - characteristic time for dissolution
ϕ - porosity
χ_k - volume fraction of species k
ψ - streamfunction

Subscripts

i - initial
k - mineral phase k
ℓ - aqueous species ℓ
m - minor mineral m
r - reaction r

APPENDIX A

Mass Conservation for Moving Discontinuities

The behavior of discontinuous fronts in general flows can be seen by considering a mass balance on phase m in the streamtube shown in Figure A-1. Since the lateral surface of a streamtube is composed of streamlines, flow can only enter or leave at the ends of the streamtube.

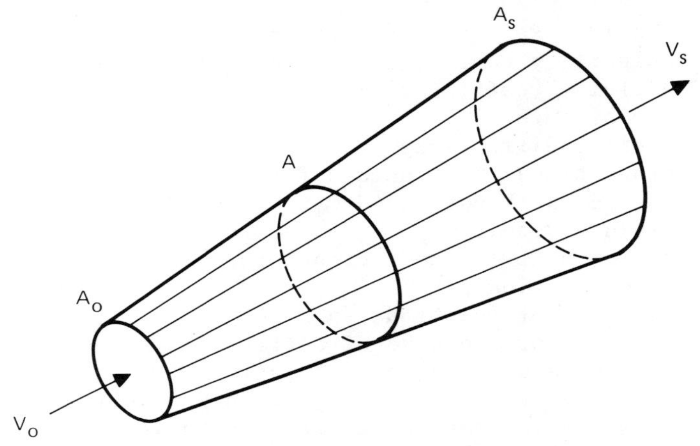

Figure A-1: Streamtube surface in a flowing fluid.

For a constant density fluid, conservation of mass requires

$$q = A_oV_o = A_sV_s = AV \tag{A1}$$

where q is the volume flow rate through the streamtube. Conservation of the total moles of mineral m in the streamtube at any instant requires

$$\int_o^S C_m A ds = \int_o^S C_{m,i} A ds - \left(\frac{\alpha_{T,m}\rho_s}{M_m}\right)\int_o^t (V_sT_sA_s - V_oT_oA_o)dt' \tag{A2}$$

where the last integral on the right is the total amount of phase m removed from the streamtube by the flow over time t. As long as there is an excess of mineral m, the molar concentration, C_m, at any instant is

$$C_m = C_{m,i} - \Xi_m = C_{m,i} - \left(\frac{\alpha_{T,m}\rho_s}{M_m}\right) \int_o^t \underset{\sim}{V} \cdot \underset{\sim}{\nabla} T dt' \qquad (A3)$$

and

$$\int_o^S C_m A ds = \int_o^S C_{m,i} A ds - \left(\frac{\alpha_{T,m}\rho_s}{M_m}\right) \int_o^S \int_o^t (\underset{\sim}{V} \cdot \underset{\sim}{\nabla} T) dt A ds \qquad (A4)$$

If we consider a flow with a steady streamtube geometry, i.e. with steady boundaries and permeability distribution, then we can interchange the order of integration in Equation A4 and combine with Equation A2 to get

$$\int_o^t \left[\int_o^S (\underset{\sim}{V} \cdot \underset{\sim}{\nabla} T) A ds - (V_s T_s A_s - V_o T_o A_o) \right] dt = 0 \qquad (A5)$$

For this to be valid, we require

$$\int_o^S (\underset{\sim}{V} \cdot \underset{\sim}{\nabla} T) A ds = (V_s T_s A_s - V_o T_o A_o) = q(T_s - T_o) \qquad (A6)$$

This result can also be derived directly from the divergence theorem (Hildebrand, 1965) and the condition $\underset{\sim}{\nabla} \cdot \underset{\sim}{V} = 0$ without reference to conservation of component m.

If mineral m becomes depleted in a region between S_1 and S_2, then in that region its dissolution reaction will have stopped at $\Xi_m = C_{m,i}$. For times greater than the time for mineral m to be depleted between S_1 and S_2

$$\int_o^S C_m A ds = \int_o^S C_{m,i} A ds - \left(\frac{\alpha_{T,m}\rho_s}{M_m}\right) \int_o^{S_1} \int_o^t (\underset{\sim}{V} \cdot \underset{\sim}{\nabla} T) dt A ds -$$

$$\int_{S_1}^{S_2} C_{m,i} A ds - \left(\frac{\alpha_{T,m}\rho_s}{M_m}\right) \int_{S_2}^{S} \int_o^t (\underset{\sim}{V} \cdot \underset{\sim}{\nabla} T) dt A ds \qquad (A7)$$

Interchanging the order of integration, subtracting from Equation A2, and combining with Equation A6, this relation can be rearranged to give

$$\int_{S_1}^{S_2} [C_{m,i} - (\frac{\alpha_{T,m}\rho_s}{M_m}) \int_{o}^{t} (\underset{\sim}{V} \cdot \underset{\sim}{\nabla} T)dt]Ads = 0 \qquad (A8)$$

APPENDIX B

The Mathematical Model

Figure 3a shows a two-dimensional folded sheet of water-saturated rock (sandstone) embedded between two impermeable shale units. A Cartesian coordinate system, having its origin at the Earth's surface, has its (vertical) z-axis pointing downward. At z = 0, the surface temperature θ_o is specified, while at depth $z=z_o$ the heat flux,

$$Q = k_s \frac{d\theta^{(s)}}{dz},$$

is given consistent with the normal geothermal gradient. Here k_s is the thermal conductivity of the shale, and $\theta^{(s)}$ is the temperature in the shale. The upper and lower boundaries of the sandstone have positions given by $z = z_T(x)$ and $z = z_B(x)$, respectively, where x is the horizontal coordinate.

Two-dimensional steady flow in the porous layer is described by Darcy's law:

$$u = - \frac{k_p}{\mu} \frac{\partial p}{\partial x} \qquad (B1)$$

$$w = - \frac{k_p}{\mu} (\frac{\partial p}{\partial z} - \rho g) \qquad (B2)$$

$$\frac{\partial u}{\partial x} + \frac{\partial w}{\partial z} = 0 \qquad (B3)$$

where the velocity components (u,w) act in the (x,z) directions, p is the pressure, μ is the viscosity of water, ρ is the density of the water and k_p is the medium permeability.

The density depends on temperature through a linear law,

$$\rho = \rho_o \{1-\alpha(\theta-\theta_o)\} \ . \tag{B4}$$

Using the Boussinesq approximation, the variable ρ enters Equations (B1)-(B3) only in the buoyancy term ρg (note that such variations have already been neglected in Equation (B3)). Thermal effects are governed by the energy equation

$$u \frac{\partial\theta}{\partial x} + w \frac{\partial\theta}{\partial z} = \kappa \left(\frac{\partial^2\theta}{\partial x^2} + \frac{\partial^2\theta}{\partial z^2}\right) \tag{B5}$$

where κ is the thermal diffusivity of the fluid-filled porous medium.

In the impermeable rock (shale) heat transfer occurs only by thermal conduction, so

$$\frac{\partial^2\theta^{(s)}}{\partial^2 x} + \frac{\partial^2\theta^{(s)}}{\partial^2 z} = 0 \quad \text{for } z < z_T(x) \text{ and } z > z_B(x) \tag{B6}$$

where $\theta^{(s)}$ denotes the temperature of the shale.

Finally, the boundary conditions are:

$$\theta^{(s)} = \theta_o \ \text{at } z = 0; \tag{B7}$$

while

$$k_s \frac{\partial\theta^{(s)}}{\partial z} = Q \ \text{at } z = z_o \ . \tag{B8}$$

At the shale-sand contacts, both the temperatures and the heat fluxes match,

$$\left.\begin{aligned} \theta &= \theta^{(s)} \\ k\left[\frac{\partial\theta}{\partial x}\frac{dz_T}{dx} - \frac{\partial\theta}{\partial z}\right] &= k_s\left[\frac{\partial\theta^{(s)}}{\partial x}\frac{dz_T}{dx} - \frac{\partial\theta^{(s)}}{\partial z}\right] \end{aligned}\right\}, \quad z = z_T$$

$$\left.\begin{aligned} \theta &= \theta^{(s)} \\ k\left[\frac{\partial\theta}{\partial x}\frac{dz_B}{dx} - \frac{\partial\theta}{\partial z}\right] &= k_s\left[\frac{\partial\theta^{(s)}}{\partial x}\frac{dz_B}{dx} - \frac{\partial\theta^{(s)}}{\partial z}\right] \end{aligned}\right\}, \quad z = z_B \qquad \text{(B9)}$$

where k is the thermal conductivity of the fluid-filled porous sandstone. Since the boundaries of the water-saturated porous medium are impermeable to fluid flow, the normal component of velocity vanishes there so that

$$w - \frac{dz_T}{dx} u = 0 \text{ at } z = z_T$$

and (B10)

$$w - \frac{dz_B}{dx} u = 0 \text{ at } z = z_B .$$

Introducing a stream function ψ, with

$$u = -\frac{\partial\psi}{\partial z}, \qquad w = \frac{\partial\psi}{\partial x} . \qquad \text{(B11)}$$

Equations (B1)-(B4) can be combined to obtain

$$\frac{\partial^2\psi}{\partial x^2} + \frac{\partial^2\psi}{\partial z^2} = -\rho_o g\alpha \frac{k_p}{\mu}\frac{\partial\theta}{\partial x} . \qquad \text{(B12)}$$

Likewise, Equation (B5) can be rewritten as:

$$\frac{\partial\psi}{\partial x}\frac{\partial\theta}{\partial z} - \frac{\partial\psi}{\partial z}\frac{\partial\theta}{\partial x} = \kappa\left(\frac{\partial^2\psi}{\partial x^2} + \frac{\partial^2\psi}{\partial z^2}\right) . \qquad \text{(B13)}$$

The equations may be put into non-dimensional form with the following change of variables

$$X = \frac{x}{L} \quad , \qquad Z = \frac{z}{H} \quad . \tag{B14}$$

$$\Psi = \frac{\psi}{U_o h} \quad , \quad T = \frac{\theta - \theta_o}{\Delta T}$$

where L is a measure of the horizontal length scale of the fold, h is the maximum thickness of the fold, H is the total change in elevation over L and

$$\Delta T = QH/k_s$$

and

$$U_o = \frac{\kappa}{h} \tag{B15}$$

This notation differs from that of Davis et. al., (1985) in recognizing two different vertical length scales. An intrinsic velocity scale is used to scale the flow equations since either the natural convection or forced flow may be zero in a particular example. Using uppercase letters for non-dimensional quantities, system (B6)-(B13) becomes,

$$\frac{\partial^2 \Psi}{\partial Z^2} + S^2 \frac{\partial^2 \Psi}{\partial X^2} = - \text{RaS} \frac{\partial T}{\partial X} \tag{B16}$$

$$\frac{\partial^2 T}{\partial Z^2} + S^2 \frac{\partial^2 T}{\partial X^2} = S \left(\frac{\partial \Psi}{\partial X}\frac{\partial T}{\partial Z} - \frac{\partial \Psi}{\partial Z}\frac{\partial T}{\partial X}\right) \tag{B17}$$

$$\frac{\partial^2 T^{(s)}}{\partial Z^2} + S^2 \frac{\partial^2 T^{(s)}}{\partial X^2} = 0 \text{ for } Z < Z_T(x) \text{ and } Z > Z_B(x) \tag{B18}$$

$$T^{(s)} = T_o \quad \text{at} \quad Z=0 \tag{B19}$$

$$\frac{\partial T^{(s)}}{\partial Z} = 1 \quad \text{at} \quad Z=Z_o \tag{B20}$$

$$\left.\begin{aligned} &T = T^{(s)} \\ &S^2 \frac{\partial T}{\partial X} \frac{dZ_T}{dX} - \frac{\partial T}{\partial Z} = \frac{k_s}{k}\left[S^2 \frac{\partial T^{(s)}}{\partial X} \frac{dZ_T}{dX} - \frac{\partial T^{(s)}}{\partial Z}\right] \end{aligned}\right\}, \quad Z = Z_T \tag{B21}$$

$$\left.\begin{aligned} &T = T^{(s)} \\ &S^2 \frac{\partial T}{\partial X} \frac{dZ_B}{dX} - \frac{\partial T}{\partial Z} = \frac{k_s}{k}\left[S^2 \frac{\partial T^{(s)}}{\partial X} \frac{dZ_B}{dX} - \frac{\partial T^{(s)}}{\partial Z}\right] \end{aligned}\right\}, \quad Z = Z_B \tag{B22}$$

In place of conditions (B10) we have the equivalent form

$$\begin{aligned} &\Psi = 0. \quad \text{at } Z = Z_B \;. \\ &\Psi = -\, Pe_T \text{ at } Z = Z_T \end{aligned} \tag{B23}$$

Here Ra is the Rayleigh number,

$$Ra = \frac{\rho_o g \alpha k_p Q H^2}{k_s \kappa \mu} \quad ,$$

Pe_T is the thermal Peclet number, $Pe_T = \frac{q_o}{\kappa}$, (B24)

and S is a parameter that measures the slope of the fold,

$$S = \frac{H}{L} \quad . \tag{B25}$$

From Equation (B23) we see that the contribution of the throughflow to Ψ will be of order Pe_T. In the absence of a throughflow Equation (B16) shows that Ψ will be of order RaS. Thus the convective derivative on the right hand side of Equation (B17) will be of order $(RaS + Pe_T)S$. In many geologic situations the slope of a bed is small ($S \cong 0.3$ or smaller) so that S^2 ($\cong 0.09$) is very small. Hence, we make the approximations

$$S^2 << 1 \quad (RaS + Pe_T)\, S << 1 \tag{B26}$$

and ignore all terms having the factor S^2 and the right hand side of Equation (B17). This means that for the relatively slow flows considered only vertical heat conduction is important in establishing the temperature field.

Solutions for the approximate system (B16)-(B23) can be obtained by direct integration. These solutions are:

$$T^{(s)} = \begin{cases} Z & \text{(upper shale)} \\ Z - (1 - \frac{k_s}{k})[Z_B(X) - Z_T(X)] & \text{(lower shale)} \end{cases} \tag{B27a}$$

$$T = \frac{k_s}{k} Z + (1 - \frac{k_s}{k})\, Z_T(X) \quad \text{(sand)} \tag{B27b}$$

$$\Psi = \frac{1}{2} RaS\, (1 - \frac{k_s}{k})\, \frac{dZ_T}{dX}\, [Z_B(X) - Z][Z - Z_T(X)] - Pe_T \frac{[Z_B(X) - Z)}{[Z_B(X) - Z_T(X)]} \tag{B27c}$$

With the solution in this form, it can be seen that the patterns of flow are determined by the geometry of the sand-shale contacts. Changes in the values of the physical parameters affect only the magnitude of the velocity with a change in sign occurring when k_s/k changes from a value less than unity to a value greater than unity. When k_s/k is equal to one, there are no horizontal temperature gradients and the flow is entirely due to the hydraulic throughflow.

Introducing characteristic velocity scales for the convective flow and hydraulic flow defined as

$$U_c = \frac{1}{2}\,\mathrm{RaS}\,\left(1 - \frac{k_s}{k}\right)\left(\frac{h}{H}\right)^2\left(\frac{\kappa}{h}\right)$$

and (B28)

$$U_h = \frac{q_o}{h}$$

the expression for Ψ can be rewritten in dimensional form as

$$\psi = U_c H G_1(X,Z) + U_h H G_2(X,Z) \tag{B29}$$

where

$$G_1(X,Z) = \left(\frac{H}{h}\right)\frac{dZ_T}{dX}\,[Z_B(X)-Z][Z-Z_T(X)]$$

$$G_2(X,Z) = \left(\frac{h}{H}\right)\frac{[Z_B(X)-Z]}{[Z_B(X)-Z_T(X)]}$$

The dimensionless functions $G_1(X,Z)$ and $G_2(X,Z)$ give the influence of the layer geometry on the flow field. The velocity components are:

$$u = -\,U_c G_{1,Z}(X,Z) - U_h G_{2,Z}(X,Z)$$

(B30)

$$w = U_c S G_{1,X}(X,Z) - U_h S G_{2,X}(X,Z)$$

where

$$G_{1,Z}(X,Z) = -2\left(\frac{H}{h}\right)\frac{dZ_T}{dX}\left[Z - \frac{1}{2}(Z_B + Z_T)\right]$$

$$G_{2,Z}(X,Z) = -\left(\frac{h}{H}\right)\frac{1}{(Z_B-Z_T)}$$

$$G_{1,X}(X,Z) = \left(\frac{H}{h}\right)\left\{\frac{d^2Z_T}{dX^2}(Z_B-Z)(Z-Z_T) + \frac{dZ_T}{dX}\left[(Z-Z_B)\frac{dZ_T}{dX} + (Z-Z_T)\frac{dZ_B}{dX}\right]\right\}$$

$$G_{2,X}(X,Z) = \left(\frac{h}{H}\right)\left\{\frac{1}{(Z_B-Z_T)}\frac{dZ_B}{dX} - \frac{(Z_B-Z)}{(Z_B-Z_T)^2}\left[\frac{dZ_B}{dX} - \frac{dZ_T}{dX}\right]\right\}$$

The reaction rate due to flow through a temperature field is calculated as

$$\dot{\Xi}_T = \left(\frac{\rho_s \alpha_{T,1}}{M_1}\right) \underset{\sim}{V} \cdot \underset{\sim}{\nabla\theta} \tag{B31}$$

where

$$\underset{\sim}{V} \cdot \underset{\sim}{\nabla\theta} = u \frac{\partial\theta}{\partial X} + w \frac{\partial\theta}{\partial Z}$$

may be calculated from Equation (B30) and the dimensional form of Equation (B27b).

REFERENCES

Blackwell, R. J., (1962), Laboratory Studies of Microscopic Dispersion Phenomena, Soc. Pet. Eng. J., March, pp. 1-8.

Carnahan, B., Luther, H. A., and Wilkes, J. O. (1969), Applied Numerical Methods, Wiley and Sons, New York.

Cathles, L. M. (1981), Fluid Flow and Genesis of Hydrothermal Ore Deposits, Econ. Geol., 75th Anniv. Vol., pp. 424-457.

Cathles, L. M. (1983), An Analysis of the Hydrothermal System Responsible for Massive Sulfide Deposition in the Hokuroku Basin of Japan, Econ. Geol., Monograph 5, pp. 439-487.

Davis, S. H., Rosenblatt, S., Wood, J. R., and Hewett, T. A. (1985), Convective Fluid Flow and Diagenetic Patterns in Domed Sheets, Amer. J. Sci., Vol. 285, March, pp. 207-223.

Eckert, E. R. G., and Drake, R. M. (1972), Analysis of Heat and Mass Transfer, McGraw-Hill, New York.

Franks, S. G., and Forester, R. W. (1984), Relationships Among Secondary Porosity, Pore-Fluid Chemistry and Carbon Dioxide, Texas Gulf Coast, Clastic Diagenesis (ed. D. A. McDonald and R. C. Surdam), AAPG Memoir 37, pp. 63-80.

Galloway, W. E. (1984), Hydrologic Regimes of Sandstone Diagenesis, Clastic Diagenesis (ed. D. A. McDonald and R. C. Surdam), AAPG Memoir 37, pp. 3-14.

Hildebrand, F. B. (1965), Advanced Calculus for Applications, Prentice-Hall, Englewood-Cliffs, N.J.

Helgeson, H. C., Brown, T. H., Nigrini, A., and Jones, T. A. (1970), Calculation of Mass Transfer in Geochemical Processes Involving Aqueous Solutions, Geochim. Cosmochim. Acta, Vol. 34, pp. 569-592.

Helgeson, H. C. (1979), Mass Transfer Among Minerals and Hydrothermal Solutions, Geochemistry of Hydrothermal Ore Deposits, 2nd Edn. (ed. H. L. Barnes), Wiley-Interscience, N.Y., pp. 568-592.

Loucks, R. G., Dodge, M. M., and Galloway, W. E. (1984), Regional Controls on Diagenesis and Reservoir Quality in Lower Tertiary Sandstones along the Texas Gulf Coast, Clastic Diagenesis (ed. D. A. McDonald and R. C. Surdam), AAPG Memoir 37, pp. 15-46.

Hoffman, A. (1972), Chromatographic Theory of Infiltration Metasomatism and Its Application to Feldspars, Amer. J. Sci, Vol. 272, January, pp. 69-90.

Lund, K., and Fogler, H. S. (1976), The Prediction of the Movement of Acid and Permeability Fronts in Sandstone, Chem. Eng. Sci., Vol. 31, pp. 381-392.

Pittman, E. D. (1979), Porosity, Diagenesis, and Productive Quality of Sandstone Reservoirs, SEPM Spec. Pub. 26, pp. 159-173.

Prigogine, I., and Defay, R. (1954), Chemical Thermodynamics, translated by D. H. Everett, Longmans Green, London.

Schmidt, V., and McDonald, D. A. (1979), The Role of Secondary Porosity in the Course of Sandstone Diagenesis, SEPM Spec. Pub. 26, pp. 175-207.

Walsh, M. P., Bryant, S. L., Schecter, R. S., and Lake, L. W. (1984), Precipitation and Dissolution of Solids Attending Flow through Porous Media, AIChE J., Vol. 30, No. 2, March, pp. 317-328.

Wood, J. R., and Hewett, T. A. (1982), Fluid Convection and Mass Transfer in Porous Sandstones--A Theoretical Model, Geochimica et Cosmochimica Acta, Vol. 46, No. 10, pp. 1707-1713.

Wood, J. R., and Hewett, T. A. (1984), Reservoir Diagenesis and Convective Fluid Flow, Clastic Diagenesis (ed. D. A. McDonald and R. C. Surdam), AAPG Memoir 37, pp. 99-110.

Wood, J. R. (1984), Advective Diagenesis: Calculation of Thermal Mass Transfer Coefficients in the Systems SiO_2-H_2O and $CaCO_3$-H_2O, COFRC TM84001387, November.

PERMEABILITY VARIATIONS OBSERVED AT THE FACES OF CROSSBEDDED SANDSTONE OUTCROPS

Fred I. Stalkup

ARCO RESOURCES TECHNOLOGY
ATLANTIC RICHFIELD COMPANY
PLANO, TEXAS

ABSTRACT

This paper examines the nature of permeability variation in crossbedded intervals of six sandstone outcrops. Crossbedded intervals are commonly found within the total cross section of sandstone reservoirs and represent an important element of reservoir anatomy. These outcrops are both of fluvial and shallow marine origin. They are approximately three to twelve ft thick and were sampled over 24- to 80-ft lateral distances by taking cores on close spacing from the faces of the outcrops. Samples were taken on one-ft centers along vertical lines, or coring stations, that were spaced two to four ft apart laterally.

A coring station is an analogue of a well drilled through the crossbedded interval. The overall character of permeability variation seen at an individual coring station did not typically persist for more than 5 to 20 ft laterally, although a particular feature such as an interval of higher or lower permeability might extend over a greater distance. Statistical analysis of subintervals identified from the coring station permeability profiles and qualitative interpretation of the x- and y-direction semivariograms for each outcrop show these crossbedded outcrops to be stratified into 0.5 to 3 ft thick laterally discontinuous layers of different mean permeability that are from 6 to 40 ft or greater in lateral dimension. The ratio of mean permeabilities of adjacent discontinuous layers usually was in the range 1.3 to 2.5. No high permeability "streaks" with permeabilities 5 or 10-fold greater than surrounding permeabilities were observed in any of the outcrops. The limited lateral persistence of coring station permeability profile character and the apparent discontinuous nature of layering indicate a stochastic rather than deterministic description of permeability variation may be required.

Within a discontinuous layer there may be smaller-scale structure to the permeability variation, but it could not be resolved with the spacing of cores in this study. Instead, an approximate model, suggested by the semivariograms, is that permeability varies randomly about the mean within a discontinuous layer.

The mean permeability calculated from samples at an individual coring station is an estimate of average permeability for the crossbedded interval at that location. For all outcrops, these estimates usually did not differ by more than a factor of 1.4, and statistical analysis failed to show any statistically significant differences among the coring station means.

Standard deviation is a measure of the variation of permeability samples about a mean. Estimates of standard deviation made from samples at individual coring stations varied as much as three fold from one station to another, and there appeared to be roughly equal probability of obtaining an estimate anywhere in this range. This behavior casts doubt on the usefulness of an estimate from only one well for characterizing stratification of thin crossbedded intervals in the region around the well.

INTRODUCTION

Permeabilities measured in core plugs taken from a well core typically show a large degree of variation. Fig. 1 shows permeabilities measured on a core from an oil producing reservoir that illustrate this statement. Because a well core represents only a negligible volume of the reservoir, some logical questions to ask are "How meaningful are the permeabilities observed in the core for characterizing the reservoir? Are they largely random such that another core a few inches or feet away would show a completely different pattern of variation; or do the observed pattern and magnitude of permeabilities persist out into the reservoir, and if so, for how great a distance?"

Outcrop coring is one method for studying the spatial variation of permeability. During the summers of 1959, '60, and '61, T. L. Polasek and C. A. Hutchinson of Arco's

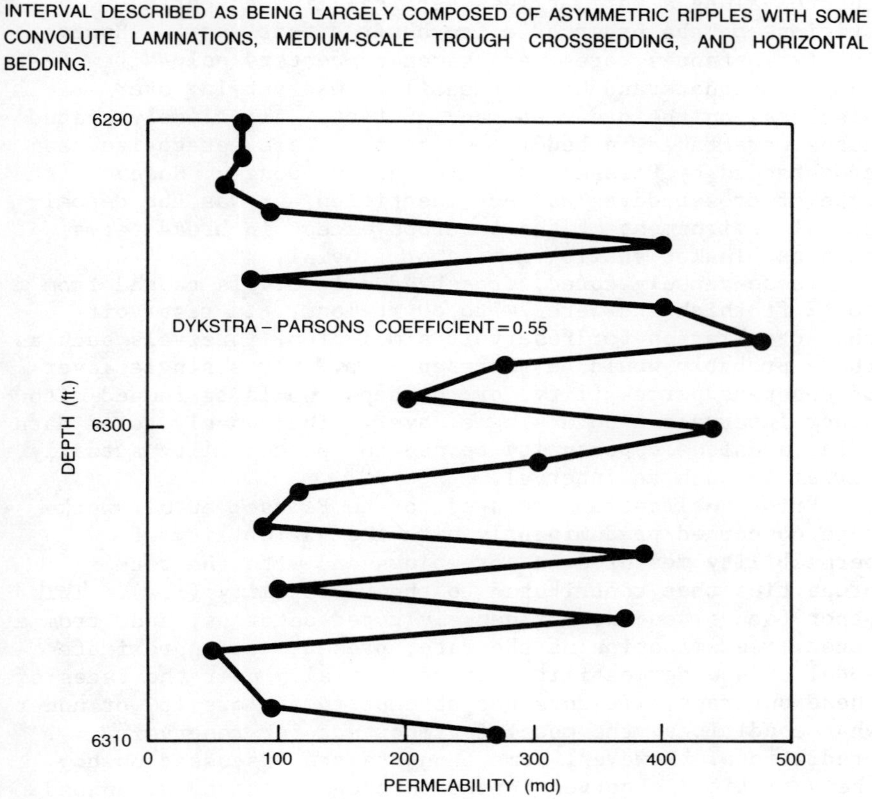

FIGURE 1. Permeability variations in a well from a producing reservoir.

Exploration and Production Research (then the Atlantic Refining Company) and C. F. Dodge, a geological consultant, carried out a program for studying permeability variations in sandstones by observing and coring some 31 outcrops. Most sampling was conducted by drilling into the face of the outcrop along a more or less vertical line at coring stations on the order of a few hundred feet apart. However, in six instances cores were taken on centers only a few feet apart to understand how permeability was varying over distances on the order of tens of feet. All closely spaced cores were taken in bedding units that were categorized as crossbedded by Polasek, Hutchinson, and Dodge. However, the type of crossbedding was not identified nor was the depositional environment of the outcrops except in broad terms, such as blanket shallow marine or fluvial.

These densely cored, crossbedded intervals ranged from 3 to 12 ft thick and were 24 to 80 ft long. In reservoir characterization for reservoir simulation, intervals such as these probably would be represented by only a single layer of constant permeability, or, perhaps, would be lumped with other intervals into a single layer. The densely cored data offer a unique opportunity to see how permeability actually varies in such an interval.

Prior publications related to the Polasek et al. work were concerned predominantly with the variability of permeability measured in core plugs and with the rock properties that contributed to the variability (1,2). This paper focuses on the six densely cored outcrops, and, from a recent reexamination of the data, presents an approximate model of how permeability varies spatially over the faces of these outcrops. It does not attempt to explore if, or under what conditions, the model is important for reservoir predictions. However, some thoughts are discussed on how the variations observed in the outcrops might be accounted for in a reservoir description.

ORIGINAL DATA COLLECTION PROCEDURES

In all six of the densely cored outcrops, cores were taken in a more or less vertical line at different stations spaced laterally across the face of the outcrop. This gave a gridwork of samples. Each station may be thought of as a "well" and the cores as samples from different intervals in the "well". At a given coring station, core samples were taken one ft apart vertically. Coring stations were spaced from 2 to 4 ft apart laterally depending on the particular outcrop under study.

Cores were taken with an air-driven coring tool. At each coring point a 12-in. long core of one-in. diameter was taken. The outer two inches were discarded to prevent surface weathering from influencing results unduly. The remaining core was cut into ten, one-in. segments on which air permeabilities were measured, both in the field and later in the laboratory on selected samples. These permeabilities were then averaged to give the permeability reported for that coring location.

Unfortunately, the original permeability measurements and even much of the original data analysis appear to be lost. The only documentation remaining are descriptions of the work and conclusions, without supporting analysis, at different stages of completion that are found in several published papers and company reports (1,2). In these reports, the absolute permeability measurements are not given; only the range of permeability at each coring location is reported. For example, permeability at a given coring point might be reported to be in the range of 400 to 600 md.

ANALYSIS AND RESULTS OF ANALYSIS

The nature of permeability variation in these outcrops was examined in the following manner:

1. Permeability profiles were compared from one coring station to another to determine how far across the outcrop similar profiles persisted.

2. From this comparison, the outcrop face was divided into subintervals that appeared to have different permeabilities.

3. These subintervals were tested by statistical techniques to determine if their mean permeabilties were, in fact, different. This was necessary to be sure that apparent differences were not caused simply by the statistics of sampling, i.e., by having a relatively small number of samples per subinterval (4 to 30) and variability from sample to sample.

4. Semivariograms (defined later) were calculated for each outcrop and compared with semivariograms calculated for several hypothetical distributions of what permeability variation might be like in these outcrops. The purpose was to see which hypothetical distribution(s) the actual semivariograms were more nearly like and to see if any additional insights could be gained from semivariogram behavior.

5. Statistical tests were made to determine if the mean permeability calculated from samples at each coring station varied from one coring station to another over the face of the outcrop. These means are estimates of the permeability of the total crossbedded interval at that location, so the statistical tests showed if the mean interval permeability varied over the face of the outcrop.

6. The standard deviations of permeability calculated from samples at each coring station were examined to see how useful such an estimate from only one coring station would be for characterizing permeability variability for the entire outcrop (i.e., if only one "well" had been drilled).

These analysis procedures and accompanying findings are described more fully below:

Comparison of Permeability Profiles and Initial Identification of Subintervals

Figs. 2 and 3 show permeability profiles at each coring station for two of the outcrops, Woodbine 1 and Rock Springs. In these figures, permeability at the mid point of reported ranges is plotted. Two judgmental criteria were used to distinguish permeability trends from what otherwise might be random fluctuations and to correlate these trends from well to well. The first criterion applied to a single station. Here, an interval of high or low permeability was identified if permeabilities at several coring points within the interval consistently increased or decreased to a high or low value (as opposed to a single fluctuation to a high or low value). The second criterion was concerned with the appearance or shape of the permeability profiles from one station to another. If the appearance of the profile or a segment of the profile persisted over several coring stations, this was taken as evidence of a trend even if an individual segment of the profile, such as an indicated thin

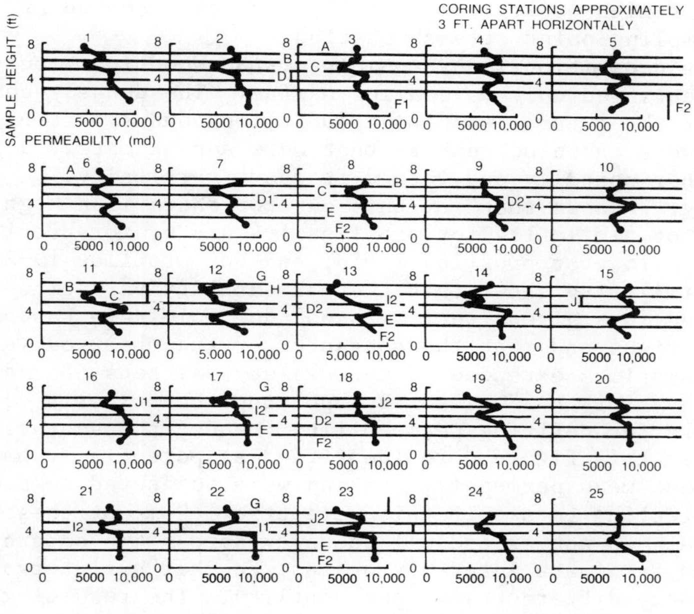

FIGURE 2. Woodbine I Outcrop: permeability profiles at coring stations and intervals for statistical testing identified from profile trends.

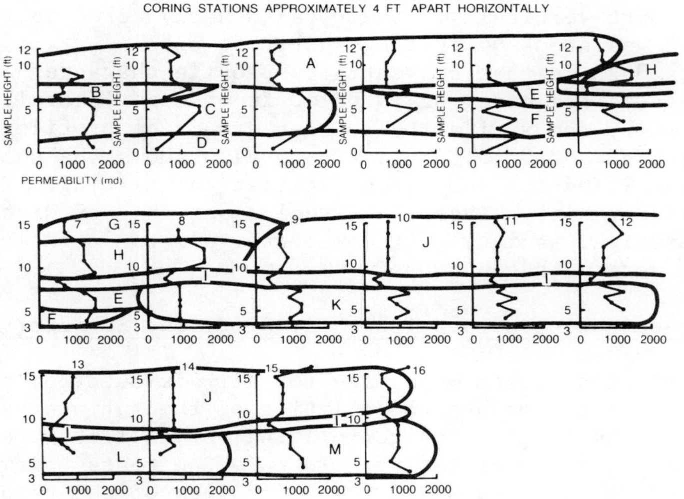

FIGURE 3. Rock Springs Outcrop: permeability profiles at coring stations and intervals for statistical testing identified from profile trends.

layer of high or low permeability, was represented by only one sampling point at each station.

The depositional environment of the Woodbine 1 outcrop was identified only as shallow marine. The densely cored part of the outcrop was about 80 ft long and 7.5 ft high. Cores were taken on centers that were approximately 3 ft apart horizontally and 1 to 2 ft apart vertically.

Fig. 2 shows that for this outcrop there is a high degree of similarity for profiles 1-4 and 6, for profiles 7-13 and 15, for profiles 16-18, and for profiles 19-23. This similarity shows that the permeability variation observed in a given "well" was not due solely to random fluctuations and that the character of the observed variation sometimes extended as much as several tens of feet or more away for the "well". By applying the criteria described above for defining permeability trends, the subintervals shown in this figure were identified as potentially having different mean permeabilities and were subjected to statistical testing to see if this were true. Some of these subintervals do not extend all the away across the outcrop, but instead terminate laterally against another subinterval of apparently different mean permeability. The resulting visual appearance is one of "discontinuous" layering.

Fig. 3 shows the coring station permeability profiles for the Rock Springs outcrop. Depositional environment was identified to be continental or coastline. Bedding was described as discontinuous crossbedding. Cores were taken on centers that were about four ft apart horizontally and one ft apart vertically. The total densely cored surface exposure was about 64 ft long and 12 ft high.

Fig. 3 shows that permeability profile character appears to extend on the order of 10 to 40 ft or more laterally, as evidenced by the similarity between groups of profiles such as 1-3, 6-8, and 9-15. This figure also shows the subintervals identified for subsequent statistical testing. This subdivision, also, gives the visual appearance of discontinuous layering, as many of the subintervals don't extend all the way across the outcrop.

Determination of Statistically Different Subintervals

Statistical tests were made for all six outcrops to determine if the mean permeabilities of the subintervals identified by tracing and correlating permeability trends were statistically different. To make the tests, permeabilities were assumed to be at the mid points of the reported ranges.

First, the data for a given outcrop were examined to see whether a normal or log normal distribution function should be used in making the tests. Although it is commonly assumed that permeability is distributed log normally, Lambert's work (3) shows that specific data may sometimes be described better by a normal distribution (and occasionally by an exponential distribution).

Figs. 4a and b show examples of normal and log normal cumulative distribution plots for Rock Springs data. On these graphs, both permeability and the logarithm of permeability are plotted vs the cumulative percent of total samples that had that level of permeability or less. Actual data are represented by the plotted points. The solid lines represent assumed normal and log normal distributions that have the mean permeability and standard deviation calculated from all samples. Mean permeability of these distributions is found at the 50% cumulative probability. Standard deviation, a measure of permeability variability among all samples, is the difference in permeabilities at the 84.1 and 50% probability levels, as shown in Fig. 4a (or the difference of the ln permeabilities for the log normal distribution). The data points were tested against the assumed distributions by the Kolmogorov-Smirnov method (4,5). A description of this test and details of its application are given in Appendix A.

The Kolmogorov-Smirnov test did not reject either distribution for any outcrop, and subsequent statistical tests were made for both distributions. Although statistical test results did not always agree for the two distributions, generally they did; and conclusions were based on those instances where calculations for both distributions were in agreement.

Three statistical techniques were used to test for permeability differences between subintervals: 1) the Tukey-Kramer test for multiple comparison of means (6), 2) comparison of confidence intervals around the mean permeability, and 3) the T test for two-sample direct comparisons. The Tukey-Kramer method is described in Appendix B. Methods for calculating confidence intervals and applying the T test are readily found in textbooks (4,7). Only summary descriptions of the three techniques are given below.

The Tukey-Kramer (T-K) method is a technique for comparing all pairs in a group of means to see which pairs are different, while preserving the same confidence level for all comparisons. (Confidence level may be thought of as the probability that differences found in the testing are not due to chance.) The method involves calculating a

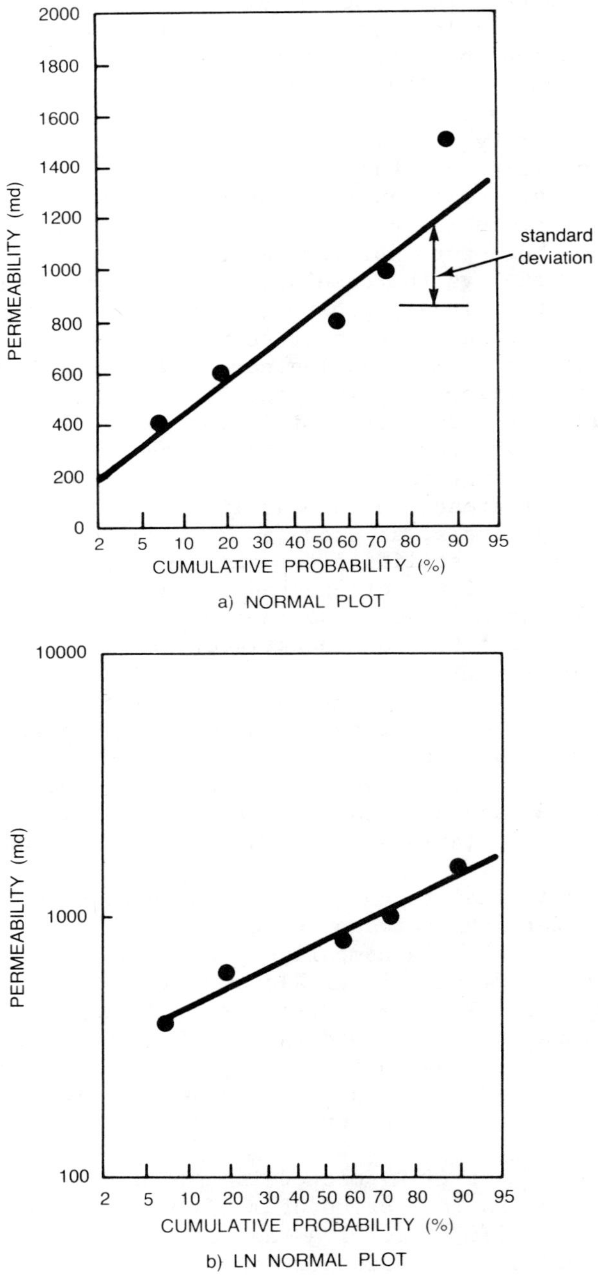

FIGURE 4. Cumulative distributions for Rock Springs permeability data.

confidence interval for the difference between two sample means, i.e., calculating a range of values within which the difference can be expected to lie at the confidence level tested. If this range does not include the value zero, the means are statistically different at that confidence level.

The T-K method assumes that the populations from which the samples were taken to calculate each mean have the same true standard deviation (the statistical assumption of variance homogeneity). When this assumption appeared to be violated (see Eq. B4, Appendix B), samples were regrouped to try and eliminate those subintervals that were causing the problem. The T-K test was performed on the remaining subintervals. Of course, excluded intervals could not be compared by the T-K method at the homogeneous variance condition. In these instances, comparison of confidence intervals around means and the T test were employed as well.

Comparison of confidence intervals involves calculating a permeability range within which the mean permeability is known to lie at a given confidence level. If the confidence intervals for two mean permeabilities don't overlap, the means can be said to be different at that confidence level. Fig. 5 shows confidence intervals calculated for the mean permeabilities of the Fig. 3 (Rock Springs) subintervals. Confidence intervals for subinterval A and C means do not overlap at the 99% confidence level for either permeability or the logarithm of permeability, and the difference between the mean permeabilities for these two subintervals is statistically different at the 99% level.

In addition to its use in situations where the homogeneous variance assumption was questionable, confidence interval comparisons were made when a difference between two subintervals was found by the T-K method for only one of the probability distributions. When there was some overlap of confidence intervals, the T test, described below, was used.

The T test is suitable for directly comparing the means of two subintervals. Although all possible pairs in a group of means can be tested by this method, it is not recommended because in this technique multiple comparisons involving the same mean use up degrees of freedom, which contributes to a cumulative decrease of the confidence level (4). However, some limited testing of means by this method can be done with a high confidence level (e.g., a given mean is only tested against one other mean).

A limitation of the T test is that samples from both subintervals must be from populations with the same standard deviation.

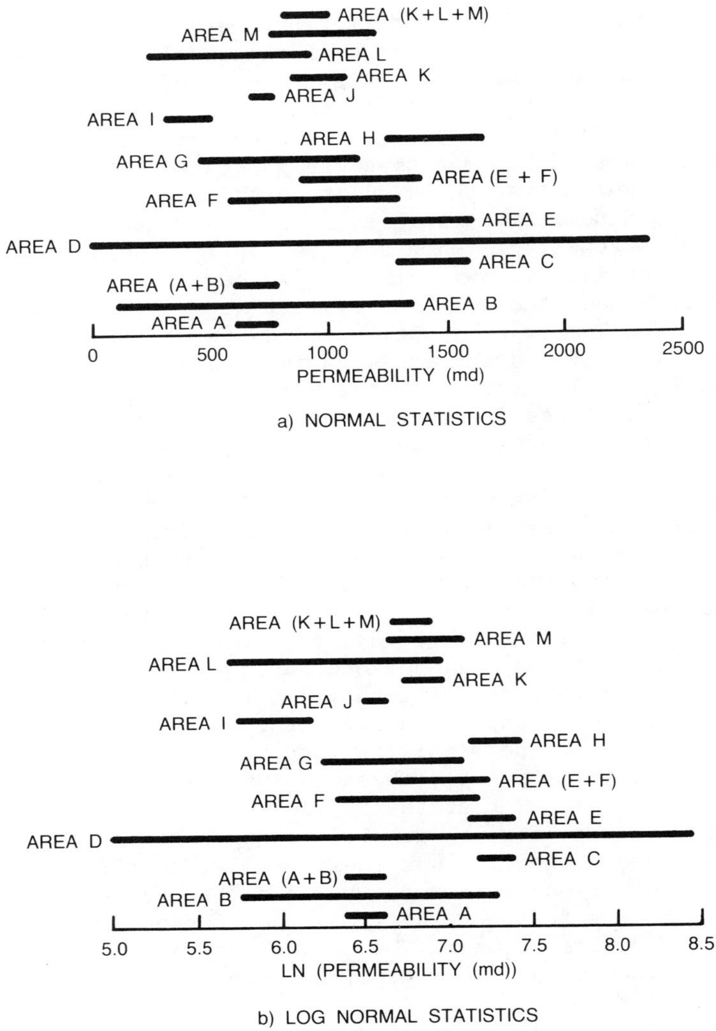

FIGURE 5. 99% confidence intervals for mean permeabilities of different areas of the Rock Springs Outcrop.

Figs. 6 through 11 show the final subintervals for all six outcrops determined to have statistically different mean permeabilities. Figs. 8 through 11 are for the Almond, Frontier, Sundance, and Fox Hills outcrops, all of which are of shallow marine origin. In these figures, the letters indicate those subintervals initially defined for statistical testing by comparing and analyzing coring station permeability profiles. The numbers in parentheses show the mean permeabilities calculated for these initial subintervals. A solid line between two subintervals indicates the T-K test or confidence interval test (usually both) showed the mean permeabilities to be different at a confidence level of at least 95% for both the normal and log normal distributions. A wavy boundary indicates that one distribution showed a T-K test difference and both distributions showed a T test difference. A dashed line indicates that neither distribution showed a difference according to the T-K test, but both indicated a significant difference according to the T test. No boundaries between two letters indicates subintervals that were initially defined for testing but failed to show statistically significant differences between their means.

The Dykstra-Parsons coefficient is sometimes used to characterize permeability variation in reservoir engineering (8). It is calculated by making a log normal cumulative distribution plot similar to Fig. 4b, except that permeability is plotted vs the percent of total samples with that level of permeability or greater. The Dykstra-Parsons coefficient is found by dividing the difference in permeabilities at the 84.1 and 50% probabilities by the value at 50%. These values are taken from the best straight line through the data. For interest, the Dykstra-Parsons coefficent for all samples from each outcrop is indicated on Figs. 6 through 11. Although these coefficients are smaller than customarily found for permeabilities measured in reservoir cores, they only reflect a small interval of one type of bedding.

Figs. 6 through 11 show that for all outcrops there is permeability layering within the relatively thin cross-bedded units examined. The layers are on the order of one to three feet thick, and some are not continuous across the face of the outcrop. Only for the Frontier outcrop do the layers appear to be continuous for the most part across the face. The ratio of mean permeabilities of adjacent discontinuous layers usually was in the range 1.3 to 3.5. No high permeability "streaks" were found in any of the outcrops (i.e., no continuous layers with mean permeabilities 5- to 10-fold higher than surrounding layers).

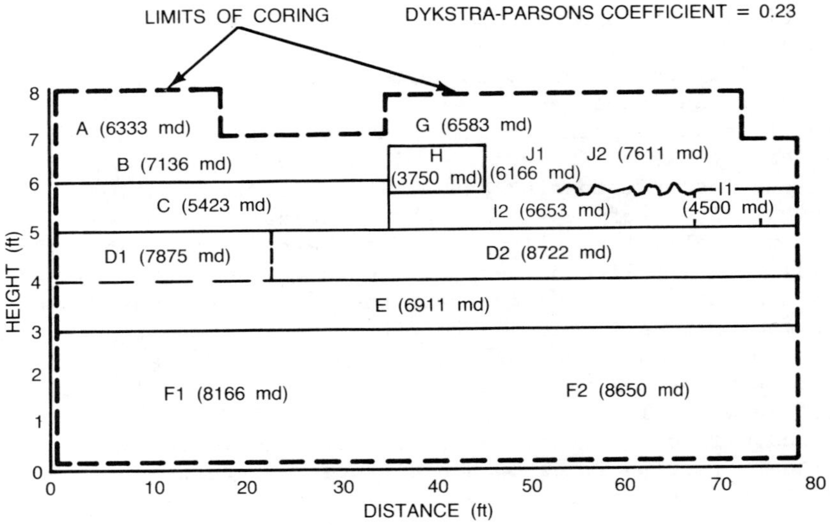

Figure 6. Woodbine 1 Outcrop: adjacent layers with statistically different permeabilities.

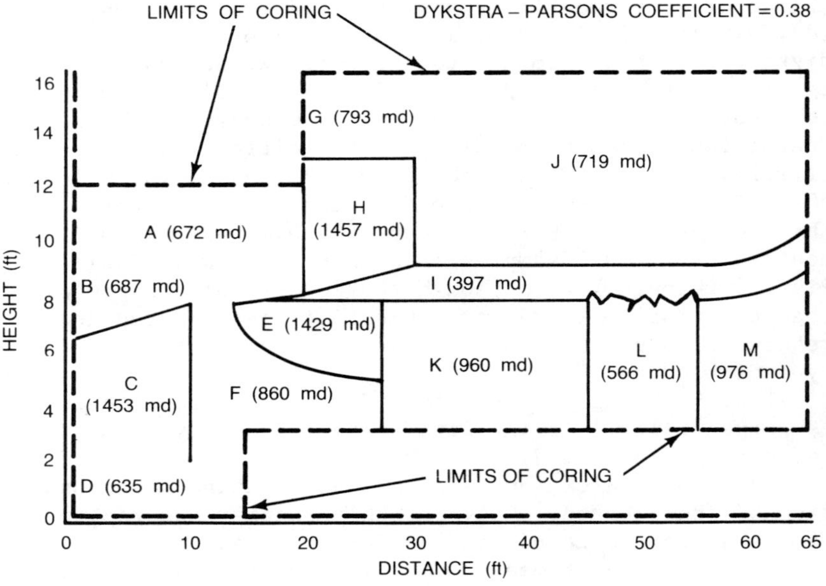

FIGURE 7. Rock Springs Outcrop: adjacent layers with statistically different permeabilities.

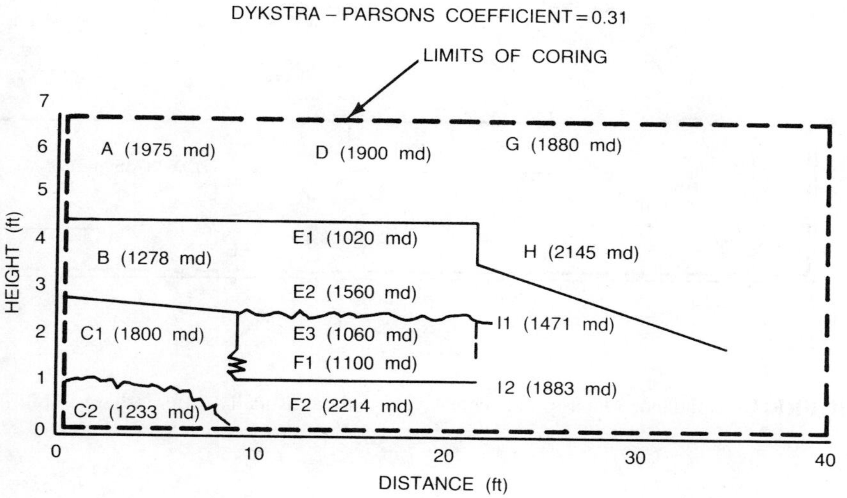

FIGURE 8. Almond Outcrop: adjacent layers with statistically different permeabilities.

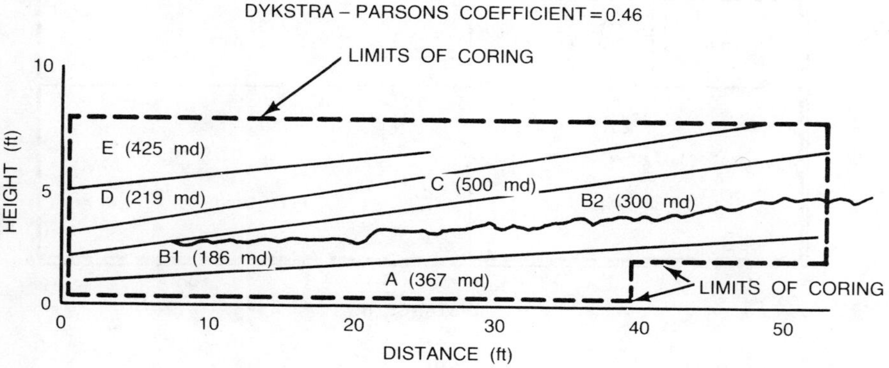

FIGURE 9. Frontier Outcrop: adjacent layers with statistically different permeabilities.

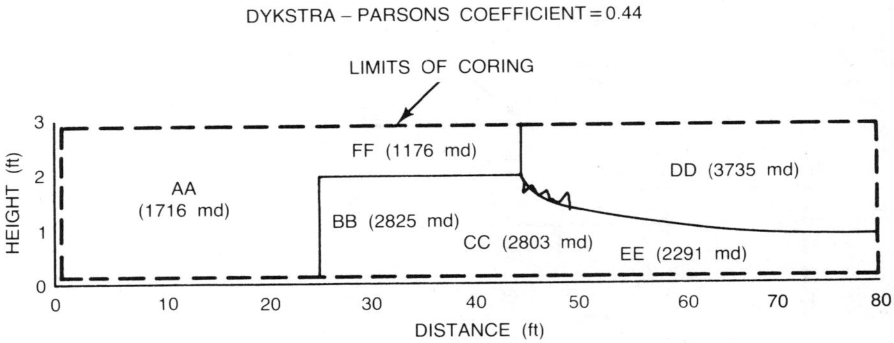

FIGURE 10. Sundance Outcrop: adjacent layers with statistically different permeabilities.

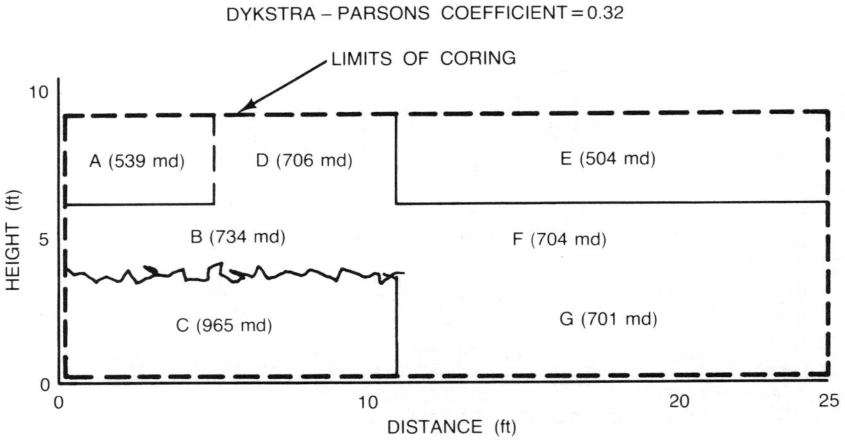

FIGURE 11. Fox Hills Outcrop: adjacent layers with statistically different permeabilities.

Analysis With Semivariograms

In geostatistics semivariograms can provide useful information about the structure of a deposit for which a quantity such as mineral richness is to be estimated. Semivariograms for the outcrop permeabilities were examined to see if they might also provide useful information about the nature of permeability variation in the outcrops.

Figs. 12a and b show how a semivariogram is calculated (9). In this example, permeability values are known at the points shown as black dots on a regular grid (Fig. 12a). Values aren't known at every point on the grid, however. To calculate the semivariogram in the x direction, the square of the difference between permeability samples is first evaluated for all samples separated by a distance, h. For example, the difference in permeabilities between points 1 and 2 is squared, between points 2 and 3, etc. The sum of all the squares is then divided by twice the total number of data pairs at spacing h, and the results plotted as the ordinate on a graph where the abscissa is the distance between sample pairs (Fig. 12b). Next, the same thing is done for all samples separated by the distance, 2h (e.g., points 1 and 3, 2 and 4, 3 and 5, etc.), and the results plotted vs the distance 2h. This is repeated for increasing spacing between sample pairs (e.g., 3h, 4h, etc.). A similar plot is made for samples in the y direction.

Fig. 12b (solid curve) shows a common shape for the semivariogram found in nature (9). Values increase with increasing distance between sample pairs until a distance, a, is reached at which the variogram levels off. This type of behavior shows that at small spacings the samples are similar, and their differences are small. Put another way, for this type of semivariogram, as the distance between sample pairs decreases, samples become more alike. With increasing spacing, samples become less alike until at spacing, a, they become completely independent of each other. The distance, a, is called the range of influence. The value of the semivariogram where it levels off is called the "sill".

Sometimes, however, the semivariogram appears to extrapolate to a positive value as sample spacing becomes smaller. In other words, the difference between two sample values doesn't appear to go to zero as the spacing between sample pairs decreases to small values. This type of behavior is shown by the dashed line in Fig. 12b. It indicates a degree of randomness to the sample values at small spacings and is called the "nugget effect" in geostatistics. Behavior of this kind will be referred to as a "nugget effect" in

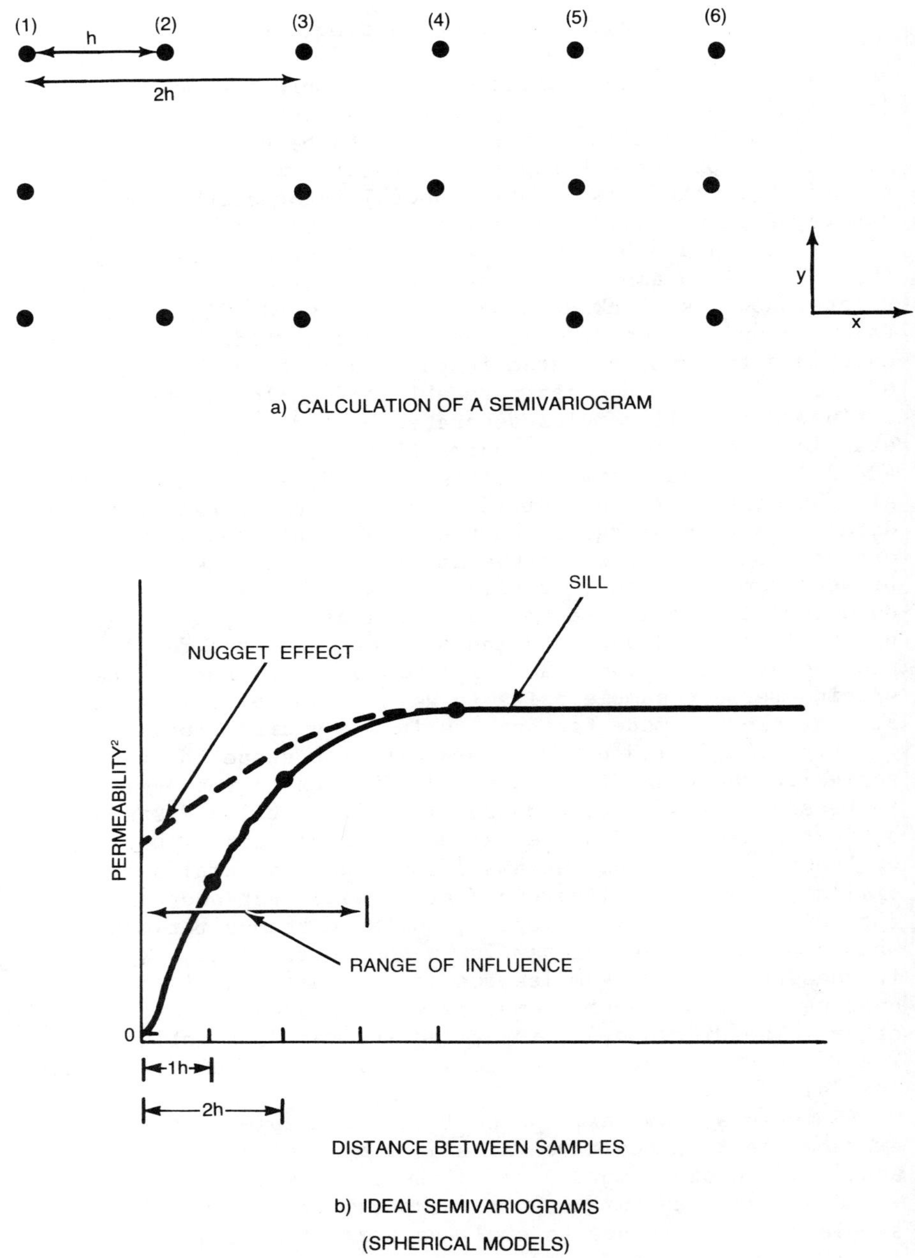

FIGURE 12. Concepts and Terminology of Semivariograms

the remainder of this paper. The standard deviation of this "nugget effect" randomness is approximated by the square root of the extrapolated semivariogram value at zero sample pair spacing.

Figs. 13a through e show semivariograms calculated for several hypotheses of how permeability could be distributed in the outcrops. These calculations were made to see if comparison of actual and hypothetical x- and y- direction semivariograms might be a useful means for inferring the nature of permeability variation in these outcrops. The hypothetical semivariograms were constructed by assigning permeabilities to the points on a 16x12 grid (x,y) for different models of permeability variation and then calculating the semivariogram for these grid- point values. For example, a layer of given mean permeability was modelled by assigning several rows of grids to represent the layer. Then, values were assigned to each grid point in the layer by drawing randomly from a normal distribution with that mean and with a standard deviation that characterized random variation within the layer. The grid points were assumed to be four ft apart laterally and one ft apart vertically, which approximated how sampling was done in the outcrops.

Table 1 shows how one model for discontinuous permeability layering was constructed. The semivariogram for this model is shown in Fig. 13d.

Fig. 13a is for a completely random distribution of permeabilities (no layering) that come from a normal distribution. Fig. 13b is for a situation where there are continuous layers of different mean permeability. The layer means come from a normal distribution with the same standard deviation as a), but within a layer, permeabilities are distributed randomly about the layer mean with one-half this standard deviation. Fig. 13c is for a situation where the layer mean permeabilities are the same as 13b, but the standard deviation within a layer varies from layer to layer.

These hypothetical semivariograms show no range of influence down to the smallest sample spacing for both x- and y-direction semivariograms. They also show what appears to be a pure nugget effect, i.e., they are essentially flat, except for fluctuations, over the entire range of sample pair spacing. This behavior reflects the random assumption for grid-point permeabilities within the continuous layers. The value of the x-direction semivariogram extrapolated back to zero spacing between sample pairs is equal to the square of the assumed standard deviation within layers, and the corresponding value for the y-direction semivariogram approximates the square of the assumed standard deviation of

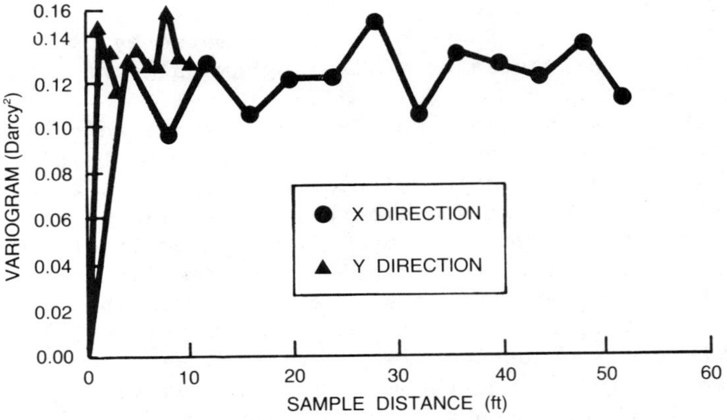

a) RANDOMLY DISTRIBUTED PERMEABILITY

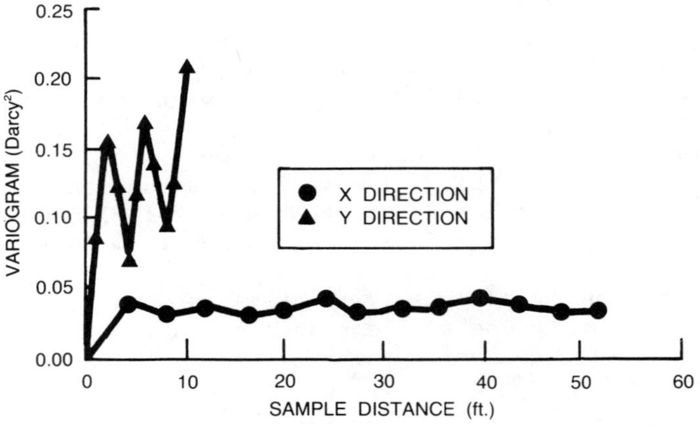

b) LAYERS OF DIFFERENT MEAN PERMEABILITY (STANDARD DEVIATION WITHIN LAYERS = 1/2 STANDARD DEVIATION BETWEEN LAYERS)

FIGURE 13. Hypothetical semivariograms

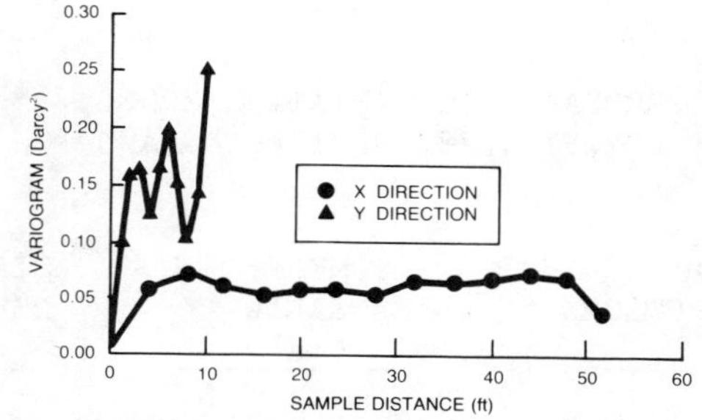

c) SAME LAYER PERMEABILITIES AS b) BUT VARIABLE STANDARD DEVIATION WITHIN LAYERS

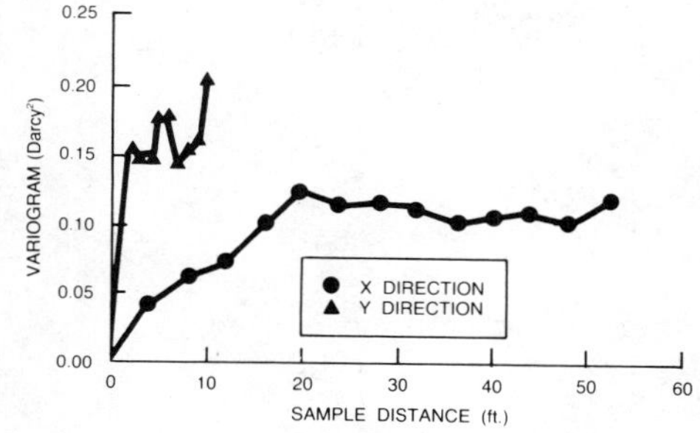

d) LONG THIN BLOCKS OF DIFFERENT PERMEABILITY (STANDARD DEVIATION WITHIN BLOCKS = 1/2 STANDARD DEVIATION BETWEEN BLOCKS

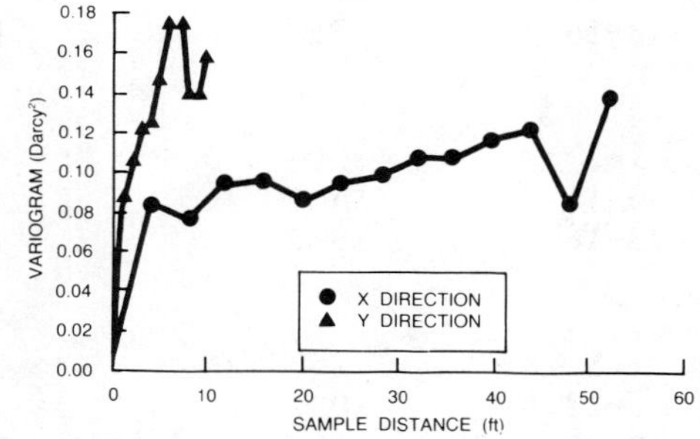

e) LONG THIN BLOCKS OF DIFFERENT PERMEABILITY (VARIABLE STANDARD DEVIATION WITHIN BLOCKS)

FIGURE 13. Hypothetical semivariograms. (cont.)

TABLE I

PERMEABILITY VARIATION MODEL FOR FIG. 13d SEMIVARIOGRAMS

ROW	COLUMN	MEAN PERMEABILITY (md)	STANDARD DEVIATION (md)
1,2	1-5	1400	175
1,2	6-11	480	175
1,2	12-16	1000	175
3,4	1,6	750	175
3,4	7-12	650	175
3,4	13-16	840	175
5,6	1-5	1000	175
5,6	6-12	1400	175
5,6	13-16	1000	175
7,8	1-6	480	175
7,8	7-12	650	175
7,8	13-16	750	175
9,10	1-5	1000	175
9,10	6-11	1400	175
9,10	12-16	840	175
11,12	1-5	750	175
11,12	6-12	1000	175
11,12	13-16	1400	175

the layer means. In the y direction, the smallest spacing of sample pairs in these hypothetical models was only 1/2 the layer thickness, so most sample pairs are across-layer pairs. Layer-to-layer variation of standard deviation within layers seems to have little effect on the variogram of the Fig. 13c model.

Figs. 13d and e show semivariograms for models where layers of different mean permeability are not continuous across the model. The mean permeabilities of the discontinuous layers had a standard deviation that was approximately the same as the standard deviation of layer means in the continuous layer models of Figs. 13b and c. In Fig. 13d the distribution of permeabilities within a layer had 1/2 this standard deviation. In the model for Fig. 13e the standard deviation of samples within a layer varied from layer to layer.

The x-direction semivariogram in Fig. 13d differs from those of Figs. 13a through c by appearing to have a range of influence of about 20 ft. This is on the order of the length of the discontinuous layers in the permeability model for this figure. It may reflect that at smaller sample pair spacings, differences are calculated predominantly for samples within a discontinuous layer, but with increasing spacing, differences are calculated more for samples that are in different discontinuous layers. There is still a nugget effect at small sample spacings, again reflecting the assumption of permeability randomness within a discontinuous layer. The y-direction semivariogram shows essentially no range of influence. Again, the smallest y-direction spacing between sample pairs is 1/2 the layer thickness, so few differences are calculated for sample pairs within layers.

Finally, although the model for Fig 13e has laterally discontinuous layers, it differs from the Fig. 13d model by having layer-to-layer variation of the within-layer standard deviation, thicker layers relative to minimum spacing for sample pairs, and some layer lengths that are a greater fraction of model length. For this model there was more opportunity to sample pairs that were within layers in the y direction, and a range of influence appears to be indicated by the y-direction semivariogram. The x-direction semivariogram for the most part increases continuously, although there is some suggestion that beyond about 40 ft it may begin to fluctuate about a sill value. If so, the indicated y- and x-direction ranges of influence appear to be rough measures of thickness of the thicker layers (y direction) and length of the longer layers (x direction). Both variograms have a nugget effect of about the same value, perhaps because of the greater testing of within-layer pairs in the y direction due to the greater layer thickness.

It is noteworthy that in all figures except 13a, the x- and y-direction semivariograms show a noticeable separation at small sample pair spacings because of the assumption of layering and the assumption that the standard deviation of samples within layers is smaller than the standard deviation of the layer mean permeabilities.

Figures 14 and 15 show the actual semivariograms for the Woodbine 1 and Rock Springs outcrops. To minimize spurious results, variogram values were not plotted when there were fewer than 30 data pairs.

The x-direction Woodbine 1 semivariogram (Fig. 14) appears to be increasing gradually over the entire range of sample pair spacing, suggesting there might be a range of influence that hasn't been reached after 35 ft. There does not appear to be a y-direction range of influence at the smallest sample pair spacing. Neither semivariogram appears to be trending to zero as sample pair spacing decreases, indicating x- and y-direction nugget effects. There is a distinct separation of the variograms at small spacings.

The Rock Springs x-direction semivariogram (Fig. 15) rises to an apparent sill value after about 10 to 15 ft (i.e., it appears to level off) but then decreases to a lower value. The significance of this behavior isn't clear, although conceivably it might indicate some type of periodic variation in discontinuous layer mean permeabilities over the lateral distance sampled. This variogram appears to have a range of influence of about 10 to 15 ft and appears to be trending to a nugget effect.

The Rock Springs y-direction semivariogram appears to have a range of influence of about 3 ft and also trends to a nugget effect. Conceivably, this behavior may indicate layers that are thicker relative to minimum sample pair spacing than for the Woodbine 1 outcrop. The x-direction values are considerably smaller than the y-directon values at small sample pair spacings, possibly reflecting a stratified character of the rock.

By comparing the actual and model semivariograms, some inferences about the nature of permeability variation in these outcrops are possible. The word "inference" is used rather than "conclusion" because the general shapes and relationships of the semivariograms probably are not unique to a particular permeability distribution. For example:

o The separation of the x and y semivariograms at small sample pair spacing suggests a layered structure to permeability. It also suggests the standard deviation

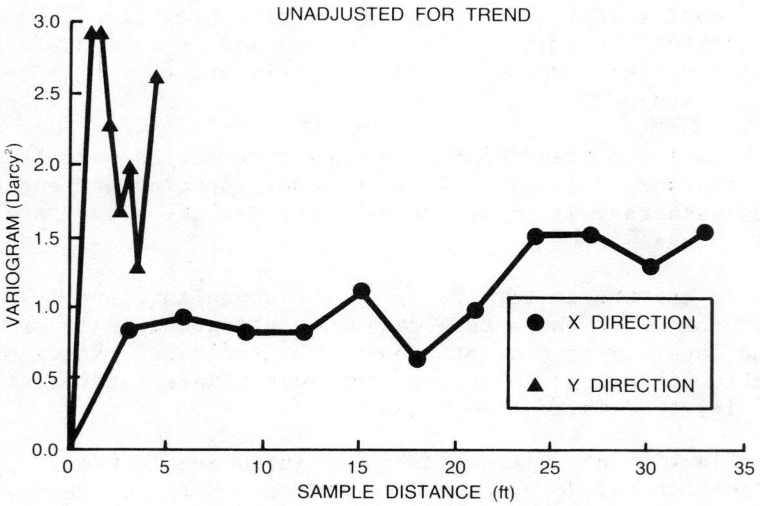

FIGURE 14. Semivariogram for Woodbine I data.

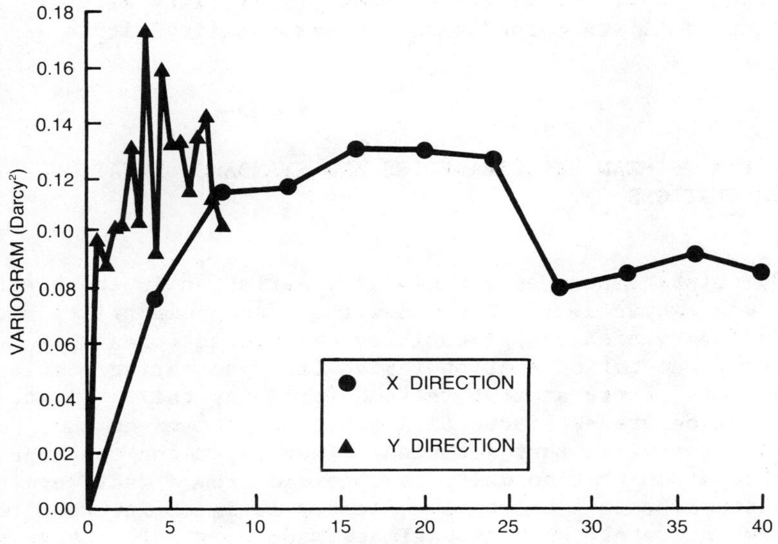

FIGURE 15. Semivariogram for Rock Springs Data.

of samples taken predominantly within layers (x direction) is smaller than the standard deviation of samples taken predominantly across layers (y direction).

- The apparent trend of both x- and y-direction semivariograms to finite values as sample spacing decreases suggests randomness to permeability variation over small distances.

- There appears to be a range of influence in the x direction for the actual data suggesting that some of the layering is discontinuous. The indicated ranges of influence appear to support the approximate lateral size of layers deduced from Figs. 6 and 7.

- The lack of an apparent range of influence in the y direction for Woodbine 1 data and the small apparent y direction range of influence for Rock Springs data suggest layer thicknesses that are small or on the same order as vertical sample spacing.

The semivariograms for the other outcrops generally showed behavior similar to Woodbine 1 and Rock Springs and generally were supportive of the concept of laterally discontinuous layers of different mean permeability with an apparent small-scale randomness to permeability within layers.

VARIATION OF MEAN PERMEABILITIES AND STANDARD DEVIATIONS AT CORING STATIONS

One final aspect of permeability variation in these outcrops was examined. This was how the mean permeability and standard deviation of permeability calculated for samples from only one coring station varied from one coring station to another. There are two reasons for doing this. First, these values are estimates of average properties of the entire interval at that location. They represent the type of information that normally is obtained from a well core to calculate average properties. Second, it is of interest to see how accurately such an estimate made from only one well characterizes the region around the well.

Fig. 16 shows mean permeabiities calculated at each coring station on the Rock Springs outcrop and the 99% confidence intervals for these means. Although visually there is

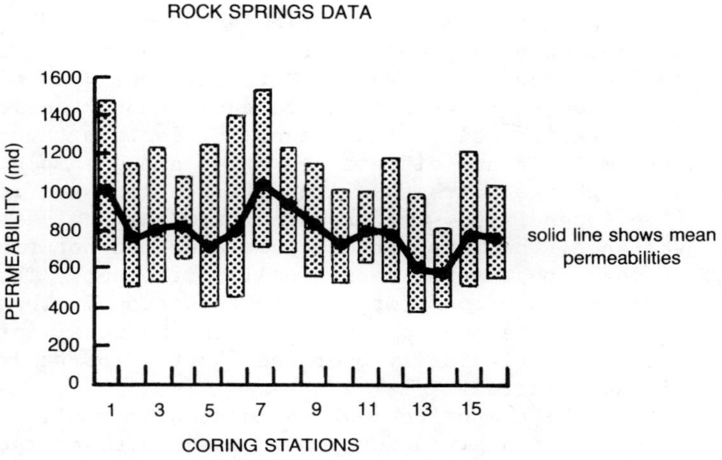

FIGURE 16. Confidence intervals for Geometric Mean Permeability at Coring Stations.

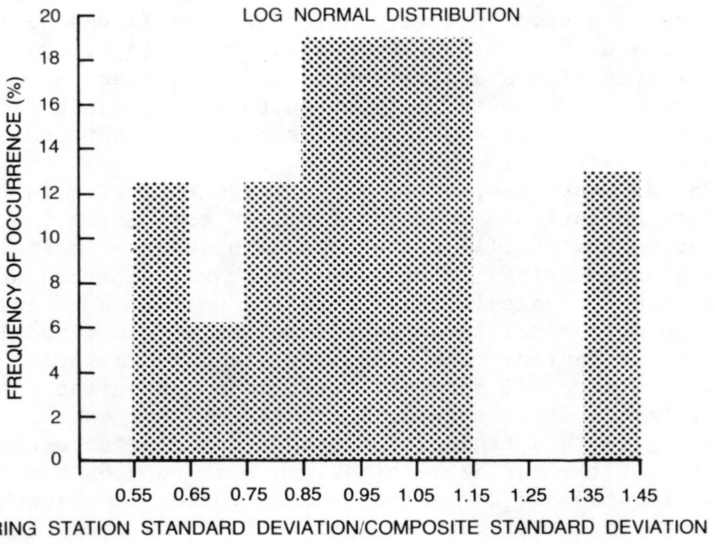

FIGURE 17. Rock Springs Outcrop: distribution of standard deviations calculated at different coring stations.

some variation among the calculated means and perhaps the suggestion of a trend, there is substantial overlap of all the confidence intervals. Moreover, analysis of variance indicated there were no differences among the means that were statistically significant. This is a technique to determine if any differences exist among sample means, although the pairs that are different are not identified (See Appendix B). As expected from the analysis of variance, the Tukey-Kramer test also did not find any pairs of means that were statistically different. The same behavior was found for the other outcrops. Thus, for all outcrops, the mean permeability of the total interval was uniform statistically over the 25-ft to 80-ft lateral distances cored.

If the composite mean calculated from all permeabilities measured on a given outcrop is assumed to characterize average permeability for that outcrop, then the estimate from samples taken at a single coring station was within ±15% of the composite mean over 60% of the time. It was within ±25% of this mean most of the time. This observation seemed to be true when there were as many as 6 to 10 samples per coring station. Thus, when there were six or more samples from the crossbedded interval, there was a relatively high chance for estimating from single-well data to within ±15% of the composite mean permeability that characterized the outcrop. However, there was a small chance that the estimate could be as much as 25 to 45% in error.

There is always a question as to which mean, arithmetic, geometric, or harmonic, is most suitable for fluid flow calculations. This is a separate issue that is not addressed in this report.

For all outcrops, there was a large variation in the standard deviation calculated from one coring station to another. This is illustrated by the histogram in Fig. 17 for the Rock Springs outcrop, which shows the frequency of occurrence for calculating a standard deviation at any one coring station that is a given fraction of the composite standard deviation. The composite standard deviation is calculated from all permeabilities measured on the outcrop and is taken here simply as a reference. For example, according to this figure, the standard deviation estimated from 12% of the coring stations was 0.55 to 0.65 the composite standard deviation, and estimates from another 12% of the coring stations were 1.35 to 1.45 the composite standard deviation. This behavior was generally true for the other outcrops, where individual estimates could differ as much as three fold. There appeared to be roughly equal probabilities for estimating values anywhere within this variation.

DISCUSSION OF RESULTS

A few thoughts are in order concerning the relevance of outcrop study to reservoir characterization. Permeability variation in a sedimentary rock is caused by such factors as grain size, grain size distribution, sedimentary structures, and other attributes determined by the original depositional processes and by the rock alterations after deposition caused by diagenetic processes. The diagenetic history of an outcrop, including surface processes, is undoubtedly different to various degrees from what the diagenetic history of the same deposit would have been if it had been buried to depth and remained there as a reservoir. So a question to be asked is: "What do permeability variations observed in outcrops mean with regard to variations in reservoirs?" An underlying assumption in trying to develop reservoir characterization from outcrop studies is that the original framework determined by depositional processes still predominantly controls permeability <u>structure</u> of the rock. In other words, if the outcrop were now at depth rather than at the surface and had undergone a somewhat different course of diagenesis, the spatial structure of permeability (although not necessarily the magnitudes or even the degree of contrasts) would be approximately the same.

The outcrops discussed in this report represent crossbedded intervals that most likely would only be a fraction of the total reservoir thickness. They represent only a portion of depositional sequences ordinarily observed in reservoirs of interest and represent only one type of sedimentary structure that is found in reservoir rocks. Ordinarily, if the well were cored, there would only be a few core plug samples from such an interval on which to measure permeability (perhaps one sample per ft or less). In reservoir simulation, such an interval ordinarily would be represented by no more than one layer of constant properties or might even be lumped with other intervals for simulation purposes.

Permeability Structure

The data clearly show there is structure to permeability variation in these outcrops and that variation is not completely random. The differences in the x- and y-direction semivariograms speak to this as do the results of statistical significance tests comparing adjacent areas of the outcrops.

The model that emerges from outcrop studies for such an interval is one of small-scale permeability layering, where the mean permeability of a given layer is different from the mean permeability of adjacent layers. These layers are on the order of 0.5 to 3 ft thick and from 6 to greater than 40 ft in lateral dimension. Because of a layer's limited lateral extent, it may butt up against another layer of different permeability as well as lie immediately above or below other layers of different permeability.

The ratio of mean permeabilities of adjacent discontinuous layers usually was in the range 1.3 to 2.5. No high permeability "streaks" were found.

Semivariogram behavior suggests that within a layer there is some degree of randomness to permeability variation. In no case did the x or y-direction semivariograms appear to be trending to zero as spacing between sample pairs decreased. This is not to say there might not be some additional structure to permeability variation on a scale that is smaller than the smallest sample spacing; samples taken within a layer may be sampling randomly from this smaller-scale structure. Crossbedding by its nature is a structure of alternating laminae of various widths, grain sizes, and permeabilities (10,11). Moreover, the coring station permeability profiles of Figs. 2 and 3 suggest that permeability changes continuously, rather than abruptly, to higher or lower values and that within a layer of high or low mean permeability, there perhaps is a building to and receding from the peak or trough value. Thus, the model of a layer in which permeabilities are distributed randomly on a small-scale about the layer mean probably is only an approximation of a more complex reality.

The standard deviation of permeability within the discontinuous layers appears to be less than the composite standard deviation for all permeability samples from an outcrop. For all outcrops, the overall within-layer standard deviation was on the order of 0.6 to 0.75 the composite standard deviation.

Over the 24-to-80 ft lateral distances sampled in these outcrops, no statistically significant differences were found among the total interval mean permeabilities estimated at the different coring stations.

Implications for Reservoir Modeling

The analyses of this paper indicate that core data from a single well or wells drilled on normal spacings give only a limited insight into the nature of permeability variation in reservoir intervals analogous to the ones studied in these outcrops. Although the outcrop data suggest that the total interval mean permeability may be estimated with confidence from well data much of the time, these data also show that details of the permeability profile at a well do not persist for very great lateral distances away from the well. It seems unlikely that the thin layers found in the outcrops could be identified with much confidence from core plug samples taken at one-ft intervals. Although the outcrop samples were spaced one ft verticaly, enough samples were taken laterally to establish statistical significance. Moreover, even if these layers could be identified in a single well, they almost certainly could not be correlated from well to well. A deterministic approach to describing such an interval doesn't seem plausible.

Another potential approach for characterizing this small-scale layering from single-well data is a stochastic one. It might be analogous to the stochastic method described by Haldorsen for characterizing shales (12). In such an approach, the layers and their mean permeabilities would be assigned stochastically within the gross interval and over the interwell distance. This would require suitable distribution functions relating layer dimensions to frequency of occurrence of these dimensions as well as characterization of the permeability distribution between and within layers.

Conceivably, the assignment of mean permeabilities to stochastic layers might be related to the standard deviation of permeability determined from a well core. A justification for doing this is the empirical observation that the mean permeabilities of the discontinuous outcrop layers generally seemed to fall in the range defined by the 10 and 85% probabilities of the composite data cumulative distribution. For example, with reference to Fig. 4a, this observation would imply that layer mean permeabilities should generally be in the range of about 450 to 1200 md for a gross interval with the type of cumulative distribution shown in this figure. Unfortunately, the large variation in standard deviations estimated from samples at each coring station seems to limit the usefulness of this type of data for stochastic characterization. Perhaps, sampling more densely than one per ft would help, but this remains to be seen. It also is not clear in stochastic layer modeling how to treat the apparent permeability variation within layers or even if it is desirable to account for this additional small-scale variation.

The distribution of layer thicknesses and lengths for a stochastic description cannot be determined from single-well data; this type of information would have to come from outcrop studies. Figs.18 and 19 show cumulative distributions for layer thickness and length derived for adjacent layers found to be statistically different. All the outcrops are represented in these figures (e.g., Figs. 6 through 11). Figs 18 and 19 probably mix apples and oranges to some degree, particularly the figure for cumulative distribution of length. Intuitively, one would expect to find more discontinuous layering and shorter layer lengths for the higher energy environments of deposition and would expect that different distributions would characterize high- and low-energy environments. In fact, the only outcrop identified as having been deposited in a fluvial environment (Rock Springs) appears to be more discontinuous than most of the others. However, considering that the depositional settings of these outcrops are only identified broadly and that the type of crossbedding is not identified at all, lumping of all the data seems justified at this stage as an initial approximation for layer distributions in crossbedded intervals.

Fig. 19 should predict lengths that are too small because many layers identified in these outcrops extended beyond the area of sampling. There is no way to know how far they may have gone, although the fact that some layer terminations are evident within the lateral distances sampled suggests that layers generally don't continue for distances that are appreciable compared to the distance over which the outcrops were sampled.

Another question to be raised concerning Fig. 19 is what to do about the layer dimension normal to the face of the outcrop? Until further studies are able to quantify the directional dependence of layer dimension, e.g., parallel or transverse to the depositing current, it seems justified to assume that Fig. 19 approximates this dimension also. This is because the outcrops probably were sampled at various orientations to the ancient current direction, and thus the data of this figure probably represent some average of lengths reflecting different orientations. Sketches of some of the outcrops suggest that different orientations relative to the direction of crossbedding were, in fact, sampled.

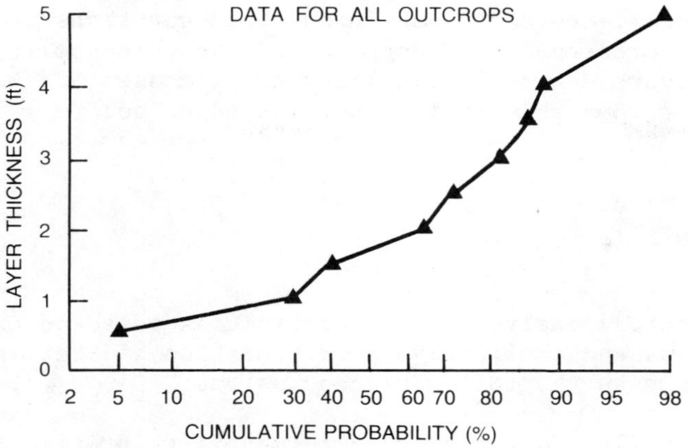

FIGURE 18. Cumulative probability distribution for layer thicknesses.

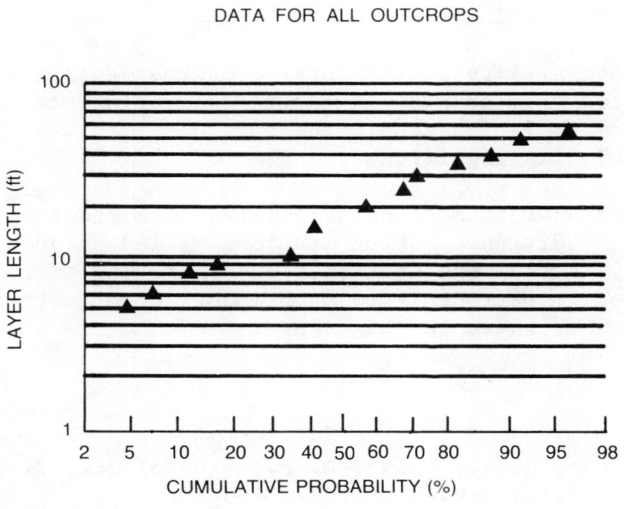

FIGURE 19. Cumulative probability distribution for layer length.

Obviously, there are many unanswered questions concerning the importance or unimportance of the permeability variations reported here on displacement processes and how to account for them when it is important to do so.

CONCLUSIONS

Statistical analysis of permeabilities measured on these six outcrops and qualitative interpretation of their semivariograms gave the following conclusions:

- The detailed character of a permeability profile found at any one coring station usually did not persist for more than 5 to 20 ft away from the coring station.
- The crossbedded intervals are stratified into 0.5- to 3-ft layers of different mean permeability that are from 6 to 40 ft or greater in lateral dimension.
- If there is smaller-scale structure to permeability variation within a layer, it could not be resolved with the spacing between samples in this study; semivariogram behavior at small sample spacings indicates some random variation within layers.
- Mean permeability of the total crossbedded interval estimated from samples at individual coring stations did not show variation across the outcrops that was statistically significant.
- Standard deviation of permeability estimated from samples at individual coring stations varied as much as three fold from one station to another, and there appeared to be roughly equal probability of obtaining an estimate anywhere in this range.

REFERENCES

1. Hutchinson, C. A., Dodge, C. F., and Polasek, T. L. (March 1961) J Pet. Tech., 223-230.

2. Polasek, T. L. and Hutchinson, C. A. (1967). Proc. Seventh World Petr. Cong., 397-407.

APPENDIX A

THE KOLMOGOROV-SMIRNOV TEST

The Kolmogorov-Smirnov one-sample test is a goodness-of-fit test that determines whether or not the hypothesis that data come from a given type of distribution can be rejected. It is concerned with the agreement between an observed cumulative distribution of sample values and a specified continuous distribution function(4,5). The one-sample test is based on the maximum absolute difference between the empirical cumulative distribution and the specified theoretical distribution. To determine whether this difference is larger than can be expected from chance, the maximum difference is compared with a critical value of D found in statistical tables for the appropriate degrees of freedom in the data. Because the outcrop data remaining today are grouped into intervals, the K-S procedure described by Tate and Clelland for grouped data was followed(5).

APPENDIX B

ANALYSIS OF VARIANCE AND THE TUKEY-KRAMER TEST FOR MULTIPLE COMPARISON OF MEANS

To test a group of means by analysis of variance, the mean square between samples (MSBS) and the mean square within samples (MSWS) are calulated. This is done by Eqs, 1 and 2:

$$SSBT = \sum_{j=1}^{k} \left[\left(\sum_{i=1}^{n_j} K_{i,j} \right)^2 / n_j \right] - \left(\sum_{j=1}^{k} \sum_{i=1}^{n_j} K_{i,j} \right)^2 / N_T \quad (1)$$

where SSBT = sum of squares between samples

Mean square between samples = MSBS = SSBT/(k-1).

$$SSWS = \sum_{i=1}^{n_j} \sum_{j=1}^{k} (K_{i,j})^2 - \sum_{j=1}^{k} [\sum_{i=1}^{n_j} (K_{i,j})^2/n_j] \qquad (2)$$

where SSWS = sum of squares within samples

Mean square within samples = MSWS = SSWS/(N_T-k).

where $K_{i,j}$ = permeability of measurement i in sample j
n_j = total number of measurements in sample j
N_T = the total number of measurements in all samples
k = the total number of means (samples) being compared.

If MSBS/MSWS > F_{table} for k-1 degrees of freedom in the numerator and N_T-k degrees of freedom in the denominator, there is a difference among at least two of the sample pairs at the probability level tested. Here, F_{table} is the tabulated F statistic.

The Tukey-Kramer method is a way to determine which pairs in a group of means are different when the number of data points in each sample is not equal(6). This method involves calculating the within-sample mean squares for all samples and then estimating the confidence interval, CI, for the difference in a pair of sample means by Eq. 3.

$$CI = \overline{K}_{j1} - \overline{K}_{j2} \pm q\left(\overline{\nu}\ (n_{j1}^{-1} + n_{j2}^{-1})/2\right)^{1/2} \qquad (3)$$

where $\overline{K}_{j1}$, $\overline{K}_{j2}$ = permeability means (or means of ln permeability) for samples j1,j2

$\overline{\nu}$ = MSWS

n_{j1}, n_{j2} = number of data points in samples j1 and j2

q = a statistic found in appropriate tables(7)

q is determined for the appropriate number of total samples and degrees of freedom (total data points minus total number of means to be compared) and for the confidence level at which the comparisons are being made. If the confidence limits calculated by Eq. 1 include zero, K_{j1} and K_{j2} are not statistically different at the confidence level tested.

The T-K method assumes normality and homogeneity of variances (i.e., $\sigma_{j1} = \sigma_{j2}$, where σ is the true standard deviation of the population as opposed to s, the sample standard deviation). The method is thought to be robust with respect to departures from normality but may lack robustness with respect to variance heterogeneity(6).

The Bartlett test is one method to test for homogeneity of variances(4). It compares the quantity calculated in Eq. 4 with the Chi-square statistic.

$$\chi^2 = \sum_{j=1}^{k} (n_j - 1) \ln(\overline{\nu}) - \sum_{j=1}^{k} (n_j - 1) \ln (s_j^2) \quad (4)$$

where sj = standard deviation of sample j
k = total number of samples (means) being compared

If $\chi 2$ is greater than the chi-square statistic found in chi-square tables for k-1 degrees of freedom and the confidence level being tested, variances are considered to be heterogeneous.

Data published in the open literature show that very simple permeability-porosity relationships hold for unconsolidated sands. From other published data, and from Sohio data, we find that straightline relationships hold in rocks which we infer to be "simple mixtures". For these rocks, permeability can be predicted easily using a porosity measurement from well logs.

Where several petrological variables (large amounts of cement, wide range in grain size, diagenetic alteration) have effect, the linear trend disappears and the scatter is symptomatic of "complex mixtures". This scatter may just be due to a conbination of many linear trends, each dependent on individual petrological variables. Porosity alone does not serve as a good predictor of permeability. A better predictor can be found if petrological variables can be included.

The purpose of the work reported here is to compute permeability, k, estimates in wells where no core data are available. We do this by establishing predictive equations for permeability in cored wells where both well logs and laboratory measurements are available. These equations are then applied in the uncored wells. The process utilizes both discriminant analysis and multiple linear regression. The equations are established independently for layered subsections of the producing Sadlerochit reservoir at Prudhoe Bay. Because a primary use of predicted permeability is for waterflood monitoring, a special effort has been made to replicate the core permeability data, both in terms of interval averages and foot-by-foot variations.

Predicted values are based on the permeability of core samples to air; no corrections for experimental or stress effects have been applied. Error analysis is based on comparison of predicted to core permeability values with special emphasis placed on the prediction of high permeability zones. As the reader will see in the section on error analysis, evaluation of what is "an adequate or a better predictor" is not simple and at this point remains somewhat subjective. It is not that there is a lack of statistical parameters available, but rather that the goals of permeability prediction are not the same for all users of the data. Should we strive for accurate average data at the expense of all other considerations? Should low permeability values be eliminated in order to better predict high values? Should we attempt to match the core data at every depth? Or should we edit erratic core data caused by inadequate sampling? All efforts at prediction rest on some kind of choice or assumption with regard to these matters.

III. PERMEABILITY, POROSITY, AND PETROLOGY

Permeability values commonly encountered in petroleum reservoirs range from a fraction of a millidarcy to several darcys. We examine the relations of ϕ and k in Figures 1 through 6. All k-ϕ plots are scaled so that a factor of ten change in k is the same length as 0.1 change in fractional porosity. (The scales will always be the same but the axis values will shift to suit the data.)

Unconsolidated Sand Packs

Beard and Weyl (1973) made 48 sieved sand samples using sand from two Texas rivers. Data shown in Figure 1 are from

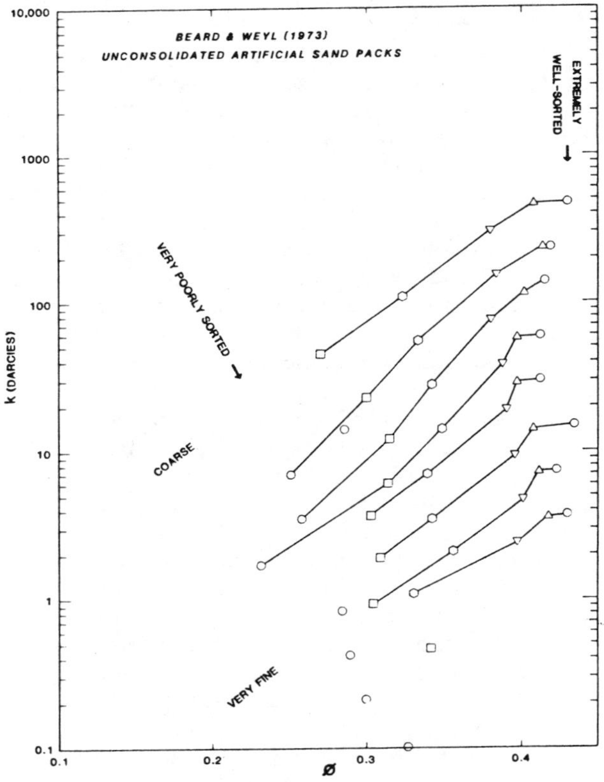

Figure 1. k-ϕ plot, unconsolidated artificial sand packs, by grain size and sorting (Beard and Weyl, 1973). Reprinted by permission of the American Association of Petroleum Geologists.

sand samples which were wet-packed in vials 1.8cm in diameter and 5.0 cm in length. Median grain size ranges from 0.840 mm for the coarse sample to 0.074 mm for the very fine sample.

Permeability increases by about a factor of two for each increment of grain size, whereas the mean grain diameter of adjacent size intervals increases by the square root of two. This is also in accord with simple models: flow is proportional to the square of the radius of the opening. Except for a few very fine, poorly sorted samples, the values of k exceed one darcy. The upper limit for a well-compacted, unconsolidated sand pack is about 500 darcies.

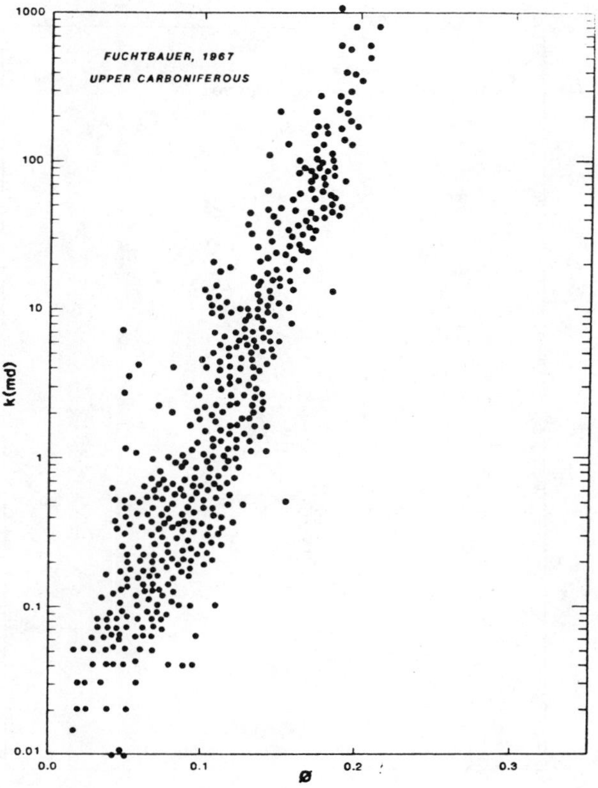

Figure 2. k-ϕ plot, Upper Carboniferous samples (Fuchtbauer, 1967). Reprinted by permission of Elsevier Science Publishing Company, Inc.

For the extremely well-sorted samples, porosity is independent of grain size, as it should be for packing of uniform spheres. The Beard and Weyl data show clearly that k increases with grain size. For those samples which are not well-sorted, an increase in the coarse grain content results in somewhat decreased ϕ as it increases k. We will see this pattern repeated in consolidated samples.

Effects of Clays and Cement

Figures 2 and 3 demonstrate a linear relationship between log (k) and ϕ. Such trends are often seen, but are not well understood. They are significant because there is little variability in log (k) when ϕ is known. In the case of the Upper Carboniferous samples (Figure 2), Fucht-

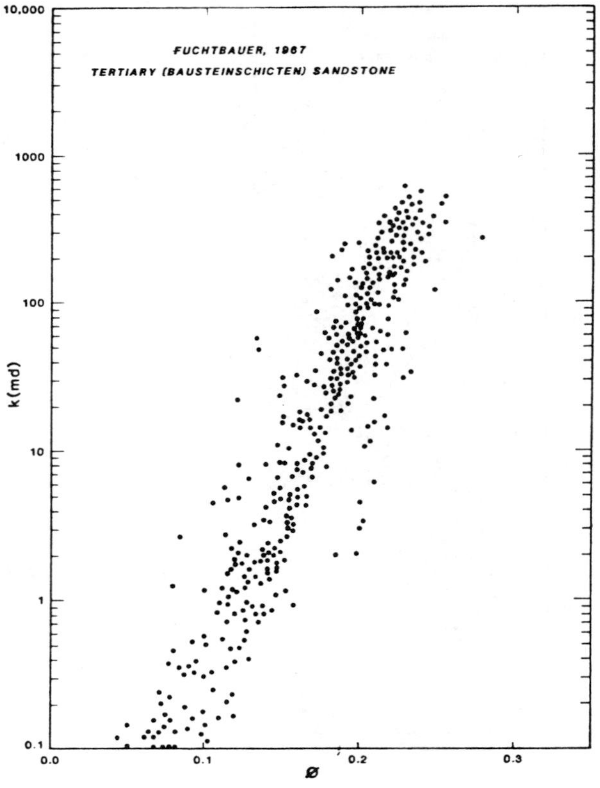

Figure 3. k-ϕ plot, Tertiary sandstone (Fuchtbauer, 1967). Reprinted by permission of Elsevier Science Publishing Company, Inc.

bauer (1967) states that the progressive decrease in k is due to the high clay content. In the case of the Bausteinschicten sandstone (Figure 3), calcite cement, increasing clay content, and decreasing grain size are all possible contributors to the decreasing k-ϕ trend.

Bos (1982) describes results from a Shell exploration well which encountered 1) clean sandstone, 2) sandstone with pores filled with kaolinite and 3) laminated sandstone, part clean and part filled with kaolinite (indicated as "laminated" in Figure 4).

Here again we see the linear relationship between log (k) and ϕ, with pore-occluding clays reducing both k and ϕ such that the kaolinite data fall on the line.

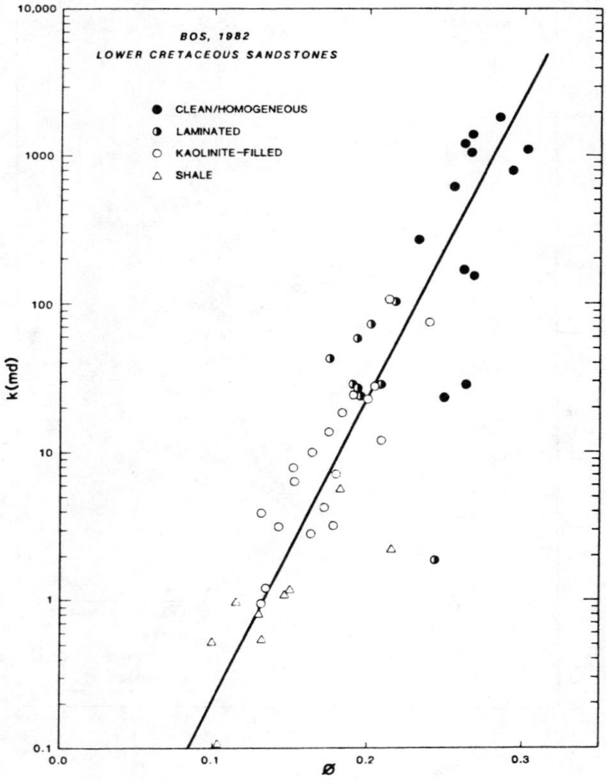

Figure 4. k-ϕ plot, Lower Cretaceous sandstones: clean, laminated, and kaolinite-filled (Bos, 1982). Reprinted by permission of the Society of Professional Well Log Analysts.

Effect of Coarse Grain Size

Shenhav (1971) classified samples by grain size; his results are shown in Figure 5. The sandstones are cemented by calcite and dolomite. Note the progressive increase in the k-ϕ slope as grain size increases. The trends are not linear as observed in other data sets. Shenhav also plotted k-ϕ according to depositional facies, but the plots (not shown here) are indistinguishable one from another. Similar results are obtained for selected non-vugular carbonates (Figure 6). The k-ϕ trends increase in both slope and offset as particle size increases.

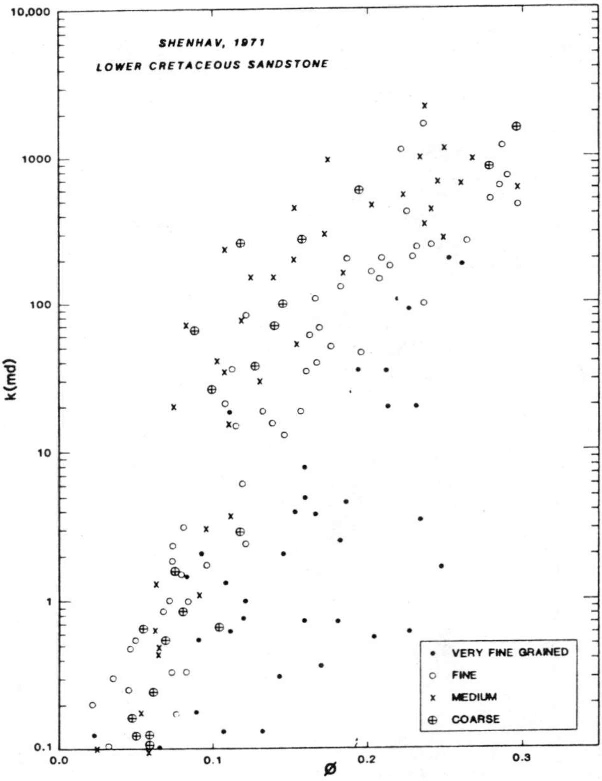

Figure 5. k-ϕ plot, cemented Lower Cretaceous sandstones, by grain size (Shenhav, 1971). Reprinted by permission of the American Association of Petroleum Geologists.

Petrological Effects Summary

The k-ϕ effects discussed above can be seen in plots for the seven zones in the Northwest Fault Block of the Prudhoe Bay Field (Figure 7). The uppermost (Zones 4B and 4A) and lowermost (Zones 1B and 1A) zones exhibit trends which are linear, or nearly so. We infer that these trends are mainly controlled by the cement and clay content. The central zones, especially Zone 3, contain a high gravel content and consequently the data are more scattered. Here, k is abnormally high in the poorly cemented, high gravel samples. Except for Zone 2A, the plots show a clean right-hand edge, indicating that the presence of extremely fine-grained material is nil.

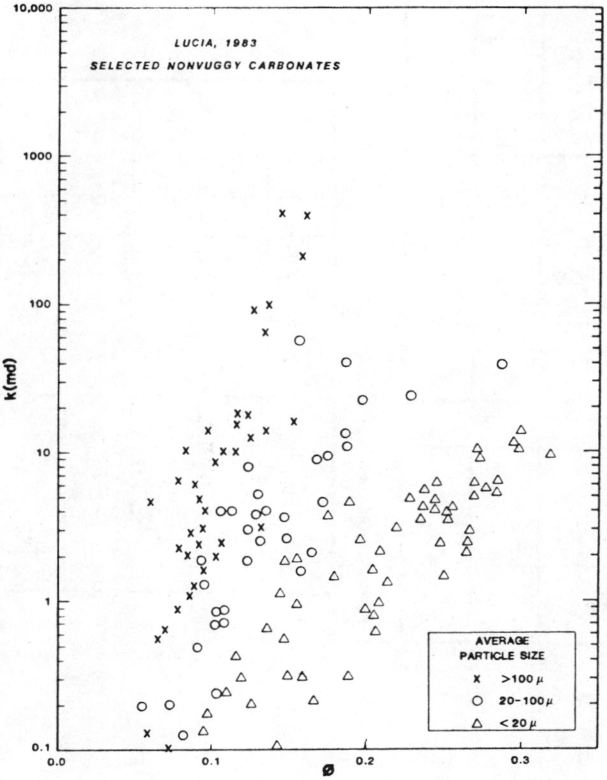

Figure 6. k-ϕ plot, non-vugular carbonates, by particle size (Lucia, 1963). Reprinted by permission of the Society of Petroleum Engineers of AIME.

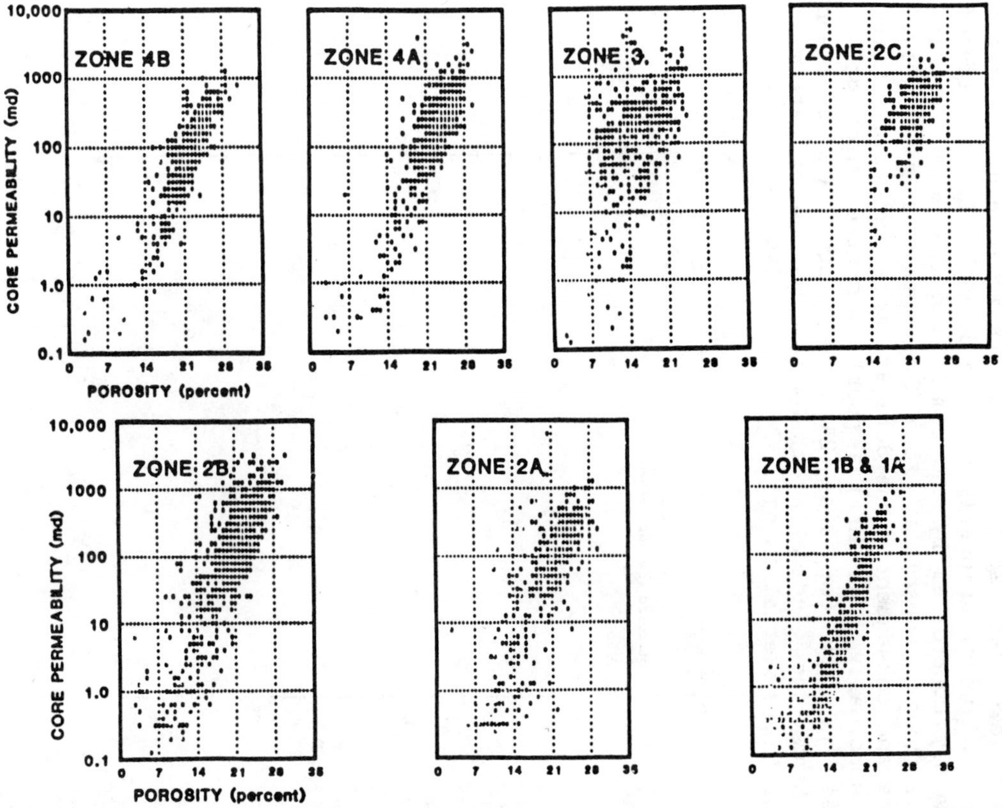

Figure 7. Permeability-porosity plots by zone, NWFB, Prudhoe Bay.

There is, at present, no theoretical model which accounts for the influence of petrological variables on k-ϕ trends. Undoubtedly, part of the problem results because k and ϕ are macroscopic properties of the pore space, yet we are attempting to explain k-ϕ relationships in terms of grain geometry and mineralogy (non-pore space). This problem may be overcome in the future if pore geometry observations become a standard technology.

Now that the petrological effects on permeability-porosity relationships have been reviewed, we will discuss how this information is applied in regression and discriminant analysis.

IV. GENERAL PROCEDURES

This NWFB permeability study utilized core data from twelve cored wells in or nearby the fault block shown in Figure 8. Core data required include the percentages of calcite cement and gravel from core description, and also laboratory measured permeability and porosity. Core porosity is used only to calibrate well log derived porosity.

Before presenting a description of the general procedures, the zonation/layering system at Prudhoe Bay should be noted. Based on a combination of petrologic and stratigraphic factors, the producing Sadlerochit reservoir is divided into several zones. These include, starting at the base, Zones 1A and 1B (very fine to fine-grained sandstone with abundant interbedded shales); 2A, 2B, and 2C (fine-grained sandstones with locally abundant shales and locally conglomeratic); Zone 3 (sandy conglomerate with few shales); and Zones 4A and 4B (fine-grained sandstone with locally abundant shales. We utilized a layering system which further subdivides these zones into layers. These zones and/or layers are utilized for all discriminant and regression analysis. They provide the means to group like lithologies for prediction purposes.

The general procedural flow of this permeability estimation method consists of five steps (illustrated in Figure 9):

1. Establish independent variables. Here we define the well log variables x_1, x_2, ... x_n. These are in fact based on a very limited number of actual well logs -- induction, density, gamma ray, and sonic -- plus several variables of location. These well logs are corrected, filtered, and combined in whatever ways we think will enhance predictive resolution. Clearly these

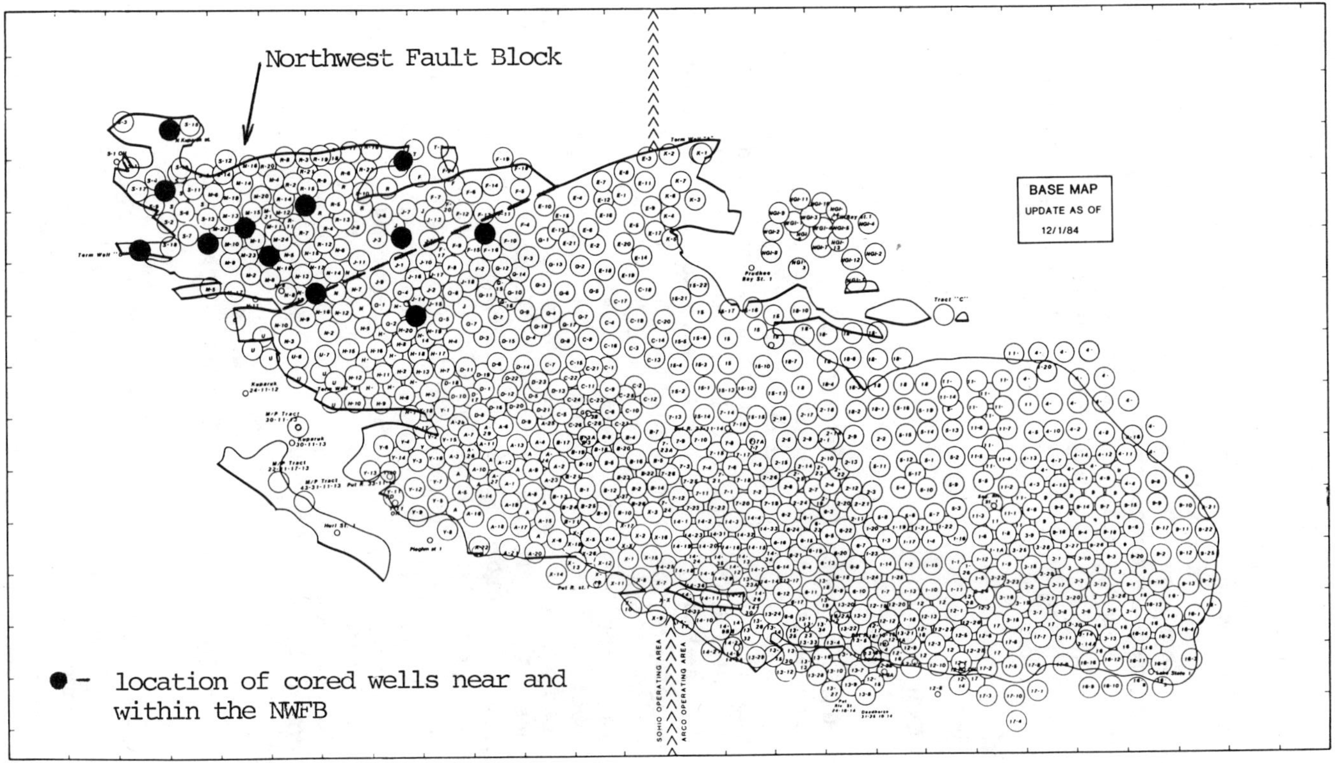

Figure 8. Prudhoe Bay location map showing NWFB area and cored well locations utilized in the multiple variable linear regression permeability study.

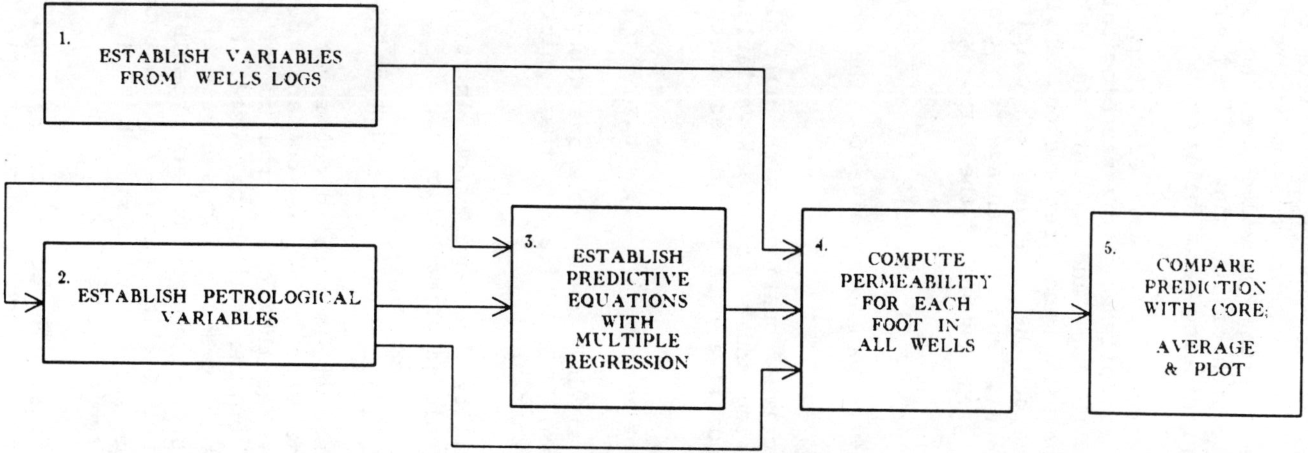

Figure 9. Flow chart for the general procedures used in the permeability prediction study.

input variables are not "independent" from one another in the physical sense nor are they meant to be; rather we rely on the sorting provided by the statistical software to find which of the many defined variables do the best job of predicting permeability.

2. Establish petrological variables. Predictors for gravel and calcite are established using petrological observations and the well log "independent" variables. Both discriminant analysis and regression analysis are employed to predict the presence of calcite cement, presence of gravel, and percentage of gravel. The equations from regression are of the form,

$$y_1 = \% \text{ gravel} = a_o + a_1x_1 + a_2x_2 + \ldots + a_nx_n$$

Statistical software packages solve for the coefficients a_i.

3. Establish predictive equations with multiple regression. The well log and petrological variables are used to establish predictive equations based on core permeability k_c,

$$\log k_c = b_o + b_1x_1 + b_2x_2 + \ldots + b_nx_n + b_{n+1}\, y_1$$

Data from all available cored wells are grouped together by zone or layer. Statistical software packages solve for the coefficients b_i.

4. Compute permeability for each foot in all wells to determine the predicted permeability k_p,

$$\log k_p = b_o + b_1x_1 + b_2x_2 + \ldots + b_nx_n + b_{n+1}\, y_1$$

In practice this step is carried out several times for the cored wells. Each time we compare the core and predicted values and then change the number or type of input variables or the methodology until the match is deemed satisfactory based on an analysis of residuals. Obviously, no such comparisons are possible for the uncored wells, but quality checks on the uncored wells are mandatory as errors and busts do occur which have not appeared in the cored wells.

5. Compare predictions with core; average and plot. We rely on graphical techniques to compare predicted with observed core data. Several types of comparisons are

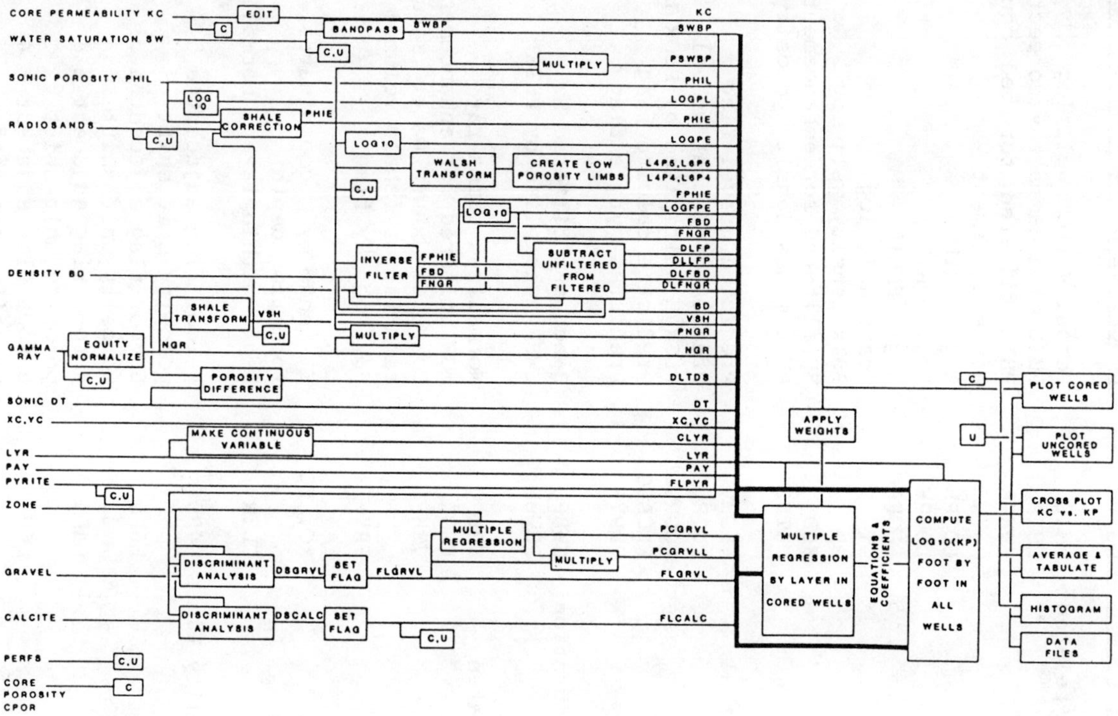

Figure 10. Logic diagram of computations for permeability prediction in the NWFB, Prudhoe Bay.

TABLE I. Description of variables used as input for regression analysis.

Category	Mmemonic	Description
Porosity-related	PHIL	Porosity from well log
	PHIE	PHIL with gamma ray correction
	FPHIE	PHIE filtered for bed boundary enhancement
	LOGPL	Log_{10} PHIL
	LOGPE	Log_{10} PHIE
	LOGFPE	Log_{10} FPHIE
	DT	Sonic well log
	BD	Bulk density well log
	FBD	BD filtered for enhancement
	DLTDS	PHIL minus density porosity
Petrological	FLGRVL	Predicted gravel flag
	PCGRVL	Predicted percent gravel
	PCGRVLL	FLGRVL multiplied by PCGRVL
	FLCALC	Predicted calcite flag
	FLPYR	Predicted pyrite flag
Water Saturation	SWBP	Band-pass filtered Sw
	PSWBP	SWBP multiplied by PHIE
Clay-related	NGR	Normalized gamma ray
	FNGR	NGR filtered for enhancement
	VSH	Computed bound water saturation factor (shale volume)
	PAY	Pay/Non-pay flag
Location	XC, YC	X and y coordinates at formation depth
	CLYR	Depth-within-layer, normalized
Low Porosity Limbs	L4P5	4 foot flag adjacent a 5 porosity unit shift
	L6P5	6 foot flag adjacent a 5 porosity unit shift
	L4P4	4 foot flag adjacent a 4 porosity unit shift
	L6P4	6 foot flag adjacent a 4 porosity unit shift
Enhancement	DL...	Difference between filtered and unfiltered curves

used because none seems truly adequate by itself. Also we compute both geometric and arithmetic averages by layer and by zone to anticipate the possible needs of different users of the data. Finally plots at one inch to 20 feet were provided for detailed inspection and at one inch to 100 feet for correlations and for quick inspection.

Having discussed the general procedures, we now describe some algorithms and techniques which are peculiar to the NWFB predictors. A detailed logic diagram for the NWFB computations is shown in Figure 10. It is arranged in the same order as Figure 9 so that Figure 9 can be used as a "map" for Figure 10. We will touch on some, but not all, of the computational features indicated in Figure 10. The independent variables are defined in Table I.

V. WELL LOG AND PETROLOGICAL VARIABLES

As stated earlier, our objective is to find predictive equations which will estimate k values. In zones where a linear trend holds, the prediction of k from ϕ is a relatively straightforward process. In zones where there is no linear trend, porosity alone is not a good predictor of k. Earlier we had been examining core data only. The prediction of k from well logs encounters problems related to the different scale of measurements afforded by core and well logs,

1. A core plug may not be representative of the rock (the sampling problem).

2. Well logs may not represent the rock (wellbore effects).

3. Well logs are spatially averaged more than core data (the resolution problem).

4. Statistical predictors of k will estimate the mean values, underestimating the higher values and overestimating the lower values.

These caveats must be kept in mind and sometimes must be accounted for by editing the core data or by filtering the well log data.

Selection of Cored Wells

Cored wells both within and closely surrounding the area in question should be included in the data set for regression. When a sufficient amount of core data exists, as with the Prudhoe Bay Field, improved permeability estimation may result by dividing the cored wells into groups for regression purposes. This division should be based on areally grouping cored wells with similar k-ϕ relationships.

In the NWFB, permeability distributions differ between the eastern and western NWFB cored wells and moreover, there is a greater density of permeability above one darcy in the east. Figure 11 illustrates these dissimilarities. Additionally, calcite cement is more prevalent in the western area. For these reasons, the two areas received a separate, but similar treatment in the prediction of permeability.

Two cored wells were included in both eastern and western regressions. A line through these two wells defined the division of uncored wells for the application of the appropriate regressions. This approach was intended to avoid abrupt changes in permeability prediction in wells on opposite sides of this east-west boundary. A better approach might be to search systematically for an optimum boundary between areas with distinct petrological characteristics.

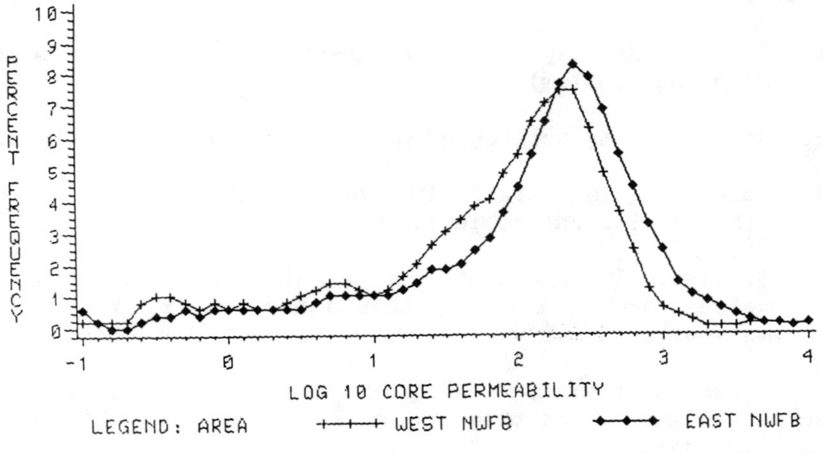

Figure 11. Core permeability distributions for the eastern and western NWFB areas.

Shale-Corrected Porosity

As indicated in Figure 10, the sonic porosity, PHIL, is corrected for shale content, and the resulting porosity, PHIE, is used extensively to compute or control the computation of other variables. For example, it is used to compute a porosity-water saturation product PSWBP, to create the low porosity limbs, and to form filtered logs.

A gamma ray based variable, VSH, is computed and represents a fraction of pore space occupied by water bound to shales; hence, porosity is reduced accordingly,

PHIE = PHIL*(1-VSH)

The equations for determining PHIL from sonic logs are derived by regression and are thus subject to overprediction in shaly rock and underprediction in clean sands. The VSH term does not correct for these statistical vagaries; it is a correction to account for bound water.

Where radioactive sands are present, the transform is not applied so that PHIE = PHIL. Both PHIE and PHIL are included as independent regression variables.

Band-passed Water Saturation

Inclusion of water saturation, S_w, in permeability regression is problematic due to changes in fluids in a zone from well to well. We have noticed, however, that the high frequency characteristics of S_w tend to correlate with permeability changes. For this reason, a high-pass filter was applied to the Archie water saturation which eliminates the low frequency transition between oil and water. Only the filtered S_w was utilized in regression analysis; use of the original S_w could produce erroneous results due to fluid type changes within the same zone or layer.

Low Porosity Limbs

Some high permeability streaks have a close spatial relation to shales. We have also noted a few high permeability intervals adjacent to low-porosity cemented sands. In either case, these high permeability streaks are near intervals which have relatively low porosity. The typical porosity profile of these occurrences contains an abrupt change between high and low porosity as opposed to a gradual change.

Three theories attempt to explain these high permeability occurrences. First, dissolution by fluid movement along permeability barriers will result in higher permeability. Secondly, extraction of carboxylic acids from shale into an adjacent sandstone could destroy carbonate cements. And lastly, a radical change between low and high energy depositional environments may have created high primary permeability. For lack of a better method of predicting permeability streaks of this nature, flag-type curves (1 yes, 0 = no) were computed to indicate proximity to shales and low porosity intervals. These flags were then included as regression variables.

The method of creating the low porosity limbs follows a technique employed by Lanning and Johnson (1982). In that paper, Walsh transformations of well logs are employed for the purpose of identifying rock units. Very simply, the Walsh transformation creates a square wave curve (constant valued for several feet before a shift in value) which overlies the original curve.

Figure 12 illustrates the Walsh transformation. The solid line traces in tracks 1 and 2 of that plot are two different Walsh transformations of the same porosity curve (dashed lines). In both cases, the Walsh porosity is constant for 4 foot intervals. They differ, however, in the placement of the shifts, which occur 2 feet apart. Abrupt changes in porosity in either or both of the Walsh transformations flagged the point from which a low porosity limb was generated.

Both 4 and 5 porosity unit changes on the Walsh porosity were used as minimum requirements for this pick, and flags were set either 4 or 6 feet in the direction of high porosity. For example, the curve L6P5 sets a 6 foot flag adjacent to low porosity from the location of each 5 p.u. change in the Walsh transformations. This example is also shown in Figure 12 by the solid shaded blocks. There are four curves of this type created. It is left up to the statistical software to select the appropriate curve for use in aiding the high permeability prediction based on optimum sensitivity and magnitude. Track 3 of Figure 12 presents core permeability from which the reader may note the coincidence between high permeability and the statistically created low porosity limbs.

Filtered Curves as Independent Variables

Most well log data have a resolution of several feet due to spatial averaging and tool movement through the formation. However, core data are obtained at one foot intervals so core resolution is greater than well log resolution. In order to predict the extreme values present in core permeability data, we needed to enhance the high frequency content of the permeability predictors. Two procedures were used.

CURVE NUMBER	LEFT MARGIN LIMIT	RIGHT MARGIN LIMIT	TRACK NUMBER	CURVE MODIFIER	LINE TYPE	NAME	TOOL	UNITS
91.0	.5	0.	10.	LIN	DSDA	PHIE	SCRV	FRAC
1091.0	.5	0.	10.	LIN	DSOL	WPHI	WAL1	FRAC
91.0	.5	0.	20.	LIN	DSDA	PHIE	SCRV	FRAC
1092.0	.5	0.	20.	LIN	DSOL	WPHI	WAL2	FRAC
1094.0	20.	0.	20.	LIN	DSOL	L6PS	LIMB	1YON
121.0	10000.	.1	30.	LOG	SDOT	K-H	CORE	MD
121.1	10000.	.1	30.	LOG	SPEC	K-H	CORE	MD

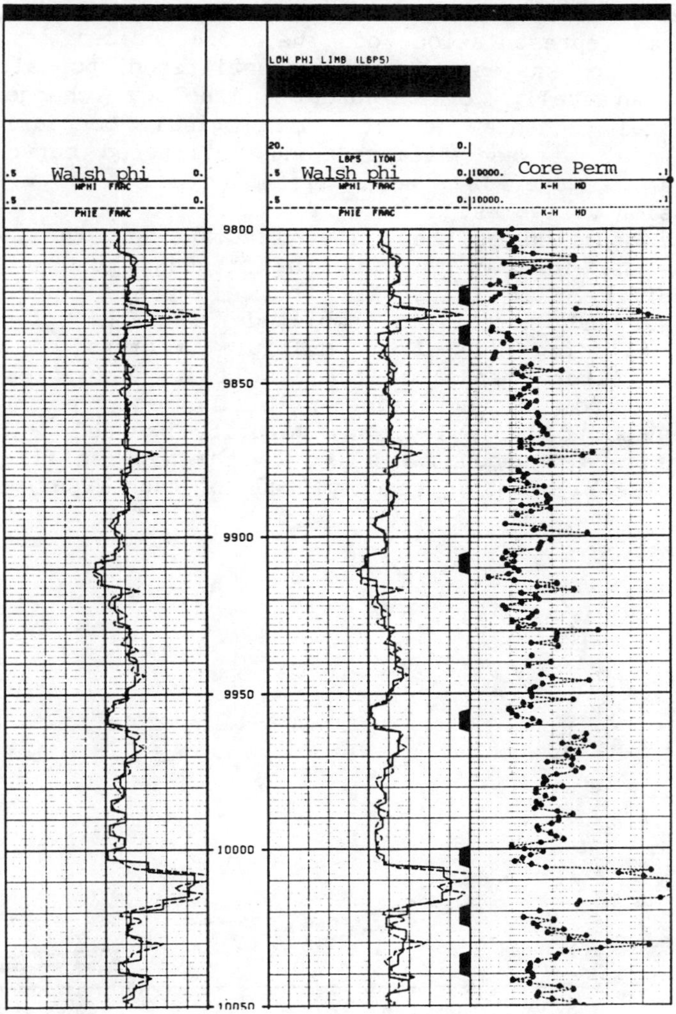

Figure 12. Two Walsh transformations (solid lines) of porosity (dashed lines) are shown in tracks 1 and 2. The solid-shaded blocks are low porosity limbs as derived from the Walsh transformations. Permeability is plotted in track 3 where high k_c is often associated with the low porosity limbs.

First, the high frequency content of predicted permeability increases with the number of regression variables. Second, three of the regression variables (porosity, bulk density, and normalized gamma ray) were inverse filtered to increase bed resolution and then were included in the regression.

The filtered porosity, for example, provides a more accurate representation of the core porosity in areas adjacent to shales than does unfiltered porosity which varies gradually over abrupt lithology changes. For additional enhancement at lithological boundaries, the differences between filtered and unfiltered curves (DLFP, etc. in Figure 10) were also included as independent regression variables.

The addition of the filtered curves and the difference curves has aided in matching the frequency content of core permeability. To illustrate this, Figure 13 presents the smoothed and normalized amplitude spectrum of (1) log (k_c), (2) log (k) from multiple regression utilizing various filtered and unfiltered well log variables and (3) log (k) from a regression utilizing only porosity, the square root of porosity, and porosity squared. Throughout the entire frequency range (up to the Nyquist frequency of 500 cycles/kilofoot), the amplitude of the NWFB predictor

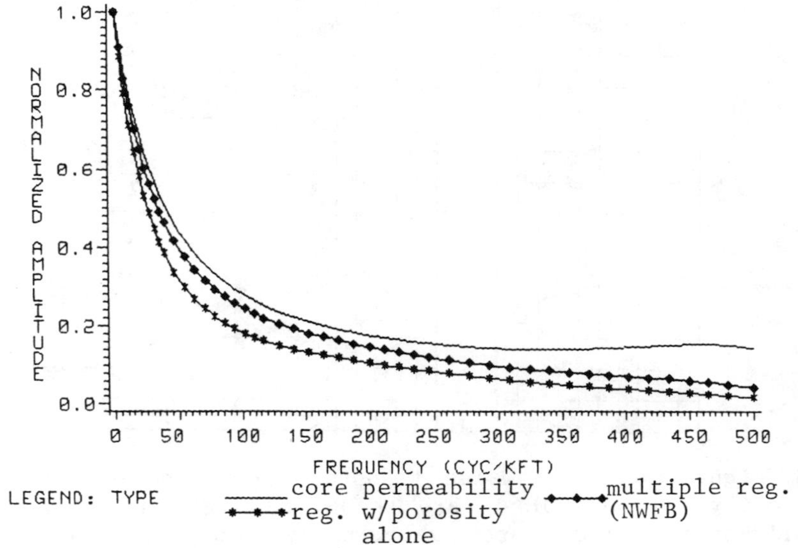

Figure 13. Amplitude spectra of permeabilities from core, multiple variable linear regression, and regression with porosity alone.

more closely matches that of the core permeability. (The reader should note that a close match of the core permeability amplitude spectra does not alone indicate accurately predicted permeabilities, but only indicates similar frequency content.)

Petrologic Variables

Equations to predict the petrologic controls on permeability were derived through discriminant and regression analysis. Flag-type curves (1 = yes, 0 = no) were created for both the presence of calcite cement and gravel. Additionally, a prediction of percent gravel was made from well log and log-derived data. No attempt was made to predict percent calcite cement because simply its presence, even in small amounts, related to permeability reduction. Also, in core samples where calcite is present, variation in the values noted is small (mean of 5% and standard deviation of 5.0) and difficult to predict.

After some adjacent zones were grouped according to similar calcite occurrences, discriminant analysis was performed for the presence of calcite. Those results produced a calcite flag curve which can be calculated from well log data. Of all the core samples where calcite cement was noted, the computed calcite discriminator correctly flagged 65.4% (123/188). Conversely, of the core data with no calcite cement, 96.3% (2042/2121) were predicted accurately.

The predictor for gravel presence was derived in a similar manner. Of the cored wells utilized, the low percentage of gravel in the upper Zone 4B (1.8%) and lower Zone 1 (0.2%) warranted exclusion of these zones from gravel discriminant analysis. When the equations to predict gravel presence are applied to the cored wells, correct flags were set in 93% of the cases where gravel exists. The gravel flag, FLGRVL, is used as a regression variable and is also used as a switch to allow a gravel percent prediction only when the gravel presence flag is "on" (variable PCGRVLL in Figure 10).

An increase in gravel content tends to decrease porosity and increase permeability. For this reason, percent gravel estimators (PCGRVL) were created. Through regression analysis, equations were derived to predict the percentage of matrix composed of gravel. Correlation coefficients from the percent gravel regessions were low, ranging from 0.43 to 0.59. A partial explanation for these low correlations lies within the core description data: not all core data were described by the same individual. No attempt was made to normalize these observations.

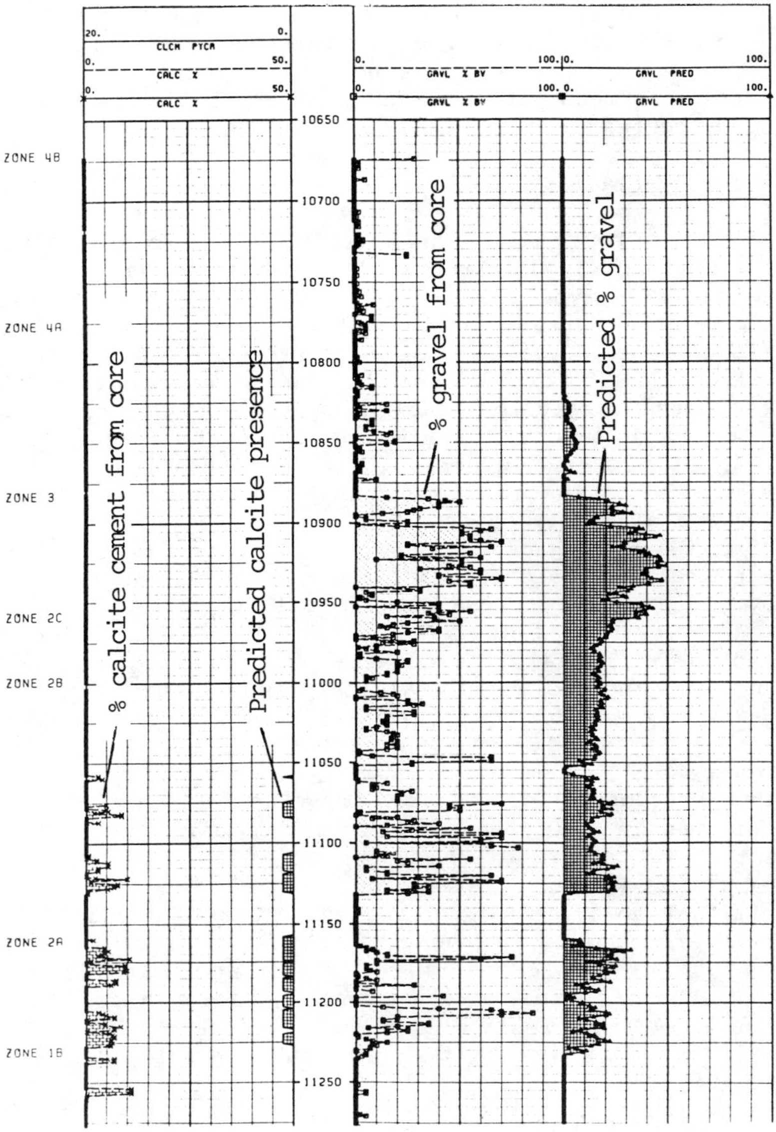

Figure 14. NWFB cored well plot including: % calcite cement from core, predicted calcite presence, % gravel from core, and predicted % gravel.

An example of calcite and gravel predictions are shown in Figure 14. The left side of the first track presents the percentage of cement that is typed as calcite. The predicted calcite indicator (1 = yes, 0 = no) is shown on the right side of that track. The percent gravel described from core and predicted percent gravel are plotted in the second and third tracks, respectively. Although extreme values are not reached by the gravel predictor, the trend is similar to core data.

VI. REGRESSION CONSIDERATIONS

Regression by Layer

Initial trials at permeability regression were worked by zone. This was an obvious step to take upon examination of permeability-porosity plots by zone (Figure 7). By a further division of zones into layers, some different patterns were apparent between layers within a zone. In particular, the base of zone 4A (layer 17) becomes more conglomeratic with a k-ϕ crossplot markedly different from the rest of zone 4A. Along with the obvious increase in correlation coefficients, regression by layer also provides more accurate averages for use in simulation grids which use the more finely spaced layering system. A mechanical division of zones into layers is obviously not the ideal procedure for grouping like lithologies. Subdivisions based on facies characterization would be more appropriate.

Weighted Regression

A major difficulty in permeability estimation is the underprediction of high permeabilities, and although of less concern, overprediction of low values. Since y-on-x regression tends to estimate the mean, there is a lower density of predicted extremes than in the core data. Because of this effect, the use of the logarithm of k compounds the problem. For example, if $\log_{10}k_c$ spans -1 to 3 (0.1 to 1000 md), regression results may have a span of -.03 to 2.3 (0.9 to 200 md). Thus the high values are especially underpredicted.

The importance of the high permeabilities necessitated a weighting scheme in order to expand the range of predicted values. Typically both high and low permeability extremes were weighted. Although low permeabilities are inherently weighted from the use of $\log_{10}k$ in the regression, weighting of both low and high values produced better results in the high permeability range.

Least squares weighting of only the high values forces a regression line up at the high end of a log k-ϕ plot, whereas the additional weighting of low values pivots the regression line and results in more accurate foot-by-foot predictions of high as well as low permeabilities. Additionally, weighting of both extremes provides more accurate geometric averages than would result from the weighting of high permeabilities only. It is interesting to note that allowing error in the independent regression variables has the same effect as weighting both high and low permeabilities. That is, both procedures produce the pivoting effect: an increased prediction of high values and a decreased prediction of low values.

Figures 15a, b show core permeability on the y-axis and predicted permeability on the x-axis for layer 10 in east NWFB. In Figure 15a, non-pay was excluded from the regression. Note that although core permeability (k_c) ranges up to 3000 md, predicted permeability (k_p) reaches only 500 md. In Figure 15b, non-pay was included and both low and high permeabilities were weighted. As compared with Figure 15a, permeabilities above 500 md are being predicted and where k_c is greater than 500 md, the predicted values are higher. Also, when both high and low values were weighted, more accurate permeabilities were computed than when only the highs were weighted.

The problem is compounded when using log k rather than k itself as the independent variable, because it causes undue weighting of the lower values, as discussed by Etnyre (1984). A formal method of weighting is suggested for log k, but the method raises some questions and has not been tested. We proceeded to use an empirical method of our own devising. Weighting was applied independently by layer depending upon the $\log_{10}(k)$ distribution within each layer. This is best illustrated by an example. In Figure 16, a frequency plot of $\log_{10}(k)$ shows the original and revised permeability distribution for layer 17. The revised portion of the plots indicates the additional core permeabilities included for regression. Factors of either 1.5, 2.0, or 3.0 were applied, with the highest weights applied to the highest permeabilities. Factors were chosen to preserve whatever log normality was present.

(a)

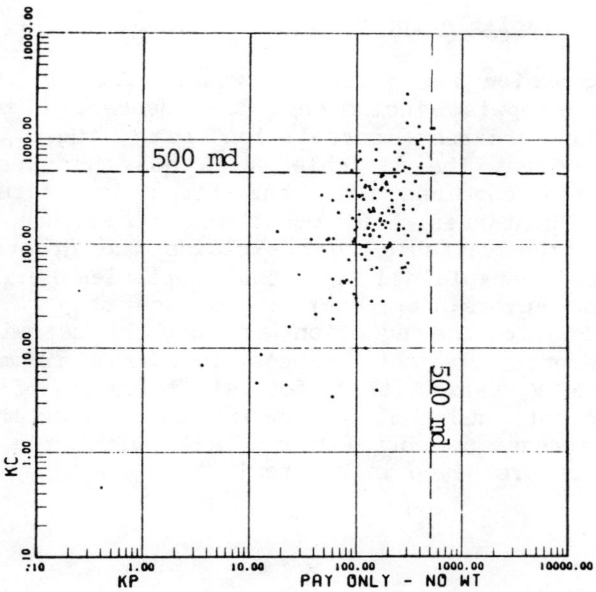

(b)

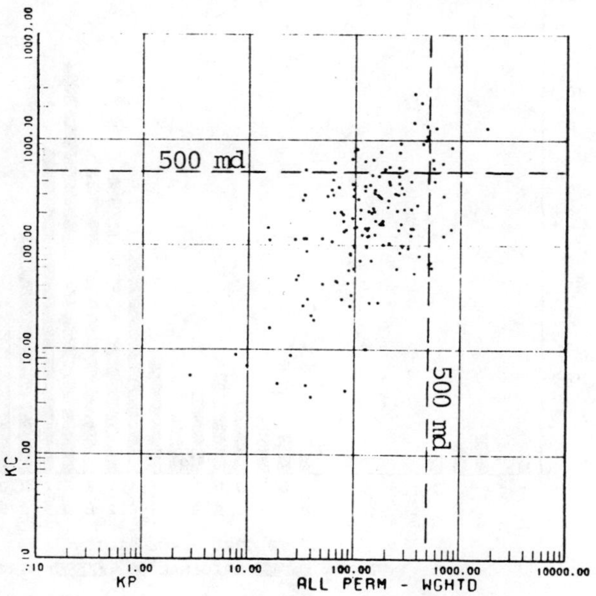

Figure 15. Core versus predicted permeability (a) without weighting and without non-pay, and (b) with weighting of permeability extremes and non-pay included in the regression.

Order of Variable Inclusion

Regression was performed within SPSS by the method of forward stepwise inclusion. The independent variable which has the greatest correlation with $\log_{10}k_c$ is entered first. Then, the variable which has the greatest correlation when combined with the first is included, and the process continues. In other words, as Nie et al (1975) state, "the variable that explains the greatest amount of variance unexplained by the variables already in the equation enters the equation at each step". Variables are entered into the equation as long as certain statistical requirements are met. These include minimum requirements for the F ratio (test for significance of a regression coefficient) and the tolerance. Examples of the increase in the regression correlation (R^2) with the addition of variables are given in Figure 17.

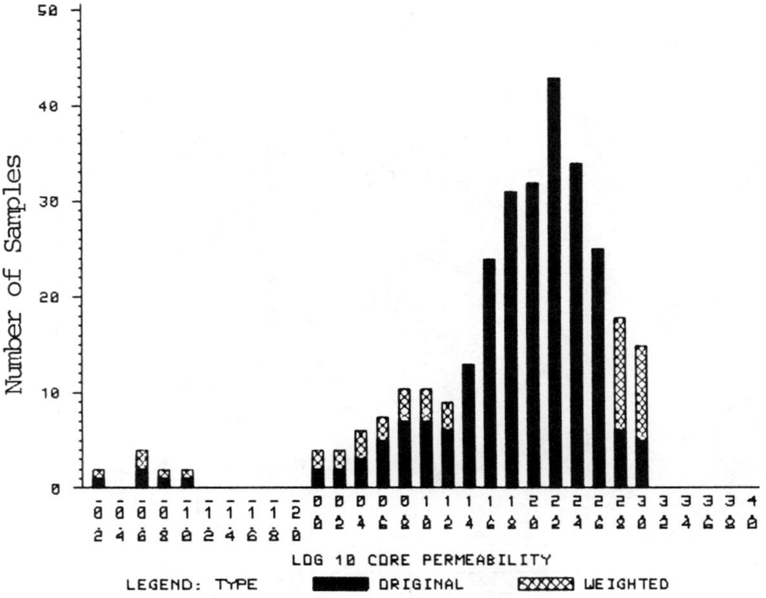

Figure 16. Example of permeability weighting scheme. The plot illustrates the distribution of the original core permeability data and the additional data values used in the regression as a result of weighting permeability extremes.

A number of observations can be made by considering the order in which variables are included in the regression. Almost invariably (30 of 33 cases) the first variable used is one of the porosity variables. This is not surprising because permeability is usually predicted from porosity alone. Porosity-related variables are selected in 11 of 33 cases as the second variable and in 8 of 33 cases as the third variable. Thus two-thirds of the second and three-quarters of the third variables are other than porosity variables and are a mixture of petrological, saturation, shale indicating, location, limb-type, and enhancement variables. The usage of non-porosity well logs as the second and third variables underscores the importance of

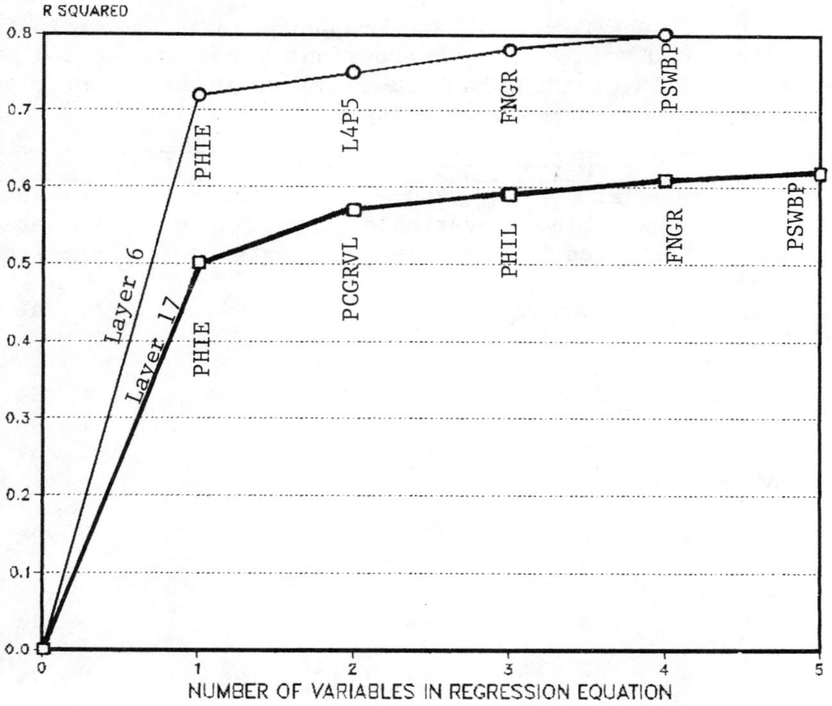

Figure 17. Examples of increase in R^2 with addition of variables in the regression equation. The variable which entered the equation at each step is noted on the plot for the two examples sited.

broadening the type and number of well log variables which are input to the regression analysis. Table II lists the percentage of each variable type included in the regression at progressive steps.

Layers 13 through 16 are typically the most conglomeratic. In these intervals, the calcite flag was used once and band-passed water saturation was used twice as the most significant variable. The three gravel variables were not used much; probably the gravel predictions were simply not accurate enough. Presumably they were replaced by water saturation.

Trend information is obviously important in permeability prediction. The location variables (depth-within-layer variable CLYR and the east-west variable XC) were used often as lower-ranking variables.

TABLE II. Breakdown of the frequency in which variables were included into the regression equation, based on category. (91% of the time, the first variable to enter the regression equation was a porosity-type variable.)

	Percentages of Variable Use			
Category	1st variable entered	2nd variable entered	3rd variable entered	Remaining variables entered
Porosity-related	91%	33%	24%	24%
Petrological	3%	24%	12%	8%
Water Saturation	6%	0%	12%	8%
Clay-related	0%	12%	9%	20%
Location	0%	18%	18%	20%
Low Porosity limbs	0%	9%	9%	7%
Enhancement	0%	3%	15%	13%

VII. ERROR ANALYSIS

To assess errors, the predicted data, k_p, are compared with the core data, k_c, and with a field-wide predictor, k_{fw}, which is based totally on porosity. We have found no simple statistic or single graphical technique which neatly summarizes the errors inherent in the permeability predictor. Instead we rely on several techniques, mostly graphical, to summarize the uncertainties and to make comparisons.

We have mentioned that well log plots are a primary means of finding errors and improving techniques during the construction of a predictor. The final plots also offer a quick way to see how well the predicted log tracks the core data. In addition, the frequency content of k_p and k_c have already been compared in the amplitude spectrum of Figure 13.

In the following discussion, crossplots, frequency distributions, averages, and residuals are used to compare k_p, k_c and k_{fw}. Data are considered together for the entire NWFB and on a zone-by-zone basis for each cored well.

Prediction of High Permeability Streaks

The cored wells were examined to see how well the predicted permeability curve duplicates "high permeability streaks". Thirty-one such streaks were picked, based on the criteria of (a) 3 or more contiguous core points averaging more than 800 md (with two exceptions) and (b) good permeability contrast with points above and below the streak. A qualitative comparison of shape was assigned to each streak. If it appeared that the predicted curve clearly indicated the streak, then the comparison is rated "good", if the predicted curve gives little or no indication of the streak it is rated "poor".

The prediction of high permeability streaks is successful in 18 of 31 cases - 17 with good shape comparison and generally quite close in average permeability plus one fair shape comparison within a factor of two in average value. From this exercise, it appears that high permeability streaks will be unambiguously predicted with 60% success, about 20% will be indicated in less than satisfactory fashion, and 20% not predicted at all.

We can also ask how many high permeability streaks will be predicted where there are none actually present. The criteria for selecting a false predicted streak were analogous to those for a core-based streak. There were 12 such false predictions in the cored wells.

The data can be summarized in this way: thirty predictions of high permeability streaks were made (18 real and 12 false). Hence 60% of total predictions are valid, 40% are not. Incidentally, the number of misses (5 fair and 6 poor) is about equal to the number of false predictions. In other words, we can expect to predict the right number of high permeability streaks, but up to 40% will be in the wrong location.

Distribution of Predicted Permeabilities

Not only have we more closely duplicated the frequency content of the core permeablility as seen in Figure 13 but also the permeability distributions have been duplicated more closely (as opposed to a predictor based on porosity alone). Figure 18 illustrates the similarities between core and the NWFB predicted permeability distribution, and also the improvement of that distribution over the previous porosity-based predictor for the same well, particularly in the high permeability range.

By investigating the differences between core and predicted permeabilities, further verification of the improvement of high permeability prediction is seen. Where

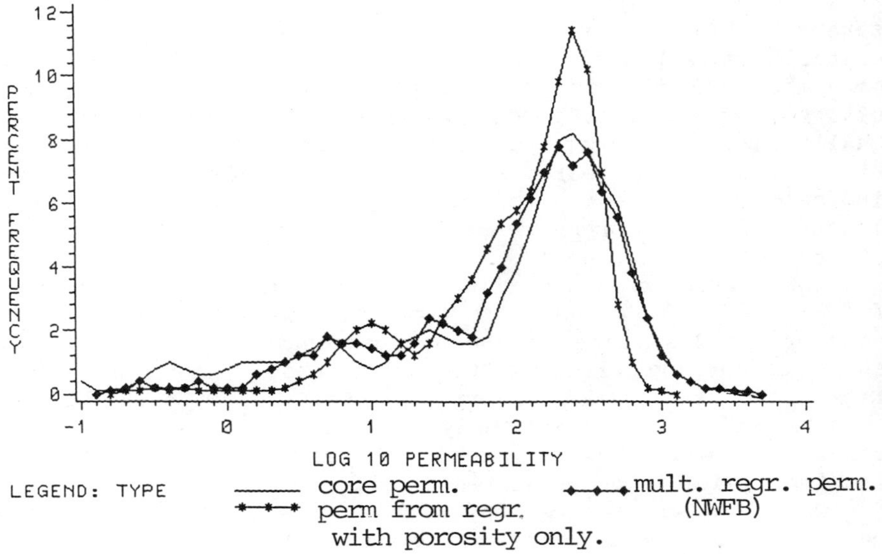

Figure 18. Frequency distributions of permeability from core, multiple variable linear regression, and regression with porosity alone.

k_c is greater than 500md, the average of (k_c-k_p) is 485 md for all NWFB cored wells; whereas the average of (k_c-k_{fw}) for the same wells is 671 md.

Foot-by-foot Permeability Prediction

In order to check the quality of permeability prediction of all regressions (all layers) combined, a crossplot of core permeability vs. predicted permeability was generated for all zones, all cored wells in the east and also in the west NWFB. For comparison purposes, k_c vs. k_{fw} (Figure 19a) and k_c vs k_p (Figure 19b) for the same eastern NWFB cored wells are shown. These clearly indicate the improvement of high k underprediction and low k overprediction.

Also shown on Figures 19a and b are the associated R^2's for the data; and more importantly, the residual mean square,

$$\text{RSDMS} = \frac{1}{N-2} \sum_{i=1}^{N} (\log k_{ci} - \log k_{pi})^2$$

for each set of data. The reduction of the residual mean square from the fieldwide predictor to the NWFB predictor is 31%.

Figure 20 summarizes the multiple variable regression quality for a cored well. On the left side of each figure, a scale bar for the number of core permeability points is given for each zone. The next two columns show the core (solid line) and predicted (dashed line) distributions on both logarithmic and linear scales. Histograms of the k_p/k_c ratio are also shown on a logarithmic scale along with the number of core samples in each zone. Finally, both geometric and arithmetic averages are given for k_c and k_p. Overall, permeability distributions between core and predicted values are similar in shape and the k_p/k_c histograms peak near the value 1.0.

Plots of permeability data with depth present another means of visually analyzing the predicted permeabilities. In Figure 21, both core and predicted permeabilities are plotted on logarithmic and linear scales. Errors in the prediction of high permeability values are more obvious on the linear scale, where errors in low core permeabilities appear insignificant. As in the percent gravel prediction, extreme values are not always reached, but trends and averages are similar. The core data from the well in Figure 21 were utilized in the construction of the regression equations. To more justly evaluate the predictions, a test well can be studied. A test well is one which was not

(a)

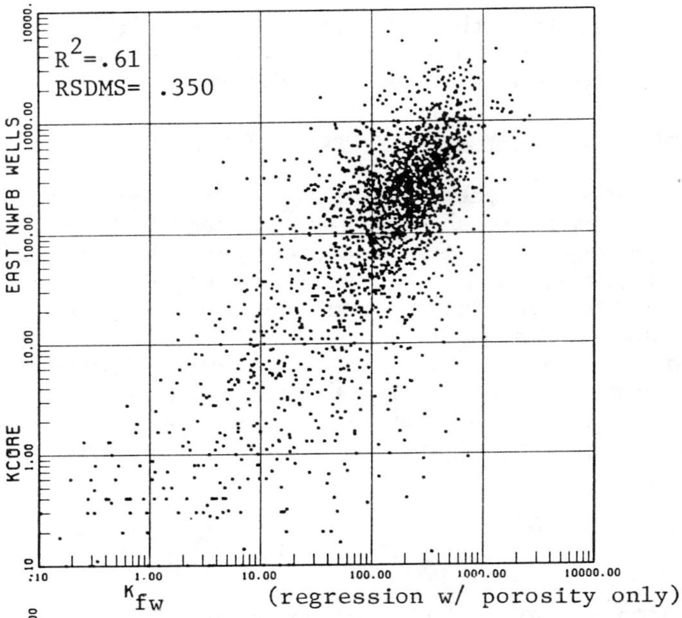

(b)

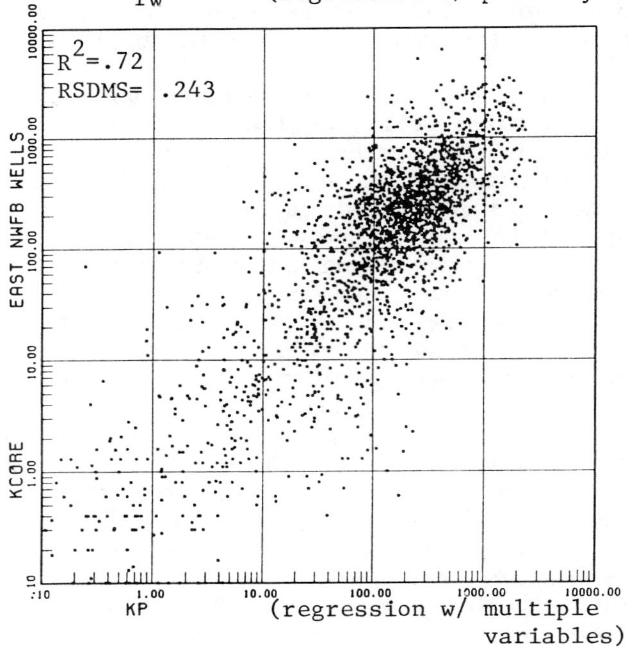

Figure 19. Plots of k_c versus (a) permeability from regression with porosity alone and (b) permeability from multiple variable linear regression.

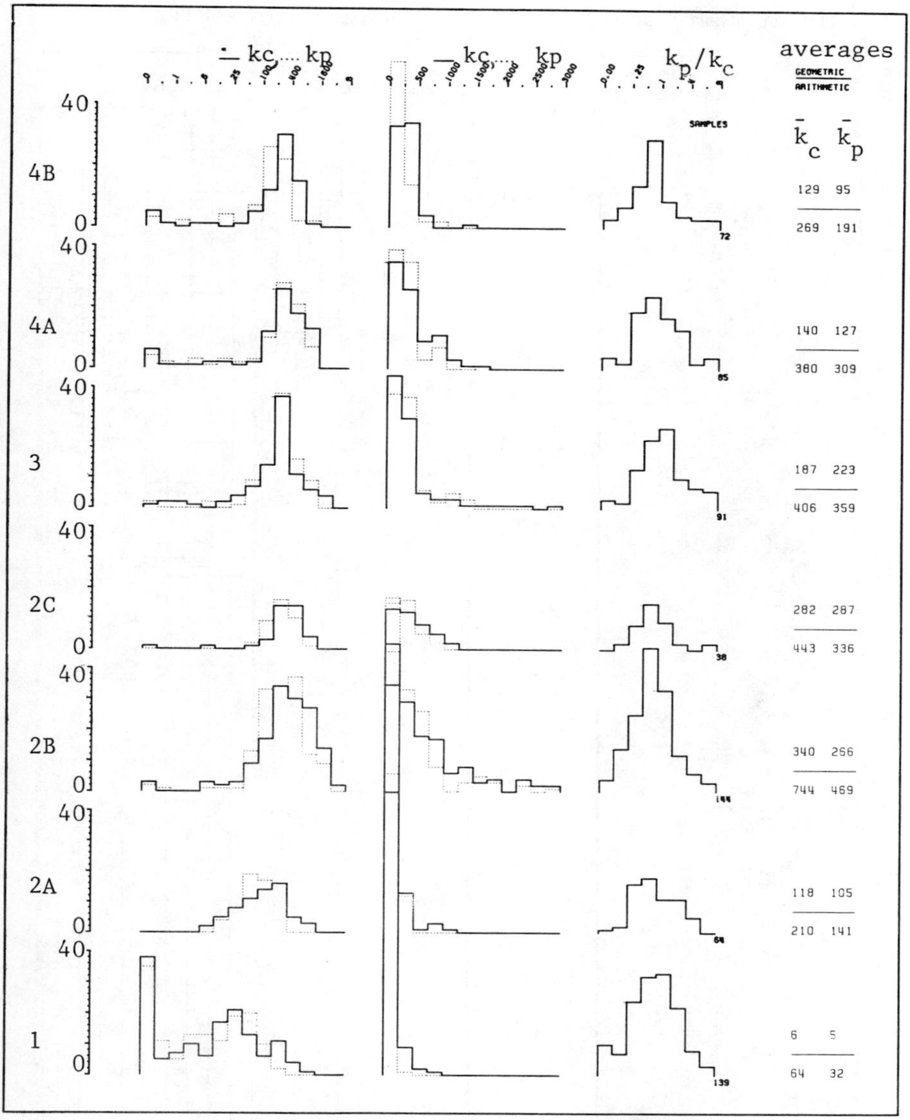

Figure 20. Permeability comparisons for a NWFB cored well. The first two columns present core and predicted permeability distributions on logarithmic and linear scales. The third column shows histograms of the ratios of predicted to core permeabilities. The last column presents geometric and arithmetic averages for both core (left) and predicted (right) permeabilities.

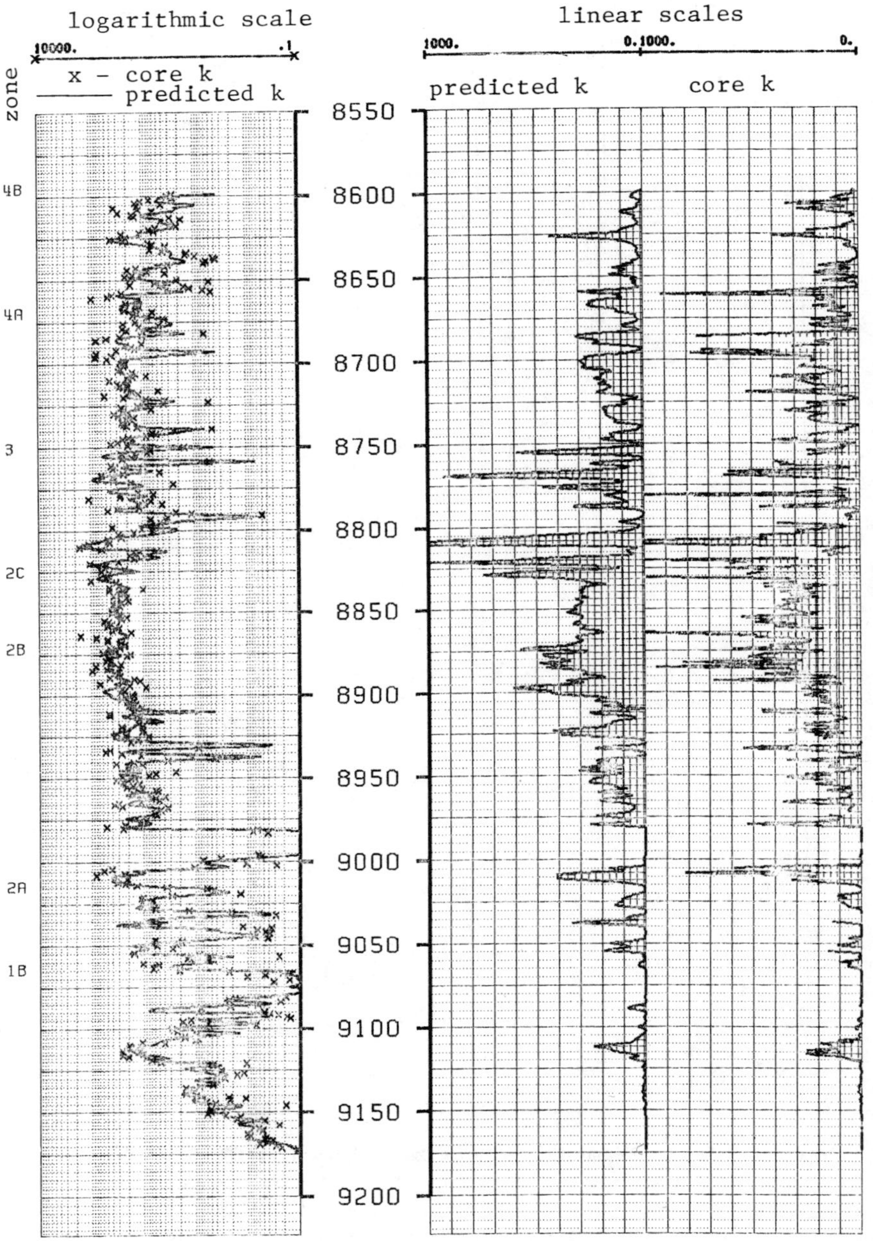

Figure 21. Logarithmic and linear scale plots of core and predicted permeabilities for a NWFB cored well which was included in the regression. Permeability increases to the left in each track.

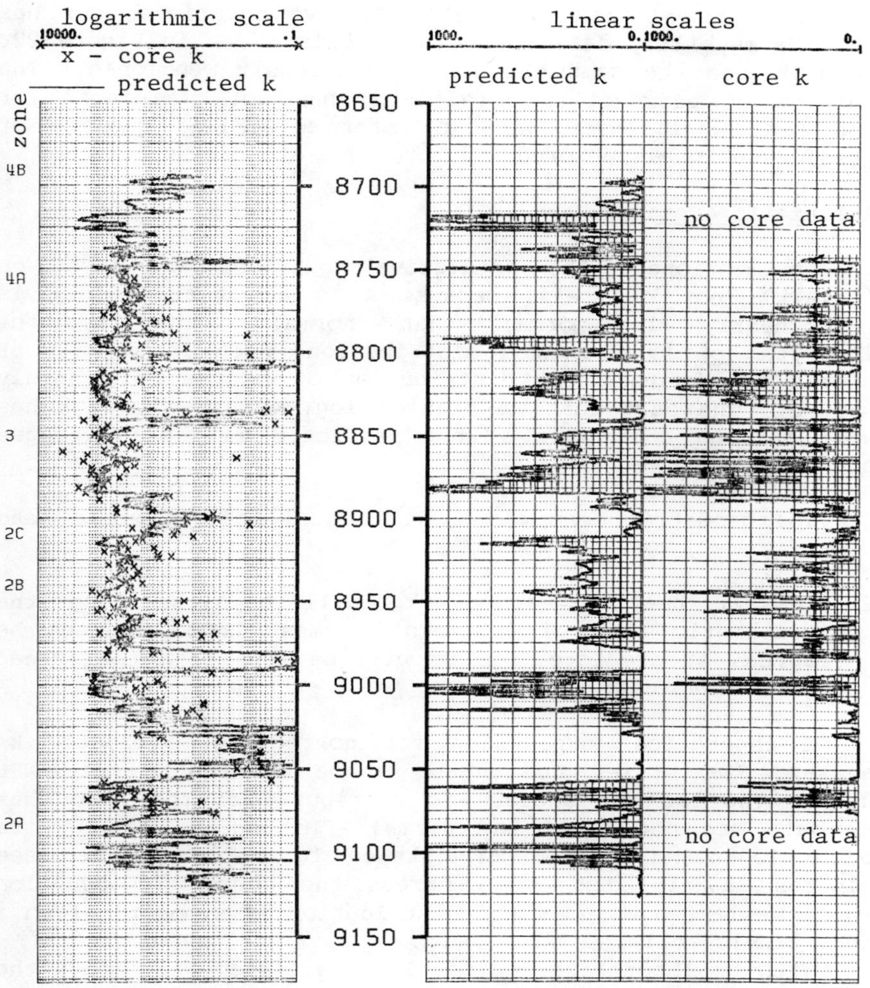

Figure 22. Logarithmic and linear scale plots of core and predicted permeabilities for the NWFB test well case. Permeability increases to the left in each track.

utilized in establishing the predictive equations, but has core permeability data. Such a well is shown in Figure 22. Here we see the same effects as previously mentioned. The quality of permeability prediction has not diminished for the test well; thus, we can infer a similar quality of prediction in all uncored wells.

Arithmetic Averages

As mentioned, the predictive equations are based on $\log_{10}(k)$, not k itself, because k is a variable which is more nearly "log normal" than "normal". That is, the logarithm of k assumes a distribution which is normal, or nearly so, whereas the distribution of k itself is highly skewed. The fact that we predict $\log_{10}k$ rather than k has some serious consequences when predicted and core averages are compared:

1. The geometric mean of k_p will equal the geometric mean of k_c.

2. The arithmetic mean of k_p will be <u>less than</u> the arithmetic mean of k_c. This comes about because the spread of predicted values will be less than the spread of core values.

A theoretical expression for correcting arithmetic k_p averages was tried and found to be inadequate for NWFB data. The lack of success is attributed to the fact that k_c data are not truly log normal. Histograms of k_c data on a logarithmic scale are skewed to the low side (see Figures 11, 16, and 20), whereas the correction equation which we tested assumes the distribution is symmetrical on a logarithmic scale.

With no theoretical correction available, we compare the averages of k_c and k_p. Figure 23 shows the layer arithmetic averages for east and west regressions. It can be seen that k_c averages are 15% higher than k_p in the east and 19% higher in the west. To correct the data, k_p should be multiplied by 1.15 and 1.19 in the east and west, respectively. The correction is only approximate as some layers deviate considerably from the best-fit line.

Other corrections to core data which need to be applied are the Klinkenberg correction and a correction for overburden pressure. Typically, they require that the core data should be multiplied by a factor ranging from 0.7 to 0.8 for Prudhoe Bay. Incomplete core cleaning could conceivably produce air permeability values which are too low, but this possibility has not been examined.

VIII. COMMENTS/CONCLUSIONS

The advantages of multiple regression methods to establish permeability prediction equations follow:

1. Regression methods are familiar to many workers and have gained wide acceptance for reservoir description applications.

2. The statistical methods are coded in accessible software packages.

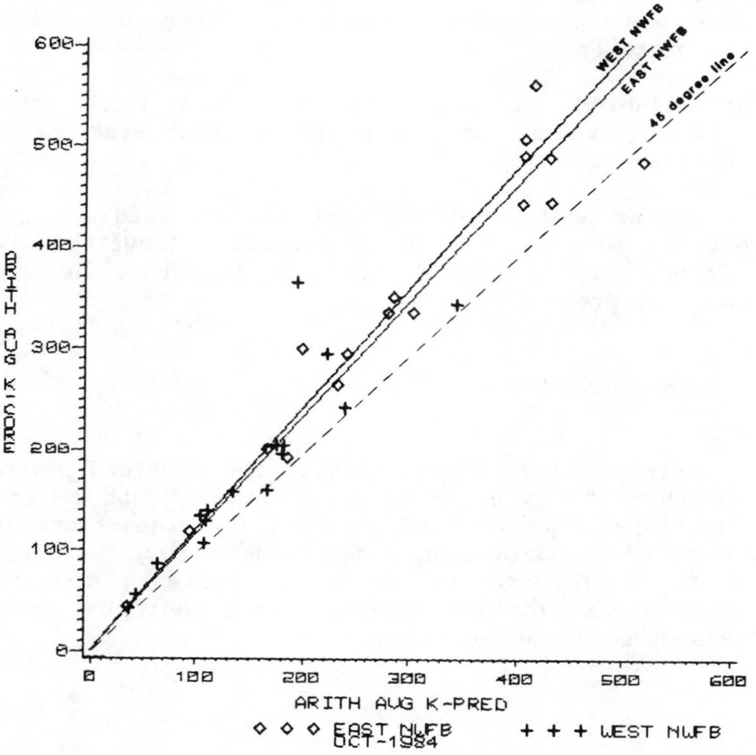

Figure 23. Artihmetic means of core versus predicted permeabilities, by regression layer for the eastern and western NWFB cored wells; least squares fit lines have slopes of 1.15 and 1.19, respectively.

3. Regression methods provide an accurate estimate of averages.

4. When porosity alone does not provide a good estimate of permeability, other variables may enter the equation to improve the prediction.

However, as we have discussed, there are disadvantages to the procedures we have pursued for the NWFB:

1. Their application requires availability of a comprehensive data base of well log and petrological data (rather than porosity alone).

2. Overprediction of low values and underprediction of high values result from the regression with log k and must somehow be compensated in order to get good estimates of the extremes.

3. Establishing the predictive equations is a laborious task and it must be done again for each area where core control exists.

No matter what predictive methods are used, we need to improve our methods of quality control. Inevitably, as we have found, quality control questions depend on the goals of permeability prediction.

IX. ACKNOWLEDGMENTS

J. Perrin and E. Piekenbrock of SPC provided petrological guidance for the permeability study. Bob Metzger of SAPC initiated work on NWFB permeability predictions in the summer of 1983 and offered suggestions during the course of the work. R. Mohundro of SPC and J. Burke of McCord-Lewis Energy Services provided valuable input regarding frequency analysis of core and log data.

X. REFERENCES

Beard, D. C. and P. K. Weyl (1973). Influence of Texture on Porosity and Permeability of Unconsolidated Sand, Bull. AAPG, v. 57 n. 2.

Bos, M. R. E. (1982). Prolific Dry Oil Production from Sands with Water Saturations in Excess of 50%, A Study of a Dual Porosity System, paper BB in SPWLA Trans.

Etnyre, L. M. (1984). Practical Application of Weighted Least-Squares Methods to Formation Evaluation. Part I: The Logarithmic Transformation of Non-linear Data and Selection of Dependent Variable. The Log Analyst, v. XXV n. 1.

Fuchtbauer, H. (1967). Influence of different Types of Diagenesis on Sandstone Porosity, Seventh World Petroleum Congress, Proc v. 2.

Lanning, E. N. and D. M. Johnson (1983). Automated Identification of Rock Boundaries: An Application of the Walsh Transform to Geophysical Well-Log Analysis, Geophysics v. 48 no. 2.

Lucia, F. J. (1983). Petrophysical Parameters Estimated from Visual Descriptions of Carbonate Rocks: A Field Classification of Carbonate Pore Space, J. Petr Tech., v. 35, n. 3.

Nie, N.H. et al (1975). Statistical Package for the Social Sciences, McGraw Hill.

Shenhav, H. (1971). Lower Cretaceous Sandstone Reservoirs, Israel: Petrography, Porosity, and Permeability, Bull. AAPG, V. 55, N. 12.

TABLE 1: This is the general mass species conservation equation.

425 7/11

MATHEMATICAL FORMULATION: ISOTHERMAL, MULTIPHASE FLUID-FLOW IN POROUS AND PERMEABLE MEDIA:

$$\frac{\partial}{\partial t}\left(\sum_{j=1}^{M} \phi\rho_j S_j w_{ij} + (1-\phi)\rho_s w_{is}\right)$$

$$+ \vec{\nabla}\cdot\left(\sum_{j=1}^{M} \Big(\rho_j w_{ij}\overbrace{\big(-\lambda_{rj}\,\underbrace{\vec{\vec{K}}}_{\text{PERMEABILITY TENSOR}}\cdot(\vec{\nabla}P_j+\rho_j g)\big)}^{\text{CONVECTIVE TERM}}\right.$$

$$\left.\underbrace{-\,\phi\rho_j S_j\vec{\vec{K}}_{ij}\cdot\vec{\nabla}w_{ij}}_{\text{DISPERSIVE TERM}}\Big)\right) = \sum_{j=1}^{M}\left(\phi S_j r_{ij} + (1-\phi)r_{is} + Q_{T_{ij}} - Q_{P_{ij}}\right)$$

HERE,

j:1,...,M (PHASES)

i:1,...,N (COMPONENTS)

TABLE 2 THE PERMEABILITY TENSOR:

PERMEABILITY FOR A POROUS MEDIUM HAVING ORTHOGONAL PRINCIPAL AXES IS A SECOND-RANK SYMMETRIC TENSOR IN 3-D SPACE:

$$\vec{\vec{K}} = \begin{bmatrix} K_{11} & K_{12} & K_{13} \\ K_{21} & K_{22} & K_{23} \\ K_{31} & K_{32} & K_{33} \end{bmatrix} \qquad (*)$$

FOR RESERVOIR SIMULATION PURPOSES NORMALLY WE RESTRICT OUR ATTENTION TO A CARTESIAN COORDINATE SYSTEM IN 3-D WHERE:

$$K_{11} \equiv K_{xx},\ K_{12} = K_{21} = K_{xy},\ K_{22} = K_{yy},\ K_{13} = K_{31} = K_{xz},\ K_{32} = K_{23} = K_{yz},$$

$$\text{AND } K_{33} = K_{zz}$$

WE NORMALLY ASSUME THAT THE PRINCIPAL AXES OF PERMEABILITY COINCIDE WITH THE DIRECTION OF THE COORDINATE SYSTEM. IN THIS CASE THE OFF-DIAGONALS IN (*) VANISH AND $\vec{\vec{K}}$ FOR ANISOTROPIC MEDIA BECOMES:

$$\vec{\vec{K}} = \begin{bmatrix} K_x & 0 & 0 \\ 0 & K_y & 0 \\ 0 & 0 & K_z \end{bmatrix}$$

IN ISOTROPIC MEDIA: $K_x = K_y = K_z$

425 8/11

II. THE DATA AVAILABLE FOR PERMEABILITY ESTIMATION

A. The Scale Problem

Fluid Mechanics provides us with the tools to derive the details of a fluid's behavior within the pores. For example, we may use the Navier-Stokes equations for the flow of a viscous fluid to determine the velocity distribution of the fluid in the pores. However, in a reservoir simulation, we are more interested in what happens at the macroscopic than at the microscopic scale. In order to properly formalize the transition from microcopic to macroscopic level, is essential to define the minimum volume over which an average can be performed. This volume is often called the Representative Elementary Volume (REV). The REV should be smaller than the entire flow domain, it should be larger than the size of a single pore, and contain a sufficient number of pores to permit the meaningful statistical average required in the continuum concept. With porosity for instance, the REV is the volume ΔVo at which the ratio $\Delta V_p/\Delta V$ (ΔV total volume, ΔV_p volume of the pores) starts varying erratically, because of the individual effects of grains and pores (Figure 1).

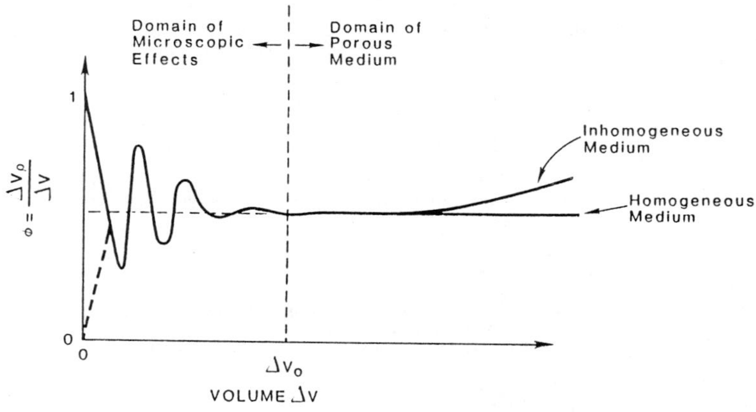

FIGURE 1 : DEFINITION OF POROSITY AND REPRESENTATIVE ELEMENTARY VOLUME (From Bear, 1968)

By averaging properties on volumes, we can describe flow within a porous medium by means of partial differential equations. These equations are, at the macroscopic level, direct consequences of fluid-mechanics equations at the microscopic level. For instance, Darcy's law can be derived from the Navier-Stokes equations (Hubbert, 1957).

In theory, the macroscopic parameters can be expressed in terms of the microscopic information. In practice, they are deduced from numerical experiments in which the various phenomena related to these parameters are observed. Any instrument used to measure macroscopic properties has a "window" of a certain size through which it is in contact with the porous medium. The instrument "sees" or averages a certain property over a certain porous medium around it.

B. Data Representative of Different Scales

H. Haldorsen (1985) mentions that, in the databases, it is a common practice to associate data to volumes they are not representative of. For instance, plug values are very often associated to one-foot long intervals, although the permeability of the plug may or may not be representative of the one-foot interval. Let us briefly summarize the orders of magnitude given by H. Haldorsen (1985):

There are about 133 plugs contained in a one-foot core interval. "Whole core" samples of varying lengths are representative of another volume, which can be two hundred times the volume of a plug.

In wells which are cored and logged, it is common to establish a correlation between the sonic travel-time and the core porosity measured at the same depth. The volume of investigation of the sonic log does not include the volume of the core, and is about one thousand times the volume of a plug.

Permeability estimates are also derived from the pressure response during pressure build-up testing. The radius of investigation of such a test can easily be several thousand feet, and the volume tested normally contains more than one petrophysical zone. When a pressure build-up result is compared to the average core-permeability at a given well, the volume of investigation of the build-up can be ten billion times larger than the volume of the cores going into the calculation of the average core permeability!

C. Permeability Averaging

In order to adequately represent the disparate movement of displacement fronts through a reservoir, it is imperative that proper knowledge of layer horizontal permeabilities exists.

Cardwell and Parsons (1945) define the equivalent permeability of a reservoir segment as "the permeability of a homogeneous segment of the same dimensions that would pass the same flux under the same pressure drop". The problem of estimating the equivalent permeability of a heterogeneous layer whose thickness is considerably greater than the core sampling interval involves an averaging scheme based on some assumptions about the nature of the heterogeneity of the formation. The two common models based on the supposition of a flow either perpendicular or parallel to parallel layers lead to the harmonic and arithmetic mean respectively (Cardwell and Parsons, 1945). The harmonic average is always smaller than the arithmetic average. It represents the "worst" configuration, while the arithmetic average indicates the "best one". Matheron (1967) has proved that the equivalent permeability always lies between these two extreme cases. The equivalent permeability K over a volume V is usually calculated from the elementary permeabilities k_i inside V using the generic expression:

$$f(K) = \text{Arithmetic average of the } f(k_i)$$

where the function f determines the nature of the average:

f = identity function for arithmetic average
f = inverse function for harmonic average
f = logarithm function for geometric average

Matheron (1967) explains that the more alternate routes the flow can find around the bad permeabilities, the higher the equivalent permeability is likely to be. For instance, the equivalent permeability is closer to the arithmetic average in three dimensions than in two dimensions, because the flow can find more ways around the bad permeabilities in three than in two dimensions.

TABLE 3 FACTORS TO CONSIDER WHEN COMPARING CORE PERMEABILITIES AND PERMEABILITIES DERIVED FROM PRESSURE TRANSIENT TESTS

FACTORS TO CONSIDER	EFFECTS
CORE PERMEABILITIES ARE NOT CORRECTED FOR THE KLINKENBERG EFFECT.	K_{CORE} TOO HIGH
CORE PERMEABILITIES ARE NOT CORRECTED FOR OVERBURDEN PRESSURE	K_{CORE} TOO HIGH
CORE PERMEABILITIES ARE NOT CORRECTED FOR TEMPERATURE EFFECTS.	$K_{CORE} \neq K_{CORE}$ IN SITU
CORE PERMEABILITIES MAY BE UNREPESENTATIVE OF IN-SITU CONDITIONS DUE TO CONTACT WITH REACTIVE FLUIDS, MUD-FILTRATE INVASION , SWELLING/DISPERSIVE CLAYS AND LABORATORY PREPARATORY PROCEDURES.	$K_{CORE} \neq K_{CORE}$ IN SITU
IN-SITU FRACTURES UNDETECTED IN THE LAB.	$K_{PBU} > K_{CORE}$
UNCONSOLIDATED SECTIONS CHANGE COORDINATION NUMBER DUE TO CHANGE IN PORE-PRESSURE.	$K_{CORE} \neq K_{CORE}$ IN SITU
K_{PBU} IS ACTUALLY $K_{ABS} \times K_{ro}$ (S_o), $(K_{ro} < 1)$	$\bar{K}_{CORE} > K_{PBU}$
MISSING CORE DATA (UNCONSOLIDATED SECTIONS), HIGHEST PERMEABILITY SECTIONS ARE LOST AND THEREFORE NOT INCLUDED IN THE AVERAGE.	$\bar{K}_{CORE}$ TOO LOW
PBU-ANALYSIS TECHNIQUE MAY BE ERRONEOUS. IN PARTICULAR, FOR WELLS WITH LOW (HP/H)-RATIOS AND HIGH PERMEABILITIES, SPHERICAL/HEMISPHERICAL/LINEAR AND RADIAL FLOW GEOMETRICS SHOULD ALL BE CONSIDERED.	$K_{PBU,\ RADIAL,} \neq K_{PBU,}$ (SPHERICAL / HEMISPHERICAL / LINEAR
CORE AVERAGING TECHNIQUES : GEOMETRIC, ARITHMETIC, HARMONIC TECHNIQUES YIELD DIFFERENT AVERAGES.	$\bar{K}_{HARM} < \bar{K}_{GEOM} < \bar{K}_{ARITH}$
THE MORE PROFOUND PROBLEM OF COMPARING AN AVERAGE MACROSCOPIC MEASURE TO A MEGASCOFIC MEASURE.	$K_{MEGA(PBU)} = f(K_{I,\ CORE});\ I = 1,\ N$
WHEN COMPARING $(KH)_{CORE}$ TO $(KH)_{PBU}$: THE ACTUAL PRODUCTION INTERVAL MAY BE DIFFERENT FROM THE ESTIMATED PRODUCTION INTERVAL. THIS CAUSES A COMPARISON BETWEEN AVERAGE CORE - PERMEABILITY IN ONE SECTION (ESTIMATED) TO AN AVERAGE IN ANOTHER SECTION (ACTUAL), IT IS REASONABLE TO EXPECT THAT THESE TWO VALUES DIFFER IN MAGNITUDE.	$(KH)_{CORE} \neq (KH)_{PBU}$

425 9/11

$\bar{K}$ = CORE AVERAGE

K_{BU} = PERMEABILITY FROM PRESSURE BUILD -UP TEST

In the present work, we are interested in averaging horizontal permeabilities along a well. This pattern is similar to the pattern of a flow parallel to parallel layers of stratification. The equivalent permeability is probably close to the arithmetic average. This has been confirmed by pressure build-up tests which showed that, at Prudhoe Bay arithmetic averages correlated better than geometric and harmonic averages with these tests. Consequently, in the mapping of layer permeabilities at Prudhoe Bay, we used the arithmetic averages.

D. Measurement Errors on the Data

We mentioned above that core measurements and permeability from pressure build-up analysis were very hard to compare because they were representative of volumes of different orders of magnitude. The other problem is that core-data are laboratory measurements, while pressure data are "in-situ" data. Core-measurements are not representative of the in-situ conditions. The most important problem to consider are differences due to overburden pressure and gas slippage effects. Many other important effects should also be considered (Table 3).

When the core is removed from the formation, all the confining forces are removed, and the rock-matrix expands, changing the fluid-flow paths inside the rock. Since some formations are more compressible than others, a significant amount of data is required to develop empirical correlations which will permit the correction of surface permeabilities for overburden pressure. And any applied correction will still be an approximate correction.

Usually, air-permeability is measured in the laboratory. Air-permeability differs from liquid-permeability. Klinkenberg (1941) proposed an explanation to this phenomena and provided an approximate relationship between air permeability and liquid permeability. Again, the liquid permeability obtained from this relationship is an approximation.

We must also mention that, although permeability from pressure build-up analysis is representative of the in-situ conditions, it is still an approximation: in particular, some assumptions have to be made about the thickness of the production interval, and about the nature of the flow.

E. Permeability Values Predicted From Other Data

On non-cored wells, permeability has to be predicted from other variables, the most useful ones being those measured on the logs. The most often sought after relationship has always been permeability and porosity. In the early attempts (up to about 1933) of the oil industry to model the reservoirs, no distinction was even made between permeability and porosity, implying proportionnality between the two variables.

A simple consideration of theoretical possibilities of the structures of porous media shows that a universal relation between porosity and permeability cannot exist. Two porous media of the same porosity can have different permeabilities. The volume of the pores certainly has an influence on permeability, but their interconnection is a dominant factor too. Hence, the idea to use other variables for predicting permeability. A correlation that can be looked for is between structure and permeability. The meaning of "pore structure" is somewhat arbitrary. One may understand that this means the "pore size distribution" as obtained from thin-plate measurements of the "granulometry of the pores" (Matheron, 1967) or from experimental capillary pressure curves. For unconsolidated sands, Beard and Weyl (1973) showed that initial permeability is dependent on grain size, sorting and porosity. An attempt has been made to introduce further parameters, such as "angularity", "roundness" (Tickell and Hiatt, 1938) or orientation of the grains. All these studies give a better understanding of the factors controlling permeability, but they also show that it is an illusion to look for a "universal" relation between permeability and other variables. Matheron (1967) recalls that permeability is a <u>tensor</u> (see Table 2), and as such cannot be explained entirely by scalar variables such as porosity, pore size distribution or sorting.

It seems that the relations between permeability and other variables should be interpreted using a statistical rather than deterministic formalism. One can predict the <u>average</u> permeability corresponding to a give pore structure, but in each particular case the actual permeability will vary about this average in a statistical manner. Hence, the advantage of the "regression approach" which tries to predict a <u>conditional average</u>, or expectation of permeability corresponding to a given set of parameters (Draper and Smith, 1980). Nevertheless, the previous empirical studies are important for guiding the choice of the dependent variables. Other important studies are those representing the porous medium by a simplified model, and from this model find the exact relations between permeability and the other parameters.

F. Kozeny Theory (1927)

Kozeny's work is today a widely accepted explanation for the permeability as conditioned by the geometrical properties of a porous medium. Kozeny's theory represents the porous medium by an assemblage of channels of various cross sections but of definite length. The Navier-Stokes equations are solved simultaneously for all channels passing through a cross-section normal to the flow in the porous medium. Finally, the permeability is expressed in terms of the specific surface of the porous medium, which is a measure of a properly defined (reciprocal) hydraulic radius. However, certain aspects are neglected, notably by assuming that a cross-section normal to the channel there is no tangential component of the fluid velocity. The theory therefore neglects the influence of conical flow in the constrictions and expansions of flow channels, just as capillaric models do. The resulting expression for permeability is:

$$K = \frac{C\,\Phi^3}{\tau\,S^2}$$

where C is Kozeny's constant, S is the surface area per unit volume, Φ is the porosity, and τ is the tortuosity.

G. Modern Empirical Permeability Correlations

Most of the modern permeability relationships were derived from Kozeny's equation. They are generally developed for use with log-derived values of porosity, Φ, and irreducible water saturation, Swirr, and include a modification term to accomodate hydrocarbons of different gravity and phase, or density. As an example, the Tixier relationship (Tixier, 1949):

$$K^{0.5} = C\,\frac{\Phi^3}{Swirr}$$

The constant C is dependent on the hydrocarbon nature. Tixier suggested C = 79 for gas. Raymer proposed a correction factor C which is a function of capillary pressure and irreducible water saturation. This factor corrects the irreducible water saturation at a given capillary pressure so that it is equal to the water saturation at the common reference used in Tixier's relationship. Note that Tixier's method is not statistical. It can be used when there are no core-measurements of permeability available. When some core data are available, it is important to check how core-permeability behaves in relation to the other variables. In that case Kozeny's or Tixier's relationship can be a starting point in establishing a relation. The various parameters of these equations should be modified to fit the data as well as possible. For instance, Brown and Husseini (1977) found through multiple linear regression that the optimum permeability prediction for the Shaybah field (carbonate) was:

$$K = \frac{57\ \Phi^{0.86}}{Swirr^{1.26}\ P_c^{0.89}}$$

where use of the equation is restricted to the intervals of irreducible water saturation (P_c is the capillary pressure). Allen (1982) obtained the following relationship between (overburden and Klinkenberg effect corrected) core permeability and several independent parameters:

$$\mathrm{Log}_e(Kcorr) = b_0 + b_1\ \rho_b + b_2\ \Phi_n + b_3\ \Delta t + b_4\ GR$$

where Kcorr is equal to 0.62 $Kair^{1\ 04}$, ρ_b is the bulk density, Φ_n the neutron porosity, Δt the sonic travel-time, and GR the gamma-ray response.

H. The Multidimensional Histogram

The approach of the multidimensional histogram (Schlumberger, 1978) is based on the same assumption as multivariable linear regression: at any depth level in a given well where there is a set of log data, for instance bulk density, sonic travel-time and gamma-ray, the probability is high of having statistically the same permeability compared to that derived from cores in cored wells for which bulk density, sonic travel-time and gamma-ray have the same values. Thus, in a non-cored well, a good estimate of permeability would be the (frequency-weighted) average of all core permeabilities corresponding to the same values of bulk density, sonic travel-time and gamma-ray. In fact, this estimate is statistically nothing else than the conditional expectation of permeability, when bulk density, sonic travel-time and gamma-ray are fixed. It can be used when there are enough data to build the frequency-weighted average. Linear regression is generally used because there are not enough data: the lack of data is replaced by an extra assumption of linear dependence of the conditional expectation relative to the log-measurements (this assumption is theoretically valid only for multi-normal distributions). The multidimensional histogram is very appropriate when there are enough data to calculate each conditional average relative to intervals of the log-data which will be used for permeability prediction.

I. Permeability From Other Log Variables

Various studies, only applicable to the fields they were run on, have found different kinds of relations between permeability and log variables. For instance, Ogbe and Bassiouni (1978) used an interesting approach to predict permeability from formation resistivity. Rabe (1957) related permeability to gamma radiation, while Kolodizie (1980) used a linear equation to estimate permeability from the normalized SP log.

J. Conclusion

The goal of this overview was to show that all the values of permeability, cored ones as well as predicted from logs, are estimates of what the real value is: for instance, cored data are dependent on the sample size, predictions from logs are never exact. And adding to the uncertainty is the averaging problem. This uncertainty must be taken into account. The geostatistical formalism, which is going to be described below, provides the tools to deal with these approximated values.

III. THEORETICAL CONSIDERATIONS ABOUT GEOSTATISTICS

The theory of Geostatistics has already been presented by many authors (Journel and Huijbregts, 1978, Matheron, 1971). The section below emphasizes the points which are relevant to petrophysical variables in general, and permeability in particular. The main reference for this overview is Matheron, 1971.

A. The Framework of Random Functions

Many earth-science variables present two main characteristics: there is some randomness in their behavior, but at the same time there is some continuity.In particular, knowing the value at one point x gives some information about the values in the neighborhood of x (continuity), but not enough to exactly predict what these values are (randomness).

Geostatistics takes into account the randomness by considering the value z(x) of the variable z at a point x as the realization, or the outcome, of a random variable Z(x). The continuity element is given by the variogram $\gamma(h)$ which associates to each distance h a measure of the difference between values a distance h apart:

$$2\,\gamma(h) = \text{Variance}\,(Z(x+h) - Z\,(x))$$

An experimental variogram is first calculated from the data, then fitted by a simple mathematical function. This function then represents the degree of continuity of the variable under study.

B. The Support Effect

The main problem Matheron (1971) wanted to solve in his original works was not the simple "point interpolation" problem, but rather the "support effect". Mining engineers used to estimate the average ore grade of a block by the grade of the core sample centered on the block. Then they applied their cut-off grade to the histogram of the estimated blocks, which was nothing else than the histogram of the samples (Figure 2).

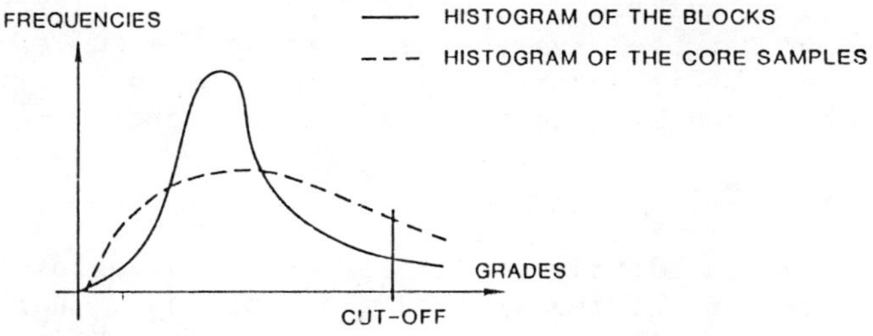

FIGURE 2: ILLUSTRATION OF THE "SUPPORT EFFECT."
THE AREA AT THE RIGHT OF THE CUT-OFF IS THE ESTIMATED RECOVERY: IT IS OVERESTIMATED IF THE HISTOGRAM OF THE CORE SAMPLES IS USED TO REPRESENT THE HISTOGRAM OF THE BLOCKS.

As a result, the block-recovery (percent of blocks with a grade above the cut-off) was always overestimated, because the method did not take into account the fact that the variance of the blocks is always smaller than the variance of the samples. Matheron calls this the "support" effect, the support being defined as the volume on which a given variable is measured. The support is nothing else than the volume over which the measuring instrument averages the measured property (see above). To correct for the support effect, Matheron proposed to estimate the average grade of the blocks:

$$z(V) = \frac{1}{V} \left(\int_V z(x)\, dx \right)$$

by a weighted average of a certain number of samples in, but also around the block.

$$z^*(V) = \Sigma\, \lambda^\alpha\, z(x_\alpha)$$

In order for $z^*(V)$ to be an "unbiased" average, the sum of the λ^α has to be equal to one. Under this condition, $Var(Z^*(V) - Z\,(V))$ can be developed according to the λ^α and to the variogram (known, through the model fitted to the data). By minimizing this expression in the λ^α, one gets the Kriging system. Notice that Kriging is nothing else than a multivariable regression of $Z(V)$ on the $Z\,(x_\alpha)$ (Draper and Smith, 1980). The regression is under the condition that the sum of the λ^α is equal to one. Because of this constraint, only the covariance between differences, that is the variogram, is needed. In fact,

$$\mathbf{Var}\ (\mathbf{Z(V)} - \Sigma\, \lambda^\alpha\, \mathbf{Z(x_\alpha)})$$

can be calculated for the difference between $Z(V)$ and any kind of weighted average, as long as the sum of the weights is one. It is interesting to evaluate this variance when the weighted average is itself an integral:

$$Var\ (Z(V) - Z(v))$$

($Z(v)$ can evidently be interpreted as a weighted average with the sum of the weights equal to one). Here $Z(V)$ may be for instance the average on a core, while $Z(v)$ is the average on a plug inside the core. As a result, knowing the variogram of the plugs only, one is able to estimate how the plug is representative of the whole core. $Var(Z(V) - Z(v))$

is called the "extension variance" of v into V. To calculate this variance, one needs to know the exact position of v inside V. This is generally unknown with plugs, for the laboratory does not usually provide this information. In that case, the extension variance can be replaced by the "dispersion variance" $\sigma^2(v|V)$, which is the average value of the extension variance when v varies randomly inside V. $\sigma^2(v|V)$ can be given another interpretation. Assume V is divided into n disjoint volumes v_i all equal to v: $V = U(v_i)$, and calculate $z(v_i)$ on each volume. Then $\sigma^2(v|V)$ is an estimate of the experimental variance of the n values $z(v_i)$. Matheron proved the remarkable "additivity relationship", for $v_3 \supset v_2 \supset v_1$:

$$\sigma^2(v_1|v_3) = \sigma^2(v_1|v_2) + \sigma^2(v_2|v_3)$$

The most simple application of this relationship is the case where v_1 is the size of small samples (such as core plugs), v_2 the size of bigger samples (such as full-diameter cores), and v_3 is the volume of the whole field:

$$\sigma^2(v_2|v_3) = \sigma^2(v_1|v_3) - \sigma^2(v_1|v_2) < \sigma^2(v_1|v_3)$$

which is exactly what was shown by the histogram of core samples versus block samples: the variance decreases when the size of the volume increases. What is remarkable is that Geostatistics provides the theoretical framework in which these empirical relations can be formalized.

Unfortunately, the additivity relationship are only valid for "additive" variables. In other words, the support effect relations are only valid when the value of the variable over a block is the arithmetic average of the individual values inside the block:

$$z(V) = \frac{1}{V}\left(\int_V z(x)\,dx \right)$$

We saw earlier that this is not true for permeability. But we also saw that the following approximation is often made:

$$f(z(V)) = \frac{1}{V}\left(\int_V f(z(x))\,dx \right)$$

Which means that the variable f(z(x)) is additive. Codreanu (1968) used for f the logarithm function, and applied the additivity relation to the logarithm of permeability: v_1 was the volume of cores, v_2 the volume of pressure build-ups an v_3 the volume of the whole field. Codreanu used the ratio between $\sigma^2(v_2|v_3)$ and $\sigma^2(v_1|v_3)$ as a measure of the heterogeneity of permeability at the scale of the distance between wells.

C. Error Cokriging

Cokriging is a method which has been successfully used in mining problems (Journal and Huijbregts, 1978) and which is very promising for petroleum applications. In lead-zinc orebodies, it can happen that one of the variables is better sampled that the other. Lead and zinc are usually well correlated, and the knowledge of the zinc (resp. lead) grade gives some information about the unknown lead (resp. zinc) grade. Thus, when estimating the value of the lead grade L(x) at a point x, it is good to use not only the available $L(x_\alpha)$ at the surrounding data points (as Kriging would do), but also the available values $Z(x_\beta)$ of the zinc grade:

$$L^*(x) = \Sigma \lambda^\alpha L(x_\alpha) + \Sigma \nu^\beta Z(x_\beta)$$

This method can be used for instance to combine the information provided by the core porosity (known at cored wells only) with the sonic travel time (known at cored and logged wells) in order to map porosity field-wide. What has been used for permeability mapping at Prudhoe Bay is Error Cokriging, which is a little simpler than the general Cokriging.

We know that the data available for permeability mapping do not have the same degree of reliability. Each datum $Y(x_\alpha)$ can be considered as equal to:

$$Y(x_\alpha) = Z(x_\alpha) + \varepsilon(x_\alpha)$$

where $Z(x_\alpha)$ is the exact (unknown) value of permeability, and $\varepsilon(x_\alpha)$ is a random error. The variance of $\varepsilon(x_\alpha)$ can be different for each value of permeability. When mapping permeability, the goal is to estimate the variable Z(x) using the information provided by the variable Y(x): this is a Cokriging problem! Each Z(x) will be estimated by a weighted average of the $Y(x_\alpha)$ at the surrounding wells:

$$Z^*(x) = \Sigma \lambda^\alpha Y(x_\alpha) = \Sigma \lambda^\alpha (Z(x_\alpha) + \varepsilon(x_\alpha))$$

Var $(Z^*(x) - Z(x))$ can be expressed using the weights, the variogram and the variance of the errors $\varepsilon(x_\alpha)$, and then minimized according to the weights, which produces the "Error Cokriging" system (see for instance Delhomme, 1974).

So, knowing what the variogram is, and what the uncertainty is on each data point, Error Cokriging can estimate permeability. Note that this possibility to weigh the data according to their reliability gives to Error Cokriging a definite advantage over the other mapping methods.

IV. A PRACTICAL APPLICATION

Permeability has been mapped on 7 lithological units of the Sadlerochit formation of the Prudhoe Bay field (see a similar work by Wadman et al., 1979). As an example, we describe below the work performed on zone 4A, which is about 80 feet thick.

The dominant lithology of zone 4A is fine grained sandstone believed to have been deposited in a braided fluvial environment. Some coarser grained sandstones and conglomerates are frequently present near the base of zone 4A. Also present in subordinate amounts are siltstone/shale units (internal communication, G. Geehan). The data available for permeability mapping were:

.Arithmetic averages of core-permeability on 62 cored wells (only wells with a recovery better than 50% were used). The non-pay intervals were not used in the average.

.Arithmetic averages of predicted permeabilities on non-cored wells, obtained using different kinds of predictors. In two areas of the field, two kinds of predictors were obtained, mainly using variables measured on the logs. For instance, for one area known as the "North-West fault block" (NWFB), a foot-by-foot regression was based on 8 cored wells and applied to 9 non-cored wells. The regression equation changed from one zone to the other. On zone 4A, the dependent variables were:

* Φ_t and $\log(\Phi_t)$, where Φ_t is porosity predicted from travel time
* Φ_c and $\log(\Phi_c)$, where Φ_c is Φ_t corrected for shales
* Φ_b the bulk density as obtained from the logs
* the subsea depth

* the gravel percent. The gravel percent is determined in two steps. Its presence/absence is first established by discriminant analysis. On zone 4A, presence of gravel was established. Then, if present, the gravel percent is estimated by linear regression for various log variables.

The predictor in the other area (FS3 area) was built in a similar fashion, but was less sophisticated than the NWFB one. All this work was performed by W. Wendt, S. Sakurai, and P. Nelson (see their paper in this volume). The regression was always performed on the logarithm of foot-by-foot permeabilities, then the antilog was taken and averaged over the pay-intervals of the zone.

The third kind of predictor was used for all the non-cored wells which were neither in NWFB nor FS3. It is based on a field-wide multiple regression of the logarithm of permeability against porosity, its square root, and its square. The work was performed by T. Lane.

A. Geostatistical Analysis of the Data

The analysis should be performed on data which are as reliable as possible. The 62 average core-permeabilities were used. Basic statistics are:

Minimum	:	98 millidarcies
Maximum	:	1317 millidarcies
Mean	:	444 millidarcies
Standard deviation	:	303 millidarcies

Figure 3 shows a map drawn from these data using an elementary mapping technique. The map does not pretend to be precise, but shows that permeability systematically increases towards the North-East of the field.

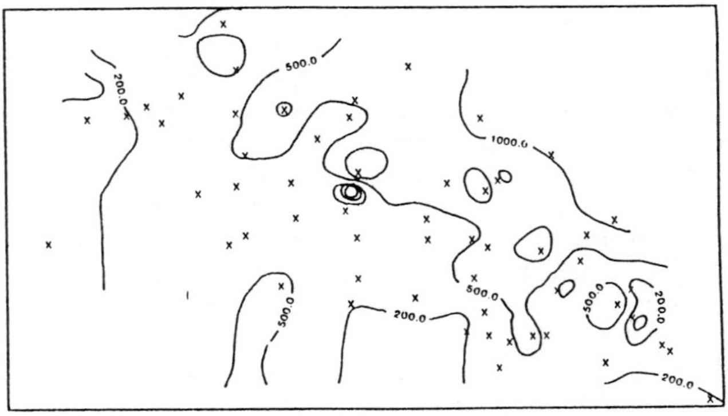

FIGURE 3: MAP OF PERMEABILITY OBTAINED FROM THE CORED WELLS ONLY.
THE DISPLAYED CONTOUR LINES ARE: 100, 200, 500, 1000 MILLIDARCIES (LOGARITHMIC INTERVAL)

425 3/11

The variograms shown on Figure 4 are in the NW-SE and NE-SW directions (tolerance 45°).

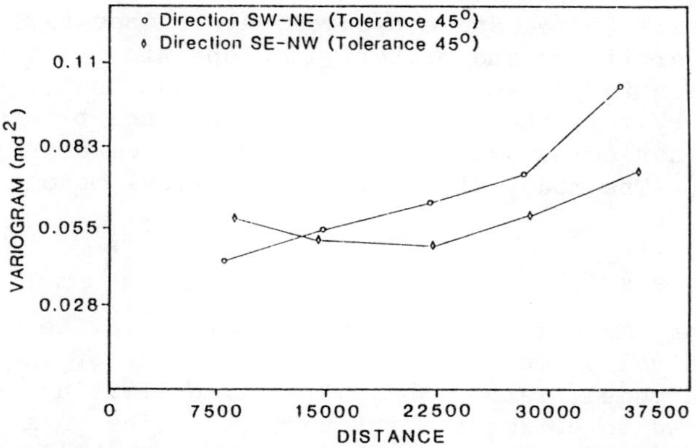

FIGURE 4: VARIOGRAMS (SW-NE AND SE-NW) CALCULATED ON THE LOGARITHM OF PERMEABILITY ON CORED WELLS, ZONE 4A.

They are calculated on the decimal logarithm of permeability, which makes them less sensitive to extreme permeability values. The lag of the variograms is 7500 feet, which is the average distance between cored wells. The two variograms behave the same way until a distance of about 15000 feet, where the SW-NE systematically increases under the influence of the trend while the SE-NW stays flat. The SE-NW variogram indicates that when the trend is not affecting the plot, there is no visible structure for permeability: two wells 30000 feet apart are not better correlated than two wells 7500 feet apart. This means that 7500 feet is greater than the scale of variation of permeability values: at 7500 feet, any kind of correlation is already lost between permeability values. We now need to geologically interpret the trend and the variograms:

* The trend increases towards the "Lower-Cretaceous Unconformity" (LCU), which bounds the field at the North-East. It is believed that, after deposition, diagenetic fluids came from the LCU and modified permeability. They enhanced permeability close to their source, while having little effect away from it. This is in agreement with the trend.

* Zone 4A is a braided streams environment. The original braided streams were probably only a few hundred feet wide, which might explain the short scale variability of permeability. We decided to choose a range of 2500 feet for the variogram. 2500 feet is about the average distance between wells (cored and non-cored) at Prudhoe-Bay. This choice is arbitrary and probably too optimistic: the scale of variation is probably still smaller than that. What we just mean by this choice is that the distance between wells, cored and non-cored, does not allow us to see any structure. The model chosen was a spherical model, with a 0.06 sill.

B. The Geostatistical Mapping

Kriging (Matheron, 1971) can directly provide estimates of average values on blocks. For the blocks of the simulation model, rather than directly estimating averages we preferred to estimate point permeabilities on a very dense grid (800 ft X 800 ft). This way, any kind of average on any kind of block could then be calculated, by just averaging the points inside the block.

Let us recapitulate the number of available data for mapping:

* 62 cored wells
* 27 FS3 wells
* 9 NWFB wells
* 595 wells with the field-wide predictor

We assumed that there was no error on average core-permeabilities. This means that the arithmetic average of the plugs inside a well is considered as representative of the true equivalent permeability. This is certainly not true, but we decided to neglect this error compared to the large errors on permeabilities predicted from logs. For these permeabilities, an estimate of the error variance had to be determined for each type of estimator: field-wide, FS3, and NWFB.

For the field wide estimator, the logarithm of actual permeabilities was plotted against the logarithm of predicted permeabilities, for the 62 cored wells. It was found that the predicted permeabilities were systematically too low. This happened because the regression was first built on the logarithm, and then the predicted values were obtained by taking the antilog. We corrected for this bias by readjusting the predicted values to the actual ones. The method was to recalculate a regression line between the 2 variables. This gave 2 results: the bias was corrected for, and an estimate of the error variance was obtained, through the variance of the residuals of the regression. The same kind of correction was applied to the FS3 and NWFB data, with a small difference: only 6 and 8 cored wells were available on each area which was not enough to build the regression line. We had to use datas from other zones of the Sadlerochit unit to calculate the regression. The final error variance obtained for field-wide, FS3 and NWFB predictors were respectively 0.03, 0.02 and 0.04. Figure 5 shows a cross-plot of the corrected estimates versus the core permeabilities on cored wells.

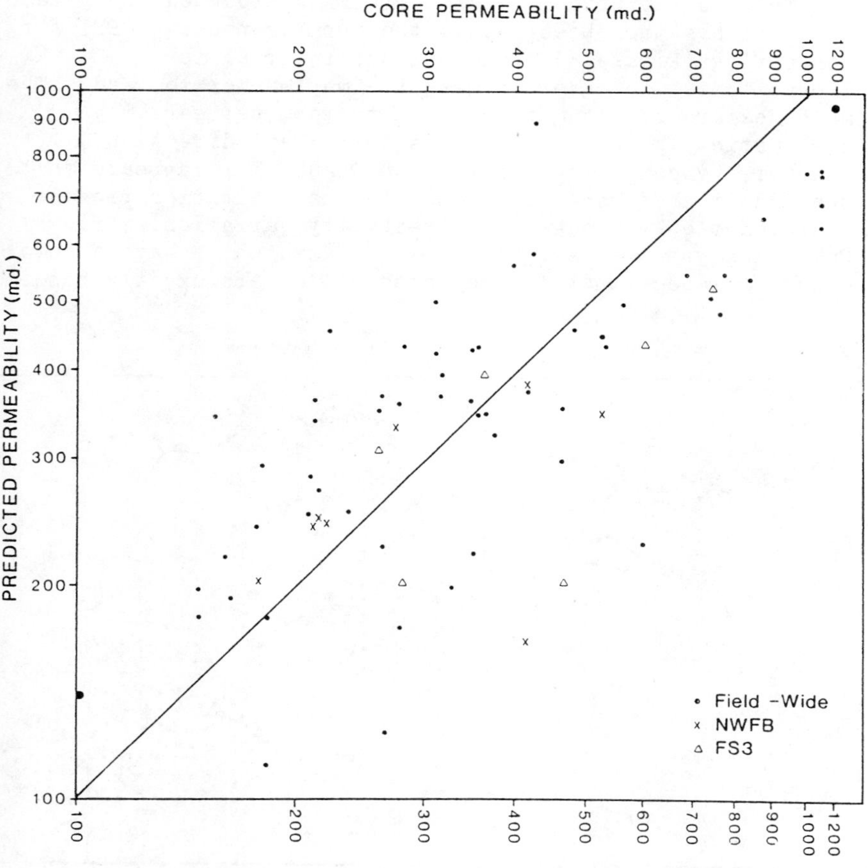

FIGURE 5. CROSS PLOT OF "PREDICTED PERMEABILITY" VERSUS "CORE PERMEABILITY" OBTAINED ON THE CORED WELLS AFTER CORRECTING FOR THE BIAS.

• = FIELD WIDE x = NWFB Δ = FS3

Using these variances, and the variogram model described above, the logarithm of permeability was interpolated on a set of grid nodes, each grid value being estimated by its 48 nearest neighbors. There are multiple reasons for mapping the logarithm rather than the raw values:

* By taking the antilog of the estimated logarithms one is sure to obtain positive estimates of permeability.
* The error variances and the variogram calculated earlier were based on the logarithms, which makes them less sensitive to extreme values than variances calculated on the raw permeabilities. In order to use them, we need to map the logarithms first.

The geostatistical package BLUEPACK produced the grid. The grid has then been fed to the mapping package CPSI for contouring (Figure 6). The contour interval is "logarithmic" in order to make the map easier to read. The main feature of the map is still there: permeability systematically increases in the North-East direction. However, by comparing Figure 6 to Figure 3 it appears that the addition of non-cored wells to the estimation process improved the knowledge of permeability variation. Error Cokriging works a very simple way: data which have a zero error variance, that is the cored wells, are exactly honored

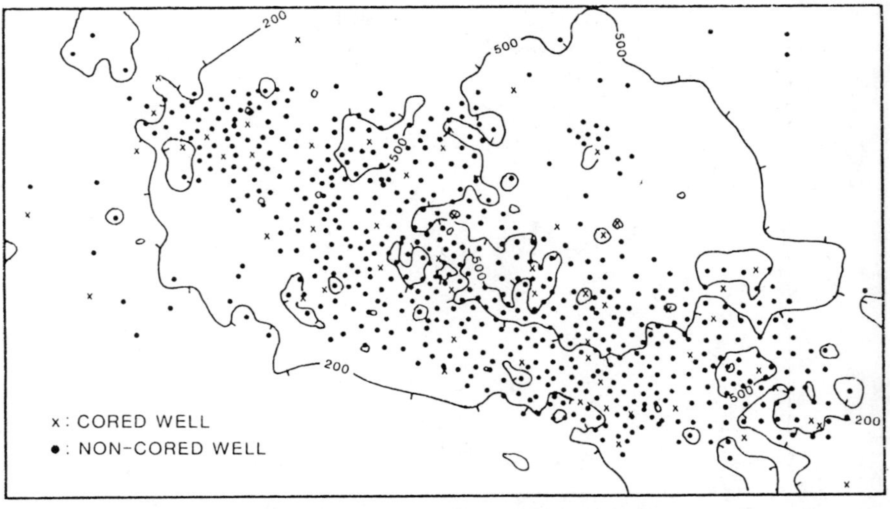

FIGURE 6 : MAP OF PERMEABILITY OBTAINED FROM ALL THE WELLS (CORED AND NON-CORED) BY "ERROR COKRIGING"

THE DISPLAYED CONTOUR LINES ARE: 100, 200, 500, 1000 MILLIDARCIES ("LOGARITHMIC" INTERVAL).

by the map. Data which have a non-zero variance are generally not honored: they are honored only if their value is similar to the values at the surrounding points. If not, they are corrected by the Cokriging algorithm, so that they look more like what happens in the neighborhood.

C. Conclusion

This paper had two main objectives: first of all insist on the fact that permeability data are never perfectly accurate. Second, that Geostatistics provides some tools to take the reliability of the data into account. In fact, it was shown that Geostatistics can do much more than that: for instance, it can quantify the degree of representativity of one sample taken inside a bigger volume.

Some improvements might still be brought to the mapping exercise. For instance, we:

* made the wrong assumption that the core data were perfect.
* jumped from the logarithm of permeability to its antilog in a rather unconsiderate manner.
* provided, through kriging, a very smoothed version of what reality is. Conditional simulations (Journel and Huijbregts, 1978) might have probably been a better tool for creating a "possible" version of reality rather than a smoothed one.

Nevertheless, this study shows that geostatistics can be used in a completely applied environment: The maps were adjusted somewhat due to core/build-up transforms and then fed into the simulator model. Thus, the costly data from cored and non-cored wells were combined in a sensible manner for mapping. Hopefully this study will prove that simplistic mapping of complicated data sets is not acceptable any more, and that the tools are here to improve the way interpolation is performed.

REFERENCES

Allen, J.R. (1982). Prediction of Permeability from Logs by Multiple Regression, SPWLA (Europe)

Bear, J. (1968). Dynamics of Fluids in Porous Media, Elsevier, New York

Beard, O.C., and Weyl, P.K. (1973). Influence of Texture on Porosity and Permeability of Unconsolidated Sands, Bull. AAPG, Vol. 57, No. 2, pp 349-369

Brown, A., and Husseini, S. (1977). Permeability from Well Logs, Shaybah Field, Saudi Arabia, SPWLA 8th Logging Symposium, June 5-8, 1977

Cardwell, W.T., and Parsons, R.L. (1945). Average Permeabilities of Heterogenous Oil Sands, Transactions AIME, Vol. 160, pp 34-42

Codreanu, D. Un Critere pour l'Evaluation de l'Heterogeneite des Gisements de Petrole a l'Echelle Centimetrique et Regionale, Communication n.A8, 3rd Colloque ARTFP, Reference I.F.P. 16055

Delhomme, J.P. (1974). La Cartographie d' une Grandeur Physique a partir de Donnees de Differentes Qualites, Meeting of the International Association of Hydrogeologists, Montpellier, France, Vol. 10, pp 185-194

Draper, N.R., and Smith H. (1980). Applied Regression Analysis, Wiley

Haldorsen, H.H. (1982). Reservoir Characterization Procedures for Numerical Simulation, Ph.D. Dissertation, The University of Texas at Austin

Haldorsen, H.H. (1986): The Problem of Scale in Reservoir Engineering (in this volume)

Hubbert, K. (1957). Darcy' Law and the Field Equations of the Flow of Underground Fluids, reprinted in "The Theory of Ground-Water Motion and Related Papers", by K. Hubbert, Hafner Publishing Company, pp 261-300, 1969

Journel, A.G., and Huijbregts, C. (1978). Mining Geostatistics, Academic Press, London

Klinkenberg, L.J. (1941). The Permeability of Porous Media to Liquids and Gases, Drilling and Production Practices, p. 200, API

Kolodizie, S., Jr. (1980). Analysis of Pore Throat Size and Use of the Waxman-Smiths Equation to Determine OIP in Spindle Field, Colorado, SPE Paper 9382, 55th Annual Conference of the SPE, Dallas, Sept. 21-24

Kozeny, J. (1927). Uber Kapillare Leitung Des Wassers im Boden Sitzungsberichte, Royal Academy of Science, Viena, Proc. Class I, Vol. 136, pp 271-306

Matheron, G. (1967). Elements pour une Theorie des Milieux Poreux, Masson, Paris

Matheron, G. (1971). The Theory of Regionalized Variables and its Applications, les Cahiers du Centre de Morphologie Mathematique de Fontainebleau, Ecole Nationale Superieure des Mines de Paris, France

Ogbe, D., and Bassiouni, Z. (1978). Estimation of Aquifer Permeabilities from Electric Well Logs, The Log Analyst, Sept-Oct 1978, pp 21-27

Rabe, C.L. (1957). A Relation Between Gamma Radiation and Permeability, Denver-Julesberg Basin, Trans. SPE, Vol. 210, p 358

Schlumberger (1982). Formation Evaluation

Tickell, F. G., and Hiatt, W.M. (1938). Bull AAPG, Vol. 22, p 1272

Tixier, M.P. (1949). Evaluation of Permeability from Electric-Log Resistivity Gradients, Oil and Gas Journal, Vol. 48, June 16, p 113

Wadman, D.H., Lamprecht, E.E., Mrosovsky, I. (1979). Joint Geologic/Engineering Analysis of the Sadlerochit Reservoir, Prudhoe Bay Field, SPE paper 7531

Wendt, W., Sakurai, S., Nelson, P. (1986). Permeability Prediction From Well Logs Using Multiple Regression (in this volume)

CODING AND EXTRAPOLATING EXPERT INFORMATION FOR RESERVOIR DESCRIPTION

Clement Kostov
Andre G. Journel

Department of Applied Earth Sciences
Stanford University
Stanford, California

ABSTRACT

Reservoir description typically makes use of a considerable amount of qualitative ("soft") information and expert interpretation. Computer-generated maps, unlike hand-drawn maps, do not account for that soft information thus requiring time-consuming, ad-hoc corrections. A Bayesian-type formalism is proposed whereby data, hard and soft, are input as prior distributions which are then interpolated through a least squares algorithm to yield estimates of posterior distributions, from which various estimators and confidence intervals are derived. A synthetic case study is presented.

I. INTRODUCTION

A constant problem in reservoir description is that of estimating the values of a variable away from the available information, which is either of a "hard" type (e.g. actual core measurements at wells locations) or of a "soft", qualitative type (e.g. interpretative geology). Volumetric computations and numerical simulation models rely on the accuracy of such estimates.

Besides actual core measurements, which typically are extremely scarce, logging data and production history provide a detailed description of the subsurface in the immediate vicinity of the borehole. Further away, geological models, well tests and surface seismic capture broader, spatially averaged features of the variable to be mapped. From that information, the geologist or engineer builds up an "expert" understanding of the spatial phenomenon and expects the contour map, whether hand-drawn or computer-generated, to conform with it.

Most existing interpolation algorithms do not account for qualitative information other than through "ad-hoc" corrections a-posteriori or "fake" hard data. The "soft-kriging" technique presented hereafter is a two-step procedure. First, expert information is coded into prior cumulative distribution functions (cdf) at certain locations, then these distributional data are combined with actual hard data by a least-squares type algorithm, yielding estimates of the posterior distributions, from which various confidence intervals can be deduced.

The traditional approaches including kriging are reviewed and their shortcomings outlined. The Bayesian-type "soft-kriging" technique is then presented together with a case-study based on synthetic data.

II. EXPERT INFORMATION : THE GAS-OIL CONTACT EXAMPLE

The mapping of the gas-oil contact (GOC) provides typical examples of information contained in interpreted exploration and production data, which is not routinely incorporated into traditional interpolation algorithms.

Hard data for the depth of the GOC come from well-log measurements, see figure 1-a. These well-log measurements may have different degrees of reliability due to log calibration and interpretation, however they can be considered as hard, error free data, compared to other types of information.

A. Inequality constraints

In some cases, only constraints on the range of possible values, instead of a precise measurement, are available for the depth of the GOC, for instance :

- because drilling stopped short of meeting the GOC, see figure 1-b, the information on the GOC-depth appears as a lower-bounded inequality, the lower bound being the maximum depth of the well.

- because of blurred log data, as in figure 1-c, where the GOC marker was missed, say due to high clay content. At location x_c the information can be coded as an upper-bounded inequality, the bound being the depth of a marker known to lie below the GOC. More generally, expert information may define the GOC depth as being within an interval $[z_1, z_2]$. A hard error-free datum would correspond to a zero-width interval.

B. Time dependency of the GOC

Another important aspect of the GOC which yields inequality-type data is its time dependence. The initial GOC prior to production is thought of as a plane gas-liquid interface, which subsides and grows more irregular as production goes on. In areas where the dominant feature is gravity drainage, the GOC level cannot rise: hence if at time t_0 the GOC depth is 3,000 m, at any later time t, the depth must be greater than 3,000 m, i.e. in the interval [3,000 , ∞]. However, not all values within that interval are equally likely, and an expert guess for the most likely depth value at time t_1 may be obtained by extrapolating in time rates of subsidence and/or correlating subsidence and production rates.

C. Gas influx mechanisms

In some reservoirs, the presence and geometry of flow barriers such as shales is of paramount importance to gas flow, and hence to the evolution in time of the GOC. D.M. Chang and H.H. Haldorsen (1984) list the numerous mechanisms controlling gas excess production at the Prudhoe Bay field, including coning, underrunning, solution gas and fault-related movements. Understanding of these mechanisms as well as a geological model for shale geometry and spatial distribution provides local three dimensional models for the surface of the GOC. This important expert information, often the only one available away from wells, must be accounted for in the GOC mapping.

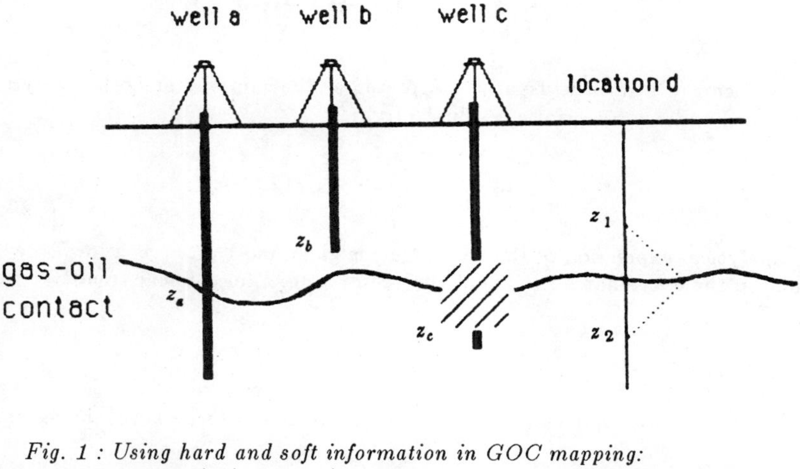

Fig. 1 : Using hard and soft information in GOC mapping:
well a - hard datum : $z(x_a) = z_a$*, (precise measurement of the GOC depth)*
well b - Lower bound : $z(x_b) \geq z_b$ *(the well did not reach the GOC)*
well c - Upper bound : $z(x_c) \leq z_c$ *(the GOC could not be detected)*
location d - Soft datum : probability distribution for $z(x_d)$ *in the interval* $[z_1, z_2]$ *(an expert information)*

III. TRADITIONAL APPROACHES

Hand-drawn and hand-corrected contour maps are the most traditional way to account for such expert information. Besides being costly and labor time consuming, this approach does not yield any measure of uncertainty for the estimated surface, nor is it clear that a geologist, however expert, can do a correct job in balancing subjective expert information with the hard data.

Another traditional approach consists in generating "fake" data at some critical locations which are then processed as if they were additional hard data. Again uncertainties, either associated with the data or resulting from the mapping procedure, are not quantified.

Among other moving-average type interpolation algorithms, the kriging technique stands out, for it provides a measure of interpolation uncertainty based on a covariance model of the spatial variability of the variable being contoured. See in particular P. Delfiner (1976).

A. The surface estimated by kriging

Let $z(x)$ be the value at location x of the variable of interest. $z(x)$ is estimated by a linear combination of the n hard data $z(x_\alpha)$ at locations x_α, $\alpha = 1,...n$:

$$z^*(x) = \sum_{\alpha=1}^{n} \lambda_\alpha(x) \; z(x_\alpha) \tag{1}$$

The surface $z(x)$ itself is usually interpreted as the sum of a smoothly varying trend $t(x)$ and a more erratic residual $r(x)$:

$$z(x) = t(x) + r(x) \tag{2}$$

The analytical expression of the trend $t(x)$ is given most often as a low order polynomial in the coordinates (x_1,x_2) of the vector x , e.g., for a linear trend :

$$t(x) = a_0 + a_1 x_1 + a_2 x_2 \tag{3}$$

with all or some of the trend parameters (i.e. the a_i 's in formula (3)),being unknown.

The weights of the linear interpolator, the $\lambda_\alpha(x)$'s in formula (1), together with the trend parameters are then obtained by multiple regression through a system of normal equations, see in particular Luenberger (1968) p. 56., also and belatedly called a "kriging system". This linear regression requires the inference and modelling of a covariance function, $Cov\ [R(x), R(x')]$, characteristic of the variability in space of the residual field $R(x)$. Kriging actually reduces to the even more traditional polynomial least-square fitting algorithm when $R(x)$ is modelled as a pure white noise.

B. Input of expert information in kriging

From equations (1) and (2), there appear two ways to introduce expert information into kriging:

1. add "fake" data : $y(x_\beta) = t(x_\beta) + w(x_\beta)$, $\beta = n+1,\ldots,n+n'$, to the initial hard data set. These fake data can be made different from the hard data by acting on the corresponding covariance values $Cov\ [W(x_\beta), W(x_\gamma)]$ and $Cov\ [W(x_\beta), R(x)]$, see Chiles (1976), and the formalism of cokriging in Journel and Huijbregts (1978), p. 324 .

2. impose particular trend expressions, possibly not of a polynomial type, in some areas. For example, a known fault surface can be introduced through a step trend function Δ, see Marechal (1984). In a 1-dimensional space:

$$t(x) = a_0 + (a_1 - a_0)\Delta_{x_0}(x) \quad i.e.$$
$$t(x) = a_0 \ , \quad \text{for } x < x_0$$
$$t(x) = a_1 \ , \quad \text{for } x \geq x_0$$

where the fault location parameter x_0 is known, but the displacement parameter $(a_1 - a_0)$ may be unknown.

3. a third way is to impose inequality constraints on the estimated values : $z^*(x_i) > z_i$, or $z^*(x_i) < z_i$ at certain locations x_i , including, but not limited to, the locations where inequality-type data are present (refer to the previous discussion on GOC) . The determination of the linear interpolator (1) is then formulated as a constrained optimization problem, see Dubrule and Kostov (1985).

C. Limitations and shortcomings of kriging

These modified kriging techniques present one limitation and one major shortcoming :

The limitation is that expert information must be introduced as scalar values, which are either the fake data values $z(x_\beta)$ or $y(x_\beta)$, or the interval bounds values z_i such that $z^*(x_i) < z_i$. The cokriging approach of Chiles (1976), amounts to considering a zero mean gaussian distribution of measurement errors. Ideally, one would like to introduce expert information as prior distributions of any shape : $Prob\ [Z(x_\beta) \leq z] = F_{x_\beta}(z)$, see hereafter.

The shortcoming, which is shared by the original kriging technique itself, is that the "kriging variance" as provided by the kriging system,

$$\sigma_K^2(x) = E[Z(x) - Z^*(x)]^2 \qquad (4)$$

does not provide any confidence interval for the estimator $Z^*(x)$, unless an unrealistic gaussian distribution for errors is assumed. More important, the frequently repeated claim that the kriging variance is a measure of local precision, see Journel and Huijbregts (1978) p. 308, is not correct, except for multinormal random fields which are almost never adequate models in practice. Indeed, the kriging variance is independent of the data values, whether hard or soft, and usually differs dramatically from the conditional estimation variance, conditional to the N data values, $E[\,[Z(x) - Z^*(x)]^2 \mid (N)\,]$, which is a characteristic of local precision. Rather the kriging variance $\sigma_K^2(x)$ is a mere index of the data configuration, used to estimate the value $z(x)$.

D. Coding expert information as cdf's.

It thus appears that an algorithm fit for handling expert information should have the following features :

1. concise representation of expert information and reasonable computational cost

2. enough flexibility to allow for input data of various types:

- hard data :

$$z(x_\alpha) = z_\alpha \;,\; \alpha = 1,\ldots,n. \tag{5}$$

- inequality-type data, possibly with infinite bounds a_β or b_β :

$$a_\beta \leq z(x_\beta) \leq b_\beta \;,\; \beta = n+1,\ldots,N = n + n^{prime}, \tag{6}$$

- distributional-type data :

$$F_{x_\beta}(x) = Prob\;[Z(x_\beta) \leq z\,] \;,\; \beta = n+1,\ldots N. \tag{7}$$

3. provide confidence intervals for the resulting estimates, or at least a measure of precision that is data-values dependent.

The inequality-type data (6) can be seen as incomplete distributional data. Indeed, $a_\beta \leq z(x_\beta) \leq b_\beta$ implies :

$$F_{x_\beta}(z) = \begin{cases} 0 & ,\text{for } z \leq a_\beta \\ unknown & ,\text{for } z \text{ in } [a_\beta, b_\beta] \\ 1 & ,\text{for } z \geq b_\beta \end{cases} \tag{6a}$$

Similarly, a hard datum of type (5) can be seen as a Dirac step-distributional function, since the condition $z(x_\alpha) = z_\alpha$ implies :

$$F_{x_\alpha}(z) = \Delta_{z_\alpha}(z) = \begin{cases} 0 & ,\text{for } z < z_\alpha \\ 1 & ,\text{for } z \geq z_\alpha \end{cases} \tag{5a}$$

Consequently,the data set (5), (6), (7) can be seen as a one-type distributional data set of size N : $F_{x_\alpha}(z)$, $\alpha = 1,...N$.

IV. THE SOFT KRIGING APPROACH

A. Presentation

Soft kriging is a procedure for estimating the posterior distribution $Prob\,[Z(x) \leq z \mid data\,]$, conditional to all the information available (hard and soft), based on least squares interpolation between distributional-type data. In short, for each threshold z , the posterior cumulative distribution function (cdf) value is estimated by a linear combination of the corresponding distributional data:

$$\begin{aligned} F_x^*(z \mid (N)) &= Prob^*\,[Z(x) \leq z \mid data\,] \\ &= \sum_{\alpha=1}^{N(z)} \lambda_\alpha(x\,;z)\, F_{x_\alpha}(z) \end{aligned} \tag{8}$$

the interpolator weights $\lambda_\alpha(x\,;z\,)$, for each location x and each threshold z , are obtained by solving a system of normal (kriging) equations calling for the covariance function :

$$C(h\,;z) = Cov\,[F_x(z)\,,\, F_{x+h}(z)\,] \tag{9}$$

Classically, this covariance, or better the corresponding variogram,is inferred from the distributional data available for the threshold z .

The corresponding ordinary kriging system is thus written :

$$\begin{aligned} &\sum_{\beta=1}^{N} \lambda_\beta(x\,;z\,)C(x_\alpha - x_\beta;z\,) + \mu(x\,;z\,) = C(x - x_\alpha;z\,) \;,\; \alpha = 1,...,N \\ &\sum_{\beta=1}^{N} \lambda_\beta(x\,;z\,) = 1 \end{aligned} \tag{10}$$

with $\mu(x\,;z\,)$ being the ordinary kriging Lagrange parameter.

B. Some remarks

(1) When only hard data of the type (5a) are present, soft kriging reduces to indicator kriging : see Journel (1984 a) for more theoretical developments, Verly and Sullivan (1984), and Isaacs (1984) for practical applications.

(2) Soft data of the type (7) and certainly inequality data of the type (6a) can be used together with hard data for the inference of the covariance models. This could represent a serious advantage in cases where hard data are scarce. Kulkarni (1984) discusses alternative weighting schemes for hard and soft data in the variogram estimator.

(3) In the presence of inequality data of type (6a), the actual number of distributional data available at any given threshold z is N minus the number of constraint intervals containing the value z.

(4) Since kriging is an exact interpolator, see Journel and Huijbregts (1978, p. 308), if soft kriging is attempted at a location x_α where preexists a hard datum z_α, the Dirac step distribution is found back,i.e.: $F^*_{x_\alpha}(z \mid (N)) = F_{x_\alpha}(z) = \Delta_{z_\alpha}(z)$

Similarly, soft kriging done at the location x_β of a soft datum yields back the input prior distribution : $F^*_{x_\beta}(z) = F_{x_\beta}(z)$

Last, soft kriging at the location of an inequality-type datum results in a posterior cdf honoring the inequality constraints :

$$F^*_{x_\beta}(z \mid (N)) = \begin{cases} 0 & \text{, for } z < a_\beta \\ \textit{estimated function of } z & \text{, for } z \text{ in } [a_\beta, b_\beta] \\ 1 & \text{, for } z > b_\beta \end{cases}$$

V. E-TYPE ESTIMATOR

From the estimated posterior cdf $F^*_x(z \mid (N))$, an estimate of the conditional expectation can be derived :

$$z^*_E(x) = E^*[Z(x) \mid (N)] = \int_{-\infty}^{+\infty} z.dF^*_x(z \mid (N)) \tag{11}$$

This estimate, also called E-type estimate, is the value z that minimizes the estimated conditional error variance $E[[Z(x)-z]^2 \mid (N)]$. It is the least squares estimator of $z(x)$ attached to the posterior cdf $F^*_x(z \mid (N))$.

The following remarks can be made :

1. discretization of the variability interval

In practice, the posterior cdf is estimated by a series of K probability values $F^*_x(z_k \mid (N))$, if K thresholds, z_k, $k=1,...,K$, have been retained to discretize the variability interval $[z_{\min}, z_{\max}]$, cf. figure 2. The integral (11) defining the E-type estimator is then approximated by the discrete sum :

$$z^*_E(x) = \sum_{k=0}^{K} z_k'(x)\,[\,F^*_x(z_{k+1} \mid (N)) - F^*_x(z_k \mid (N))\,] \tag{12}$$

with $F_x^*(z_0 \mid (N)) = 0$, $F_x^*(z_{K+1}) \mid (N)) = 1$ and, $z_k'(x)$ being the conditional mean of the class $]z_k, z_{k+1}]$ estimated for example by the mean of all hard data falling into that class.

2. class exactitude of the estimator

Because of the discretization into K classes, the E-type estimator (12) is only class-exact, that is at the location x_α of a hard datum z_α the E-type estimate identifies the mean of the class to which belongs z_α instead of the datum z_α itself. Similarly, at the location x_β of an inequality-type datum, say $z(x_\beta) \leq b_\beta$, the E-type estimate $z_E^*(x_\beta)$ would honor not the bound b_β but the lower bound of the class containing the value b_β. Let z_k be that lower class bound, then : $z_E^*(x_\beta) \leq z_k \leq b_\beta \leq z_{k+1}$. On figure 3, at location $x = 2$, the E-type estimate is 5.75, indeed lesser or equal to the lower bound of the class]5.75,6.] which contains the bound value 6 of the inequality interval [6,∞].

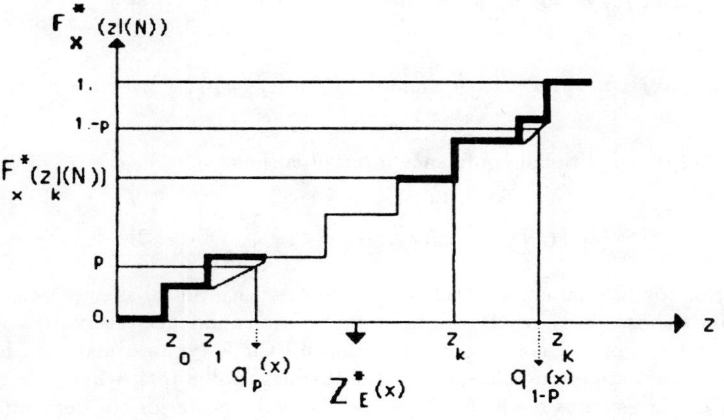

Fig. 2 : Discrete estimate of the posterior conditional cdf at location x, $F_x^*(z \mid (N)) = Prob^*(Z(x) \leq z \mid (N))$, *for K threshold values :* $z = z_k, k = 1, \cdots, K$.
$z_E^*(x)$ *is the cdf mean, an estimate of the unknown value* $z(x)$.
$q_p(x)$, $q_{p-1}(x)$ *are the* p^{-th} *and* $(1-p)^{-th}$ *quantiles of the cdf.*

3. other estimators

The availability of the posterior cdf $F_x^*(z \mid (N))$ allows consideration of non least-squares estimators, for example the median (M-type) estimator $z_M^*(x)$, equal to the median of the estimated posterior cdf. The M-type estimate is the value z that minimizes the estimated conditional absolute error $E[\mid Z(x) - z \mid (N)]$; it is the L1-norm optimal estimator attached to the posterior cdf $F_x^*(z \mid (N))$. The M-type estimator is better outlier-resistant and may be preferred to the E-type estimator for some applications. For a more detailed discussion on loss functions and optimal estimators, refer to Lehmann (1983), and Journel (1984 b).

VI. CONFIDENCE INTERVALS

Direct readings from the estimated posterior cdf $F_x^*(z \mid (N))$, cf. figure 2, provides estimates for the two opposite p and $(1 - p)$ conditional quantiles :

$$q_p^*(x) \text{ , } in \ [0.,.5] \ such \ that \ : \ F_x^*(q_p(x) \mid (N)) = p$$

$$q^*_{(1-p)}(x) \text{ ,} in \ [.5,1.] \ such \ that \ : \ F_x^*(q_{(1-p)}(x) \mid (N)) = 1 - p$$

The "(1-2p)" conditional confidence interval is thus $]\ q_p^*(x)\ ,\ q^*_{(1-p)}(x)\ ]$, with:

$$Prob^* \Big[Z(x) \ in \ \ ]q_p^*(x), q^*_{(1-p)}(x)] \Big] = 1 - 2p \tag{13}$$

Posterior conditional cdf's attached to variables with initial skewed prior distributions are themselves skewed, yielding asymmetric confidence intervals, not necessarily centered on the conditional expectation (the E-type estimator). Isaaks (1984) shows a case study on concentration of chemical pollutants, where the E-type estimator corresponds to the 80^{-th} percentile of the posterior cdf, leaving thus only 20% probability that it would be exceeded by the true unknown value (instead of the 50% probability predicted in the gaussian model).

Conditional confidence intervals are independent of the particular estimator retained to estimate $z(x)$. Thus the confidence interval in (13) is the same whether the mean (E-type) or the median (M-type) estimate has been retained, and is usually centered on neither one.

VII. A SYNTHETIC CASE STUDY

A. The data set

To test the technique, before actual applications, a 1-dimensional synthetic case study was designed. On figure 3, the abscissa axis represents the space coordinates and the ordinate axis represents the variable z.

The data set is composed of 24 equally spaced data :

1. 9 hard data at locations $x_\alpha = 1,3,5,11,14,16,20,21,24$
2. 2 inequality constraints at locations $x = 4$ and 6 ; for instance at location $x_4 = 4$, the estimate should belong to [2,4]
3. 10 open intervals with lower bound constraints at locations $x = 2,7,8,9,10,13,15,19,22,23$; for instance at $x_2 = 2$ the estimate should be greater than 6
4. 3 open intervals with upper bound constraints at locations $x = 12,17,18$; for instance at $x_{12} = 12$ the estimate should be less than 1

B. Input parameters

The values taken by the variable z were assumed to be within the interval [0.,10.], which was discretized into $K + 1 = 40$ intervals of equal widths 0.25, with bounds:

$$z_0 = 0., \; z_1 = 0.25 \;, \; \cdots \;, \; z_K = 9.75 \;, \; z_{K+1} = 10.$$

Altogether 39 thresholds were considered for soft-kriging, and the 39 corresponding correlograms of type (9) were all set equal to an exponential model of parameter 5, that is of practical correlation range 15 :

$$C(h\,;z_k) = exp\,(\,-h\,/5\,) \;\;, \text{for } all \;\; z_k \;, \; k = 1,..,39 \tag{14}$$

Estimation was performed on a regular grid of 250 nodes, spaced 0.1 unit apart, and with coordinates ranging from 0.0 to 24.9 .

All data were used for the estimation at each location. Moreover, the covariance model being independent of the discretization values z_k, the kriging weights are also discretization independent. It suffices then,in order to obtain all the weights required, to solve 250 linear systems, differing only by their right-hand sides. The assumptions of constant data configuration and covariance model independent from the discretization values, allow very important computational savings, (Journel and Huijbregts, 1978, p.309).

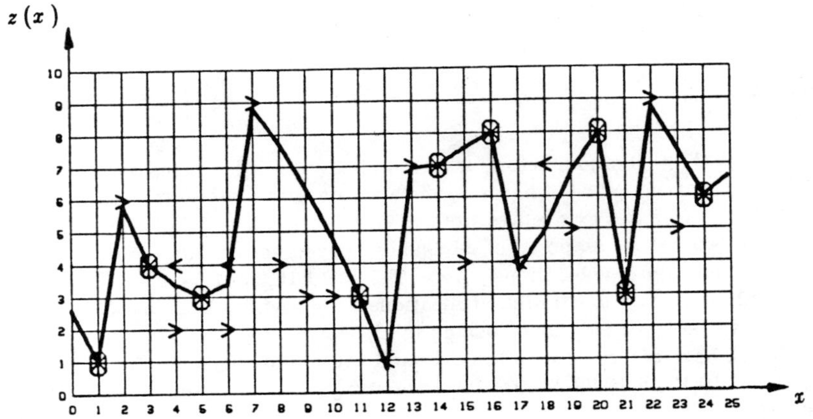

*Fig. 3 : Estimated profile honoring (1) hard data : locations indicated by *, e.g.* $z(x_1) = 1$ *; (2) inequality constraints : locations indicated by* $>$ *for the lower bounds e.g.* $z(x_2) > 6$, *and* $<$ *for the upper bounds, e.g.* $z(x_{12}) < 1$.
The estimated profile $z_E^*(x)$ *honors the inequality data to the extent of the z-discretization interval (.25), e.g.* $z_E^*(x_2) = 5.75$, *which is the lower bound of the interval* $[5.75, 6]$ *containing the constraint value 6.*

C. Discussion of results

At each of the 250 grid nodes, soft kriging provides an estimate of the posterior cdf $F_x^*(z_k \mid (N))$ discretized by the 39 thresholds z_k, see figure 2. To check for exactitude, the 24 posterior cdf's generated at the 24 data locations were plotted on figure 4. As expected, the estimated posterior cdf yields back the Dirac step distribution at hard data locations, and honors the inequality bounds where required. The apparent poor resolution of the estimated posterior cdf's (limited number of steps defining each cdf) is due to the 1-dimensional data configuration: indeed the two distributional data that bracket the location to be estimated receive most of the interpolation weights ("screen effect",cf. Journel and Huijbregts, 1978, p.331)

Figures 3 and 4 show the mean $z_E^*(x)$ of each estimated cdf, as derived from formula (12). The estimated profile honors the input constraints, both from hard and inequality type data, to the extent of the discretization interval (here equal to 0.25 along the ordinate axis). For example at location $x_2 = 2$, the inequality constraint is $z(x_2) \geq 6$ and the estimated value is $z^*(x_2) = 5.75$, equal to the lower bound of the class $]5.75\ ,\ 6.\ ]$, which includes the bound of the inequality constraint. Given the strong correlation at short distances and the hard data $z(x_1) = 1$ and $z(x_3) = 4$ it makes sense that the estimated value $z^*(x_2)$ would be at the lowest authorized limit. The strong correlation explains also the behavior of

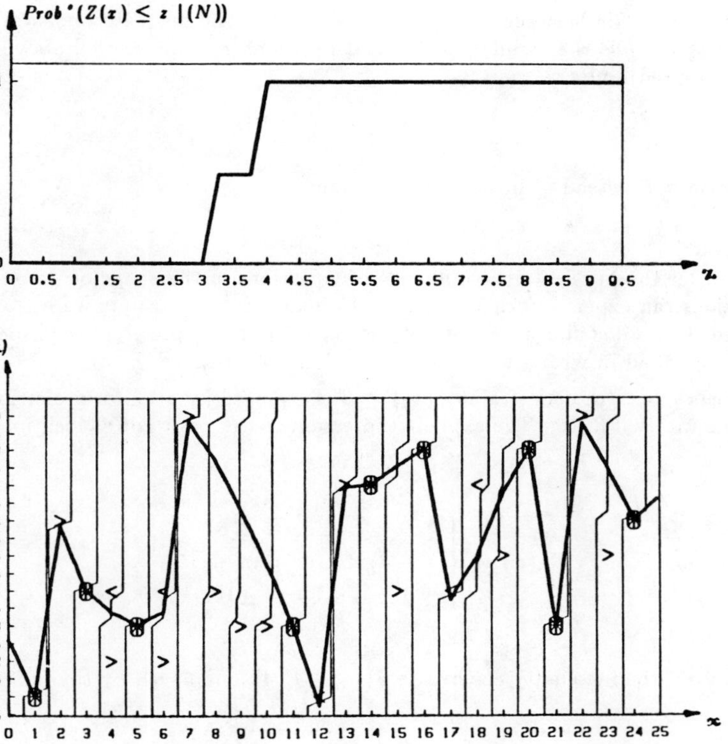

Fig. 4 : (a) Estimated cdf at location x=4. (b) Estimated profile as in figure 2, and estimated posterior cdf at each of the 24 data locations.

the estimated profile between $x_7 = 7.$ and $x_{10} = 10$. The inequality constraint at x_7, $z(x_7) \geq 9$, pulls the estimated values at locations x_8, x_9, x_{10}, well above their lower authorized limits.

D. Uniform cdf instead of inequality constraints

In order to appreciate the influence of variations in the shape of the initial distributions (an expert information) , the 15 inequality constraints were transformed into distributional data of the type (7). More precisely each inequality interval was filled in with a uniform distribution, for instance :

- for $x_2 = 2$, the constraint $z(x_2) \geq 6$ is replaced by a uniform distribution within the interval [6,10]. The explicit expression of the prior cdf at location x_2 is then :

$$F_{x_2}(z) = \begin{cases} 0 & \text{,for } z \leq 6 \\ (z-6)/4 & \text{,for } z \text{ in } [6,10] \\ 1 & \text{,for } z \geq 10 \end{cases}$$

- for x_4, the inequality constraint $z(x_4) \in [2,4]$ is replaced by the uniform distribution :

$$F_{x_4}(z) = \begin{cases} 0 & \text{,for } z \leq 2 \\ (z-2)/2 & \text{,for } z \text{ in } [2,4] \\ 1 & \text{,for } z \geq 4 \end{cases}$$

The E-type estimated profile using uniform distributions instead of incompletely defined cdf's is shown in figure 5. Again, the estimator honors all constraints, whether equalities or inequalities.

Comparison of figures 3 and 5 shows that the introduction of uniform cdf's within the constraint intervals has forced the E-type estimator away from the bounds, toward the center of the interval of authorized values, as at locations x_2 and x_{13}. Also, between abscissas x_7 and x_{11}, where there are few hard data, the introduction of uniform distribution has changed rather drastically the final estimate : $z_E^*(x_8) = 7.75$ on figure 3, versus $z_E^*(x_8) = 6.75$ on figure 5. Conversely, in the intervals $[x_3, x_5]$ or $[x_{20}, x_{25}]$, where hard data are numerous, the influence of the soft data is limited, and this is indeed a very desirable effect.

Let us insist again, that complete lack of information about a cdf should be modelled as an incompletely defined cdf, not as a uniform distribution, which is a stronger assumption.

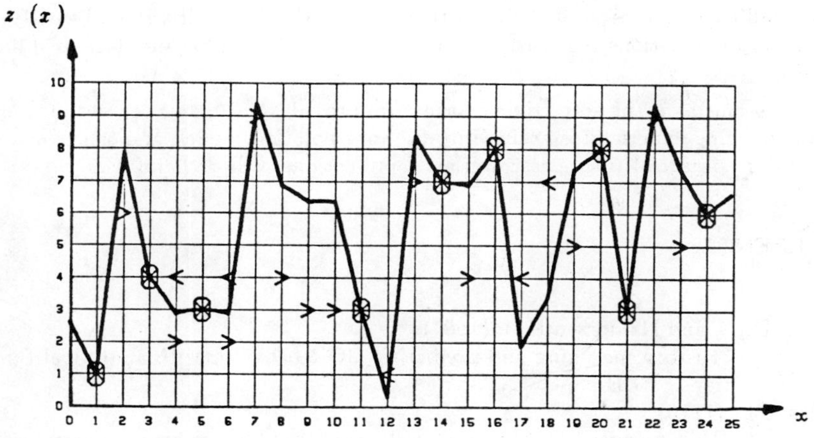

*Fig. 5 : Estimated profile honoring : (1) hard data (* locations), (2) prior uniform cdf filling in the inequality intervals of figure 2.*

VIII. CONCLUSIONS

Soft kriging provides an interpolation formalism whereby all data are considered as prior distributions, thus allowing the input of expert information at selected locations in the form of cdf's. Inequality constraints appear as incomplete prior distributions and do not call for any specific treatment. These prior distributions are then linearly interpolated through a classical least-squares procedure to yield an estimated posterior distribution at each node of the estimation grid. From these posterior distributions, optimal estimates of the unknown variable together with their confidence intervals, can be derived for mapping purposes.

The soft kriging formalism features :

- a theoretically consistent way to combine expert (soft) information including its reliability index, with both equality and inequality constraints (hard data).
- the possibility of using both inequality constraints and soft data to improve the inference of the spatial correlation model
- the derivation of a whole posterior cdf which yields optimal estimates of different kinds, not limited to the traditional least squares estimate.
- the derivation of actual confidence intervals, which are not restricted to the gaussian model, and which are data configuration and data values conditioned.

Present limitations of the formalism include :

- a yet limited record of practical applications, particularly concerning the use of expert information
- a rather heavy statistical inference, if a different covariance model is used for each discretizing threshold.

- an insufficient appraisal of the conditions, which the prior cdf's should satisfy, e. g. type of distributions, , consistency between soft information and data at different locations.

It seems as if the soft kriging formalism provides too many degrees of liberty for introducing expert information into the interpolation model, and that some guidelines , object of future research and applications, will be helpful.

REFERENCES

Chang, D.H., and Haldorsen H.H. (1984)
"History matching and predicting GOR behavior with analytical functions." SPE paper 12763.

Chiles, J.P. (1976)
"How to adapt kriging to non-classical problems : three case-studies" , in Advanced Geostatistics in the Mining Industry, ed Guarascio et al., publ Reidel, Dordrecht, pp.80-89.

Delfiner, P. (1976)
"Linear estimation of non stationary spatial phenomena", in ibid, pp. 49-68.

Dubrule, O. and Kostov C.
"An interpolation method taking into account inequality constraints", in Mathematical Geology (in press).

Isaaks, E. (1984)
"Risk qualified mappings for hazardous waste sites: A case study in distribution free geostatistics",MSc Thesis, Stanford University, 85p.

Journel, A.G. and Huijbregts, C. (1978)
Mining Geostatistics, publ. Academic Press, second print 1981, 600p.

Journel, A.G. (1984 a)
"The place of non parametric geostatistics", in Geostatistics for Natural Resources Characterization ed. Verly et al., publ. Reidel, Dordrecht, Part 1, pp. 307-335.

Journel, A.G. (1984 b)
"mAD and conditional quantile estimator", in ibid., Part 2, pp. 261-270.

Kulkarni, R. (1984)
"Bayesian kriging in geotechnical problems", in ibid., Part 2, pp. 775-786.

Lehmann, E.L. (1983)
Theory of point estimation, publ. Wiley and Sons, 506 p.

Luenberger, D.G. (1968)
Optimization by vector space methods, Wiley and Sons, 326 p.

Marechal, A. (1984)
"Kriging seismic data in presence of faults", in Geostatistics for Natural Resources Characterization, ed. Verly et al., publ. Reidel, Dordrecht, Part 1, pp.271-294.

Verly, G. and Sullivan, J. (1984)
"Multigaussian and probability krigings.Application to the Jerritt Canyon deposit." SME paper, preprint no. 84-52.

between them and more uniform properties within them, having recognizable, through not necessarily regular, dimensions.

For these reasons (the continuum nature and structured spatial arrangement), we represent the inherent variation of a flow property as a spatial stochastic process, or random field, characterized by a covariance function. We further assume the random field is statistically homogeneous--i.e. the flow property at each fixed location in space is a random variable whose expected value is a constant and whose covariance function depends only upon the separation vector (between two fixed points in space)--thereby validating analysis in terms of the spectral density function--the distribution of variance over different wave numbers.

Finally, each series of field data represents a single realization of a stochastic process. Application of stochastic theory implicity involves a probability average which can be thought of as an average over a large number of realizations or an ensemble average. To estimate the covariance function of the logarithm of permeability from field observations, we must assume equivalence of a spatial average and the ensemble average; in other words, an ergodic hypothesis (Lumley and Panofsky, 1964) must be assumed. Ergodicity means that an average over a single long record will converge to the expected value of the underlying process. The hypotheses of ergodicity and statistical homogeneity are interrelated and neither can be proved for a single realization of finite length.

A. Methods and Terminology of Data Analysis

The spectral and autocorrelation functions in this paper are computed using International Mathematical and Statistical Libraries (IMSL) subroutines and functions, which are based on Biomedical (BMD) computer programs. The indirect spectral estimation procedure is used, wherein the Fast Fourier Transform of the autocovariance function is used to estimate the spectral density because estimates of the autocovariances and autocorrelations are desired.

We assume a statistically homogeneous random field or second order stationary process, designated as $V(x)$ where x is a fixed location in space (Jenkins and Watts, 1968). We then consider estimation of the autocovariance function $R_V(\xi)$ and spectrum $\phi_V(k)$, where

$$R_V(\xi) = E\{[V(x + \xi) - E(V(x + \xi))][V(x) - E(V(x))]\},$$

$$\phi_V(k) = \frac{1}{2\pi} \int_{-\infty}^{\infty} R_V(\xi)\, e^{-ik\xi}\, d\xi.$$

The estimates are based on equally spaced observations, $V(j\Delta)$, $j=1,2,\ldots,N$. If there is no trend, the expected value $E(V(x))$ is estimated by

$$\overline{V} = \sum_{j=1}^{N} V(j\Delta)/N$$

and $R_V(\xi)$ is estimated by

$$\hat{R}_V(j\Delta) = \sum_{i=1}^{N-j} [V(j + i)\Delta) - \overline{V}][V(i\Delta) - \overline{V}]/(n-j).$$

The raw estimate of the spectrum (or spectral density) is the transform of $\hat{R}_V(j\Delta)$ given by

$$\hat{\phi}_V(h/2m\Delta) = \frac{2\Delta}{\pi} \sum_{j=0}^{m} \varepsilon_j\, \hat{R}_V(j\Delta) \cos(jh\pi/m)$$

where $\varepsilon_j = 1$, $0 < j < m$, $\varepsilon_0 = \varepsilon_m = 1/2$ and $h = 0,1,2,\ldots m$. Truncation at m of the autocovariance function smooths the raw estimate of the spectrum, which is desirable because unsmoothed estimates are not consistent estimates of the spectral density (see Jenkins and Watts, 1968). Additional smoothing is done by taking weighted averages of the above raw estimates using a Hamming window, which is a typical weighting.

We graph the logarithm of the smoothed spectral estimates because it reveals more spectral detail and yields confidence intervals of constant width in the logarithmic domain.

B. Integral and Correlation Scales

1. Background

In the stochastic models we discussed in Bakr, et al. (1978) and Gutjahr, et al. (1978) and in similar models studied by others (Dagan, 1979; Smith and Freeze, 1979), an important feature of the results is a scale associated with the input process. The input processes we are treating here are the porosity and the logarithm of the permeability.

In this section we examine the different types of scales and their estimation.

In general two kinds of scales have been used--the first is an integral scale and the second is a correlation scale. Both types of scales give a rough measure of the average distance over which one can expect to find a significant correlation between values of a random process.

Let $\rho_V(s) = R_V(s)/\sigma_V^2$ be the autocorrelation function associated with the stochastic process V(x). The integral scale, λ, is defined by

$$\lambda = \int_0^\infty \rho_V(s)ds, \text{ if } \lambda \neq 0.$$

For certain correlation functions the integral may be zero (which is the case for the "hole" covariance function,

$$R_V(s) = \sigma_V^2(1 - |s|/\ell) \exp\{-|s|/\ell\}).$$

This hole covariance function fits the data. In such a situation the <u>correlation scale</u>, ℓ, is introduced and defined by

$$\ell^2 = -\int_0^\infty s\, \rho_V(s)ds.$$

For the "hole" covariance function, $R_V(\ell)=0$.

2. Methods of Estimation

The spectal density ϕ_V at k=0 is simply given by

$$\phi_V(0) = \frac{1}{2\pi}\int_{-\infty}^{\infty}\sigma_V^2\rho_V(s)ds = \frac{1}{\pi}\int_{-\infty}^{\infty}\sigma_V^2\rho_V(s)ds = \frac{\sigma_V^2}{\pi}\ \lambda$$

and consequently the estimator suggested for λ is

$$\hat{\lambda} = \hat{\phi}_V(0)\pi/\hat{\sigma}_V^2.$$

An advantage of this estimator is that one can use the statistics of the spectral estimates, $\hat{\phi}$ (k), to make confidence statements about $\hat{\lambda}$.

A rough estimate of the correlation scale can be computed by statistically best fitting a straight line to the logarithm of the autocovariance function over the smallest five lagged distances (including zero lagged distance) and using the fitted slope. Alternately, the lagged distance at which the autocorrelation function equals 1/e is an estimator of the correlation scale. This e-fold drop of an autocorrelation function is a "rule of thumb" and is not based on mathematical theory nor models.

C. Geological Features of Investigated Rock Units

The Mount Simon Sandstone of Late Cambrian to Early Ordovician age nonconformably overlies Precambrian crystalline rocks and conformably underlies argillaceous sandstone and/or shale of the Eau Claire Formation. At and near its type section in the vicinity of Eau Claire, Wisconsin, the 234 feet thick Mount Simon Sandstone (Ulrich, 1914) consists dominantly of light yellow gray and light brown, friable, moderately sorted, medium grained, quartzose sandstone which is thick (1-3 feet) bedded and cross-bedded (Ostrom, 1970). Rare shales and pebbly sandstones provide the only lithological heterogeneity in the dominantly homogeneous unit. Feldspar averages 3% of the grains and ranges up to 22.5% (Potter and Pryor, 1961; Stauffer and Thiel, 1941) but Asthana (1969) reports an average of 18% and range up to 40.1%.

The data we study came from core samples of the upper 600 feet or less of the Mount Simon Sandstone located primarily in McLean, Livingston, and LaSalle counties in northeastern Illinois, approximately 300 miles southeast of its type section. Based on subsurface data, Templeton (1950) and Buschbach (1964) described the approximately 1600-2000 feet thick, cross-bedded, friable, very fine to very coarse grained Mount Simon Sandstone in northeastern Illinois and recognized three members within the uppermost 600 feet based on alternations of more poorly sorted, coarser grained, locally pebbly sandstone with better sorted, finer grained sandstone.

We also study the Pennsylvanian Tensleep Sandstone in Wyoming because of its lithological similarity to the Mount Simon Sandstone. Todd (1959), based on surface and subsurface examination, describes the 87-422 feet thick Tensleep Sandstone as dominantly white but buff weathering, friable to indurated, moderately to well sorted, fine to very fine-grained, quartzose sandstone, which is very thick (3-7 feet) bedded and cross-bedded. Rare thin beds of limestone and dolomite provide the only lithological heterogeneity in this dominantly homogeneous unit. Feldspar grains average 4.2% of grains and range up to 11%. The variability in induration is due to variable amounts of quartz, anhydrite and dolomitized cement.

Finally we also study Saraband logs of boreholes in the Cenozoic basinfill of the Southern Imperial Valley, just east of Holtville, California. Basinfill in intermontane basins of the southwestern U.S. in general and particularly in the Imperial Valley consists of complexly interbedded gravel, pebbly sands, sands, muddy sands, sandy or silty muds and muds. The Saraband Log is the Schlumberger trade name for a computer processed interpretation (see Poupon, et al., 1970) of sand and mud, porosity and permeability, and nature of pore fluid based, in this case, on data from a Dual Induction Laterolog, a Borehole Compensated Sonic Log, and a Compensated Neutron Log-Compensated Formation Density Log.

D. Porosity and Permeability Data of Investigated Rock Units

Porosity and (intrinsic) permeability measurements of core samples were available for the Mount Simon and Tensleep sandstones. Core samples were spaced one foot apart for most of the borehole cores but five of the 26 Mount Simon Sandstone boreholes had two foot spacing between adjacent core samples. Both vertical (perpendicular to layering) and horizontal (parallel to layering) intrinsic permeability data were available for the Mount Simon Sandstone but not the Tensleep Sandstone. In addition, core samples from some of the boreholes in the Mount Simon Sandstone were used to determine both the maximum horizontal permeability and horizontal permeability perpendicular (or normal) to the maximum. Intrinsic permeability core sample tests of the Mount Simon Sandstone were typically determined by gas (air) flow through the sample, but two boreholes also had determinations by liquid (water) flow through the sample.

Twenty-one boreholes in the Mount Simon Sandstone had continuous core recovery for sampling at one foot intervals of porosity and permeability tests; series of these data average 168 data and range from 73 to 348 data. The five boreholes with core sample determinations at two-foot data spacing provided series averaging 108 data (216 feet) and ranging from 81 to 141 data. For two boreholes in the Mount Sandstone core recovery was incomplete, with one or more short gaps of three to eight data in the series; in these cases the missing core determinations were estimated by extrapolation of straight lines from the three core sample data overlying the gap and the three core sample data underlying the gap.

From the Cenozoic basinfill in California we have Saraband logs from five boreholes within a three and one-third square mile area. We interpolated numerical values of percentage porosity from the right hand curve in the right hand track and logarithm of permeability index in millidarcies from the dashed curve in the left hand track of a Saraband log at two foot spacing between interpolation points. We use series of 144 data (subinterval thickness of 288 feet), because such a series length is similar to series from the Mount Simon and Tensleep sandstones, but we also use series of 300 and 600 data. We analyzed an average of 2602 data (5204 feet), with a range of 2300 to 3512 data per borehole.

Data from the first three boreholes/logs that we analyzed indicate a strong positive correlation both of porosity with logarithm of permeability as well as of correlation scale of a series of porosity data with the correlation scale of a series of logarithm of permeability data for the same interval. Consequently, we analyzed series only of the logarithm of permeability data for the last two analyzed boreholes.

The permeability index shown on the Saraband log is estimated from the log response of several different types of geophysical log. Commonly a geophysical tool has a separation distance between an energy emitter and receiver so that its log response at a given borehole depth is a property which inherently is an average over the separation distance and thus the permeability index is more analogous to transmissivity than to intrinsic permeability.

E. Representative One-Dimensional Arrays of Original Data

Figure 1 shows representative series of the logarithm of intrinsic permeability at one-foot data spacing of the Illinois Mount Simon Sandstone. The two series on the left (IL608 upper and lower) are from the longest single record available to us. The two series on the right (IL034 and IL038) come from boreholes separated by about one mile within the past productive Lake Bloomington Field.

All of the series in Figure 1 are better regarded as "zig-zag lines" rather than smooth curves; this distinction is made apparent by comparison of Figure 1 with Figure 2 showing representative series of the logarithm of permeability at two-foot data spacing from California borehole number 2 penetrating the Cenozoic basinfill of the Imperial Valley. In Figure 1 virtually all projections toward higher logarithm values are thin and pointed; whereas in Figure 2, many projections toward higher logarithm values are ten feet or more thick and have a "blunt" shape. Within some depth subintervals of the data series in Figure 2 there are some thin, pointed projections toward higher logarithm of permeability values indicating the variability in the parameter within these realizations of this process.

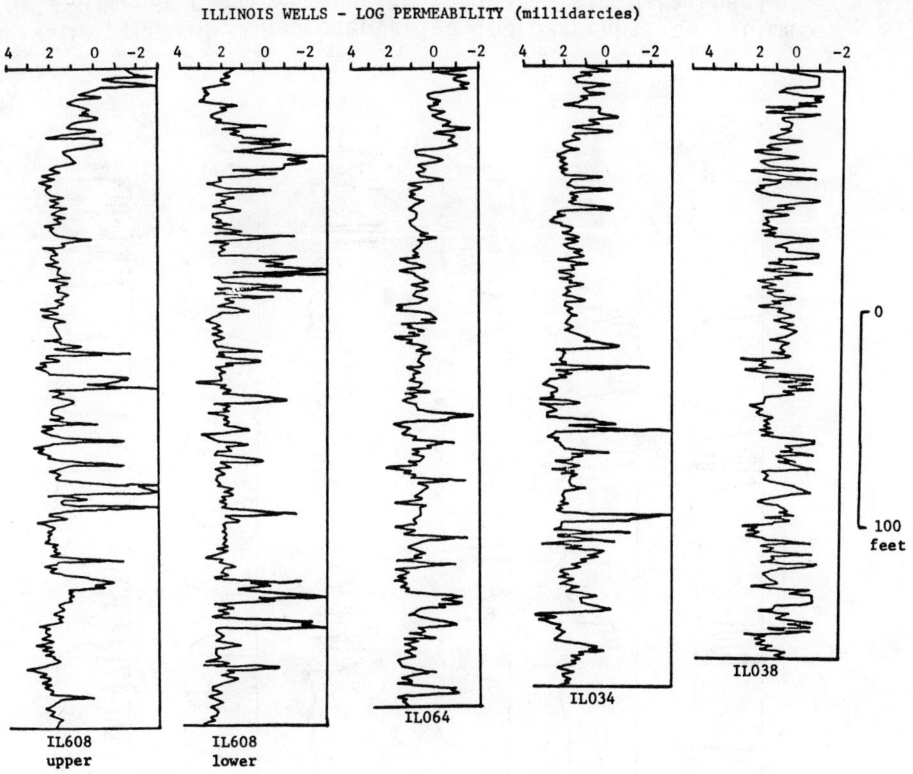

FIGURE 1. Representative log permeability series of Illinois Mount Simon Sandstone at one ft. data spacing.

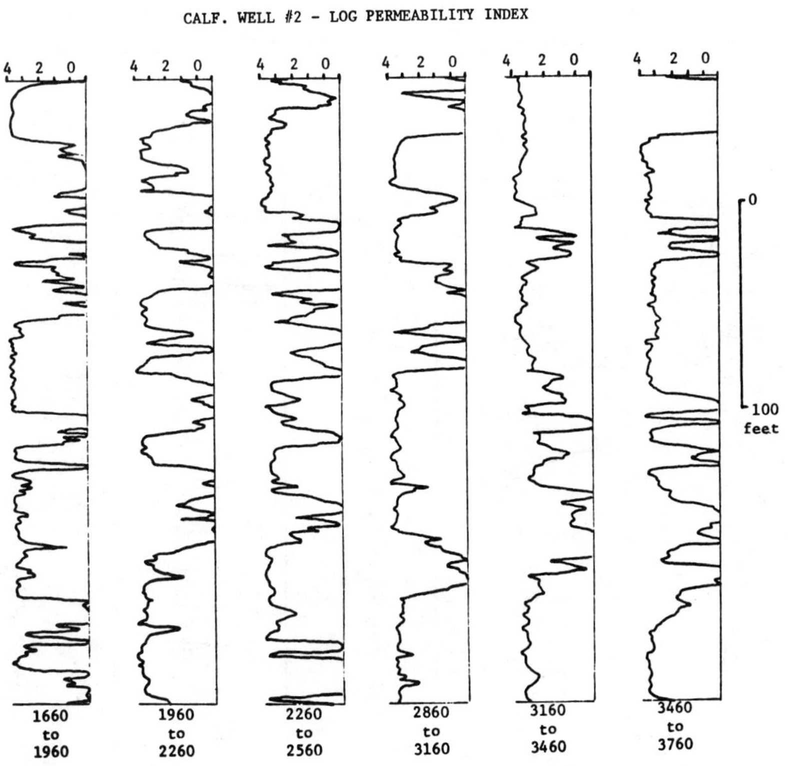

FIGURE 2. Representative log permeability estimate series of California basin fill at 2 ft. data spacing.

The series in both Figures 1 and 2 are from boreholes penetrating essentially flat (less than five degrees) lying stratigraphic units so that the series investigated are one-dimensional arrays normal to layering. Layer structure is more apparent in Figure 2 than in Figure 1.

F. Autocovariance Functions of Representative Series

As stated above, we represent the inherent variation of a flow property as a spatial stochastic process characterized by a covariance, or autocovariance, function. Figure 3 shows three autocovariance functions of series of the logarithm of permeability, each typical of one of the three geological units under investigation. The three original data series are similar in length of record and two-foot data spacing.

Obviously the autocovariance function of the California basinfill differs from those of the quartzose sandstones of Wyoming and Illinois by having an order of magnitude higher autocovariance at zero lagged distance (which is equal to the variance of the original data record) and by exhibiting a more apparent decrease as lagged distances increase. However, it is difficult to assess the similarity or dissimilarity of the autocovariance functions of the quartzose sandstones and whether these two autocovariance functions decrease as lagged distance increases. Therefore we "normalize" the autocovariance functions by dividing the autocovariance at each lagged distance by the variance of the data series. The result of this normalization is the autocorrelation function which has an autocorrelation function value of one at zero lagged distance.

II. REPRESENTATIVE AUTOCORRELATION AND LOG SPECTRAL DENSITY FUNCTIONS

We here compare and contrast selected representative autocorrelation and corresponding log spectral density functions for series of logarithm of permeability data from the geological units previously described. A label for each series consists of letter(s), either IL, WY, or C representing Illinois--Mt. Simon Sandstone, Wyoming--Tensleep Sandstone or California Cenozoic basinfill, respectively, followed by a numeral to identify individual boreholes. Following the numeral there may be additional

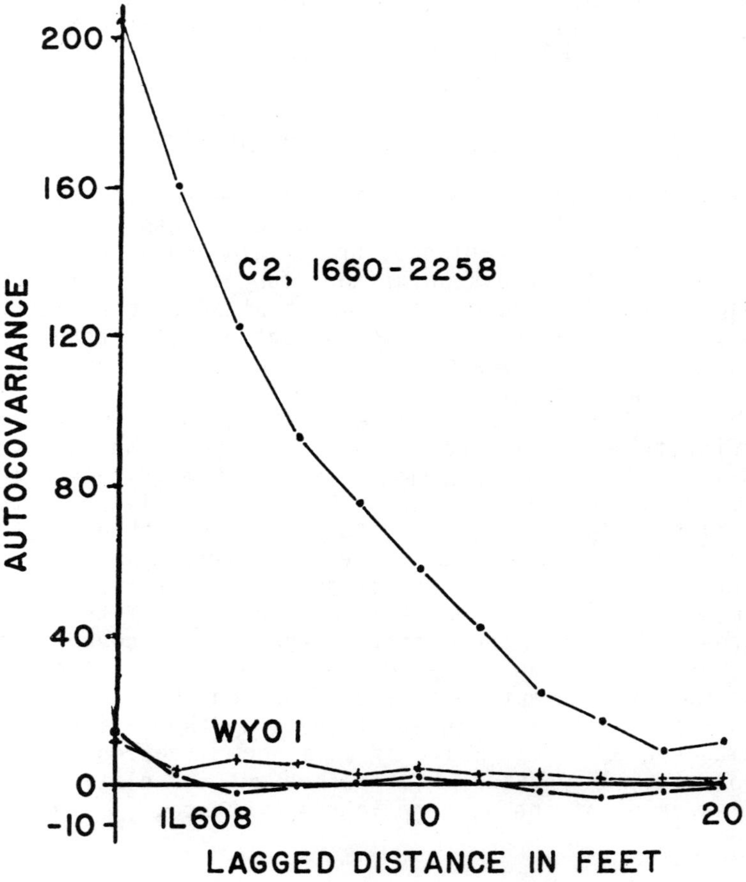

FIGURE 3. Autocovariance function of C2 (300 data), WY01 (91 data) and IL608 (304 data), each at 2 ft. data spacing.

letters and/or numbers; the California series have additional numbers indicating the depth (in feet) interval of the series consisting of the minimum depth, hyphen and maximum depth, whereas the Illinois series have additional letters indicating the nature of the data such that H represents horizontal permeability, MH represents maximum horizontal permeability or NH represents horizontal permeability normal to the maximum horizontal permeability. Where appropriate, the length of a series in terms of number of data in the series and/or data spacing in feet between spatially adjacent data in the series is indicated.

A. Illinois Lake Bloomington Field

The Lake Bloomington Field is located in McLean County, Illinois and its original data series (see two series at the right of Figure 1), autocorrelation functions and corresponding logarithm of spectra are typical of half, or slightly more, of all of our Illinois--Mount Simon Sandstone data.

Figure 4 shows the autocorrelation functions of the two series, each at one- and two-foot data spacing. The autocorrelation functions rapidly decrease and achieve our "rule of thumb" e-fold drop at one-foot spacing or at two-foot lagged distance for the series of two-foot data spacing. All four autocorrelation functions become parallel to the zero autocorrelation line with minor fluctuations about it.

Figure 5 shows the corresponding logarithm of the spectra which are generally flat, beginning at slightly negative values and becoming slightly more negative at higher frequencies. Each logarithm of the spectrum exhibits fluctuations, but all these fluctuations are insignificant in comparison to the 95% confidence interval. Moreover, the logarithms of all four spectra do not differ significantly from each other.

The autocorrelation functions in Figure 4 and corresponding logarithms of the spectra in Figure 5 approximate white noise, meaning there is no spatial correlation structure for the logarithm of intrinsic permeability series in these two boreholes, which are about one mile apart. Similar autocorrelation functions and corresponding logarithms of spectra are obtained for series of porosity data at either one- or two-foot data spacing in the Lake Bloomington and many other fields. Correlation and integral scales of these series are small, less than 1.5 feet for series at one-foot data spacing or 1.5 to 2.0 feet for series at two-foot data spacing.

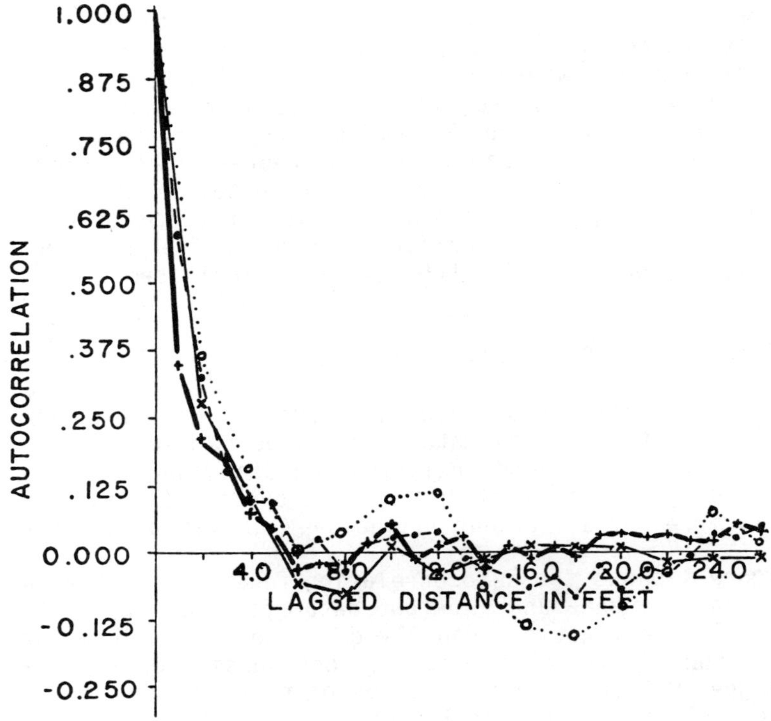

FIGURE 4. Autocorrelation functions representative of Illinois Lake Bloomington field permeability series at 1 and 2 ft. data spacing. IL 34H (+) 288 data @ 1' spacing, IL 34H (x) 144 data @ 2' spacing, IL 34H (•) 273 data @ 1' spacing and IL 38H (o) 137 data @ 2' spacing.

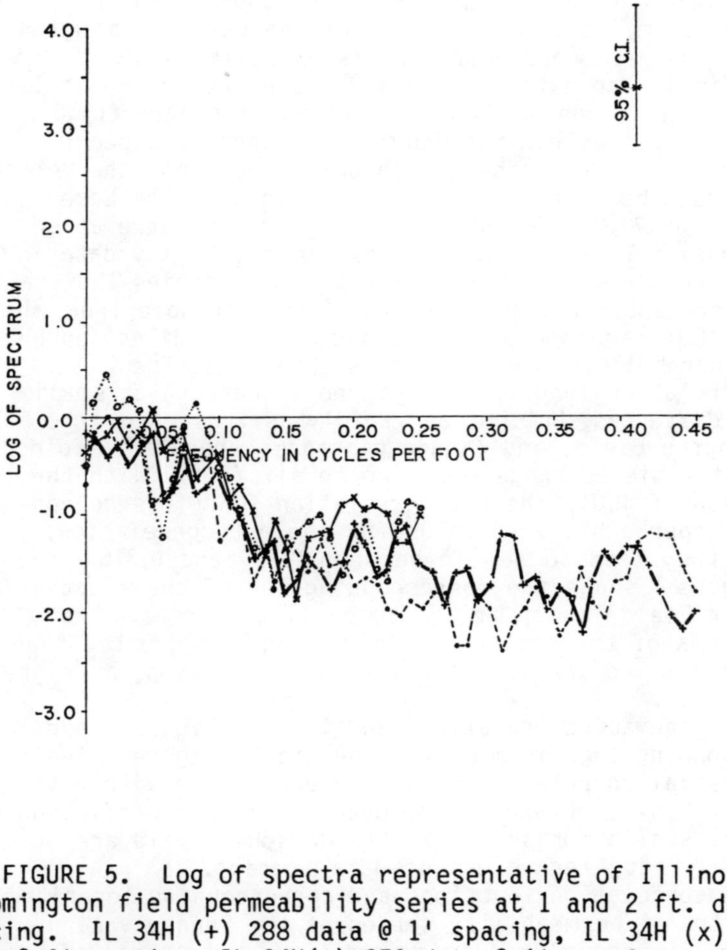

FIGURE 5. Log of spectra representative of Illinois Lake Bloomington field permeability series at 1 and 2 ft. data spacing. IL 34H (+) 288 data @ 1' spacing, IL 34H (x) 144 data @ 2' spacing, IL 34H(•) 273 data @ 1' spacing and IL 38H (o) 137 data @ 2' spacing.

B. Illinois Herscher Field

The Herscher Field is in Kankakee County, Illinois and is about 54 miles east northeast of the Lake Bloomington Field. The east northeast line is approximately perpendicular to extrapolations of isopachous contour lines of the Mount Simon Sandstone shown in Templeton (1950). If facies changes within the Mount Simon Sandstone occur perpendicular to isopachous contour lines, then the Herscher Field would be in a different facies than in the Lake Bloomington Field. Figure 6 shows four autocorrelation functions of logarithm of intrinsic permeability data in three boreholes. The borehole from which series IL09 came is located about two miles north of the borehole from which series IL01 came and about one and one-half miles southeast of the borehole from which series IL04 came. The autocorrelation functions in Figure 6 decrease at small lagged distances, but not as rapidly as those from the Lake Bloomington Field, and lagged distances where an e-fold drop is first achieved range from two to six feet. With the exception of IL01, the autocorrelation functions do not closely approach a zero value for the autocorrelation, rather they tend to remain between 0.250 and 0.375.

Figure 7 shows the corresponding log of the spectra for Herscher Field logarithm of permeability series. The logarithms of the spectra begin with values greater than 1.0 and generally decrease, with little fluctuation, at higher frequencies.

Both the autocorrelation functions in Figure 6 and corresponding logarithm of the spectra in Figure 7 indicate that spatial correlation structure does occur within the Herscher Field, regardless of data spacing. Correlation and integral scales for series in the Herscher Field are greater than 1.5 feet, regardless of data spacing.

Autocorrelation functions and corresponding logarithms of spectra of permeability series of the Pennsylvanian Tensleep Sandstone in Wyoming are similar to those of the Cambrian Mount Simon Sandstone in Illinois. Flat logarithm of spectra, indicating white noise and lack of spatial correlation structure, do occur for some series of Wyoming data, although not as commonly as in the Illinois data. More than half of the Wyoming Tensleep Sandstone autocorrelation functions and corresponding loagrithms of spectra indicate the presence of spatial correlation structure similar to that in the Illinois Herscher Field (Figures 6 and 7).

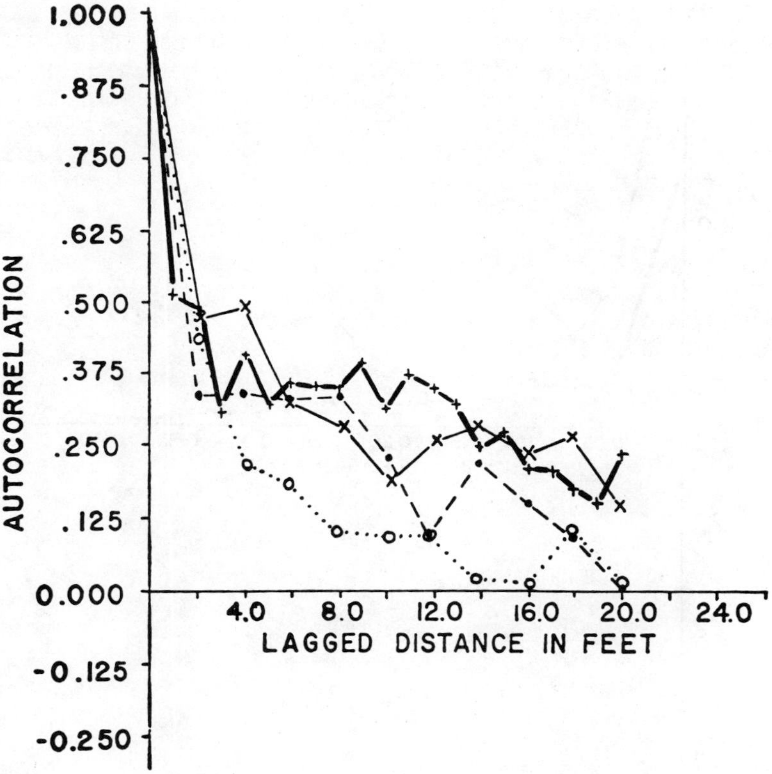

FIGURE 6. Autocorrelation functions representative of Illinois Herscher field permeability series at 1 and 2 ft. data spacing. IL 09H (+) 204 data @ 1' spacing, IL 09H (x) 102 data @ 2' spacing, IL 04MH (•) 107 data @ 2' spacing and IL 01MH (o) 128 data @ 2' spacing.

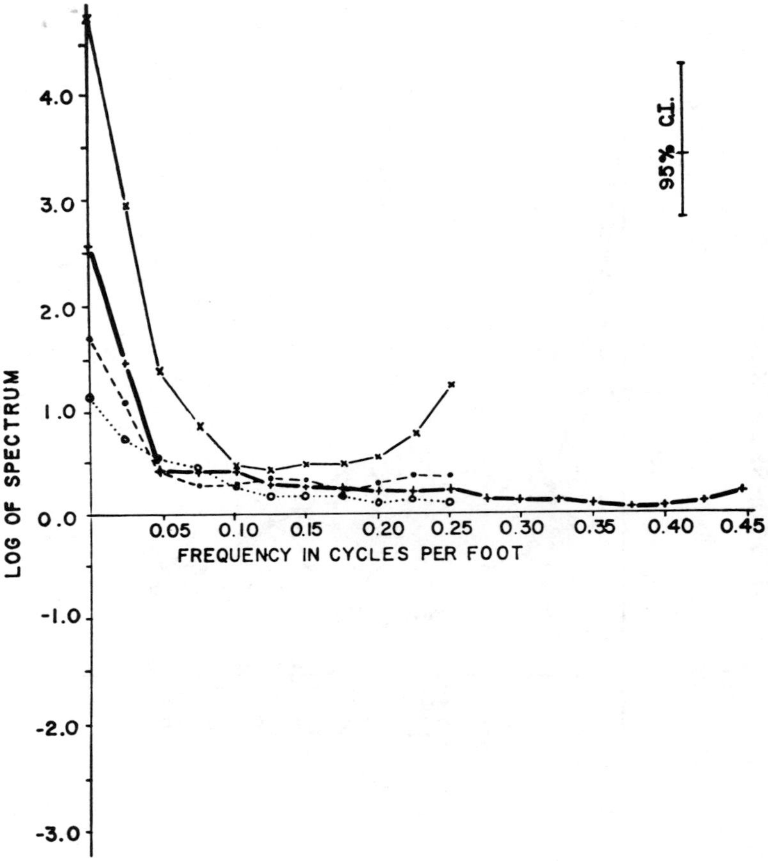

FIGURE 7. Log of spectra representative of Illinois Herscher field permeability series at 1 and 2 ft. data spacing. IL 09H (+) 204 data @ 1' spacing, IL 09H (x) 102 data @ 2' spacing, IL 04MH (•) 107 data @ 2'spacing and IL 01MH (o) 128 data @ 2' spacing.

C. California Borehole 2

In Figure 2 we presented examples of series of the original logarithm of the permeability index from borehole number 2 in the Cenozoic basinfill of the Imperial Valley of California. Figure 8 shows autocorrelation functions of four series of 144 data at two-feet data spacing, labelled by the footage depth subinterval of the series. The autocorrelation functions exhibit a rather gentle and somewhat steady (meaning few, small fluctuations) decrease to near zero autocorrelation. Initial e-fold drops occur at lagged distances ranging from six to twelve feet.

The corresponding logarithms of spectra are shown in Figure 9. They begin at values of the logarithm of the spectrum between 2 and 2.5 and decrease relatively rapidly to zero and negative values, with many fluctuations as frequency increases. None of the fluctuations of a logarithm of a spectrum are statistically significant and only at a few frequencies are there significant differences between logarithms of the spectra.

Spatial correlation structure is indicted by the autocorrelation functions in Figure 8 and logarithms of spectra in Figure 9 within the California basinfill, which is typical and representative of all series from the five boreholes penetrating this deposit.

III. GEOLOGICAL CONTROLS ON INTEGRAL AND CORRELATION SCALES

As we have said above, both the integral and correlation scales give a rough measure of the average distance over which once can expect to find a significant correlation between values of a random process such as either porosity or the logarithm of permeability. Geologists recognize and define layering (stratification) based on similarity of a host of properties, such as porosity, color, lithology, grain composition and size, etc., within a layer and dissimilarity of those properties with adjacent layers. Therefore, it appears reasonable that the average thickness of layers should correspond to, if not exert a dominant control on integral and correlation scales of a property of the medium in a one-dimensional array normal to layering. In order to test this proposed correspondence, we examined subintervals of the logarithm of permeability estimate data in California Well number 2 for which correlation and

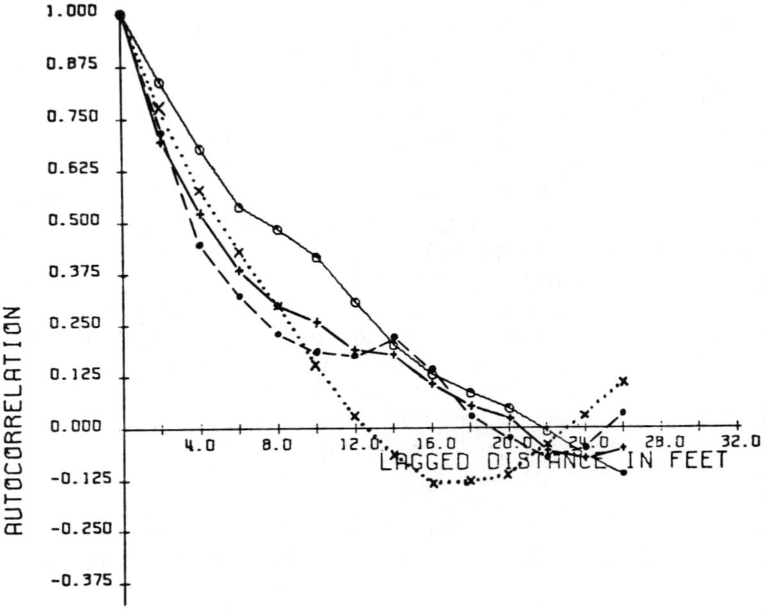

FIGURE 8. Autocorrelation functions representative of California permeability estimate series at 2 ft. data spacing. C2, 1660 - 1948 (+), 1970 - 2258 (x), 2260 - 2548 (•) and 2570 - 2858 (o).

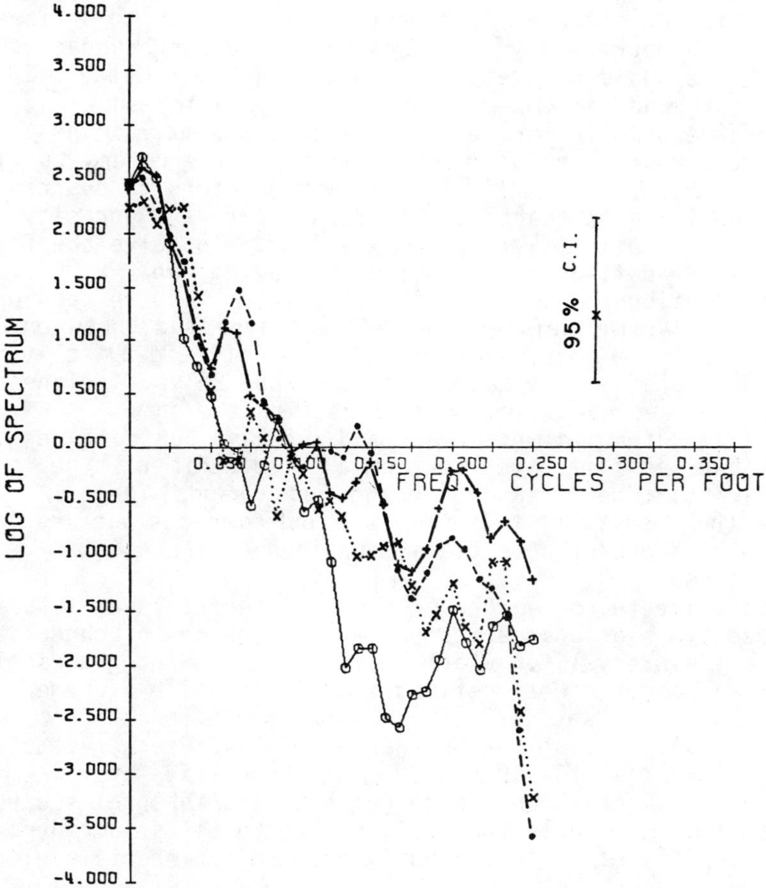

FIGURE 9. Log of spectra representative of California permeability estimate series at 2 ft. data spacing. C2, 1660 - 1948 (+), 1970 - 2258 (x), 2260 - 2548 (•) and 2570 - 2858 (o).

integral scales had been computed, and analyzed the Saraband Log for the number and thickness of layers of different sediment types within that layer set from which average thickness of layering could be computed.

Two relatively minor technical problems became apparent: definition of different sediment types in order to define boundaries between layers and, secondly, depth boundaries of subintervals did not match boundaries of layer sets. We use the right-hand track of the Saraband Log, which shows percentage of bulk volume composed of clay, matrix and porosity; Poupon, et al. (1970) indicate clay refers to clay mineral abundance, matrix (matrix rock) refers to quartz sand and includes quartz and feldspar silt, and porosity refers to volume not occupied by solids. To solve the first problem, we defined sediment type following two classifications. The first classification is more than or equal to 80% clay defines the sediment type mud, between 80% and 50% clay defines sandy mud, between 50% and 20% clay defines defines muddy sand, and less than 20% mud defines sand. The second classification retains the same definitions for sediment types mud and sand but combines the other two sediment types to form a third sediment type "mix"--a mixture of sandy mud and muddy sand. Table 1A shows the results of this analysis and compares the average layer thickness of both classifications enclosed in rectangles.

Relative to the second problem, different layer sets bounded by layer boundaries nearest to the depth boundaries of the subinterval for which correlation and integral scales had been computed were defined and within which average layer thickness was computed. Comparison of the left column of Table 1A, showing the boundaries of layer sets, with the left column of Table 1B, showing the boundaries of subintervals for which the correlation and integral scales are computed, reveals that neither the thickness nor depth of a boundary of the differently defined subintervals differ by more than 12 feet.

The average layer thickness within layer sets (enclosed by rectangles) in Table 1A should be compared with the correlation and integral scales and the 95% confidence interval range for approximately the same subintervals shown in Table 1B. The computed correlation scale commonly is between the average layer thickness calculated for the two classifications of sediment type. Moreover, the computed correlation scale is commonly underestimated by the average thickness of layers defined by the more detailed first classification of sediment types; the average thickness of layers defined by the second classification system tends to over estimate the correlation scale because this system

TABLE 1A. C2 mean sediment layer thickness within layer sets

Classification 1					Classification 2			
Depth (ft) Bounds of layer sets	M = Mud (≥80% clay) sM = sandy Mud (80-50% clay) mS = muddy Sand (50-20% clay) S = Sand (≤20% clay)				M = Mud Mx = Mix (80-20% clay) S = Sand			
		Σ thk	#	$\overline{\text{thk}}$		Σ thk	#	$\overline{\text{thk}}$
1660	M	39.5	9	4.39	M	39.5	9	4.39
	sM	68.5	23	2.98	Mx	105.0	19	5.53
	mS	36.5	23	1.59				
1957	S	152.5	10	15.25	S	152.0	10	15.25
		297.	÷62	4.57		297.	÷38	7.82
1960	M	41.0	11	3.73	M	41.0	11	3.73
	sM	81.5	21	3.88	Mx	123.0	22	5.59
	mS	41.5	20	2.08				
2257	S	133.0	10	13.30	S	133.0	10	13.30
		297.	÷62	4.79		297.	÷43	6.91
2258	M	19.0	8	2.38	M	19.0	8	2.38
	sM	68.0	19	3.58	Mx	128.5	21	6.12
	mS	60.5	23	2.63				
2546	S	140.5	14	10.04	S	140.5	14	10.04
		288.	÷54	4.50		288.	÷43	6.70
2570	M	22.0	10	2.20	M	22.0	10	2.20
	sM	51.5	18	2.86	Mx	86.0	16	5.38
	mS	34.5	15	2.30				
2846	S	168.0	6	28.00	S	168.0	6	28.00
		276.	÷49	5.63		276.	÷32	8.63

TABLE 1B. Computed correlation scale and integral scale of approximately same subinterval as Table 1A layer sets

Computed correlation scale (ℓ), integral scale (λ) and 95% confidence interval (CI)			
Depth (ft) Bounds of sub-intervals	$\ell_{\log k}$ ℓ_ϕ	$\lambda_{\log k}$ λ_ϕ	$(CI_{\log k})$ (CI_ϕ)
1660			
	6.30	4.84	(4.18 - 11.62)
	7.07	4.55	(3.92 - 10.91)
1948			
1970			
	6.87	4.72	(4.16 - 11.57)
	6.15	4.23	(3.66 - 10.17)
2258			
2260			
	5.27	5.06	(4.36 - 12.14)
	3.57	4.09	(3.53 - 9.82)
2570			
	10.48	6.27	(5.41 - 15.05)
	10.46	5.07	(4.37 - 12.17)
2858			

recognizes fewer sediment types and does not recognize extremely common boundaries between sandy muds and muddy sands. The computed integral scales for series of both porosity and logarithm of permeability agree more closely with average thickness of layers defined by either classification system than do the correlation scales. Nevertheless, the average thickness of layers defined by either classification corresponds well, better than simply order of magnitude correspondence, with the correlation and integral scales computed for subintervals of the Saraband Log. Thus, average layer thickness does exert a dominant control on correlation and integral scales computed for one-dimensional subintervals of the logarithm of permeability data in the Cenozoic basinfill.

We do not have geophysical logs or core descriptions of the boreholes in the Mount Simon or Tensleep sandstones from which we have porosity and permeability data. Figure 10c shows that correlation scales for logarithm of permeability data from boreholes in the Mount Simon Sandstone are commonly between 0.5 and 3.0 feet if data spacing was one-foot and correlation scales between 1.0 and 3.0 feet if data spacing was two-feet. Assuming thickness of layering does not changed significantly between the type section and location of our subsurface data, then the 1-3 feet thick bedding of the Mount Simon Sandstone corresponds well with correlation scales of logarithm of permeability data in one-dimensional arrays normal to bedding. The Tensleep Sandstone is lithologically similar to the Mount Simon Sandstone except for thicker (3-7 feet) bedding. Thicker bedding is perhaps reflected by comparison of the distribution of the correlation scales (see Figure 10b) of the Tensleep Sandstone, which are between 1.0 to 6.5 feet, compared to 0.5 to 3.0 feet for the Mount Simon Sandstone. However, small sample size of borehole data from the Tensleep Sandstone precludes more rigorous testing of this slight difference in lithology.

Comparison of the distribution of correlation scales based on one-dimensional arrays of logarithm of permeability data between the internally heterogeneous Cenozoic basinfill with the internally homogeneous and lithologically similar quartzose sandstones of the Mount Simon and Tensleep sandstones further indicates the correlation and/or integral scales are a property of the medium. Correlation scales of the Cenozoic basinfill (Figure 10a) are more broadly distributed, between 2.5 and 10.5 feet with a mode in the 6 to 6.5 feet class, whereas correlation scales for series similar in length of record (about 150 data) and data spacing (two-feet) are distributed between 1.5 and 6.0 feet

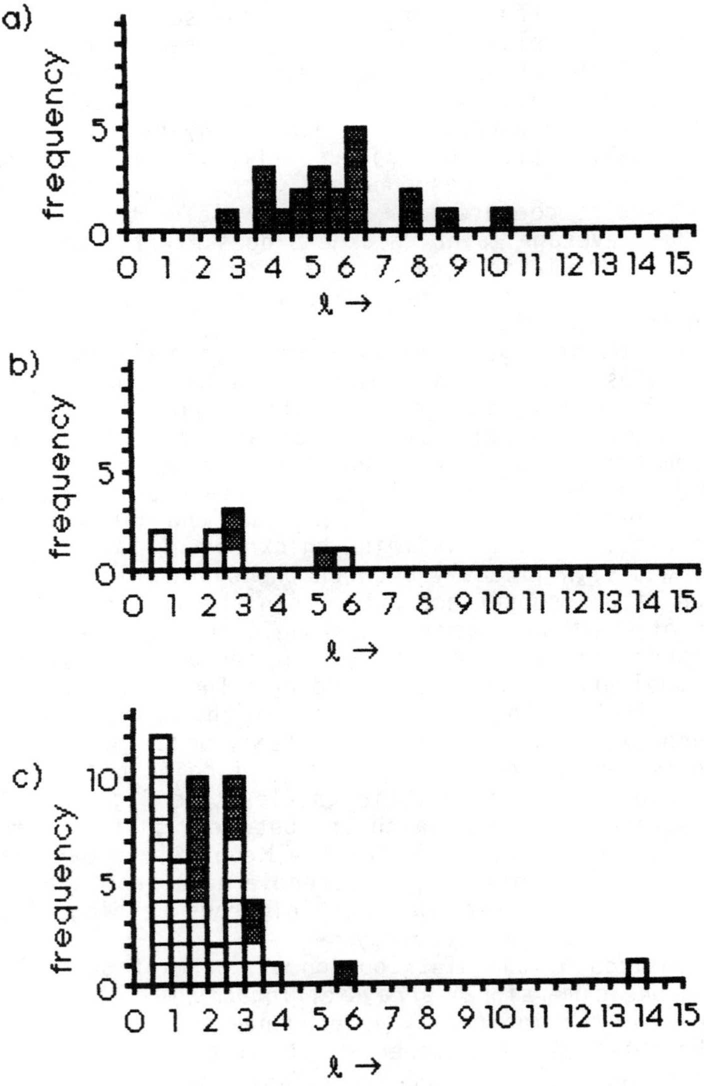

FIGURE 10. Frequency diagram of correlation scale (ℓ) of permeability series from: a) California, b) Wyoming and c) Illinois with ℓ from series of 2 ft. spacing shaded.

with a mode in the 1.5 to 2.0 feet class (Figures 10b and 10c) for the combined Mount Simon and Tensleep sandstones.

IV. SUMMARY, CONCLUSIONS AND RECOMMENDATIONS

Based on our stochastic analyses of field data series of porosity and permeability, we recommend record lengths of about 100 or more equally spaced data. Truncation of the autocovariance function to smooth the raw estimate of the spectrum should occur at 20 lags for record lengths of about 100. Truncation is related to the amount of smoothing and if longer records are available the truncation should not necessarily increase proportionately. For example for record lengths of about 580 data points we would recommend a maximum of 40 lags with possibly further reductions based upon examination of the data.

We further recommend normalizing autocovariance functions by dividing each lagged distance by the variance of the series, thereby producing the autocorrelation function which can be employed for comparison of series. We find that a Fast Fourier Transform of the autocovariance function to estimate the spectral density is satisfactory and that graphing the logarithm of smoothed spectral estimates enhances spectral detail and comparison of the log spectral density functions relative to a constant confidence interval.

We find that correlation scales derived from the autocorrelation function and integral scale derived from the logarithm of the spectral density function of a single series of field data are a property of the medium. We believe the average thickness of geological layering in the medium exerts a dominant control on both the correlation and integral scales for one-dimensional series normal to layering.

REFERENCES

Asthana, V. (1969). "The Mount Simon Formation (Dresbachian Stage) of Wisconsin," unpub. Ph.D. dissertation, Wisconsin Univ., 159 pp.

Bakr, A. A., L. W. Gelhar, A. L. Gutjahr, and J. R. MacMillan. (1978). "Stochastic Analysis of Spatial Variability in Subsurface Flows, 1, Comparison on One- and Three-Dimensional Flows," Water Resour. Res., 14, 2, pp. 263-271.
Buschbach, T. C. (1964). "Cambrian and Ordovician Strata of Northeastern Illinois," Illinois Geol. Sur. Rept. Inves., 218, 40 pp.
Dagan, G. (1979). "Models of Groundwater Flow in Statistically Homogeneous Porous Formations," Water Resour. Res., 15, 1, pp. 47-63.
Gutjahr, A. L., L. W., Gelhar, A. A. Bakr, and J. R. MacMillan. (1978). "Stochastic Analysis of Spatial Variability in Subsurface Flows, 2, Evaluation and Application," Water Resour. Res., 14, 5, pp. 953-960.
Jenkins, G. M., and D. G. Watts. (1968). "Spectral Analysis and Its Applications," Holden-Day, San Francisco, 525 pp.
Lumley, J. L., and H. A. Panofsky. (1964). "The Structure of Atmosphere Turbulence," John Wiley, New York, 239 pp.
Ostrom, M. E. (1970). "Sedimentation Cycles in the Lower Paleozoic Rocks of Western Wisconsin," Field Trip Guidebook for Geol. Soc. America annual meeting.
Potter, P. E., and W. A. Pryor. (1961). "Dispersal Centers of Paleozoic and Later Clastics of the Upper Mississippi Valley and Adjacent Areas," Geol. Soc. America Bull., 72, 8, pp. 1195-1250.
Poupon, A., C. Clavier, J. Dumanoir, R. Gaymard, and A. Misk. (1970). "Log Analysis of Sand-Shale Sequences--A Systematic Approach," Jour. Petroleum Tech., pp. 867-881.
Smith, L., and R. A. Freeze. (1979). "Stochastic Analysis of Steady State Groundwater Flow in a Bounded Domain, 1, One-Dimensional Simulations," Water Resour. Res., 15, 3, pp. 521-528.
Stauffer, C. R., and G. A. Thiel. (1941). "The Paleozoic and Related Rocks of Southeastern Minnesota," Minn. Geol. Sur. Bull., 29, 261 pp.
Templeton, J. S., Jr. 1950). "The Mount Simon Sandstone in Northern Illinois," Illinois Geol. Sur. Circ. 170, 9 pp.
Todd, T. W. (1959). "Areal Petrology of the Sacajawea and Amsden Formations, and the Tensleep Sandstone, Big Horn Basin, Wyoming," unpub. Ph.D. dissertation, Univ. Texas Austin.
Ulrich, E. P. in C. D. Walcott. (1914). "Cambrian Geology and Paleontology," Smithsonian Misc. Coll., 57, p. 354.

SIMULATOR PARAMETER ASSIGNMENT AND THE PROBLEM OF SCALE IN RESERVOIR ENGINEERING

HELGE H. HALDORSEN

SOHIO PETROLEUM COMPANY
SAN FRANCISCO, CALIFORNIA

I. INTRODUCTION

Stated simply, numerical reservoir simulation is the utilization of a mathematical model consisting of governing equations, boundary and initial conditions, constraints, and auxiliary relationships. The objective is to understand the complex chemical, physical and fluid flow processes occurring in a petroleum reservoir sufficiently well to be able to optimize the hydrocarbon recovery. The model is operated through time in such a manner that the real system is reproduced (history matched) or predicted. A grid block, node or cell system serves as the physical representation. This can be either fixed or moving, while mass, energy and momentum may be transported through it.

Reservoir description is the combined effort of discretizing the reservoir into subunits, such as layers and grid blocks, and assigning values of all pertinent physical properties to these blocks. For this task, data from several scales and sources are available, each with a given accuracy and each with a measurement averaging over a different volume (both in terms of size and location) of rock. A significant amount of interpolation, extrapolation and engineering interpretation is often necessary in the description phase since the detail required in the quantification of lateral property variability by far surpasses the detail of the sampling. This problem coupled with the scale difference between sampling supports (i.e., averaging volumes of measurements) and simulator grid block sizes commonly encountered in field scale models poses a paramount challenge in arriving at a realistic reservoir description at the scale of such grid blocks.

Since, in the process of simulation, conservation equations are imposed on all grid blocks through time, the choice of physical properties and auxiliary relationships for each block directly affects the uniqueness of a history match and also the predicted outcome of the process being modelled (primary, secondary, or EOR processes).

Today, almost without exception, decisions on commerciality, maximum efficient rate (MER), overall reservoir management, and the initiation of secondary and enhanced oil recovery projects are based on answers obtained from numerical simulation studies.

For newly discovered fields, the major business decisions such as the location and number of platforms (offshore), depletion strategy, the number and location of wells, and gas and water injection operations are made prior to production initiation. This is also the time when the least data for reservoir description are available. In many instances, only widely spaced discovery and appraisal wells are the basis for a detailed reservoir description necessitated by the system of grid blocks in the numerical model.

For developed fields with close well spacing, the variability in rock properties both vertically (in wells) and laterally (between wells) can pose a problem in that, although correlatable or deterministic features exist, no apparent structure in property variation is seen within units. Several problems are apparent:

1) How does one arrive at meaningful "layers" in a simulation model? From an engineering standpoint, what we want are "hydraulic" (or flow) units. This implies that we seek to achieve a vertical division of the reservoir into rock units which under a given driving force behave differently.

2) While the reservoir is sampled vertically only at a few locations, a detailed system of grid blocks, the majority of which are located away from the sampled locations, must be assigned values of physical properties.

This basically states the problem with reservoir characterization. There is usually an uncertainty (and few guidelines) associated with 1) the identification of hydraulic units, 2) the assignment of "average" physical properties to hydraulic units at the wells and 3) the interpolation and extrapolation of properties to grid blocks away from the wells.

Superimposed on the problems regarding the definition of hydraulic units and their average properties is often a conflict between numerical detail needs and descriptive possibility both (and particularly the former) ultimately being constrained by the maximum possible number of grid blocks (NX x NY x NZ ⩽ Limit). In order to avoid; 1) excessive numerical dispersion, 2) grid orientation effects and 3) solution instabilities and to achieve: 1) preservation of correct local behavior around wells, 2) maintainance of sharp moving fronts, and 3) correct representation of viscous, capillary, dispersive, and gravity effects, the modeller would like to operate with millions of grid blocks. Due to the maximum grid block number constraint and lack of data for such a detailed reservoir description, however, this is not possible nor practical, and a compromise between the desire for detail (in terms of the number of grid blocks) from a 1) computational, 2) reservoir descriptive and 3) cost/efficiency standpoint is sought.

Reservoir description has not reached the same level of confidence and understanding as have the fluid transport formulations and numerical methods. Therefore, today the need for quantitative geological modelling should be emphasized more and more as the sophistication of the numerical models continue to escalate.

In spite of a detailed theoretical basis for the physics and chemistry involved in a particular process, a precise mathematical model for these phenomena, adequate empirical relationships for PVT-behavior, and highly efficient computational schemes for the solution of large sets of equations, the problem remains the same. It is not possible to make useful performance predictions or to arrive at unique history matches of any transport process taking place in a porous medium unless the reservoir description is reliable.

The major steps involved in a reservoir simulation study are shown in Figure 1. Although the intermediate steps certainly are challenging and very important for arriving at a good simulator[1-4] this paper concentrates on problems and procedures associated with the first step in any simulation study, the reservoir description[5-7]. Again, the goals of reservoir description are:

1) to arrive at a meaningful subdivision of the reservoir into hydraulic units,

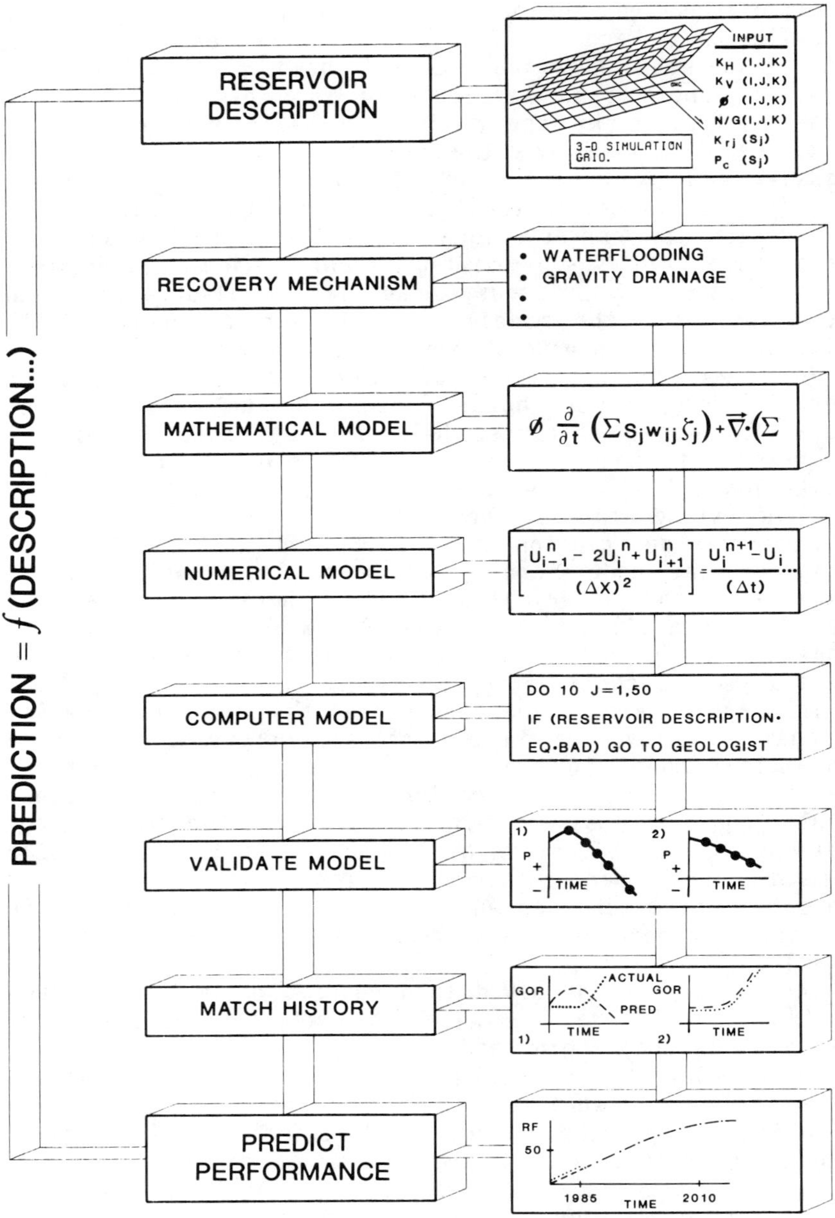

FIGURE 1: THE MAJOR STEPS IN A RESERVOIR SIMULATION STUDY.

2) further discretization of hydraulic units into grid blocks, and

3) to arrive at estimates of the parameters in the partial differential equations describing the process as a function of spatial position. The parameter estimates have the meaning of average or pseudo values at the scale of grid blocks in the discretized version of a continuous reservoir.

First the concept of a Representative Elementary Volume (REV) is revisited. Second, data sources, scales, accuracies and transforms due to scale are discussed. Current reservoir description practice is reviewed and critiqued in light of this discussion. The paper concludes with a call for more synergism between the engineering, geological and other disciplines which contribute to reservoir description for simulation purposes.

II. THE CONTINUUM APPROACH TO POROUS MEDIA AND THE SCALE PROBLEM

Reservoir description must be associated with a particular volume-scale in the reservoir. The relevant volume-scale in the context of numerical simulation is that associated with the practical size of the grid blocks. Since the various measurements or data available for reservoir description are from many different volume-scales (see Section B) and since the transport equations are intrinsically founded on the continuum approach (i.e., rock volumes associated with a mathematical point are homogeneous), it is appropriate first to review this often forgotten basic assumption and also to point out some of the dilemmas which arise.

As noted by Hubbert [8] and later Bear [9], by introducing the term "porous medium" and by considering flow of fluids through a porous medium, we have implicitly introduced the concept of a continuum. The continuum approach is first discussed below.

A. THE CONTINUUM APPROACH

Anyone who has inspected a thin-section of a porous medium will agree that it is close to impossible to describe in any exact manner the geometry of the grains and pore-walls which constitute the flow domain inside a porous medium (Figure 2A). We could, of course, in principle, solve a flow problem utilizing Navier-Stokes equations[10]

(A)

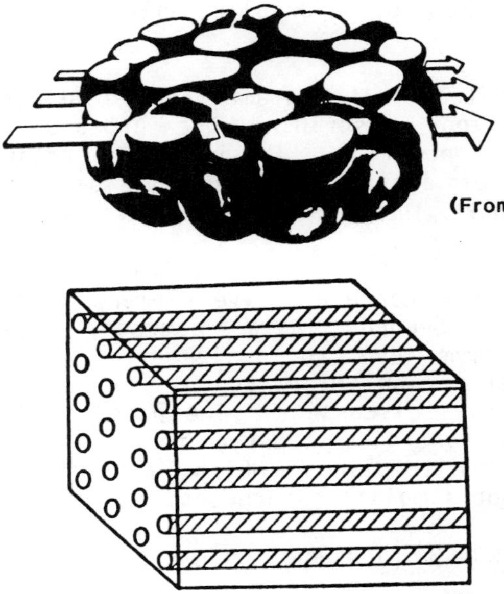

(From Keelan[11], 1977)

(B)

(From Scheidegger[12], 1960)

FIGURE 2: (A) – THE COMPLEX GEOMETRY OF SOLID SURFACES WHICH BOUND THE FLOW DOMAIN.
(B) – STRAIGHT CAPILLARIC MODEL.

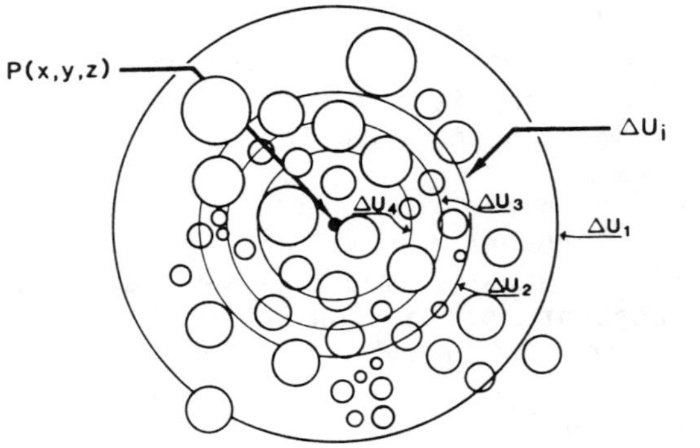

FIGURE 3: P(x,y,z) IS THE CENTROID OF ΔU_i. ΔU_i IS A ROCK VOLUME HAVING THE SHAPE OF A SPHERE AROUND P(x,y,z)

(i = 1,2,3.... indicates shrinking volumes)

to determine the velocity distribution of the fluid in the pore space satisfying a boundary condition of zero velocity at all fluid-solid interfaces. This approach, however, would be exceedingly complex except for a system of straight capillary tubes (Figure 2B) and the solutions would be of no practical value. Hence, to circumvent the difficulties associated with a microscopic reservoir description, we pass to a coarser level of averaging, to the macroscopic level. Now the question is: How much rock do we need to average over before we obtain a representative value of a rock property?

Essential to the concept of a continuum is the notion of a Representative Elementary Volume (REV). The task, therefore, is to determine if and how a rock property (associated with a mathematical point P) is a function of the amount of rock-volume associated with P. Intuitively the REV should be smaller than the entire flow domain, it should be larger than the size of a single pore, and contain a sufficient number of pores to permit the meaningful statistical average required in the continuum concept.

When a porous medium is heterogeneous, with, for example, porosity varying in space, the upper limit of the length dimension of the REV should be a characteristic length that indicates the rate at which changes in porosity take place. The lower limit is related to the size of the pores or the grains.

To illustrate the continuum approach, we borrow some of Bear's [9] discussion and first define volumetric porosity and its REV.

Let P be a mathematical point inside the domain occupied by the porous medium. Consider a volume ΔU_i (having the shape of a sphere) much larger than a single pore or grain, for which P is the centroid (Figure 3). For this volume, we may determine the ratio:

$$\Phi_i = \Phi_i(\Delta U_i) = \left[\frac{(\Delta U_V)_i}{\Delta U_i}\right] \qquad \ldots (1)$$

$(\Delta U_V)_i$ = Volume of space within ΔU_i

Repeating the same procedure, a sequence of values $\Phi_i(\Delta U_i)$, i = 1, 2, 3 ... may be obtained by gradually shrinking the size of ΔU_i around P as a centroid: $\Delta U_1 > \Delta U_2 > \Delta U_3$ ... For large values of ΔU_i , the ratio Φ_i may undergo gradual changes as ΔU_i is reduced, especially when the considered domain is heterogeneous (e.g. layers). Below a certain value of ΔU_i , depending on the distance of P from boundaries of heterogeneity, these changes or fluctuations tend to decay, leaving only small amplitude fluctuations that are due to the random distribution of pore sizes in the neighborhood of P. However, below a certain value of ΔU_o, we suddenly observe large fluctuations in the ratio Φ_i. This happens as the dimensions of ΔU_i converging on the mathematical point P, Φ_i will become either one or zero, depending on whether P is inside a pore or inside a grain. Figure 4 illustrates the relationship between Φ_i and ΔU_i.

The medium's volumetric porosity, $\Phi(P)$, at point P is defined as the limit of the ratio Φ_i as $\Delta U_i \rightarrow \Delta U_o$:

$$\Phi(P) = \lim_{\Delta U_i \rightarrow \Delta U_o} \Phi_i(\Delta U_i(P)) = \lim_{\Delta U_i \rightarrow \Delta U_o} \left[\frac{(\Delta U_V)_i(P)}{\Delta U_i}\right] \quad \ldots (2)$$

For values of $\Delta U_i < \Delta U_o$, we must consider the actual presence of pores and solid particles; in this range there is no single value that can represent the porosity at P. The volume ΔU_o is therefore the Representative Elementary Volume (REV) or the physical (or material) point of the porous medium at the mathematical point P.

A complicating factor is that the magnitude of the REV for different properties may differ. For liquids, the REV of density is an order of magnitude smaller than a single pore, whereas the REV for porosity contains a sufficient number of pores such that fluctuations in the ratio $(\Delta U_V)_i/\Delta U_i$ disappear. The REV for permeability may be very different from the REV of porosity in a heterogeneous medium (Figure 5).

Actually, ΔU_o (defined as REV earlier) should be considered only as the lower limit of the range of volumes from which we may choose the REV for a certain property. For the case of heterogeneities in fluid or medium properties, Bear et al.[16] define a characteristic length L (and a corresponding volume L^3) or three lengths L_i (in the direction of the coordinate system x_i, with a corresponding volume $L_1L_2L_3$) satisfying:

$$L_i - \left[\frac{P}{\left[\frac{\partial P}{\partial X_i}\right]}\right] \quad \ldots (3)$$

where P is the considered property. The volume L^3 (or $L_1L_2L_3$) then serves as the upper limit of the range from which we may choose the REV for property P (Figure 5):

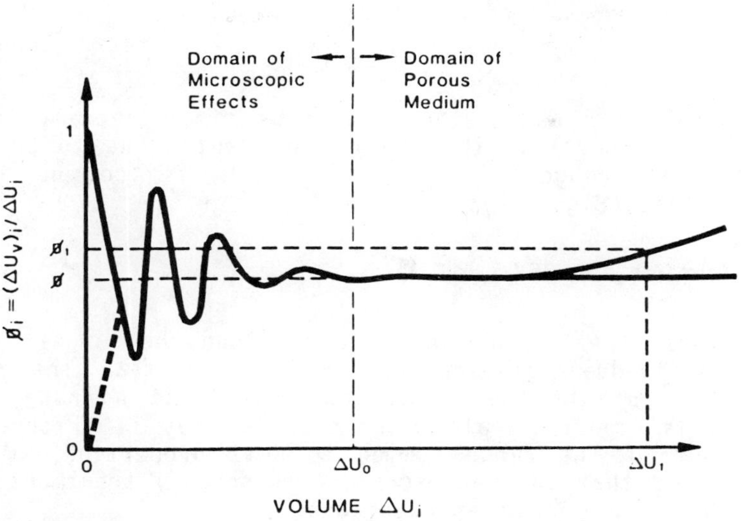

FIGURE 4: DEFINITION OF POROSITY AND REPRESENTATIVE ELEMENTARY VOLUME (ΔU_o). (From Bear[9], 1968)

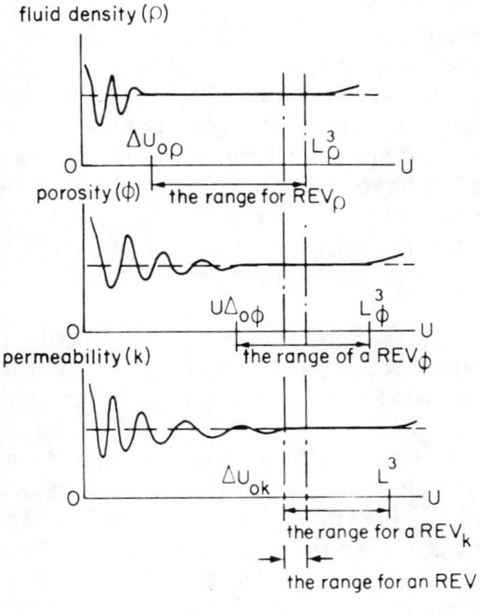

FIGURE 5: THE VARIOUS FLUID AND ROCK PROPERTIES HAVE DIFFERENT REVs. (From Bear and Braester[16], 1972)

$$\Delta U_{o,P} \leqslant REV_P \leqslant L^3 \quad \ldots (4)$$

When at a mathematical point we simultaneously consider several properties, the REV of the medium has to be chosen from the range $\Delta U_o \rightarrow L^3$ which is common to all properties (Figure 5):

$$\operatorname*{Max}_j (\Delta U_{o,P})_j \leqslant REV \leqslant \operatorname*{Min}_j (L^3)_j \quad ..(5)$$

Thus, the REV should be larger than the largest ΔU_o of the individual properties, and smaller than the smaller value L^3 of the individual properties. If no range exists which is common to all properties, we may still choose the REV out of the range common to most properties, with the knowledge that our macroscopic (megascopic) treatment of the problem is no longer an exact one.

Thus, by introducing the concept of porosity and the definition of a REV, we have replaced the actual medium composed of "holes" and sand grains by a fictitious macroscopic continuum in which way may assign values of any property (whether of the medium or of the fluids in the pore space) to any mathematical point in it.

Bachmat and Bear[65] notes that, in principle, any Arbitrary Elementary Volume (AEV) may be selected as an averaging volume and that the selection in any particular case depends only on the model's objectives. The authors, however, proceed to propose a procedure for determining a range of averaging volumes within which measurable characteristics remain constant (i.e., the derivative of the property with respect to volume is zero). It seems clear, however, that each time a new scale of heterogeneity is encountered, a new and larger REV may be defined (see Section IV).

Baveye and Sposito[63] recently reviewed the operational significance of the continuum hypothesis and concluded that the REV concept as applied to porous media is both unnecessarily restrictive and experimentally unverifyable. In its place they suggest a new relativist concept in which macroscopic or megascopic physical properties are defined as convolution products of microscopic or macroscopic properties of a porous medium with weighting functions that represent the appropriate measuring instruments.

Reservoir engineering has fully adopted the classical continuum approach with little concern about its validity on all modelling scales. The actual multiphase porous medium is replaced by a fictitious continuum; a structureless substance, to any point of which we can assign variables and parameters that are functions of the spatial coordinates of the point and of time. Sometimes the multiphase porous medium is replaced by a number of overlapping continua. Each of these represents one phase (solid, fluid 1, fluid 2, etc., Figure 6). To every point in space we may assign macroscopic properties of any of these continua. Interactions may take place between the continua (e.g., relative permeability and capillary pressure). The variables and the parameters of the various fictitious continua, averaged over a REV, enable us to describe flow and other phenomena within a porous and permeable medium domain by means of partial differential equations. Such equations describe what happens at every physical (macroscopic or megascopic) point in space and at every physical instant of time. Note that the fluid flow equations [29] are formulated on a macroscale (the scale of ΔU_0) but are invariantly solved on a megascale (the scale of large grid blocks in reservoir simulation models). Such a scale change is almost as severe as the one from grains and pores to ΔU_0.

In applying the continuum approach on the megascopic scale, we need to introduce megascopic medium parameters and coefficients. Since one grid block may contain billions of ΔU_0s, taking into account heterogeneities (structured or random) will require a statistical approach, since in many ways the task of finding average megascopic properties is similar to that of calculating, for example, permeability of a core plug from the characteristics of individual pores without having to describe in detail each individual pore. The passage from the macroscopic to the megascopic scale may be just as "painful" as the passage from the microscopic to the macroscopic. In effect (since no naturally occurring porous medium is homogeneous) porous medium properties are functions of the averaging volume (Figure 7, 8). Therefore, measurements on one scale (for example, cores) are not directly applicable on another scale (grid blocks) unless an averaging over the new larger volume is performed. Porous medium properties, hence, are always associated with a particular averaging volume. The averaging volume should always be specified along with the property value.

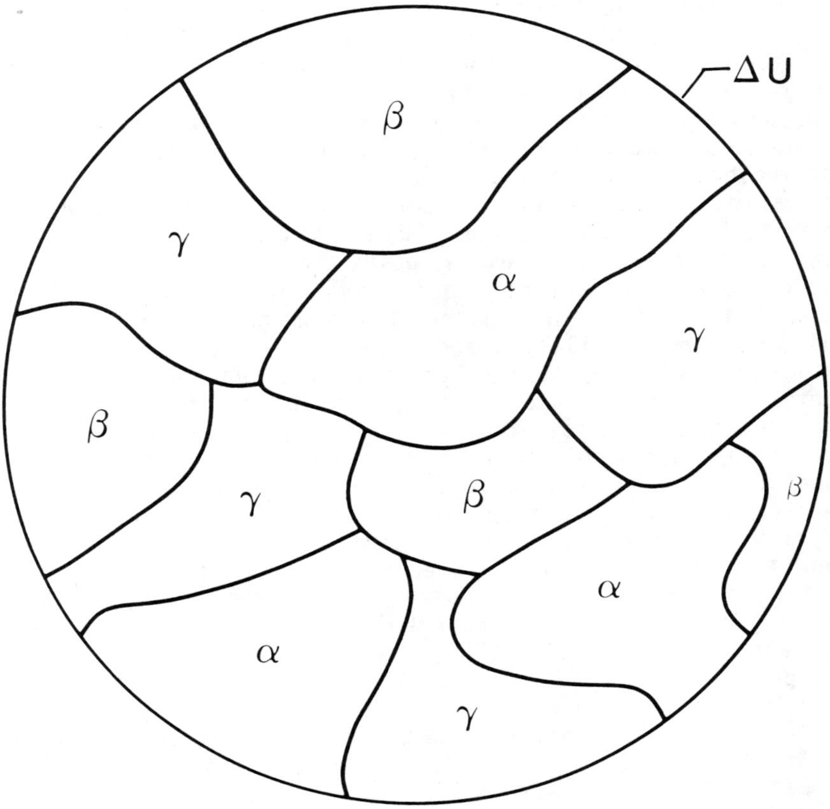

FIGURE 6: HYPOTHETICAL AVERAGING VOLUME (ΔU) WITH THREE PHASES (α, β, γ) PRESENT.
(From Hazzanizadeh and Grey[13], 1979)

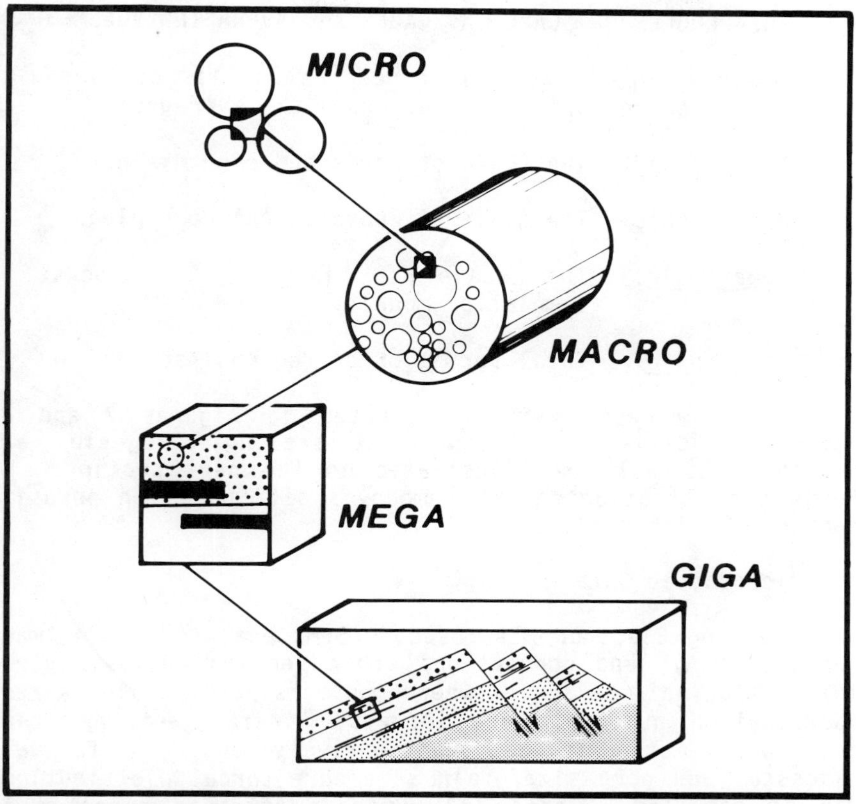

FIGURE 7: FOUR CONCEPTUAL SCALES ASSOCIATED WITH POROUS MEDIA AVERAGES. (From Haldorsen[14], 1983)

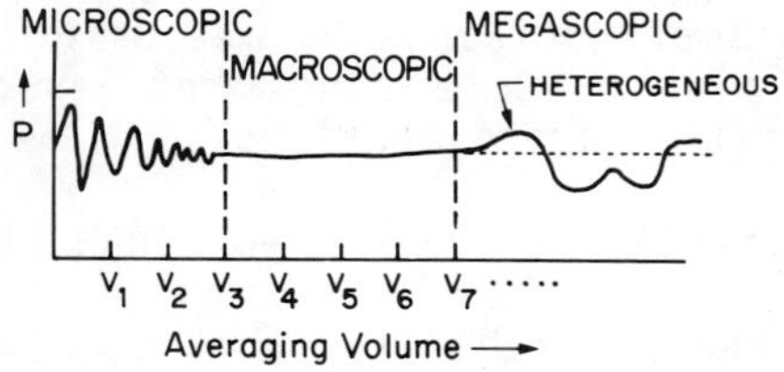

FIGURE 8: PROPERTY, P, PLOTTED AS A FUNCTION OF AVERAGING VOLUME. (From Pickens et al.[15], 1977)

B. THE PROBLEM OF SCALE, AVERAGES AND AVERAGING VOLUMES

Generally speaking, four conceptual scales of averaging volumes can be recognized for porous media averages:

Microscopic: The scale of pores and sand grains.

Macroscopic: The scale of conventional core plugs.

Megascopic: The scale of large grid blocks in simulation models.

Gigascopic: Total formation or regional scale.

These concepts were illustrated in Figures 7 and 8. Several subdivisions within these scaling categories are possible, as will be illustrated in the next section. We focus our attention on any company's data bases on porosity and permeability values.

1. Microscopic Data

Scanning Electron Miscroscope (SEM) analysis, Pore Image Anaysis (PIA), and conventional thin section analysis yield information at a scale where pore sizes, grain sizes, coordination numbers, pore throats, mineral-geometry, clay positions, etc. may be quantitatively and qualitatively assessed, and pore size/grain size/pore throat distributions may be computed. There are numerous attempts reported in the literature[30-35] where microscopic data is used to calculate macroscopic properties.

2. Macroscopic Data (Core Data)

Macroscopic data are obtained by measuring properties of conventional core plugs. Core plugs retrieved for routine core analysis typically have dimensions:

$$V_1=(L_1)(A_1)=(1.5/12)(\pi)(1/24)^2 ft^3=0.00068\ ft^3 \quad \ldots(6)$$

Let us denote the permeability (horizontal) measurements on such core plugs $K_{H,V1}$. In most companies' database, one value of permeability is associated with a certain depth (Figure 9) in a given well (e.g., at 9,434.0 ft, the permeability is equal to the measured permeability of plug No. 8, Figure 9). This permeability value exists from 9,433.5 to 9,434.5 ft in Figure 9. If, for example, a shale

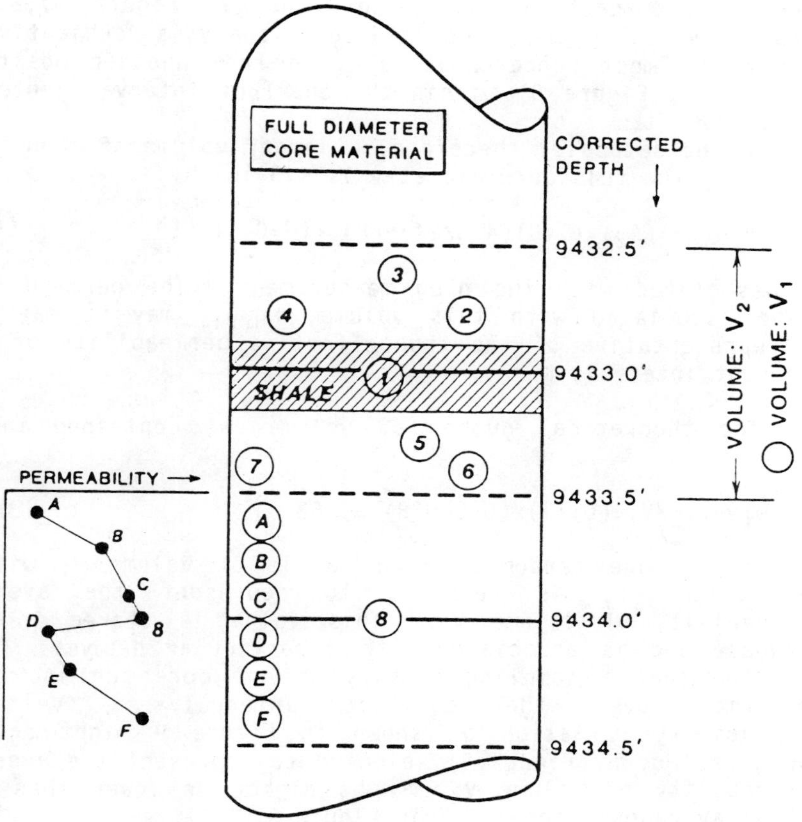

FIGURE 9: FULL DIAMETER CORE SAMPLE AND NUMBERED PLUG SITES.

occurs at a certain depth (Core No. 1, Figure 9), the laboratory doing the routine core analysis subjectively selects a "most representative" plug from another position (plugs 2-7, Figure 9) within the one-foot interval centered at 9,433.0 ft.

In the database, therefore, a larger volume of magnitude (assuming the full core diameter is 4.5 in):

$$V_2=(L_2)(A_2)=(1.0)(4.5/24)^2(\pi)\ ft^3=0.11\ ft^3 \quad ...(7)$$

is associated with the plug measurement. The permeability value associated with this volume, $K_{H,V1}$, may or may not be representative of the true effective permeability of the one-foot interval, $K_{H,V2}$.

The theoretical number of volumes V_1 contained in V_2 is:

$$N_{1/2}=V_2/V_1=(0.11)/(0.00068) = 162 \quad ...(8)$$

Hence, one randomly drawn sample of volume V_1 within the volume V_2 is selected to represent the average permeability of volume V_2. This value is stored in the database and is associated with a particular depth. If we envision that a sampling density of one core per inch was performed over the one-foot interval, a vertical permeability variation as shown in Figure 9 might occur. The core (No. 8, Figure 9) selected to represent the average permeability of volume V_2 may be higher or lower than the actual average of cores A-F in Figure 9.

The most important point to bear in mind from this discussion is: one sample among a population of 162 is drawn at random within a one-foot interval. Since severe vertical and lateral variation in horizontal permeability may occur within inches only (Figures 10A, 10B, 11), we may introduce errors in the database of core permeabilities by assuming that the selected core is representative of the average of the 162 possible cores.

A larger volume V_3 (Figure 12) and associated permeability $K_{H,V3}$ is introduced into the database when "whole core" samples of varying lengths are tested. The depth associated with this permeability value in the database is the midpoint depth of the core section studied. The volume of V_3 is:

$$V_3 = N \times V_2 \quad ...(9)$$

where N is the number of whole core feet.

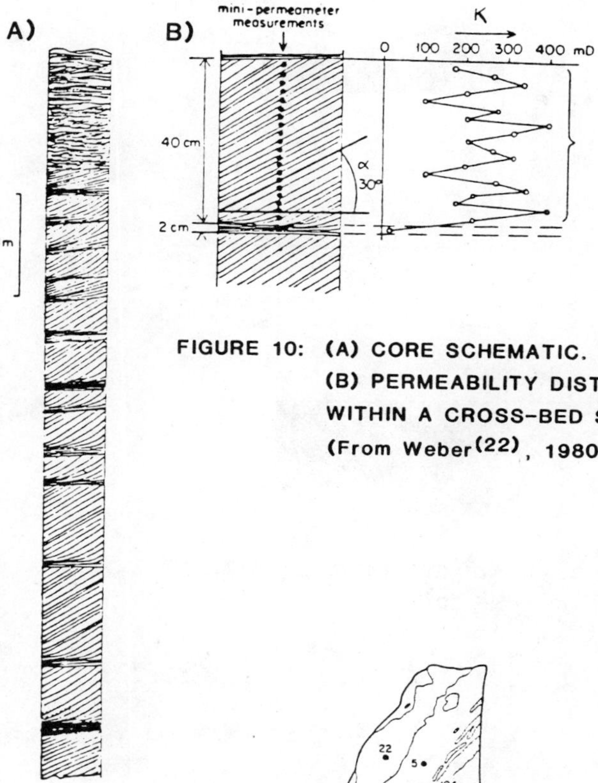

FIGURE 10: (A) CORE SCHEMATIC.
(B) PERMEABILITY DISTRIBUTION WITHIN A CROSS-BED SET.
(From Weber[22], 1980)

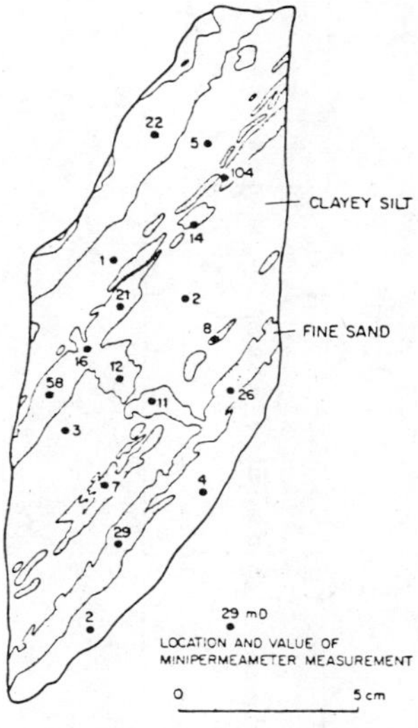

FIGURE 11: PERMEABILITY DISTRIBUTION IN INTERBEDDED BURROWED FINE SANDSTONE AND CLAYEY SILT MEASURED ON A CORE WITH THE MINIPERMEAMETER. (From Weber[22], 1980)

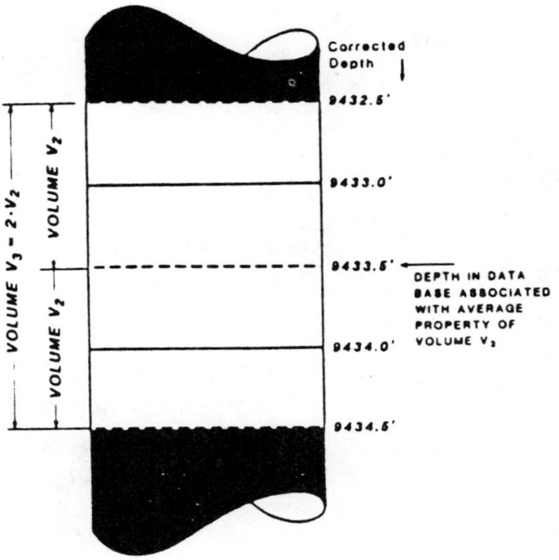

FIGURE 12: SEVERAL FEET OF FULL-DIAMETER CORE MATERIAL. (VOLUME V_3)

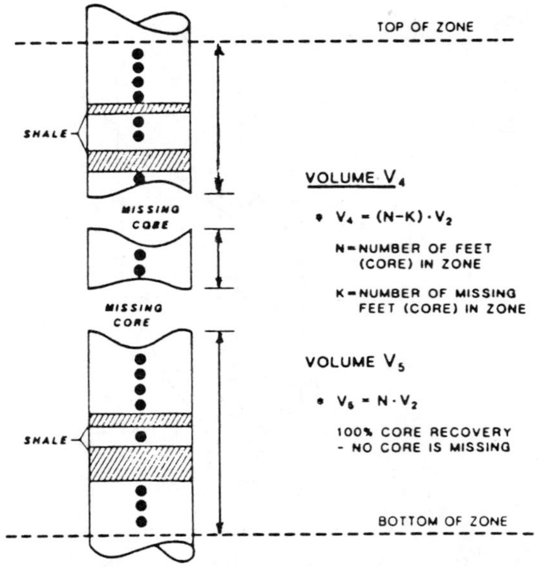

FIGURE 13: CORE AVERAGES IN A SPECIFIC ZONE AND THE PROBLEM OF MISSING DATA.

Once the size of the averaging volume becomes significantly larger than a conventional macroscopic core plug, we refer to it as being of megascopic size. Hence, volume V_3 above could be included in this category. Volumes V_4 and V_5 in Figure 13 are examples of larger megascopic volumes. In this case, we consider portions of a petrophysical zone (hydraulic unit) or the whole zone. The average permeabilities associated with megascopic volumes V_4 and V_5 are not measured directly, but are arrived at by averaging (arithmetic, harmonic, or geometric) the macroscopic core plug measurements available within V_4 or V_5.

3. Megascopic Data

a. Log Data

In wells which are cored and logged, it is common to establish a correlation between the porosity deduced from wireline logs and the core porosity measured at the same depth. Note that the core material is altogether removed from the volume of investigation of the log.

The volume of investigation of a sonic log is indicated in Figure 14. The porosity estimate deduced from the sonic - Δt is a moving average "measuring" two vertical feet of rock at a time and about 1.5 inch into the formation. The volume of rock investigated is about:

$$V_{Son}=(((8.5+3)/24)^2-(8.5/24)^2)\pi 2 \text{ ft}^3=.654 \text{ ft}^3 \quad \ldots (10)$$

The number of core plugs theoretically contained in two vertical feet of whole core is:

$$N = 2\ N_{1/2} = 2(162) = 324 \quad \ldots (11)$$

The porosity measured on one of these possible 324 cores is actually compared to the sonic porosity. The ratio of log to core volume investigated is:

$$N_{1/Son} = (.654)/(.00068) = 962 \quad \ldots (12)$$

Collins et al.[36] showed that in a statistically homogeneous reservoir, the porosity distribution must be normal or Gaussian. They also found that the standard deviation, σ, of the porosity distribution varies inversely as the square root of the volume per sample:

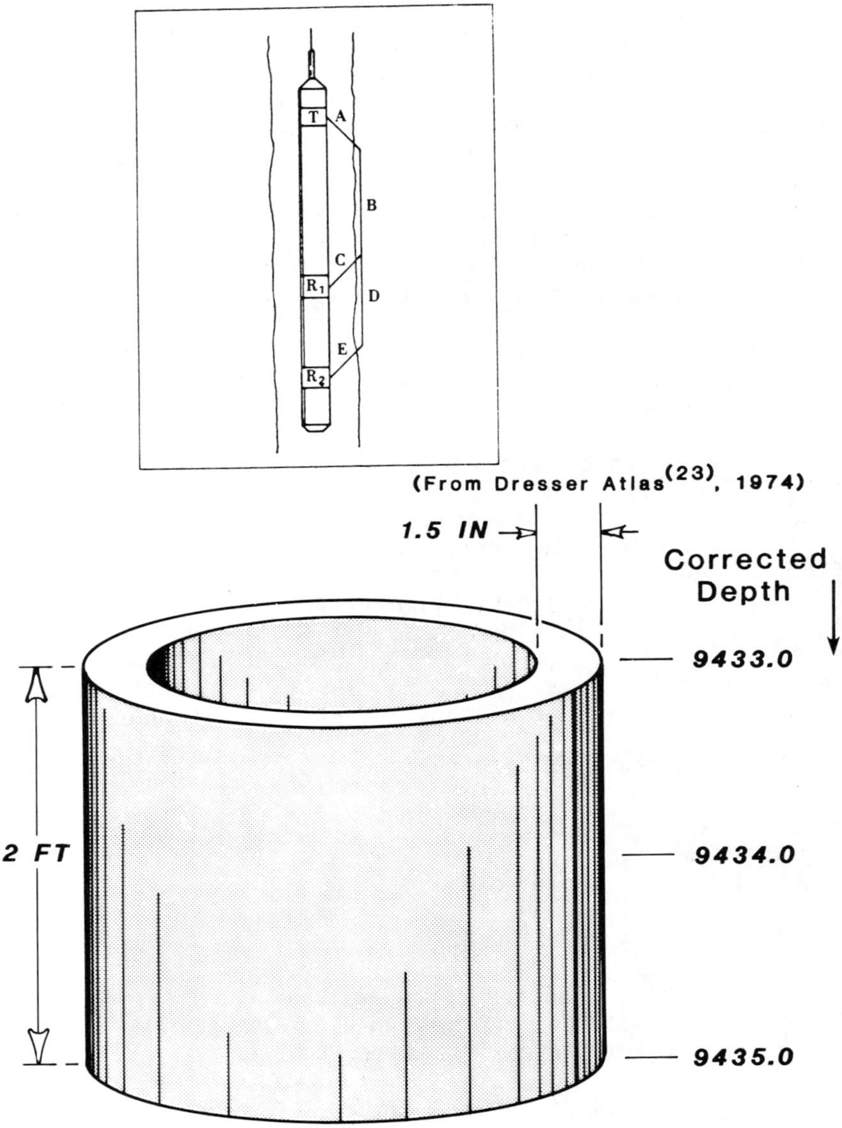

FIGURE 14: SONIC LOG ARRANGEMENT AND ITS VOLUME OF INVESTIGATION.

$$\sigma = \sqrt{\frac{\epsilon}{V} \bar{\Phi} (1 - \bar{\Phi})} \quad \text{.. (13)}$$

where,

ϵ = Characteristic Volume, L^3
V = Sample Size, L^3
$\bar{\Phi}$ = Average Reservoir Porosity, Fraction

Thus, the frequency distribution of porosity obtained using core plugs will exhibit a much greater spread than that resulting when sonic logs are used to estimate porosity. In fact, theoretically the standard deviation of porosities estimated from sonic logs, σ_{Son}, will be given by:

$$\sigma_{son} = \sqrt{V_{core}/V_{son}}\ \sigma_{core} = .03\ \sigma_{core} \quad \text{..(14)}$$

Needless to say the scale problem is again apparent in this type of approaches. There is, however, a real need for valid log to core correlations and relationships. What is needed is more attention to the problem of scale. Geostatistics seem to have a real potential for solving many of the scale problems relating to porosity.(26) Collins et al.(36) further concludes that:

i) Porosity of megascopic units should be computed from core data as the volume weighted arithmetic average.

ii) The probable error in the average porosity value is proportional to the inverse square root of total volume of cores analyzed.

iii) In the case of the particular reservoir studied (Woodbine Sandstone), every 5-cubic-foot piece of rock has a porosity within 0.5% of the average porosity (ϵ found to be 1.7 $10^{-6}ft^3$).

iv) Unique distributions of porosity do not exist. Thus, all distributions previously reported in the literature are useless for precise estimates, unless the sample sizes were reported. This dependence of the distributions on sample size also implies that it is not just the number of samples which is important in arriving at a reliable estimate of average porosity but also the total volume of samples.

b. Residual Oil Saturation Determination

A much sought after parameter in reservoir engineering is the residual oil saturation. For a given reservoir being produced under a given scenario, many different sources and scales of measurements are the basis for establishing S_{or}-estimates as shown in Figure 15. There is, of course, no reason why all the estimates should come out with the same number given 1) the vast difference in rock volume being investigated among the various techniques, 2) the variability of S_{or} due to rock-fluid interactions, and 3) the variability of the rock itself.

c. Large Scale Pressure Tests

Permeability estimates (megascopic) may be derived from the pressure response during pressure build-up testing. The radius of investigation of such a test is a function of time and formation properties (Figure 16):

$$r_{inv} = \sqrt{.00105\ [kt/\Phi\mu c]} \qquad ...(15)$$

where,

t = flowing time; hrs
k = permeability; md
Φ = porosity; fraction
μ = viscosity; cp
c = compressibility; psi^{-1}

and can easily be several thousand feet. Also, the volume tested normally contains large-scale heterogeneities. Thus, the horizontal permeability estimate inferred from the test truly is an average over a vast volume. Assuming the radius of investigation is 1,500 feet and the perforated interval is 200 feet, the volume of investigation is:

$$V_{BU} = 1.4 \times 10^9\ ft^3 \qquad ...(16)$$

Now if all the 200 feet were sand and core recovery was 100 percent, the volume of the core plugs going into the estimation of an average core permeability is:

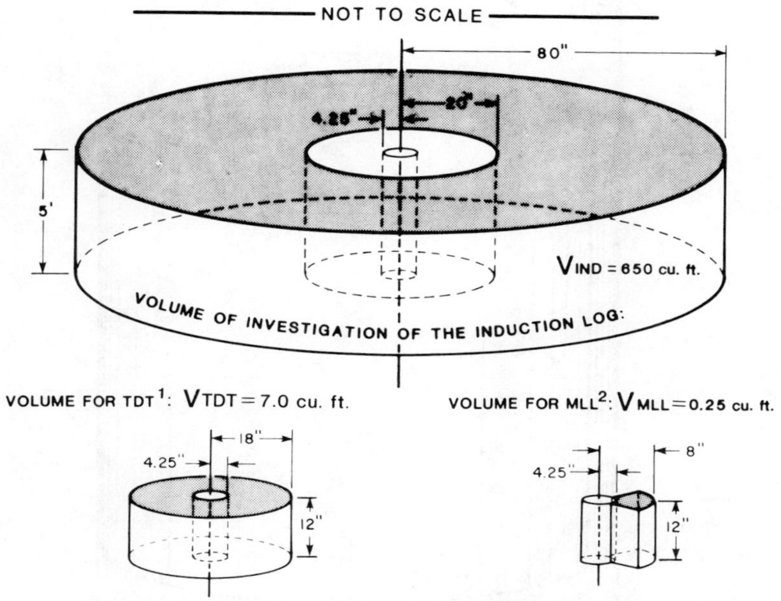

- VOLUME INVESTIGATED BY CTT[3]: $V_{CTT} = 15{,}000$ cu. ft.

- VOLUME OF A CORE PLUG: $V_{CORE} = 0.003$ cu. ft.

VOLUME RATIOS

$$\frac{V_{CTT}}{V_{CORE}} = 5{,}000{,}000 \qquad \frac{V_{IND}}{V_{CORE}} = 200{,}000 \qquad \frac{V_{TDT}}{V_{CORE}} = 2000 \qquad \frac{V_{MLL}}{V_{CORE}} = 80$$

1 TDT = THERMAL DECAY TOOL

2 MLL = MICROLATERLOG

3 CTT = CHEMICAL TRACER TEST

FIGURE 15: OVERVIEW OF AVERAGING VOLUMES ASSOCIATED WITH RESIDUAL OIL SATURATION DETERMINATION.

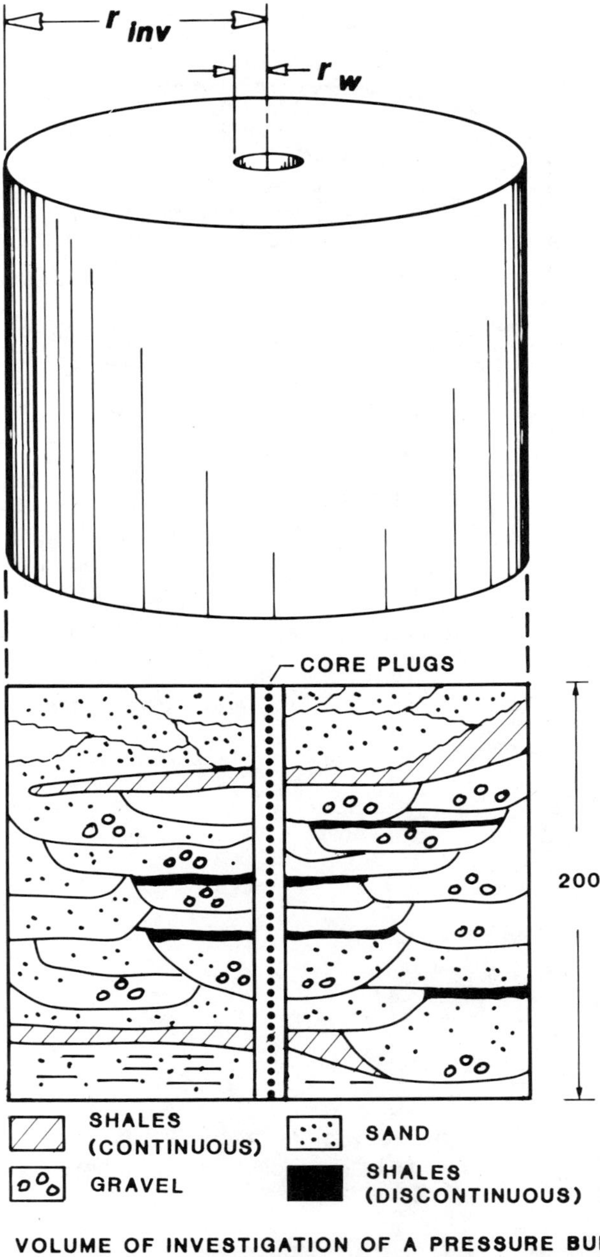

FIGURE 16: VOLUME OF INVESTIGATION OF A PRESSURE BUILD-UP TEST AND CROSS SECTION INDICATING LARGE-SCALE INTERNAL HETEROGENEITIES

$$V_{200} - 200(V_1) = .136\ ft^3 \qquad \ldots(17)$$

The ratio of the build-up volume to the core volume is then:

$$R = 10.3 \times 10^9 \qquad \ldots(18)$$

i.e., the volume of investigation of the build-up is ten billion times larger than the volume of the 200 cores going into the calculation of the average core permeability.

Still, however, the averaging of permeability, i.e., arithmetic vs geometric or harmonic averaging continues to be a hot reservoir engineering issue (Figure 17). Dubrule et al.[26] point out the many reasons why core averages and build-up permeabilities may disagree. Many authors correctly advise that conditioning of core averages with well tests is the only possible solution to this problem. If hydraulic units (Figure 18) may be identified, a relationship between the build-up permeability and the zonal core averages could be:

$$\left[\frac{(Kh)_{BU}}{\widetilde{k}_{ro}(S_o)}\right] = \sum_{j=1}^{N_z} \alpha_j \overline{K}_j h_j \qquad \ldots(19)$$

where,

$(Kh)_{BU}$ = Build-up derived permeability-thickness product, L^3

$\widetilde{k}_{ro}$ = Thickness-averaged oil relative permeability at time of test, Fraction

S_o = Thickness averaged oil Saturation, Fraction

α_j = Correction Factor for Zone j, Dimensionless (determined by Regression)

$\overline{k}_j$ = Average core Permeability in Zone j, L^2

N_z = Number of Zones

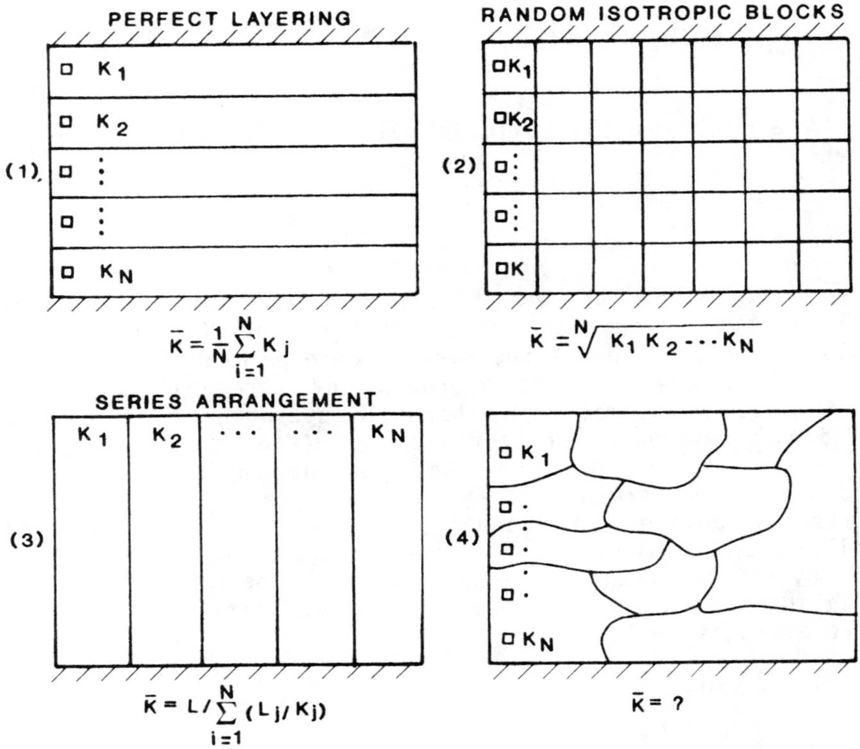

FIGURE 17: AVERAGING OF PERMEABILITY (From Ref. 37) (From Stewart et al.(37), 1980)

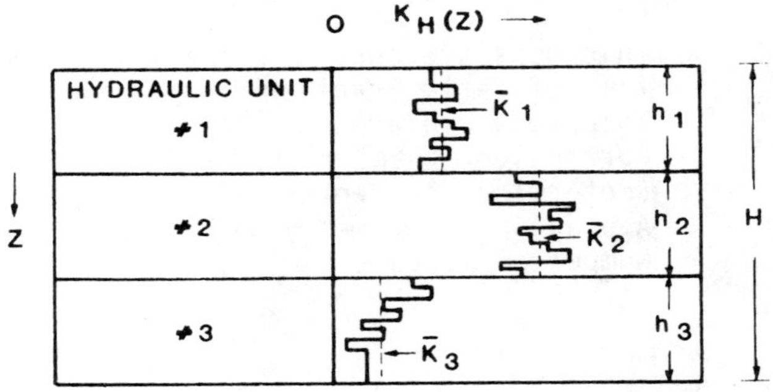

FIGURE 18: LAYERING OF PERMEABILITY (From Ref. 37) (From Stewart et al.(37), 1980)

Based on analytical work, Collins et al.[61] found that the amount of spread of the permeability distribution is dependent on the size of the samples (i.e., large spread or variance for core plug samples and less spread for build-up permeabilities). This was empirically confirmed by Codreanu[38] who compared distributions of pressure build-up derived and core derived permeabilities (Figure 19). We see from Figure 19 that both for Reservoir A and B the variance in permeability is much less for the large-scale (build-up) data. In addition, for Reservoir B (Table 1) the mean permeabilities are close (and the build-up variance very small) thus indicating an almost homogeneous system at the volume scale of build-up tests. For Reservoir A, the means are far apart and the variance in build-up permeabilities is quite significant. This indicates strong spatial variability (random or with a trend) at the scale of the volume of investigation of the build-up test.

In the case of the Woodbine Sand, Collins et al.[61] found that any piece of the reservoir of 225 cubic ft volume should have essentially the same permeability. Their other conclusions of general importance were:

i) More samples are required in very permeable reservoirs than in tighter reservoirs for the same degree of precision in average permeability determination.

ii) The probable error in average permeability decreases as the sample volume increases and increases as the average permeability increases.

The permeability derived from a pressure build-up test is given by:

$$K - \left[\frac{162.6\,(q\mu B_o)}{mh}\right] = f\,(q,\mu,B_o,m,h) \qquad \ldots(20)$$

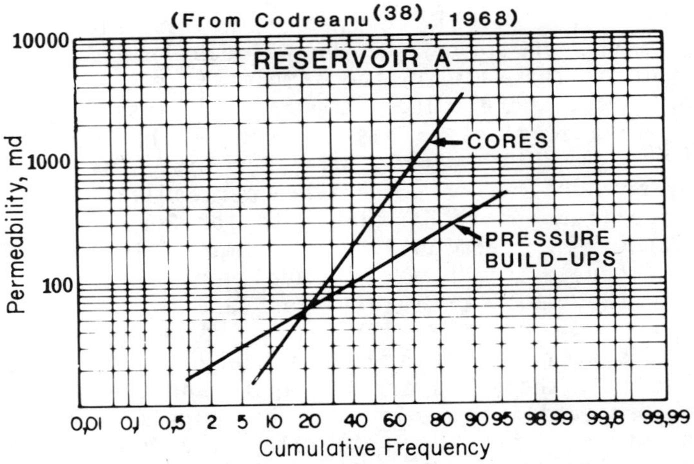

FIGURE 19A: PERMEABILITY FROM CORES AND BUILD-UPS PLOTTED AS A FUNCTION OF CUMULATIVE FREQUENCY

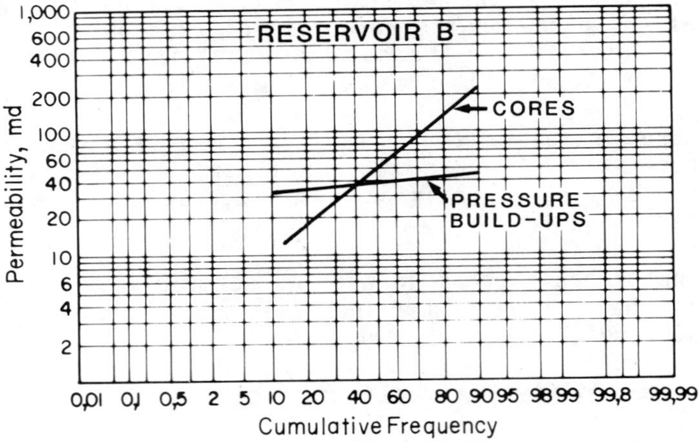

FIGURE 19B: PERMEABILITY FROM CORES AND BUILD-UPS PLOTTED AS A FUNCTION OF CUMULATIVE FREQUENCY

TABLE 1: MEAN AND VARIANCE IN PERMEABILITY FOR RESERVOIRS A AND B.

	Geometric Mean Permeability (md)	Variance (md^2)
RESERVOIR A:		
- CORES	322	4.000
- BUILD-UPS	122	0.684
RESERVOIR B:		
- CORES	48	1.440
- BUILD-UPS	37.5	0.0128

The error in this estimate, given uncertainties or errors in the parameters, is approximately given by:

$$\Delta K = \left[\frac{\partial f}{\partial q}\right]\Delta q + \left[\frac{\partial f}{\partial \mu}\right]\Delta \mu + \left[\frac{\partial f}{\partial B_o}\right]\Delta B_o$$

$$+ \left[\frac{\partial f}{\partial m}\right]\Delta m + \left[\frac{\partial f}{\partial h}\right]\Delta h \qquad \ldots(21)$$

Thus,

$$\Delta K = \left[\frac{162.6 \quad \mu B_o}{mh}\right]\Delta q + \left[\frac{162.6 \quad q\, B_o}{mh}\right]\Delta \mu + \left[\frac{162.6 \quad q\,\mu}{mh}\right]\Delta B_o$$

$$- \left[\frac{162.6 \quad q\,\mu\, B_o}{hm^2}\right]\Delta m - \left[\frac{162.6 \; q\,\mu\, B_o}{mh^2}\right]\Delta h \qquad \ldots(22)$$

Now if we have the following well-test data and associated uncertainties:

Variable	Value	Accuracy
q	1000	± 50
μ	1	±.25
B	1.3	± .1
m	84	± 10
h	100	± 25

we find:

$$K = \left[\frac{162.6 \;\; 1000. \;\; 1.3 \;\; 1.}{84. \;\; 100.}\right] = 25.2 \text{ md} \qquad \ldots(23)$$

and,

$$\Delta K = .025\,\Delta q + 25.2\,\Delta\mu + 19.35\,\Delta B_o - .3\,\Delta m - .25\,\Delta h \qquad \ldots(24)$$

which gives a maximum possible error of ΔK = 18.7 md and a minimum error of ΔK = 0.23 md. Hence, we conclude that pressure transient permeabilities also are associated with errors, and the magnitude of the coefficients in Equation (24) indicate which term contribute the most to the error. This is also where we should concentrate our efforts in terms of increasing the accuracy.

Further important aspects of absolute permeability averaging, mapping and estimation are outlined in more detail in Refs. 25, 26.

Additional megascopic information may be derived from interwell tracer tests, interference and pulse-tests. Such tests are particularly helpful in determining directional trends in permeability and also the presence of hydraulic units. The concept of a hydraulic unit is very important from a reservoir engineering (and particularly from a numerical simulation) point of view as we briefly discuss next.

III. PATHWAYS AND BARRIERS: HYDRAULIC UNITS

It was pointed out in the introduction section that one of the most important up-front descriptive efforts for simulation purposes is to identify hydraulic units. That is, we seek a vertical zonation of the reservoir which makes sense in a hydraulic or fluid-flow context. Such fluid pathways are often determined by continuous and discontinuous shales. This is discussed extensively in Refs 24 and 27 and warrants only limited discussion here. Simulation layer boundaries, should coincide with hydraulic unit boundaries, which is easily determined in the case of continuous shales. If no barriers are present and the zonation must be based on (apart from numerical and process modelling detail needs), for example, the knowledge of porosity and permeability variations with depth, statistical techniques such as the ones proposed by Testermann(39), Gill(40), and Bennion and Hope(62) are viable.

Also, vertical variograms may indicate the degree of structure in the property variation. As pointed out by Archer(29), increasing (decreasing) permeability level with depth within a reservoir should be modelled as explicitly as possible due to its bearing on recovery processes (Figure 20).

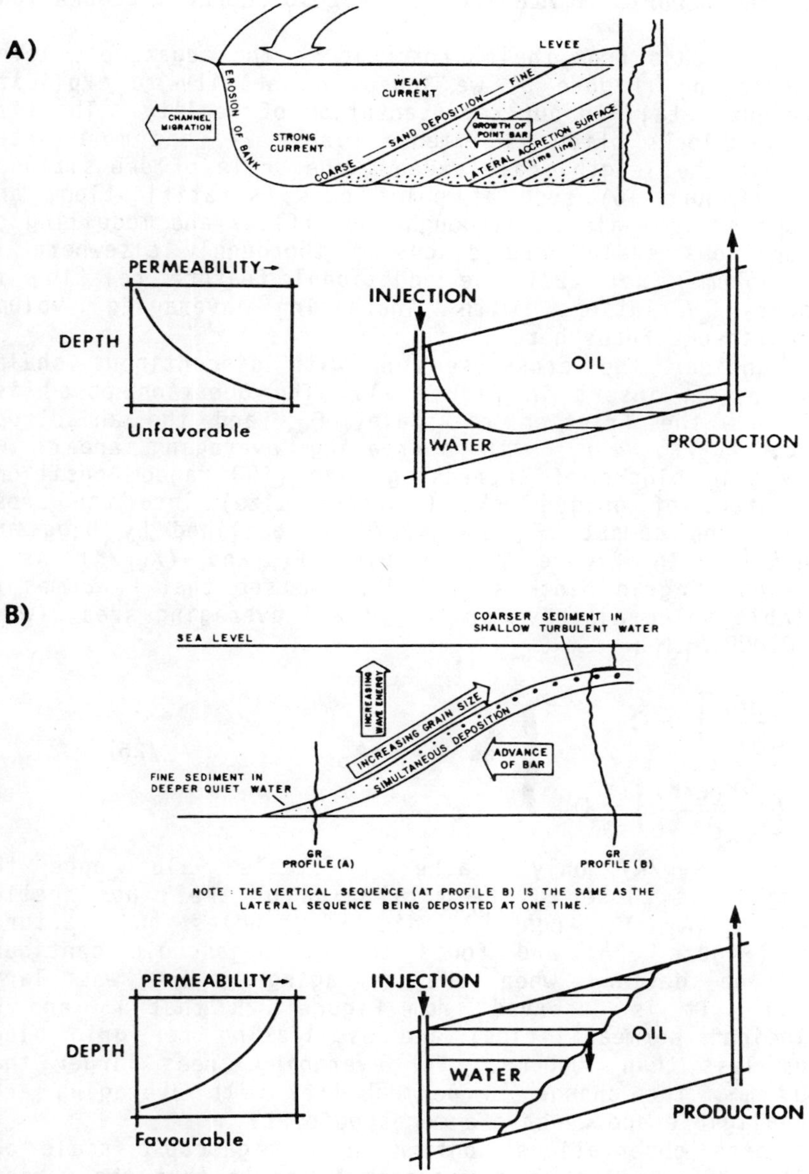

FIGURE 20: EFFECT OF PERMEABILITY DISTRIBUTION IN WATERFLOODING. (From Archer(29), 1983)

IV. MACROSCOPIC TO MEGASCOPIC: PSEUDO PROPERTY GENERATION

As we pass from the macroscopic to the megascopic level of averaging (Figure 8) we loose the ability to explicity represent detail in our representation of reality. The size of grid blocks often is immense (Table 2), and more often than not the length scale exceeds the scale of depositional units (Figure 16) such as point bars, stratification, and discontinuous shales. Although the effects and modelling of discontinous shales are discussed thoroughly elsewhere in this volume (Ref. 27) one additional feature relating to property variations with increasing averaging volume warrants some focus here.

Consider the cross section with discontinous shales shown as an insert in Figure 21. The question posed is: How does the fraction of shale, F_s, and the anisotropy ratio, K_{VE}/K, vary with increasing averaging area? We throw grid blocks of increasing size (100 random positions of centers of grid blocks for each size) into the cross section and compute F_s and K_{VE}/K as outlined by Begg and King.[41] In Figure 21 we plot F_s and (K_{VE}/K) as a function of grid block size (LH). We see that F_s comes to a stable value already for very small averaging areas (when $A_o = 0.0001 A_{TOT}$):

$$\left.\begin{array}{l} \left[\dfrac{\partial F_s}{\partial A}\right] \rightarrow 0 \\[2ex] \left[\dfrac{\partial (K_{VE}/K)}{\partial A}\right] \rightarrow 0 \end{array}\right\} \text{ as } \quad A \longrightarrow A_o \qquad \ldots(25)$$

whereas (K_{VE}/K) only reaches a stable value once the averaging is close to the total area of the cross section (when $A'_o \cong .1 A_{TOT}$). Long[65] did REV-studies on fractured rock (Figure 22A) and found that a megascopic continuum could be defined when the averaging region was large enough. It is observed from Figure 22B that K_1 and K_2 (principal permeabilities) are oscillating for grid block sizes less than $1000 cm^2$. For averaging areas larger than $5000\ cm^2$, the change in permeability with averaging area is negligible and we have a megascopic REV.

These observations confirm on a megascopic scale our earlier observation on a macroscopic scale that the various properties have different REVs. The megascopic length scale at which properties such as F_s and K_{VE}/K become stable should be termed the MEGA-REV to distinguish it from our earlier definition of a MACRO-REV.

TABLE 2: REPORTED GRID BLOCK SIZES IN SELECTED FIELD SCALE SIMULATION STUDIES

(From Haldorsen[14], 1983)

Ref.	Process	ΔX (ft)		ΔY (ft)		ΔZ (ft)		V (ft^3)		Number of Blocks
		min	max	min	max	min	max	min	max	
17	UNDERSAT. OIL RESERVOIR	1640	1640	984	8858	23	31	3.7×10^7	4.5×10^8	5606
18	GAS RESERVOIR	4921	4921	4921	4921	230	750	5.5×10^9	1.8×10^{10}	375
19	SIMULATION OF WELL TEST	800	2200	800	3400	20	200	1.2×10^7	1.5×10^9	468
20	RETROGRADE GAS RES.	775	1500	775	1500	110	200	6.6×10^7	4.5×10^8	1560
21	SOLUTION GAS DRIVE RES.	2000	2000	1500	1500	10	60	3.0×10^7	1.8×10^8	150

– ΔX, ΔY, ΔZ ARE THE LENGTHS OF THE SIDES OF THE GRID BLOCKS

– V IS THE VOLUME OF THE GRID BLOCKS

NOTE: $$\text{VOLUME RATIO} = \frac{V_{\text{AVERAGE GRID BLOCK ABOVE}}}{V_{\text{CORE}}} = \frac{2.6 \times 10^9}{0.00068} \cong 4 \times 10^{12}$$

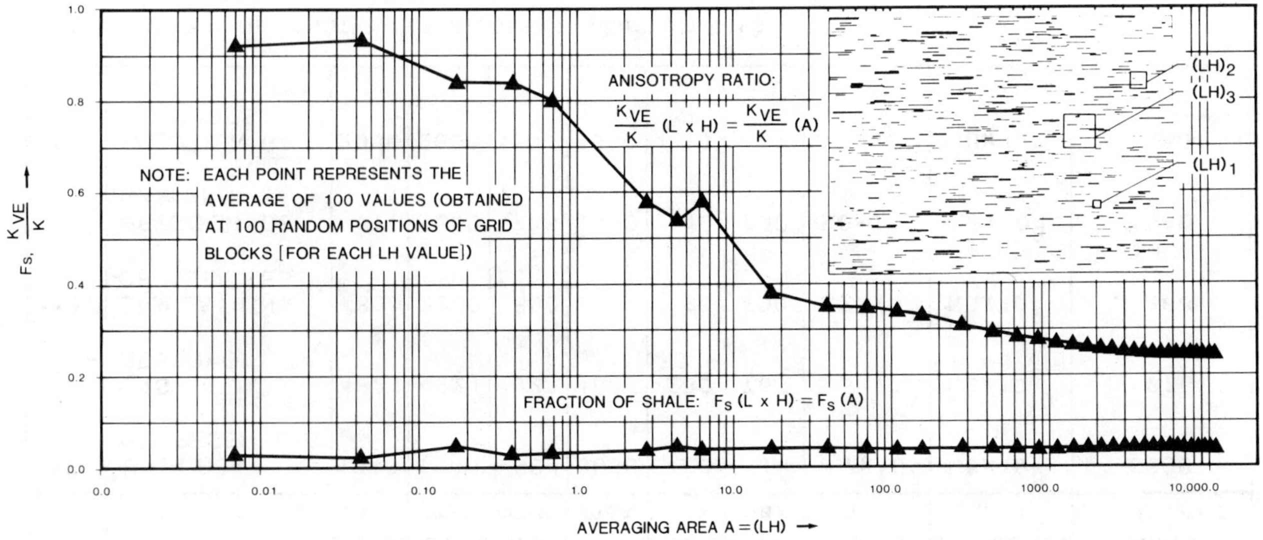

FIGURE 21: FRACTION OF SHALE, F_S, AND ANISOTROPY RATIO (K_{VE}/K) PLOTTED AS A FUNCTION OF AVERAGING AREA (LH).

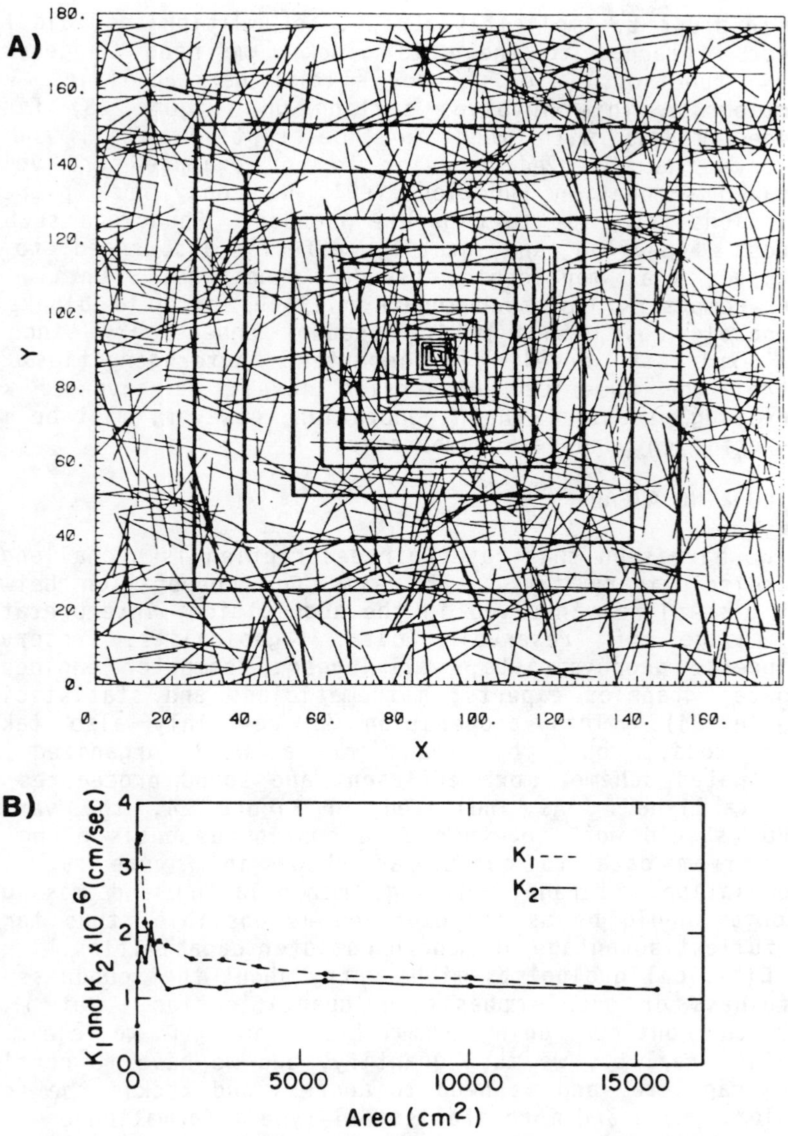

FIGURE 22: A) RANDOM 2-D FRACTURE NETWORK AND INCREASING AVERAGING AREA
B) PRINCIPAL PERMEABILITIES PLOTTED AS A FUNCTION OF AVERAGING AREA (From Long[(65)], 1985)

In order to implicitly account for multiphase fluid flow effects internal to the grid blocks, we need to generate pseudo functions (e.g. pseudo relative permeabilities, well functions). This involves setting up (Figure 23) finely gridded models with much more descriptive detail and to preserve the observed behavior in for example pore volume averaged saturation functions.(28) Not only rock property variations cause us to generate pseudos. Phenomena such as viscous fingering and gravity effects also need to be captured via megascopic pseudo functions. Until the super-computers can accommodate millions of grid blocks at reasonable cost, thus making pseudos unnecessary since we can explicity model all important heterogeneities and control the numerical dispersion level, we have to keep generating pseudos. Their generation, however, must be made more efficient and less cumbersome.

V. LOOKING AHEAD

We have seen that reservoir description is a challenging interdisciplinary task. It calls for cooperation between most disciplines involved in the acquisition, interpretation and usage of reservoir data: geologists, reservoir engineers, geophysicists, well-loggers, computer engineers, computer graphics experts, mathematicians and statisticians (Figure 24). This cooperation is certainly also taking place today, but, by means of a well organized and coordinated scheme, more efficient and sound procedures can be established. As indicated in Figure 24, the various groups should work together on a continuous basis along the lines from data to output as shown in Figure 25. The accumulation, storage, editing, manipulating and massaging of data should be as computerized as possible, thus taking the fullest advantage of modern computer capabilities.

Classical geologists might worry about the soundness and usefulness of such emphasis on quantification. But if we want to continue doing simulation, and if we claim to simulate reality, we must quantify, and we have to continue to extrapolate, and we need to address and tackle the scale problem. More and more statistical-type information on

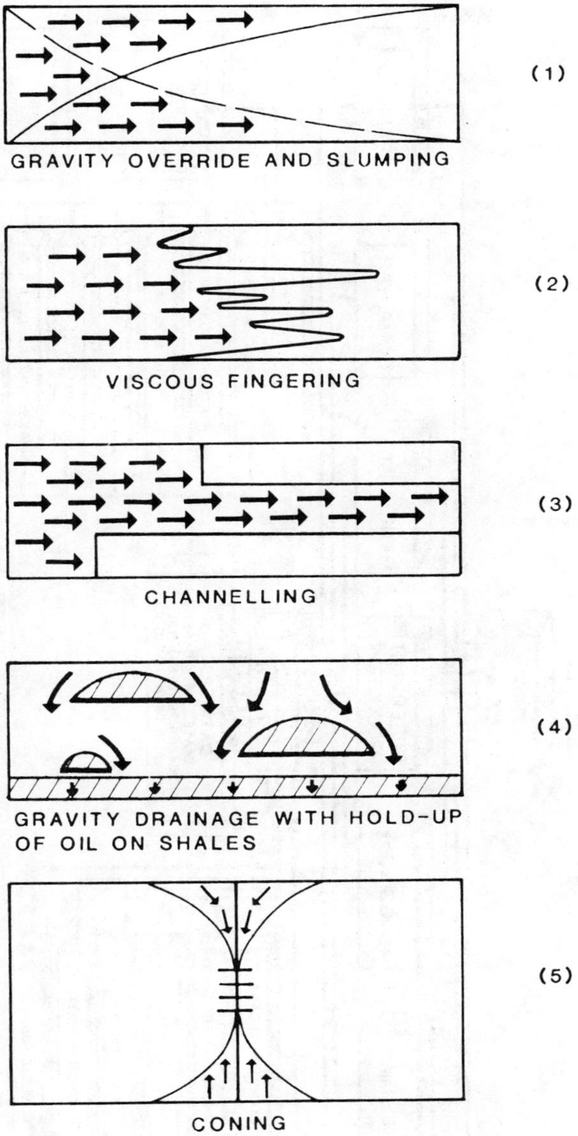

FIGURE 23: ILLUSTRATION OF SITUATIONS REQUIRING PSEUDO-FUNCTION CONSTRUCTION (DUE TO COARSENESS OF GRID, ⊟ = 1 LARGE GRID BLOCK)

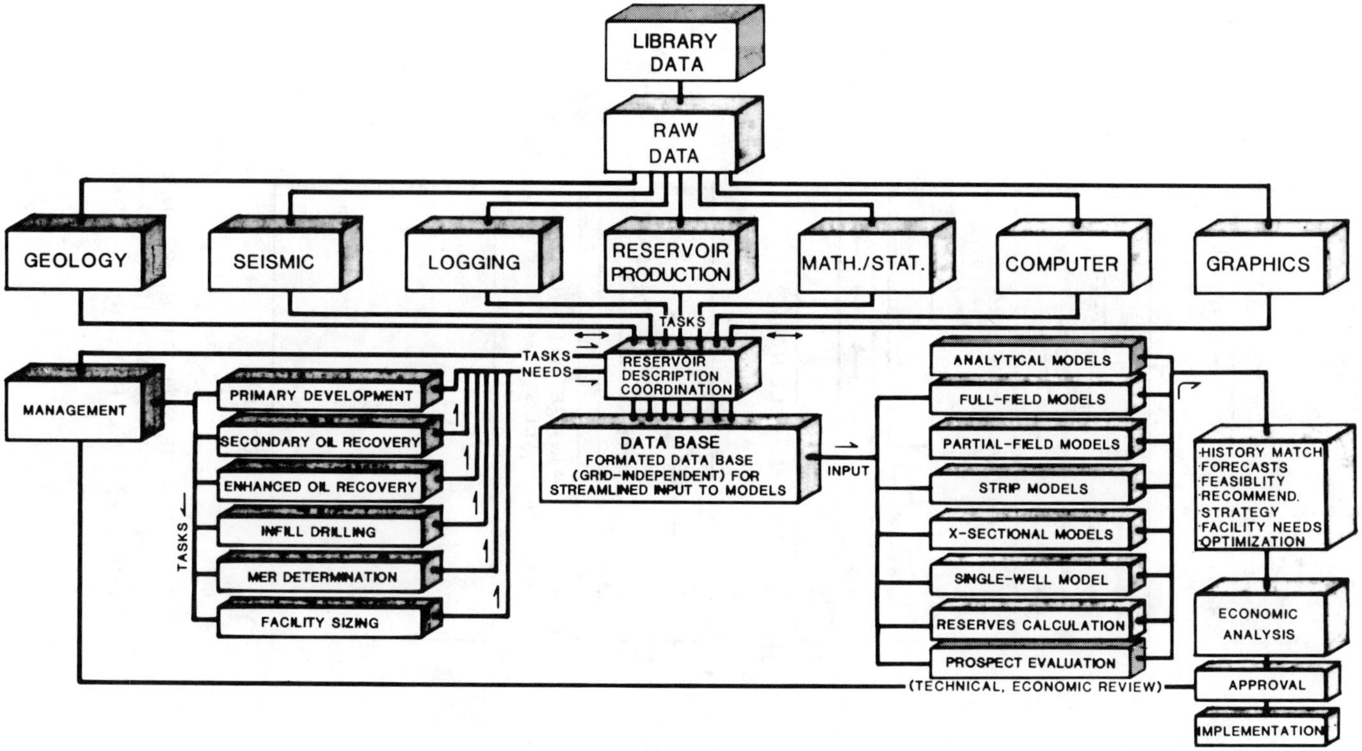

FIGURE 24:
RESERVOIR DESCRIPTION COORDINATION

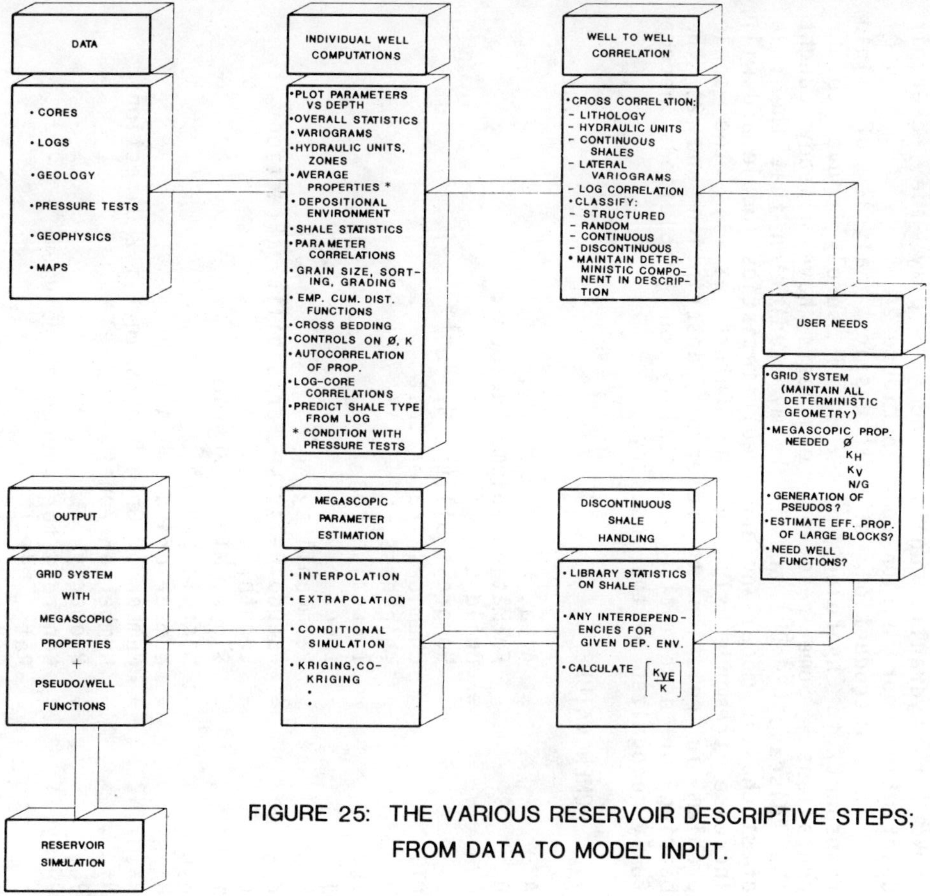

FIGURE 25: THE VARIOUS RESERVOIR DESCRIPTIVE STEPS; FROM DATA TO MODEL INPUT.

sizes and geometries of sandbodies and permeability barriers (e.g. shales) will be needed. A new breed of earth scientists will be in demand. They should 1) posses the classical geological knowledge, 2) have a strong background in statistics, 3) understand excatly what kind of data the engineer needs and how simulators work, 4) have a keen interest and understanding of the scale problem, 5) always look for hydraulic units, and 6) play the reservoir description coordination role in Figure 24.

The introduction of stochastic analysis of fluid transport to hydrology[42-52] will surely prove to be contagious. Some reservoir engineers have already caught the "disease"[53-60] and it is my prediction that the role of stochastic modelling and (geo) statistics should and will increase tremendeously in the future as 1) the scale problem in relation to reservoir description and 2) the bearing of descriptive uncertainty on predictions become fully recognized by the industry.

VI. NOMENCLATURE

A	=	Area, L^2
B_o	=	Formation Volume Factor
BU	=	Build-up
c	=	Compressibility, $MT^{-2}L^{-1}$
CTT	=	Chemical Tracer Test
h	=	Thickness, L
Ind	=	Induction Log
j	=	Index
$\underline{K}$	=	Permeability, L^2
$\tilde{k}_{ro}$	=	Pseudo Relative Permeability, Fraction
L	=	Length, L
m	=	Slope, $MT^{-2}L^{-1}$/cycle
MLL	=	Microlaterolog
$N_{A/B}$	=	Number of Volume A in B
NX,NY,NZ	=	Number of Grid Blocks in X,Y,Z direction
N_Z	=	Number of Zones
P	=	Property
$P(x,y,z)$	=	Centroid for averaging volume
q	=	Rate, L^3T^{-1}
r_{inv}	=	Radius of Investigation, L
R	=	Ratio between Volumes, Dimensionless
REV	=	Representative Elementary Volume, L^3
S_o	=	Oil Saturation, Fraction
Son	=	Sonic Log
TDT	=	Thermal Decay Tool
t	=	Time, T

X, Y, Z	=	Coordinates, L
$\Delta X, \Delta Y, \Delta Z$	=	Grid Block Dimensions, L
V	=	Sample Size, L^3
ΔU_o	=	Representative Elementary Volume, L^3
ϵ	=	Characteristic Volume, L^3
ΔU_i	=	Volume, L^3
$(\Delta U_v)_i$	=	Void Volume, L^3
$\Delta U_{o,P}$	=	Representative Elementary Volume of Property P, L^3
ΔK	=	Error in Permeability Estimate, L^2
$\alpha \beta \gamma$	=	Phases
μ	=	Viscosity, $ML^{-1}T^{-1}$
$\bar{\phi}$	=	Average Porosity, Fraction
ϕ	=	Porosity, Fraction
α	=	Zonal Correction Factor, Dimensionless
ρ	=	Density, ML^{-3}
Δt	=	Travel Time, T
σ	=	Standard Deviation

VII. ACKNOWLEDGEMENTS

I wish to express my appreciation to the management of Sohio Petroleum Company for permission to publish this paper. Special thanks are extend to Angie A. Lingad and Lorry P. Navellier for typing the manuscript and to Sohio's drafting department for their assistance with figures and tables.

VIII. REFERENCES

1) Aziz, K. and Settari, A.: PETROLEUM RESERVOIR SIMULATION, Applied Science Publishers, London, 1979.

2) Coats, K. H.: "Reservoir Simulation: State of the Art," Jour. Pet. Tech., Vol. 34 (1982) pp. 1633-1642.

3) Ewing, R.E.: THE MATHEMATICS OF RESERVOIR SIMULATION, Siam, Philadelphia, 1983.

4) Kendall, R. P., Morrell, G. O., Peaceman, D. W., Silliman, W. J., and Watts, J. W.: "Development of a Multiple Application Reservoir Simulator for Use on a Vector Computer," SPE Paper No. 11483 presented at the Middle East Oil Technical Conference, Manama, Bahrain, 1983, pp. 335-348.

5) Sneider, R. M., Tinkler, C. N., and Meckel, L. D.: "Deltaic Environment Reservoir Types and Their Characteristics," Jour. Pet. Tech., November 1978, pp. 1938-1946.

6) Harris, D. G., and Hewitt, C. H.: "Synergism in Reservoir Management - The Geologic Perspective," Jour. Pet. Tech., July 1977, pp. 761-770.

7) Harris, D. G.: "The Role of Geology in Reservoir Simulation Studies," Jour. Pet. Tech., May 1975, pp. 625-632.

8) Hubbert, M. K.: "Darcy's Law and the Field Equations of the Flow of Underground Fluids," Trans., AIME, Vol. 207, 1956, pp. 22-239.

9) Bear, J: DYNAMICS OF FLUIDS IN POROUS MEDIA, Elsevier, New York, 1968.

10) Yuan, S. W.: FOUNDATIONS OF FLUID MECHANICS, Prentice- Hall, London, 1967.

11) Keelan, D. K.: "Core Analysis," Part 2, Chapter 12 of the 1977 Edition of Subsurface Geology, compiled by L. W. LeRoy and D. O. LeRoy, Colorado School of Mines, Golden, CO.

12) Scheidegger, A. E.: THE PHYSICS OF FLOW THROUGH POROUS MEDIA, University of Toronto Press, 1960.

13) Hassanizadeh, M. and Gray, W. G.: "General Conservation Equations for Multi-Phase Systems: 1. Averaging Procedure," Advances in Water Resources, Vol. 2, September 1979

14) Haldorsen, H. H.: Reservoir Characterization Procedures for Numerical Simulation, Ph.D Dissertation, May 1983, The University of Texas at Austin.

15) Pickens, J. F., Cherry, J. A., Gillham, R. W., and Merritt, W. F.: "Field Studies of Dispersion in a Shallow Sandy Aquifer," vs DOE Invitational Well Testing Symposium, Berkeley, CA, October 19-21, 1977, Proceedings, pp. 55-62

16) Bear, J. and Braester, C.: "On the Flow of Two Immiscible Fluids in Fractured Porous Media." in Proceedings of the First Symposium on Fundamentals of Transport Phenomena in Porous Media, pp. 177-202, Elsevier, New York, 1972.

17) Chauvin, A. L., Jackson, R. F., McGee, P. R., and McMichael, C. L.: "Development Planning for the Statjor Field Using Three-Dimensional and Areal Reservoir Simulation," SPE Paper No. 8334 Presented at the 54th Annual Technical Conference and Exhibition of the SPE in Las VEgas, Nevada, Sept. 23-26, 1979.

18) Van Beek, F., and Troost, P. J. P. M.: "The Groningen Gas Field: A Case History of the Development of a Giant Gas Field," Jour. Pet. Tech., July 1979, pp. 815-820.

19) Myers, G. A., Johnston, R. D., and Mainwaring, J. R.: "Simulation of Prudhoe Bay Interference Test," SPE Paper No. 9456 Presented at the 55th Annual Technical Conference and Exhibition of the SPE in Dallas, Tx., Sept. 21-24, 1980.

20) Holst, P. H., and Zadick, T. W.: "Compositional Simulation for Effective Reservoir Management: The Brady South Weber Pressure Maintenance Project," SPE Paper No. 9321 Presented at the 55th Annual Technical Conference and Exhibition of the SPE in Dallas, Tx., Sept. 21-24, 1980.

21) Bansal, P. P., Harper, J. L., McDonald, A. E., Moreland, E. E., Odeh, A. S., and Trimble, R. H.: "A Strongly Coupled, Fully Implicit, Three-Dimensional, Three-Phase Simulator," SPE Paper No. 8329 Presented at the 54th Annual Technical Conference and Exhibition of the SPE in Las Vegas, Nevada, Sept. 23-26, 1979.

22) Weber, K. J.: "Influence on Fluid Flow of Common Sedimentary Structures in Sand Bodies," SPE Paper No. 9247, presented at the 55th Annual Technical Conference and Exhibition of the SPE in Dallas, Tx, September 21-24, 1980.

23) Dresser Atlas: Log Review 1, 1974.

24) Geehan, G. W. et al.: "Geologic Prediction of Shale Continuity," This Volume.

25) Wendt, W. et al.: "Permeability Prediction From Well Logs Using Multiple Regression," This Volume.

26) Dubrule, O., and Haldorsen, H. H.: "Geostatistics for Permeability Estimation," This Volume.

27) Haldorsen, H. H., and Chang, D. M.: Notes on Stochastic Shales; From Outcrop to Simulation Model," This Volume.

28) Rosenzweig, J. J., et. al.: "The Development of Pseudo Functions for Three Phase Black Oil Simulators," This Volume.

29) Archer, J. S.: "Reservoir Definition and Characterization For Analysis and Simulation", Preprint PD 6 of the 11th World Petroleum Congress, London, 1983.

30) Yuan, H. H.: The Influence of Pore Coordination on Petrophysical Parameters," Soc. Pet. Eng. Paper No. 10074 presented at the 56th Annual Technical Conference and Exhibition in San Antonio, Tx, 1981.

31) Heiba, A. A., Davis, H. T., and Scriven, L. E.: "Statistical Network Theory of Three Phase Relative Permeability," Soc. Pet. Eng. Paper No. 12690 presented at the Fourth SPE/DOE Symposium on EOR, Tulsa, 1984.

32) Chalkley, H. W., et al., Science, Vol. 110, 1949, p. 295.

33) Pathak, P., Winterfield, P. H., Davis, H. T., and Scriven, L. E.: "Rock Structure and Transport Therein: Unifying With Voroni Models and Percolation Concepts," SPE Paper No. 8846 Presented at the First SPE/DOE EOR Symposium in Tulsa, OK, April 20-23, 1980.

34) Dullien, F. A. L., Dhawan, G. K., Gurak, N., and Babjak, L.: "A Relationship Between Pore Structure and Residual Oil Saturation in Tertiary Surfactant Floods," Soc. Pet. Eng. Jour., August 1972, pp. 289-295.

35) Millington, R. J., and Quirk, J. P.: "Permeability of Porous Solids," Trans. Faraday Soc., Vol. 57, 1961, pp. 1200-1207.

36) Collins, R. E.: FLOW OF FLUIDS THROUGH POROUS MATERIALS, PennWell Publishing Company, Tulsa, Oklahoma, 1961.

37) Stewart, G., Wittman, S. T., and Lefevre, D.: "Well Performance Analysis: A Synergistic Approach to Dynamic Reservoir Description," SPE Paper No. 10209 Presented at the 56th Annual Technical Conference and Exhibition of the SPE in San Antonio, Tx, October 5-7, 1980.

38) Codreanu, D.: "Un Critere Pour L'Evaluation De L'Heterogeneite Des Gisements De Petrole A L'Echelle Centrimetrique et Regionale," Troisieme Colloque De L'Association De Recherche Sur Les Techniques De Forage et De Production, PAU, France, June 10-14, 1968.

39) Testerman, J. D.: "A Statistical Reservoir Zonation Technique," SPE Paper No. 286.

40) Gill, D.: "Application of a Statistical Zonation Method to Reservoir Evaluation and Digitized Log Analysis," AAPG Bull., Vol. 54, No. 5, May 1970, pp. 719-729.

41) Begg, S. H., and King, P.: "Modelling the Effects of Shales on Reservoir Performance: Calculation of Effective Vertical Permeability," SPE Paper No. 13529 Presented at the 8th SPE Symposium on Reservoir Simulation in Dallas, Tx, February 10-13, 1985.

42) Dieuhlin, A.: "The Effect of Scale in the Propagation of Pollution in an Aquifer," Ph.D Dissertation, School of Mines in Paris, France, 1980.

43) Schwartz, F. W.: "Macroscopic Dispersion in Porous Media: The Controlling Factors," Water Resources Research, Vol. 13, No. 4, August 1977, pp. 743-749.

44) Smith, L., and Freeze, R. A.: "Stochastic Analysis of Steady State Groundwater Flow in a Bounded Domain, 1: One-Dimensional Simulations," Water Resources Research: Vol. 15, No. 3, June 1979, pp. 521-528.

45) Smith, L., and Freeze, R. A.: "Stochastic Analysis of Steady State Groundwater Flow in a Bounded Domain, 2: Two-Dimensional Simulations," Water Resources Research, Vol. 15, No. 6, December 1979, pp. 1543-1559.

46) Freeze, R. A.: "A Stochastic-Conceptual Analysis of One-Dimensional Groundwater Flow in Nonuniform Homogeneous Media," Water Resources Research, Vol. II, No. 5, pp. 725-741, 1979.

47) Lippmann, M. J.: Two-Dimensional Stochastic Model of a Heterogeneous Geologic System, Ph.D Dissertation, University of California, Berkeley, 1973.

48) Newmann, S. P.: Statistical Characterization of Aquifer Heterogeneities, an Overview, Recent Trends in Hydrology-Proceedings Symposium, Lawrence Berkeley Lab., February 1979.

49) Bakr, A. A., Gelhar, L. W., Gutjahr, A. L., and MacMillan, J. R.: "Stochastic Analysis of Spatial Variability in Subsurface Flows, Part 1: Comparison of ONe- and Three-Dimensional Flows," Water Resources Research, Vol. 14, No. 2, pp. 263-271, 1978.

50) Gutjahr, A. L., Gelhar, L. W., Bakr, A. A., and MacMillan, J. R.: "Stochastic Analysis of Spatial Variability in Subsurface Flows: Part II Evaluation and Application," Water Resources Research, Vol. 14, No. 5, pp. 953-960, 1978.

51) Smith L.: "Spatial Variability of Flow Parameters in a Stratified Sand," Mathematical Geology, Vol. 13, No. 1 pp. 1-21, 1981.

52) Delhomme, J. P.: "Spatial Variability and Uncertainty in Groundwater Flow Parameters: A Geostatistical Approach," Water Resources Research, Vol. 15, No. 2, April 1979, pp. 281-290.

53) Smith, P. J. and Brown, C. E.: "Stochastic Modelling of Two-Phase Flow in Porous Media," SPE 11128 Presented at the 57th Annual Technical Conference and Exhibition in New Orleans, La., September 26-29, 1982.

54) Lambert, M.: A Statistical Study of Reservoir Heterogeneity, MS Thesis at The University of Texas at Austin, 1981.

55) Bennion, D. W., and Griffiths, J. C.: "A Stochastic Model for Predicting Variations in Reservoir Rock Properties," Soc. Pet. Eng. Jour., March 1966, pp. 9-16.

56) Potter, P. E., and Blakely, R. F.: "Generation of a Synthetic Vertical Profile of a fluvial Sandstone Body," Soc. Pet. Eng. Jour., September 1967, pp. 248-251.

57) Huhn, S. L.: A Comparison of Several Techniques to Generate a Vertical Distribution of Permeabilities and Porosities for Reservoir Modeling, MS Thesis at The University of Texas at Austin, 1982.

58) Knutson, C. F.: "Modeling of Noncontinuous Forth Union Mesa Verde Sandstone Reservoirs, Piceance Basin, Northwestern Colorado," Soc. Pet. Eng. Jour., August 1976, pp. 175-188.

59) Cekirge, H. M., and Kleppe, J.: "Stochastic Methods in Reservoir Engineering," SPE Paper No. 9608 Presented at the Middle East Conference of the SPE in Manama, Bahrain, March 9-12, 1981.

60) Knutson, C. F., Maxwell, E. L., and Millheim, K.: "Sandstone Continoity in the Mesaverde Formation, Rulison Field Area, Colorado," Jour. Pet. Tech., August 1971, pp. 911-919.

61) Collins, R. E.: Private Communication

62) Bennion, D. W., and Hope, A. C. A.: "Application of Power Spectral Analysis and Digital Filtering in Evaluating the Distribution of Rock Properties," SPE Paper No. 4121, Presented at the 47th Annual Technical Conference and Exhibition of the SPE in San Antonio, Texas, October 8-11, 1972.

63) Baveye, P., and Sposito, G.: "The Operational Significance of the Continuum Hypothesis in the Theory of Water Movement Through Soils and Aquifers," Water Resources Research, Vol. 20., No. 5, pp. 521-530, May 1984.

64) Bachmat, Y., and Bear, J.: "The Concept and Size of an REV", Paper presented at the NATO Advanced Study Institute on Fundamentals of Transport Phenomena in Porous Media, July 14-23, 1985, Newark, Delaware USA.

65) Long, J.C.S.: "Verification and Characterization of Continuum Behavior of Fractured Rock at AECL Underground Research Laboratory", Technical Report No. LBL-14975 from Lawrence Berkeley Laboratory, University of California, February, 1985.

RESERVOIR CHARACTERIZATION FOR SIMULATION, HARTZOG DRAW FIELD, WYOMING

Charles L. Hearn
John P. Hobson
and
Michael L. Fowler

Cities Service Oil and Gas Corporation
Tulsa, Oklahoma

I. ABSTRACT

The Hartzog Draw field, a major sandstone oil reservoir in the Powder River basin of Wyoming, was discovered in 1975 and unitized in 1980. The field is under waterflood and infill drilling is in progress. Enhanced recovery by CO_2 miscible flooding is under consideration. To aid in predicting waterflood and CO_2 flood oil recovery using a reservoir simulator, a comprehensive geological study was conducted.

From analysis of cores and logs in a portion of the field, geologic facies were identified and correlated. Flow units (zones of the reservoir with similar fluid flow characteristics) were mapped. Three general areas, each having similar geological properties, were identified. For each area, a prototype well was developed having average characteristics of wells in that area. The reservoir description was extended field-wide using statistical methods applied to wireline log data.

The stratigraphic and depositional model developed for Hartzog Draw indicates that layering is a major feature in the most permeable Hartzog Draw flow units. Based on the properties of the prototype wells, a reservoir description was developed for each geologic area. Simulator predictions of waterflood and CO_2 flood performance were made for flood patterns in each area. The predictions illustrate the effects of reservoir characterization on simulation results.

II. INTRODUCTION

Finite difference simulators allow a more realistic reservoir description to be used in modeling reservoir performance than do simple analytical methods. However, even these sophisticated tools require generalization and simplification of reservoir geology. Compromises must be made to accommodate available computer storage and speed, incomplete knowledge of reservoir heterogeneity, and limitations imposed by the model grid framework. Oil recovery predictions may be very sensitive to the method of representing reservoir geology in the model. This is particularly true for fluid injection projects, where reservoir heterogeneity is a predominant factor affecting oil recovery. It is important, therefore, to pay considerable attention both to the accuracy of the geological reservoir description and its representation in the simulation model.

This paper describes the reservoir characterization of the Hartzog Draw field, a major sandstone oil reservoir. As used here, reservoir characterization refers both to the development of a geological reservoir description and the assignment of simulator parameters on the basis of that description. The simulator was used to predict waterflood and carbon dioxide miscible flood recovery. Topics to be discussed include: 1) a summary of reservoir geology based on a detailed study of a limited area of the field, 2) methods used to extend this geologic description to the entire field, 3) reservoir stratigraphy, 4) simulation technique and development of model parameters, and 5) examples of the effects of reservoir description on simulation results.

III. RESERVOIR PROPERTIES AND FIELD HISTORY

The Hartzog Draw field, located in northeastern Wyoming (Fig. 1), is 22 miles long and from 1 to 4 miles wide. The development, reservoir properties, and unitization of the field described by Hunt and Hearn (1982) are summarized here. Production is from the Shannon sandstone at 9,400 ft. Fig. 2 shows net pay distribution; maximum sand thickness is 60 ft and average thickness is 20 ft. Average porosity is about 12% and average air permeability about 12 md. The reservoir contains a black oil with an original bubblepoint pressure of 1,550 psig. Initial oil in place, based on the latest volumetric determination, was 329 million STB.

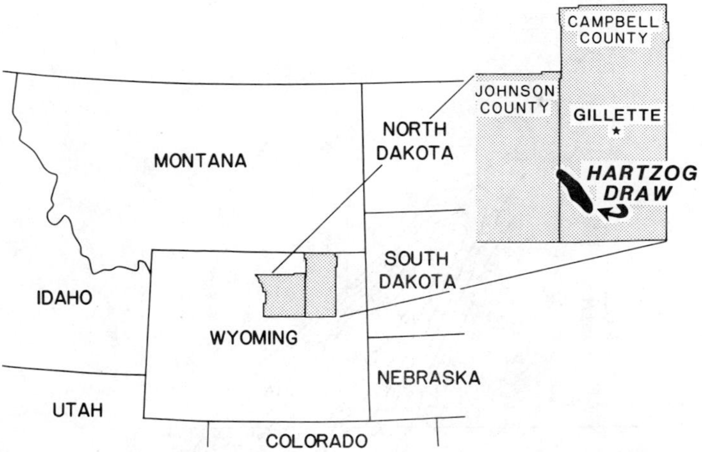

Fig. 1. Field location map. (From Hearn, et al. 1984. Copyright SPE-AIME.)

Field development began in 1975. Drilling of the original wells, on 160-acre spacing (Fig. 2), was completed in 1979. The peak oil production rate of 39,000 STB/D was reached in 1977 (Fig. 3), after which the rate declined because of reservoir pressure depletion. During primary production there was a significant difference in reservoir performance between the east and west sides of the field (Hearn, et al. 1984). In general, wells east of the long axis of the field were much better performers than wells west of the axis. As discussed later, this trend is related to the distribution of reservoir rock types, or facies.

Based on reservoir studies conducted during primary production, the field was unitized for secondary recovery in 1980. Plans were to conduct a waterflood using only the original wells on a spacing of 160 acres per well, or 320 acres per five-spot pattern. However, there was concern that waterflooding on such wide spacing would not be efficient. Early studies indicated possible limited reservoir continuity on this spacing, and laboratory tests suggested that water injectivity might be low. There also was interest in possible enhanced recovery by CO_2 miscible flooding. In 1981, a pilot infill drilling program was begun in approximately the center of the field (Figs. 4, 5). As shown in Fig. 4, the pilot program consisted of drilling nine infill producing wells and converting the original wells in the pilot to water injection. This resulted in nine 160-acre five-spot patterns. The purpose of the pilot was to evaluate waterflooding (and possible CO_2 flooding) on closer spacing, and to provide needed additional geological and engineering data.

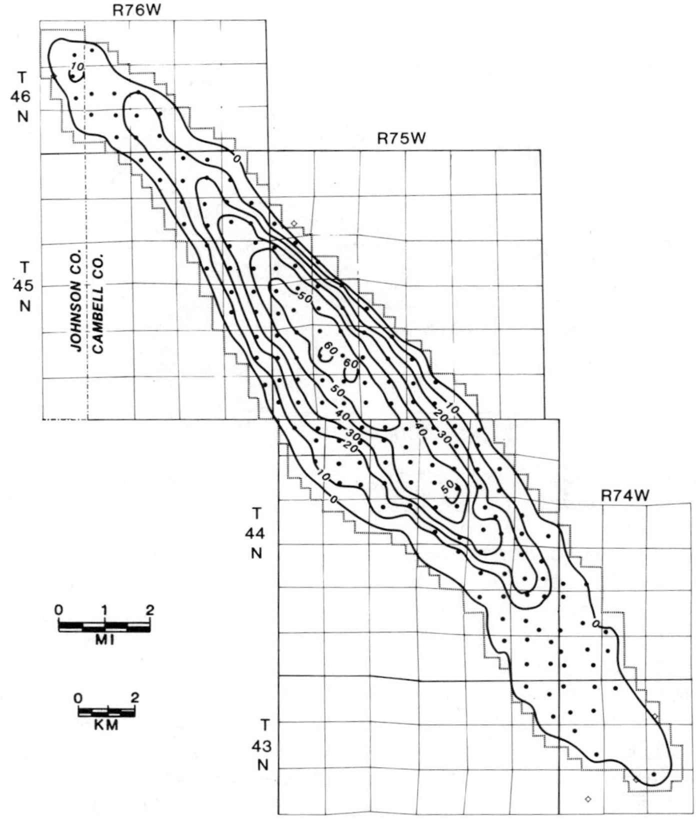

Fig. 2. Net pay isopach map. Wells shown are original wells (160-acre spacing).

As shown in Fig. 3, water injection began in 1981, and the oil production rate began to increase in 1982. Most of this early response was from the pilot patterns. The excellent waterflood response in the pilot area supported the evaluation of further infill drilling. This resulted in working interest owner approval to drill approximately 100 additional production wells and convert existing wells to water injection. Infill drilling began in 1983 and will be completed in 1985. Fig. 5 shows the well configuration upon completion of the infill drilling program. The waterflood pattern in most of the field will then be the same as in the pilot project, a 160-acre five-spot. As seen in Fig. 3, oil production rate has continued to increase in response to injection and

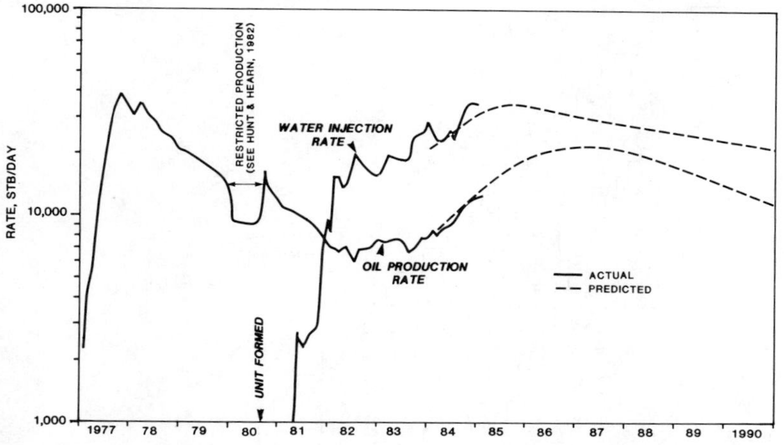

Fig. 3. Actual and predicted performance of the Hartzog Draw field.

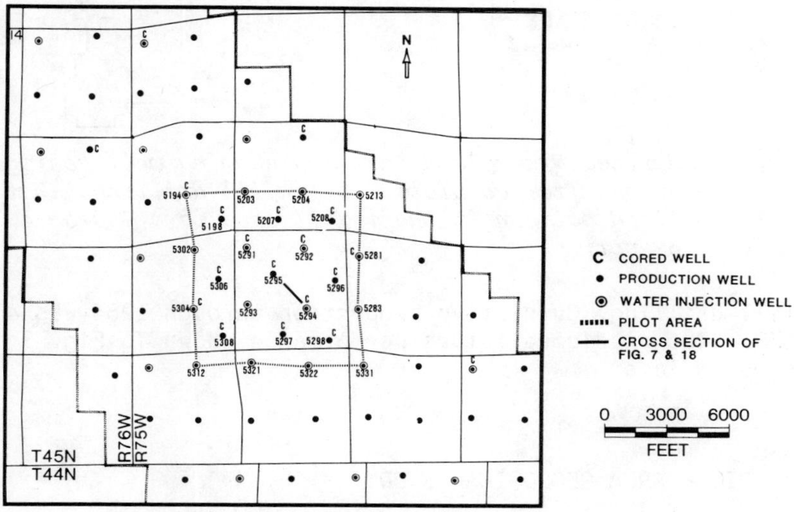

Fig. 4. Pilot area map. Nine production wells in pilot area are infill wells.

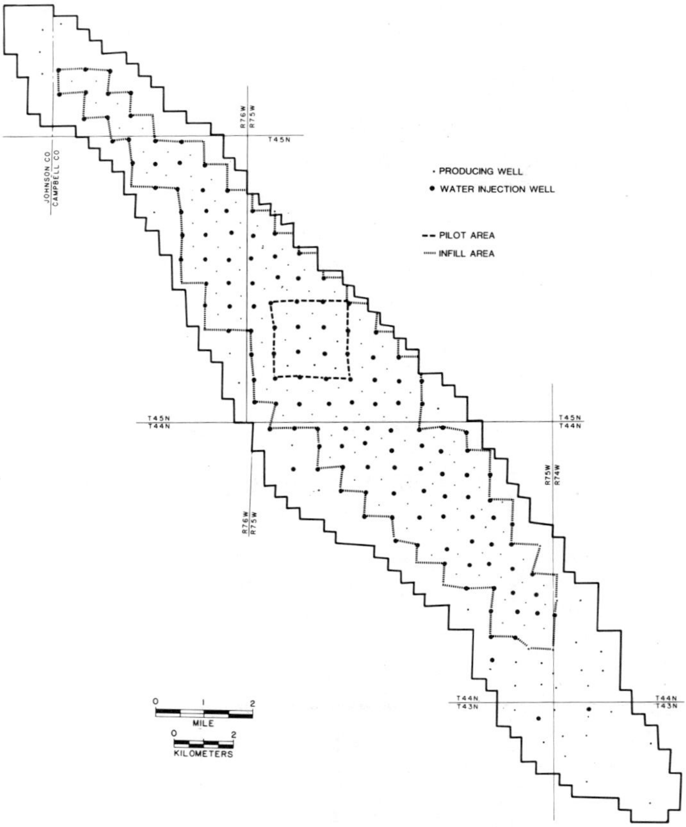

Fig. 5. Hartzog Draw field base map showing well configuration after completion of infill drilling. Waterflood pattern in the infill area is a 160-acre 5-spot.

infill drilling. Cumulative production through 1984 was 41.7 MMSTB. The predicted future performance shown in Fig. 3 is discussed later.

IV. PILOT AREA GEOLOGICAL STUDY

Nineteen cores in or near the pilot area were available, including all of the initial nine infill wells (Fig. 4). A comprehensive stratigraphic, sedimentologic, and mineralogic study was conducted using these cores and logs available on all wells. Networks of well log cross sections, correlated with core descriptions, were used to develop a detailed reservoir description of the study area, which includes the pilot patterns and one to two well locations beyond. A description of the pilot area geology given by Hearn et al. (1984) is summarized here.

A. Geology

As will be discussed further in a following section, the Shannon sandstone of Hartzog Draw was deposited as a complex of sand ridges that stood above the seafloor of the Late Cretaceous marine seaway in what is now the eastern Powder River basin. These sand ridges were formed in water that was 60 to 300 ft deep and as far as 100 miles from shore (Martinsen and Tillman, 1978, 1979; Tillman and Martinsen, 1979). The reservoir, a stratigraphic trap, is enclosed in the Cody shale.

Fig. 6 is a composite type-log illustrating the distribution of rock facies and petrophysical properties in the reservoir. The Hartzog Draw Shannon contains three sandstone lenses; the distribution of these lenses in the pilot area was discussed by Hearn et al. (1984). The main reservoir

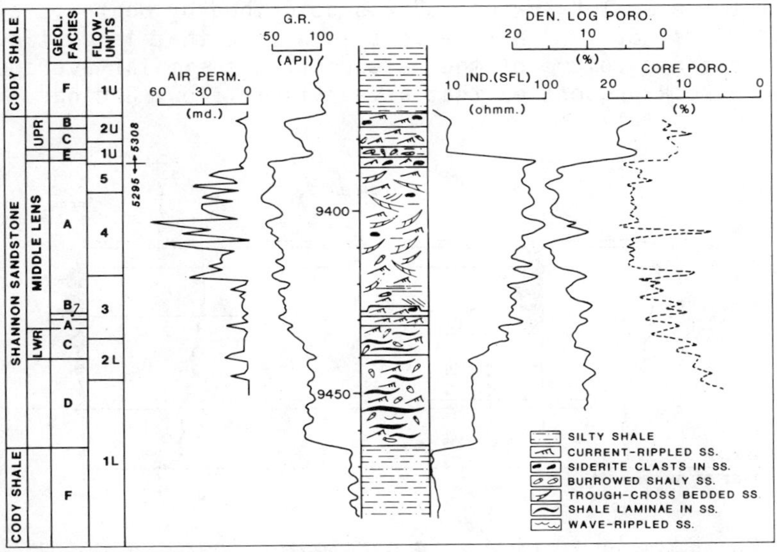

Fig. 6. Composite reservoir quality profile (based on wells 5295 and 5308). (From Hearn et al. 1984. Copyright SPE-AIME.)

rock is in the middle lens. As shown in Fig. 6, several rock facies have been identified based on those of Martinsen and Tillman (1979). Facies A, which is the most porous and permeable reservoir rock, is crossbedded, medium- to fine-grained sandstone with scattered clasts of sideritic mudstone and minor laminae of shale. Facies B consists of crossbedded and rippled coarse- to medium-grained sandstone with clasts of shale and sideritic mudstone. Facies C is horizontally laminated and ripple-bedded, slightly shaly sandstone. Facies D is a burrowed, ripple-bedded, very shaly, fine-grained sandstone. A bioturbated facies (E) separates the upper sandstone lens from the middle lens on the west side of the reservoir and also occurs in association with Facies D. The Cody shale (Facies F) overlies the reservoir sandstone and underlies Facies D.

B. Flow Units

Maps and cross sections of rock facies help in understanding the bedding characteristics and depositional environment of the reservoir. However, a zonation more directly related to reservoir properties that influence fluid flow (such as permeability) is useful for reservoir engineering purposes. To focus on variations in the reservoir that affect fluid flow, a subdivision of the Hartzog Draw pilot area into several "flow units" was described by Hearn et al. (1984) (Figs. 6, 7). A flow unit is a zone that is continuous over a defined volume of the reservoir, has similar averages of those rock properties that affect fluid flow, and has

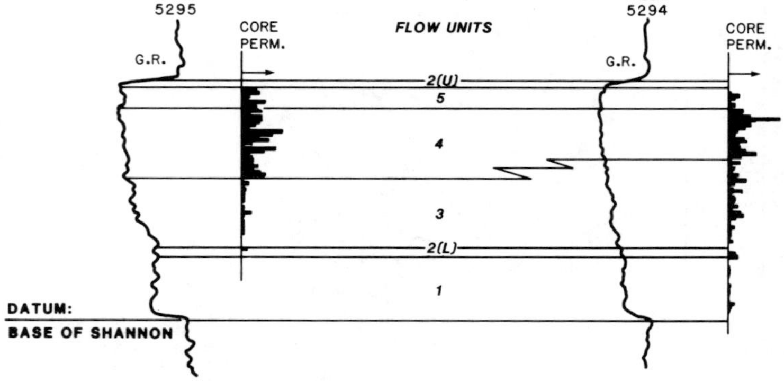

Fig. 7. Example of flow unit development between two wells in the center geologic prototype area. Location of cross section is shown in Figure 4.

similar bedding characteristics. The distribution of flow units is related to the facies distribution, but flow unit boundaries do not necessarily coincide with facies boundaries. Delineation and mapping of flow units is based on considerations of reservoir stratigraphy and vertical and lateral variation of permeability. This process is somewhat subjective, and is aided by abundant core data. In the Hartzog Draw pilot area, consistent patterns of continuity of flow units could be developed from cross section networks. Methods used in a field-wide extension of flow units are discussed later.

Fig. 6 shows the division of the composite reservoir quality profile into five flow units. Flow Units 1 and 2 are shown at both the top and bottom of the interval because they have similar properties and, in some cases, merge at the edges of the reservoir. Also, although not shown here, Flow Unit 2 was split into two sub-units, 2A and 2B, with 2A having better permeability and continuity. Fig. 7 is an example of the flow unit correlation between two wells (see Fig. 4 for location). The dependence of the flow units on permeability profile is evident in Fig. 7, although flow unit

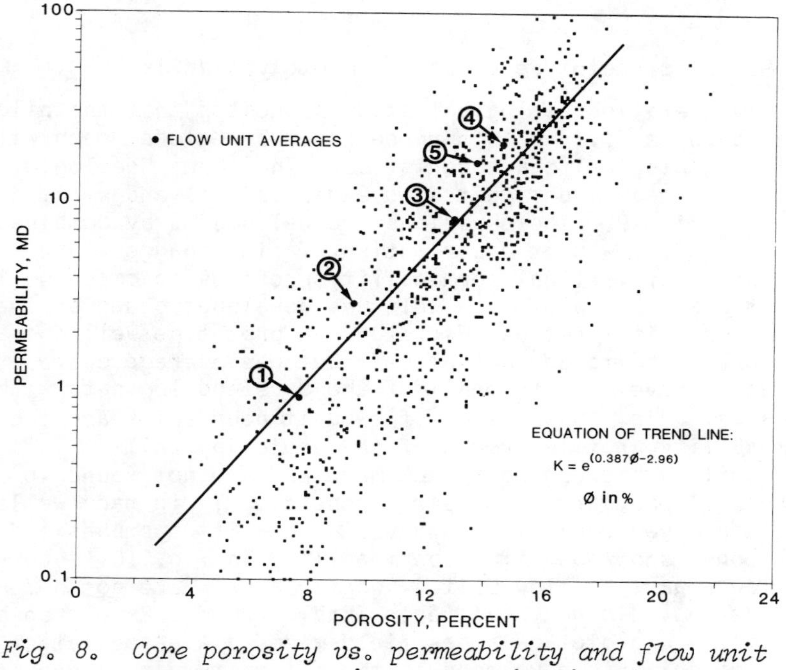

Fig. 8. Core porosity vs. permeability and flow unit average properties. Porosity is corrected for overburden pressure.

definition also depends on stratigraphy. It is emphasized that, in Hartzog Draw, the flow units are not isolated zones, but are in communication. Also, they are not homogeneous. Permeability varies both vertically (Fig. 7) and areally within a flow unit.

The flow units have different average values of permeability and porosity. Fig. 8 shows these averages in comparison to the entire population of measurements made on Hartzog Draw cores. Flow Units 4 and 5 have the highest permeability and porosity, and Flow Unit 1 has the lowest. Also shown in Fig. 8 is a trend line relating permeability to porosity.

The flow units provide a basis for determining net pay, or the thickness of reservoir which has the greatest contribution to flow of injected and produced fluids. On the basis of their permeability, porosity, and continuity characteristics, Flow Units 2A, 3, 4, and 5 were defined as net pay (Fig. 2). This resulted in a slightly greater net pay, with slightly lower average porosity, than net pay determined in early reservoir studies, which was based only on a 9% porosity cutoff (Hunt and Hearn, 1982). The net pay contains approximately 99% of reservoir flow capacity and 92% of hydrocarbon pore volume.

C. Geologic Prototype Areas and Prototype Wells

The detailed geological study suggested that the pilot area could be divided into three general regions, each with similar geological characteristics. The three "geologic prototype areas", designated as east, center, and west, are shown in Fig. 9. These regions are delineated by combined criteria of flow unit distribution, facies changes, and similarity of vertical permeability profiles in cored wells. For each area, a prototype well was developed based on the geological similarities (Fig. 10). A prototype well does not correspond to any actual well but has average characteristics developed from all of the core and log data within the area. Fig. 10 shows the flow unit boundaries and the permeability profile trends of the prototype wells. The permeability trends are approximate and are not found in every well, but are seen fairly consistently in many wells within a given geologic prototype area. It is emphasized that zones showing uniform permeability in Fig. 10 (such as the lower part of Flow Unit 4, center area), are <u>not</u> homogeneous, but did not have recognizable trends. Extension of the geologic prototype areas field-wide, discussed later, was based on criteria developed in the pilot study as expressed mainly by the prototype wells.

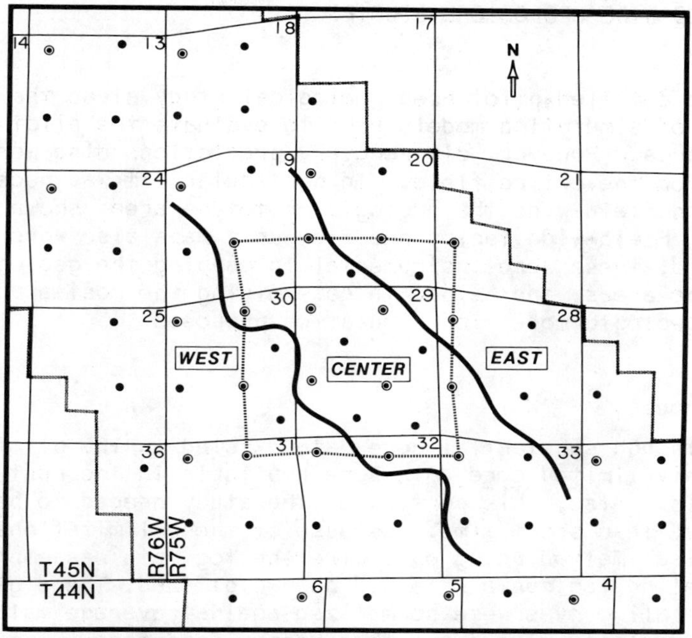

Fig. 9. Geologic prototype areas in pilot project. Each area has similar geological characteristics. (From Hearn, et al. 1984. Copyright SPE-AIME.)

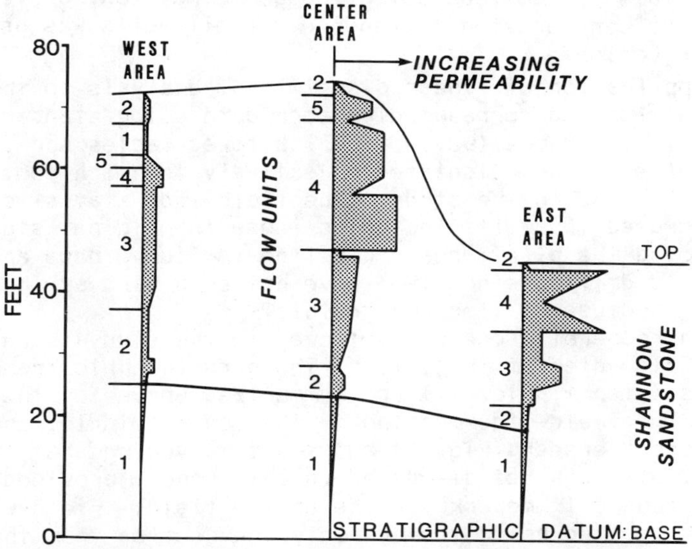

Fig. 10. Prototype wells showing flow units and permeability profile trends. (From Hearn, et al. 1984. Copyright SPE-AIME.)

V. FIELD-WIDE GEOLOGICAL STUDIES

The detailed pilot area geological study aided the development of simulation models used to evaluate the pilot performance. However, oil recovery predictions also were needed for the entire field. In particular, it was necessary to extend field-wide the geologic prototype areas shown in Fig. 9. Field-wide facies and flow unit maps also were developed; these were instrumental in mapping the geologic prototype areas, and helped in determining the most appropriate geologic model for simulation purposes.

A. Methods

Although substantial core data existed in the pilot area, only limited core data were available in the rest of the field. Also, this portion of the study needed to be completed in a short time. Because of these limitations, a statistical method using only wireline log data was employed. All wireline log curve data had been digitized. The digital data for all curves were normalized against average values for a 50-ft shale section above the Shannon Sandstone. The porosity data derived from the formation compensated density (FDC) curve were adjusted for shale content by applying appropriate corrections using the gamma ray (GR) curve. A single "clean sandstone" standard for all wells was used in the correction.

Application of linear discriminant analysis to the wireline log and corresponding core data using standard statistical methods (Davis, 1973) allowed facies and flow unit values to be calculated objectively for each foot in each well used in the study. The field-wide statistical studies used 165 wells including those in original studies in and around the pilot area. Wireline log curve data and cores for wells drilled since 1983 have not been used at this writing because of time constraints.

The general procedure followed in the study is outlined by the flow diagram (Fig. 11). The stratigraphic framework studies (step 1, Fig. 11) are summarized above for the pilot area. The field-wide mapping of the lower, middle, and upper sandstone "lenses" (Fig. 6) and the intervening shaly sandstones, although not discussed in this paper, provided a stratigraphic framework for the entire field. Field-wide studies treated below include only the section from the top of the middle sandstone lens to the base of the Shannon (Fig. 6), i.e. the upper sandstone and subjacent shaly sandstone are excluded. Maps and cross sections of facies

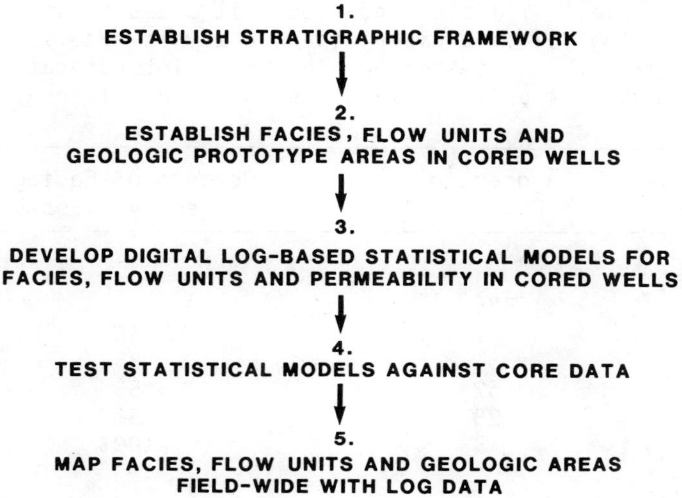

Fig. 11. General procedure followed to produce field-wide maps of facies, flow units, and geologic areas.

and flow units are based entirely on identifications using the statistical models erected with wireline log and core data from 13 cored wells (step 3, Fig. 11). The discriminant models were constructed by characterizing the geologists' foot-by-foot designations of facies and flow units in the cores in terms of corrected and normalized data from several log curves over the same intervals. The statistical models use volume of shale (GR-based), corrected FDC porosity, and SP wireline curve data. Other available log data could not be used because of missing or poor log data in some wells and necessity for having the same wireline log suites in each well. An important step was testing the accuracy of the models in reproducing the core-derived data (step 4, Fig. 11). The testing procedure consisted of foot-by-foot prediction of flow units and facies in the 13 cored wells using the models and log data and comparison of the results with the core data on the same foot-by-foot basis. Results of this test are shown in Table I. Reclassifications of Flow Units 4-5, 3, 2A, and 1 and Facies A (cross-bedded facies) are very good. Although reclassifications of Flow Unit 2B and Facies B-E are less satisfactory, maps and cross sections using the data provide geologically sensible patterns and trends as illustrated in a following section.

Flow units and facies, as discussed above, are based on the statistically-derived footage data. Field-wide extension of geologic areas rests on broader considerations of distri-

TABLE I. Test of facies and flow unit classification using wireline log data and discriminant analysis models. Results are compared, i.e. tested, against core interpretations of facies and flow units in 13 wells in the pilot area.

	Aggregate Footage	Percent of Footage Correctly Classed
Facies		
A	385	81%
B	77	35%
C	81	33%
D	62	55%
E	25	56%
F	3	100%
Flow Units		
4-5	202	85%
3	148	76%
2A	57	66%
2B	14	54%
1	71	80%

bution of the flow units, and permeability profiles for each well, as well as facies distribution, sandstone thickness, and wireline log character. The permeability profiles were produced by inserting normalized, shaliness-corrected FDC porosities for core porosity in the regression equation that defines the trend line of core porosity vs core permeability in 19 cored wells (Fig. 8). The computed permeability profiles were used in judging the ordering and average magnitudes of permeabilities in uncored wells throughout the field. Initially, two additional permeability profiles were constructed for each well by using a combination of wireline log curves and a discriminant-multiple regression analysis model and also by using a log porosity vs core permeability regression equation. Visual comparison of the three profiles with core permeabilities in the 13 cored model wells showed that the permeability curve derived through the core porosity vs core permeability regression equation gave the best match. An example well (5296, Fig. 4) comparing corrected FDC porosity, the log-based permeability profile, and core permeability and porosity is shown in Fig. 12. Whereas comparison of profiles among wells can be done with the porosity curves, knowledge of the general magnitude and variation in permeability is convenient. Permeability profiles were created

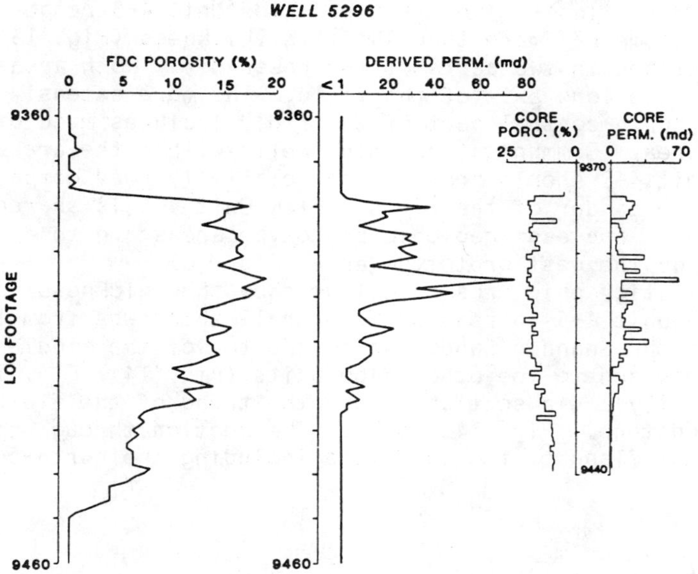

Fig. 12. Comparison of log porosity, log-derived permeability, and core porosity and permeability on a foot-by-foot basis in a well from the center geologic prototype area within the pilot area. Location of well is shown in Figure 4.

for each well in the field with adequate log data and were used in making comparison with the prototype well profiles of Fig. 10. The computed permeability profiles played a major role in final assignment of each well to the most appropriate geologic prototype area.

B. Distribution of Flow Units

Flow unit thickness maps were contoured for Flow Units 4 and 5 combined, 3, 2A, 2B, and 1, using flow unit values that were determined statistically. Flow Units 4 and 5 have generally similar average permeability and porosity characteristics (Fig. 8) and are adjacent (Figs. 6 and 7); therefore, they were combined for mapping purposes. The combination is referred to below as Flow Unit 4-5. Also, Flow Unit 2 was subdivided into 2A (better and more extensive permeability and porosity) and 2B. Flow Unit 4-5 represents the best quality reservoir, i.e. the flow unit contains both the greatest porosities and permeabilities and most extensive zones of good permeability. Therefore, Flow Unit 4-5 is used here as an illustrative example. Flow Unit 4-5 ranges from zero to somewhat more than 35 ft in thickness (Fig. 13) and is developed in two parts of the reservoir. Both areas lie east of the long axis of the field. The more extensive area in the north-central part of the field includes much of the pilot area. Communication among wells within the areas with Flow Unit 4-5 should prove to be relatively good compared to the remainder of the field. Flow Unit 4-5 is shared by the center and east geologic prototype areas and to a lesser extent by the west prototype area (Fig. 10).

Stratigraphic cross sections show the thickness variation of Flow Unit 4-5 in respect to overall thickness from the base of the Shannon Sandstone to the top of the middle sandstone and to the other flow units (Fig. 14). Fig. 14A is essentially a dip section across the trend of the field and the sandstone. Fig. 14B is a strike section through the northeast flank of the sandstone including the north-central

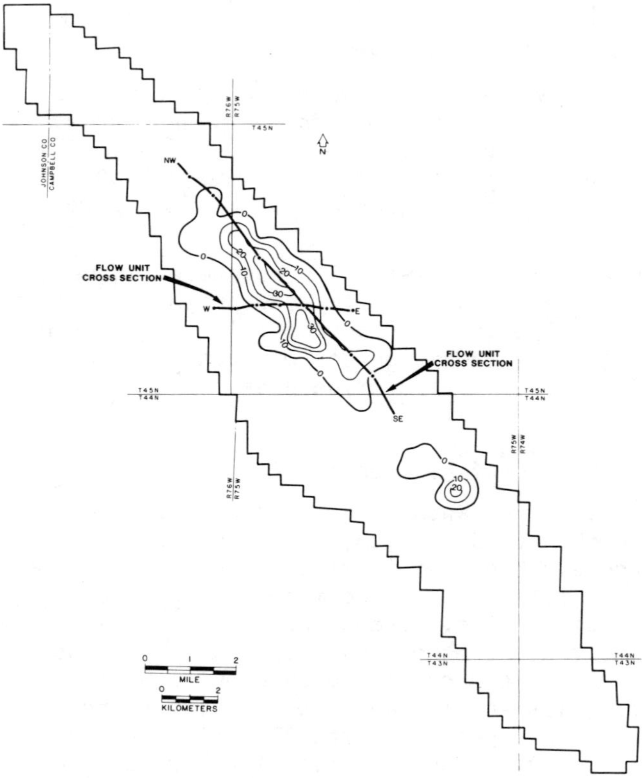

Fig. 13. Thickness of Flow Unit 4-5. Flow Unit 4-5 is enclosed in Facies A (compare with Figure 15).

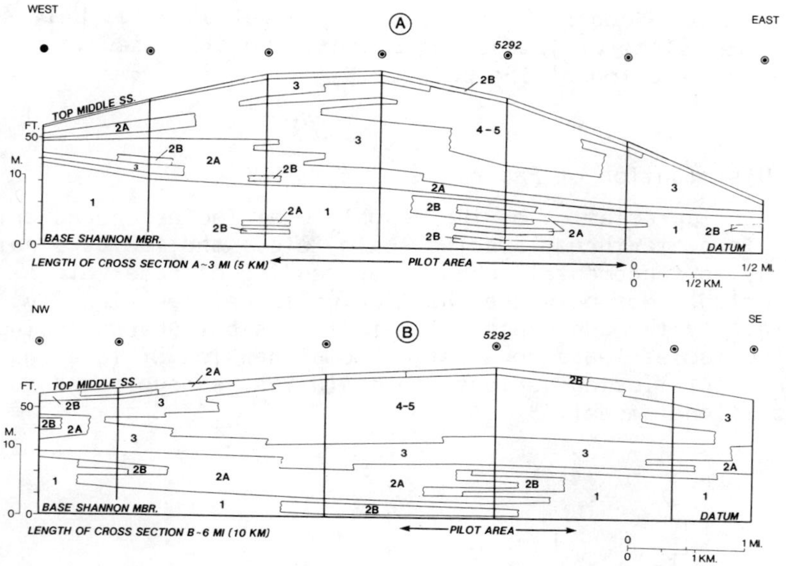

Fig. 14. Stratigraphic cross sections across (A) and parallel (B) to depositional strike and the field axis showing stratigraphy and distribution of Flow Units 1 through 4-5 in respect to Shannon thickness. Locations of cross sections are shown in Figure 13.

development of Flow Unit 4-5. As illustrated in Fig. 14A, Flow Unit 4-5 is thickest within the upper northeast flank. Flow Unit 4-5 is underlain by Flow Unit 3. Flow Unit 3 is replaced laterally by Flow Unit 2A, which in turn is replaced by 2B and 1. Although the delineation of flow unit boundaries is sometimes subjective, for engineering purposes the contacts between flow units are assumed to be abrupt, and flow units are treated as discrete reservoir zones. The areal variation of production discussed earlier is related to flow unit distribution in the middle sandstone lens. The lower part of the Shannon below the lower sandstone lens is characterized by Flow Unit 1, except in areas where Flow Unit 4-5 is thickest (Fig. 14). In these areas the lower Shannon contains Flow Units 2A and 2B. The distribution of flow units in the Shannon is related in considerable part to the distribution of facies, as discussed in the following section.

Distribution of flow units determines in large part the outlines of the geologic areas shown in Figs. 9 and 19. All but 15 wells classed in the east and center areas, for example, penetrated Flow Unit 4-5. Only two wells classed in the west area contain a thin, poorly-developed Flow Unit 4-5. Characteristics of the geologic areas are discussed in a following section of the paper.

C. Distribution of Facies

An appreciation of the depositional facies characteristics (e.g. textures, bedding scale, etc.) and facies stratigraphy is fundamental to an understanding of reservoir variation. Not only are the flow units related directly, or indirectly through diagenesis, to facies but understanding of the facies leads to a better comprehension of layering characteristics on various scales in the reservoir and to a depositional model.

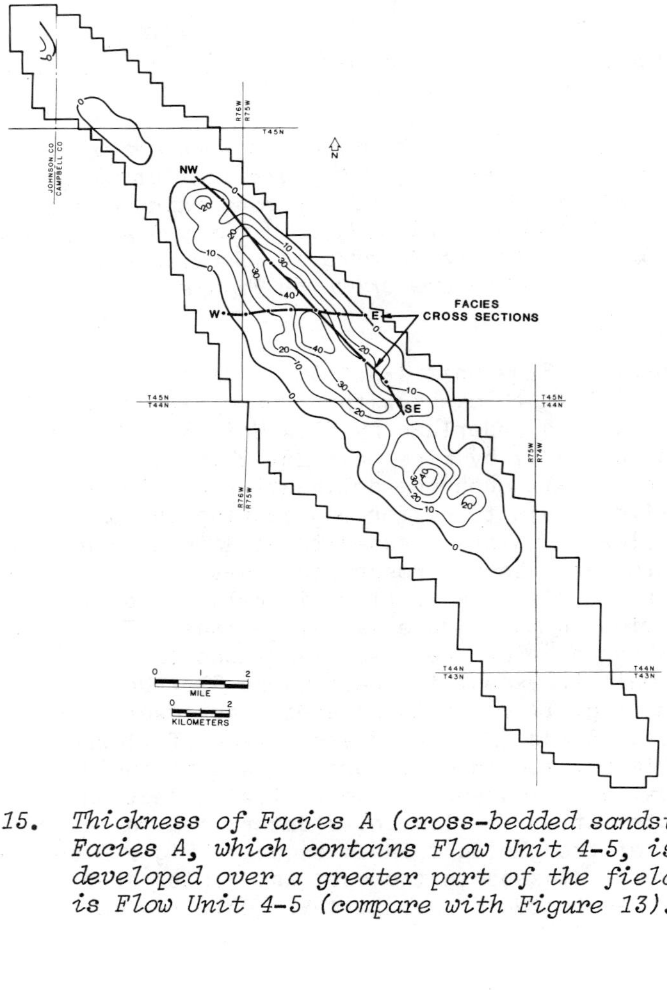

Fig. 15. Thickness of Facies A (cross-bedded sandstone). Facies A, which contains Flow Unit 4-5, is developed over a greater part of the field than is Flow Unit 4-5 (compare with Figure 13).

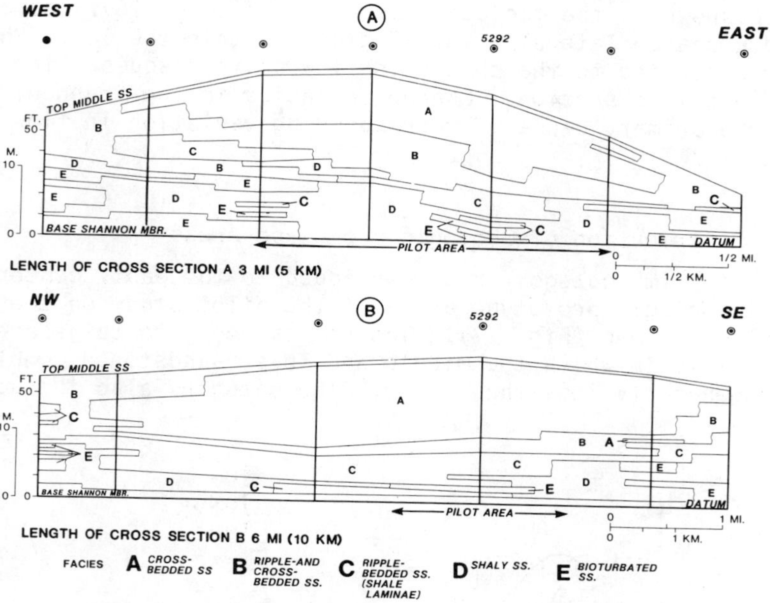

Fig. 16. Stratigraphic cross sections across (A) and parallel (B) to depositional strike and the long axis of the field showing stratigraphy and distribution of Facies A-E in respect to Shannon thickness. Facies F encloses the Shannon sandstone. Locations of cross sections are shown in Figures 13 and 15.

The thickness map of Facies A, cross-bedded sandstone, is shown as an example of facies distribution at Hartzog Draw (Fig. 15). Facies A encloses Flow Unit 4-5, but is much more extensive than the flow unit. The areal extents of Facies A and Flow Unit 4-5 can be contrasted by comparison of the respective thickness maps and cross sections (Figs. 13-16). The explanation of the contrast apparently lies in a gradual lateral and downward change within Facies A into sandstone of slightly poorer reservoir quality and continuity. Comparison in terms of vertical section and relationships to Facies B-E can be made with the cross sections Figs. 14 and 16.

Facies A is asymmetrically distributed in respect to the long axis of the field and to the trends of maximum pay thickness (Fig. 2) and thickest Shannon section except in the northern and southern parts of the field. The relationship between Shannon thickness and facies in the center of the field can be observed in the dip cross section, Fig. 16A.

Also shown in the facies cross sections (Fig. 16), Facies D and E change laterally into Facies C within the lower Shannon, corresponding to the change in flow units discussed in the previous section. Change in facies in the Shannon is the primary cause of corresponding variation in flow units.

D. Distribution of Geologic Prototype Areas

A "rim" category has been added to the east, center, and west geologic prototype areas of the pilot study on the field-wide map (Fig. 17). The rim category contains areas of the field in which the middle and lower sandstones combined are generally less than 25 ft. The category also includes an

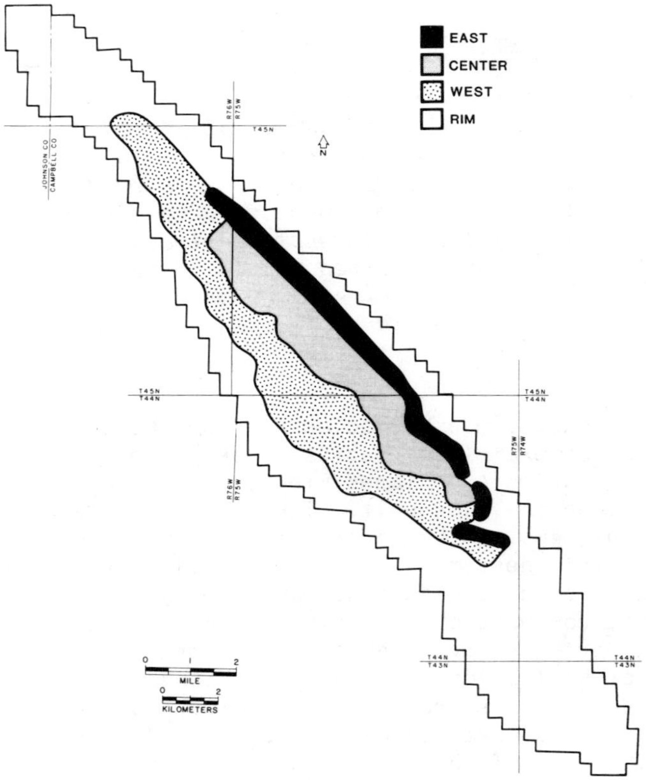

Fig. 17. Geologic area map. Identification of areas is based mainly on combined consideration of flow unit distribution, facies distribution, permeability profiles, and Shannon thickness.

area in the south part of the field where the lower sandstone is readily mappable; the lower sandstone, however, is generally 20 ft or less in thickness. Average corrected FDC porosities for pay are comparable to or less than poorer sandstones in the west area. Flow unit analysis through the discriminant analysis model shows a complex interbedding of Flow Units 2A, 2B, and 3 in the middle and lower sandstones in the rim area suggesting poor well-to-well continuity. Most of the wells in the rim category lie outside the area of infill drilling (Fig. 5).

The general procedures used in classifying the wells according to the prototype geologic areas and prototype wells of the pilot study have been discussed under Methods. Distribution of flow units, apparent vertical ordering of permeabilities, and distribution of facies in the interval from the base of the Shannon to the top of the middle sandstone (Fig. 6) were used in combination as primary parameters in creating the map (Fig. 17). Therefore, the geologic area map does not coincide exactly with other maps, e.g. the Flow Unit 4-5 thickness (Fig. 13). In the south-central part of the field, the center area trend was extended to include some borderline wells between the east and west geologic areas to produce a more sensible map pattern.

VI. STRATIGRAPHY AND DEPOSITIONAL MODEL

A. Stratigraphy

Detailed stratigraphy in the center area is illustrated in Fig. 18 for two wells a location apart in the pilot area. Flow unit stratigraphy between the wells has been shown in Fig. 7. One well is an injector (5294) and one a producing well (5295). Facies packages are not shown as such. Rather, various lithologies that affect porosity and permeability are emphasized. Shown are carbonate-cemented zones, coarser-grained sandstone with medium-scale cross bedding, finer-grained sandstone with small-scale cross bedding with and without shale laminae, very shaly silty lenticular- to wavy-bedded or bioturbated sandstone, and silty shale. Small solid triangles indicate low permeability zones as detected and correlated with log-based permeability profiling. The low permeability zones characterize and appear correlatable over much of the center area.

Considerable variation in thickness and lateral extent has been noted for many of the features shown in Fig. 18. However, their general stratigraphy suggests a model composed

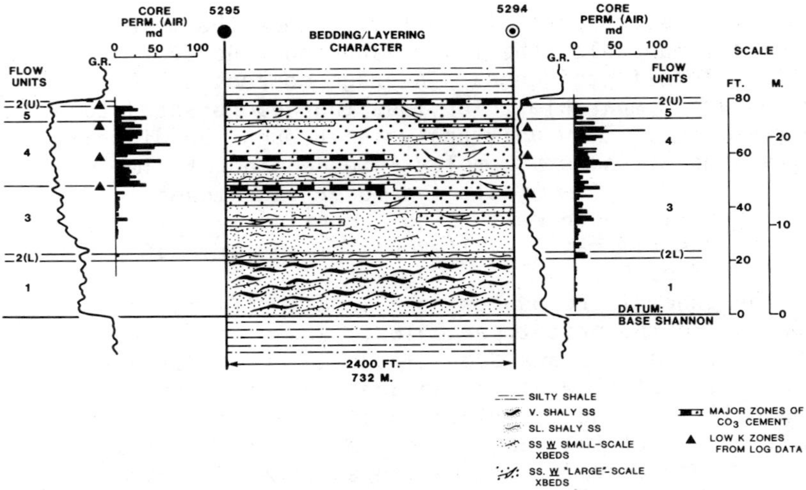

Fig. 18. Layering continuity of various lithologies and of permeability zones illustrated between two wells in the center prototype area. Flow unit stratigraphy between the wells is shown in Figure 7. Of particular interest are zones of carbonate cementation that extend between wells, as well as zones of low permeability (solid triangles) based mainly on log-derived permeability profiles. Locations of wells are shown in Figure 4.

of more or less continuous layers on the present well spacing, in vertical communication but with restricted vertical permeability. Such a model appears appropriate for the better quality reservoir sandstone in the center and east areas. A model consisting of less continuous, overlapping layers may be more characteristic of much of the west area, which also has a much lower average permeability. On a smaller scale than in Fig. 18, features such as cross bedding, laminae of ductile grains (glauconite, shale chips), lenses of sideritic mudstone and pebbles, and shale drapes or discontinuous laminae probably influence fluid flow in the Shannon.

B. Depositional Model

A possible depositional model for the Shannon (Fig. 19) aids in explaining and portraying the distribution of facies, lithologies, and the layering characteristics. The model emphasizes the offshore shelf-ridge interpretation recently documented in detail for the Shannon by Tillman and

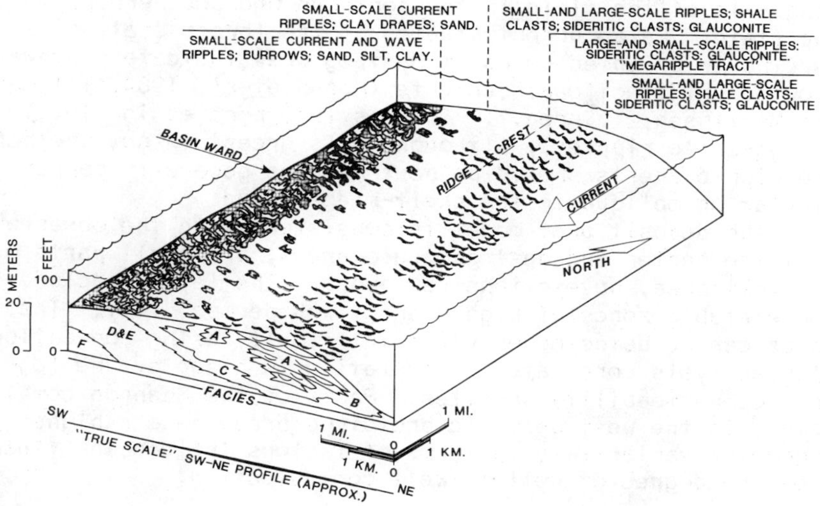

Fig. 19. Depositional shelf-ridge model for the upper part of the Shannon during deposition of Facies A. Results of deposition are sheet-like lithologic bodies of variable extent. The true-scale profile at the base of the figure illustrates the very low relief on the seafloor across the ridge complex.

Martinsen (1984; in prep.). Tillman and Martinsen (1984) studied in great detail stratigraphy of the facies in surface exposures, mainly of the western flanks of ridges, and illustrate striking lateral and vertical variations in facies often over distances of a fraction of a mile. Their studies of the surface exposures may provide an analogy for the relatively great stratigraphic variation in facies in the west geologic prototype area at Hartzog Draw. Asymmetry of form and facies has been observed both at Hartzog Draw (Tillman and Martinsen, 1979) and in the Sussex Sandstone at House Creek Field (Hobson et al. 1982). In both fields the higher energy facies and relatively simple facies stratigraphy are best developed within the upper, northeastern, basinward flanks of thicker parts of the shelf-ridge complexes. Distribution of facies in the lower Shannon at Hartzog Draw (Fig. 16) suggests that stronger currents were operating on the basinward sides of higher parts of the complexes early in their depositional and accretionary development. Apparently, the more gently inclined western flanks were relatively protected. Facies are more symmetrically distributed re the long axis of the field in the northern and southern parts of the field. Paleocurrent studies of the larger scale cross bedding in oriented cores at Hartzog Draw indicate southward flowing unidirectional currents (Hearn et al. 1984; Tillman and Martinsen, in prep.). This is incorporated in Fig. 19 (large-scale ripples), although it is uncertain now whether the ripple crests should be portrayed as generally perpendicular or oblique to the shelf-ridge trend.

The depositional model is consistent with the observation that the center and east geologic areas, and small parts of the west area, have extensive layers. The layers include correlatable zones of higher and lower permeability. These zones can be detected only with difficulty using conventional plug analysis core data. Correlation is aided by the log-derived permeability profiles. Parts of the Shannon complex, mainly in the west geologic prototype area, have a higher degree of variability in terms of factors influencing fluid flow and degree of well-to-well communication.

VII. RESERVOIR SIMULATION

Prediction of Hartzog Draw waterflood (and possible future CO_2 flood) performance required a reservoir modeling approach which would both incorporate the important geologic features controlling fluid flow, and also satisfy time and computing cost constraints. The following discusses prediction techniques and the development of simulator parameters. Although emphasis is on methods, limited simulation results are given to illustrate the effects of reservoir description on oil recovery predictions.

A. Simulation Method

The large number of Hartzog Draw wells precluded simulation of the entire field, or even groups of wells, with a sufficiently fine numerical grid to reduce numerical dispersion to an acceptable level. Also, many simulator runs were needed to investigate possible CO_2 flood operating conditions (such as the start time of CO_2 injection and the amount of CO_2 injected). To allow investigation of different operating scenarios with a practical number of simulator runs, a method described by Rester and Todd (1984) was used. A single flood pattern symmetry element was simulated for each of the three major geologic prototype areas ("east", "center", and "west") shown in Figs. 9 and 19. The prototype wells

(Fig. 10) were used as the basis for developing simulator reservoir properties for each prototype. The simulation predictions for each prototype pattern element were scaled up to represent the appropriate reservoir area. (These steps are described further below.) Considering the simplifying assumptions made, this method may not accurately predict the performance of individual wells, but should be reasonable for groups of wells if the prototype pattern descriptions are representative.

The simulator used was that described by Chase and Todd (1984), which can simulate waterflood and miscible flood performance. Predictions were made for a pattern element of one-quarter of a 160-acre five spot for each of the three prototype reservoir descriptions. The prototype results were scaled and combined as described by Rester and Todd (1984). With this method, each actual field pattern element is assigned to one of the prototypes based on the geological area map (Figs. 9, 19). The performance of any combination of field pattern elements is predicted by 1) scaling the appropriate prototype prediction for a given field element to account for differences in pore volume and flow capacity between the field and prototype elements, and 2) superposing the scaled field element predictions to obtain a prediction for the entire field or for any desired group of patterns.

B. Reservoir Description for Simulation

The Hartzog Draw reservoir contains various sedimentary lithologies, and there is considerable vertical variation of permeability (Figs. 6, 7, 10). This heterogeneity can significantly influence the vertical sweep efficiency of injected fluids. As discussed in previous sections of this paper (and by Hearn et al. 1984), the geological study suggests that a layered reservoir model is appropriate for Hartzog Draw. This is shown by correlatable features between wells (Fig. 18) and by the similarity of permeability profiles in wells within a given field area (Fig. 10). The layers are in communication, but vertical permeability is restricted by shale laminations and carbonate-cemented zones. Other sedimentary structures such as crossbedding are superimposed on this configuration.

For each of the three prototype pattern elements, a conventional layered reservoir description was developed assuming continuous layers between wells, each having uniform permeability and porosity, with no areal variation of properties. Although this is an oversimplification of actual reservoir stratigraphy, a more complex model was not justified since structural details between wells are unknown. With such a model, the ordering of layers may have significant influence on sweep efficiency. For example, the harmful effects of gravity override of CO_2 would be reinforced by a high-permeability layer at the top of the reservoir. A layered model was used for all three prototypes; although, as discussed earlier, the continuity of reservoir layers in the west geologic area is probably not as good as in the east and central areas.

Layer thicknesses of 2 to 4 feet were used to allow approximate conformance to the permeability trends of Fig. 10. This resulted in 9, 15, and 11 layers for the east, center, and west prototypes, respectively. Simulation with such relatively thin layers may model viscous, gravity, and capillary crossflow effects more realistically than thicker layers (Todd et al. 1982; Claridge, 1982). Actual reservoir sand lenses may have been more accurately represented by even thinner layers, but computing cost would have been excessive. Sensitivity studies to investigate the effect of layer thickness on simulation results would have been desirable but were not conducted because of time constraints.

Layer properties were determined separately for each flow unit of each prototype. For cored wells in a geologic area, all core permeability measurements within a flow unit were combined and sorted on permeability. The sorted data were divided into a number of groups equal to the desired number of layers for the flow unit, and the average permeability determined for each group (layer). The resulting layers were ordered, within each flow unit, in approximate agreement with the permeability profile trends of Fig. 10. This layering method combines the "positional" and "permeability ordering" methods described by Craig (1971). Core sample positions were maintained by flow units, but were ordered within each flow unit. This process tends to develop layers with a permeability variation similar to the original core data, whereas a completely positional approach usually results in layers with a lesser permeability variation (Craig, 1971).

The resulting layer properties are illustrated in Table II for the center area prototype. Since porosity varies much less than permeability, a uniform porosity was used for each flow unit. Table III shows the average properties for all three prototypes. The average permeabilities are 20 md for the east prototype, 13 md for the center, and 3 md for the west. This illustrates the transition in reservoir quality from good on the east side of the field to poor on the west side, in agreement with the geological study and production trends.

Weber (1982) and others have discussed the effect of thin, discontinuous shale laminations on fluid flow. In the prototype pattern simulations, this effect was approx-

TABLE II. Layer Properties of Center Area Prototype

Layer Number	Flow Unit	Thickness ft	Permeability md	Porosity
1	5	2.5	31.3	0.120
2	5	2.5	5.0	0.120
3	4	3.5	15.0	0.140
4	4	3.5	20.0	0.140
5	4	3.5	9.7	0.140
6	4	3.5	3.3	0.140
7	4	3.5	45.5	0.140
8	4	3.5	26.2	0.140
9	3	3.6	1.6	0.118
10	3	3.6	20.9	0.118
11	3	3.6	10.8	0.118
12	3	3.6	7.0	0.118
13	3	3.6	4.2	0.118
14	2A	4.0	0.8	0.092
15	2A	4.0	5.6	0.092

TABLE III. Average Properties of Prototype Patterns

Geologic Prototype Area	Net Thickness* ft	Average Permeability md	Average Porosity	Number of Model Layers
East	29	20.3	0.125	9
Center	52	13.4	0.123	15
West	35	3.0	0.108	11

*Flow Units 2A, 3, 4, 5

imated by a uniform reduction of vertical permeability. Had the necessary statistical information on Hartzog Draw shales been available, a more rigorous method, such as that described by Haldorsen and Lake (1984), which incorporates the continuity and spatial disposition of shales using statistical methods, may have been desirable. However, simulator runs using the center area prototype showed negligible difference in oil recovery using vertical/horizontal permeability ratios of 0.1 and 0.01 for both waterflood and CO_2 flood simulations. For a k_v/k_h ratio of 1.0, waterflood recovery was slightly increased and CO_2 flood recovery significantly decreased (these results are affected by layer ordering, so are not general). Geological core descriptions and limited laboratory data (Hearn et al. 1984) indicate that the average k_v/k_h ratio in the major Hartzog Draw flow units is considerably less than 1.0. A value of 0.1 was used for all simulation runs.

We feel that the effects of layering will dominate Hartzog Draw sweep efficiency. The most significant sedimentary structure not included in the present model probably is crossbedding. Weber (1982) and Kortekaas (1983) discuss effects of crossbedding on fluid flow and possible methods of including such effects in simulation models. In general, however, methods of characterizing crossbedding in simulators, while retaining the effects of other significant features such as layering, are not well developed. As discussed further below, we recognize that future Hartzog Draw reservoir performance may indicate the need to include other features in the model.

C. Some Simulation Results

With the geological model described above, the simulator was used to predict waterflood and CO_2 flood recovery for each of the the prototype pattern elements. Other necessary simulation parameters (relative permeability, fluid properties, etc.) were carefully developed but are not discussed here. Simulations were conducted for one-quarter of a 160-acre five-spot pattern, using a 7 x 7 areal grid and the number of layers for each prototype shown in Table III. To establish initial waterflood oil saturations, primary oil production was simulated for each pattern element using average production rates for wells in the geologic area. During injection, constant pressure well constraints (flowing bottomhole pressures of 6000 psi for injectors and 500 psi for producers) were used, approximating actual field conditions. Well skin factors were based on pressure buildup and falloff tests.

Fig. 20 illustrates the effect of the different reservoir descriptions in the three geologic prototype areas. Predicted waterflood cumulative oil production vs time is shown for each prototype, both as a fraction of original oil-in-place (OOIP) and as total barrels produced (the latter has been scaled to represent a full 160-acre pattern). After 40 years of water injection, the predicted center and west area waterflood oil production is 31% OOIP, while the east is slightly better, 35%. In terms of total barrels produced, the center area is by far the best because of its much greater thickness (Table III). Both the center and east

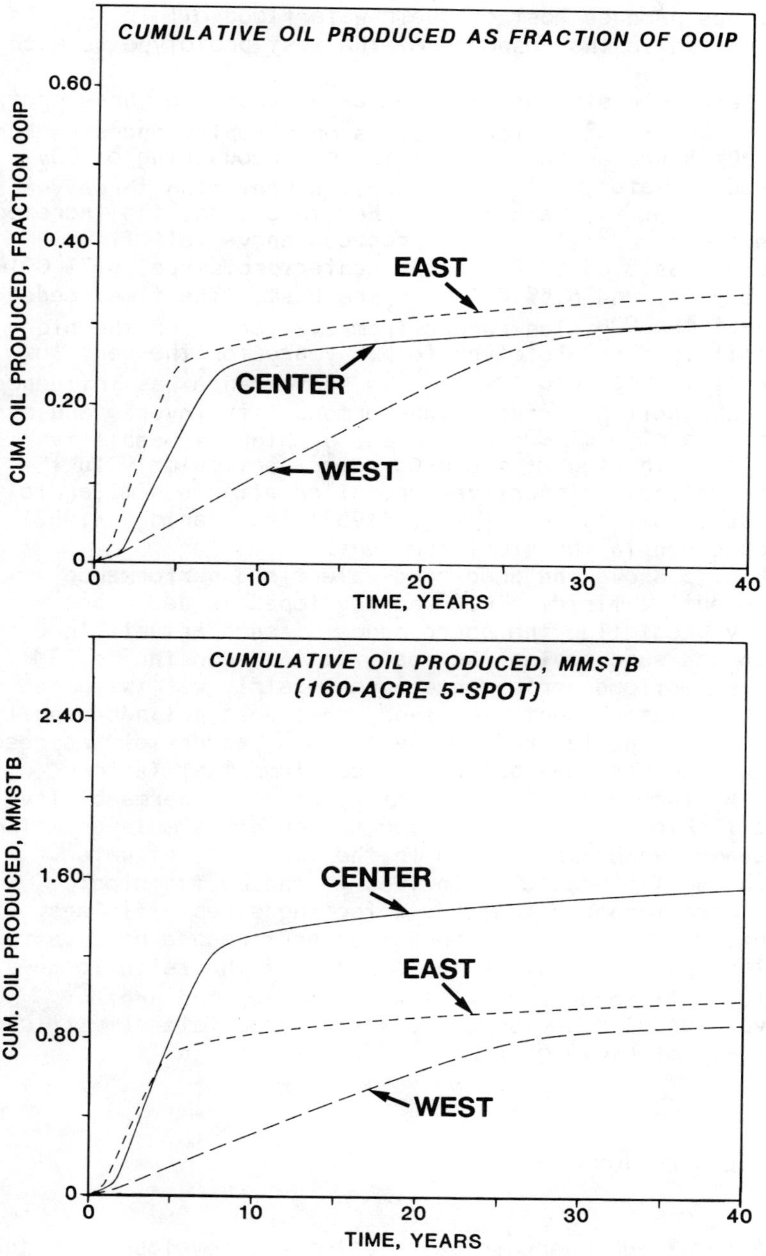

Fig. 20. Comparison of waterflood performance of three prototype patterns.

prototypes produce most of their waterflood oil in the first 10 years, while the response of the west prototype is much slower.

There were similar differences between the three prototypes in CO_2 flood performance. As an example, one case run was a 30% hydrocarbon pore volume continuous slug of CO_2, followed by water, with CO_2 injection beginning three years after start of the waterflood. For this case, the incremental oil recovery (additional oil produced above waterflood recovery) was 8.0% OOIP for the center prototype, 6.7% OOIP for the east, and 6.5% OOIP for the west. The time needed to inject the CO_2 slug ranged from 2.2 years for the high-permeability east prototype to 5.2 years for the very low permeability west prototype. CO_2 performance was influenced by CO_2 channeling through high-permeability layers, and by gravity override of CO_2 reinforced by high permeability layers near the top of the reservoir (particularly in the east prototype). Interlayer crossflow effects similar to those described by Todd *et al.* (1982) and Claridge (1982) were observed in the simulator runs.

Fig. 3 shows the predicted waterflood performance for the entire field. This was developed as described above by predicting the performance of each actual flood pattern and superposing the results. As seen in Fig. 3, actual waterflood performance agrees fairly well with the predictions at present. However, this does not necessarily confirm that the layered reservoir model accurately represents Hartzog Draw geology, but rather confirms that factors affecting throughput (permeability, relative permeability, and well skin factors) have been accurately simulated. There has been no water production in the field; first water breakthrough is predicted in 1986 in the east geologic area. Stratigraphic features affecting sweep efficiency will not be reflected in production performance until after breakthrough. The future performance of the waterflood--and CO_2 flood, if conducted--will show whether the present reservoir model represents the most appropriate simplification of actual reservoir geology.

VIII. CONCLUSIONS

A detailed reservoir description was developed for the Hartzog Draw field pilot area. Based on core and log data, rock facies and flow units were correlated and mapped. Flow units, zones of the reservoir with similar fluid flow characteristics, represent an intermediate step between the geological reservoir description and simulator grid properties. The facies and flow unit correlations aided in dividing the pilot region into three general areas, each with similar geological properties. For each area, a prototype well was developed with properties--permeability profile trends, in particular--representing wells in that geologic area. The pilot area facies, flow unit, and geologic prototype area maps were extended field-wide using statistical methods applied to wireline log data.

Reservoir simulation models must incorporate geologic features that have a significant effect on reservoir performance, and also must satisfy time and computing cost constraints. The geological study indicated that layering is a major depositional and diagenetic feature in the Hartzog Draw reservoir. A model for predicting Hartzog Draw waterflood and CO_2 miscible flood performance was developed for each geologic prototype area. Features included in the model were layer permeabilities retaining the permeability variation of the reservoir core data, layer ordering consistent with the permeability profiles of the prototype wells, and the effect of shale laminations on vertical permeability. Future reservoir performance, particularly after breakthrough of injected fluids, will indicate whether this model is the most appropriate for Hartzog Draw.

ACKNOWLEDGEMENTS

We thank Cities Service Oil and Gas Corporation and the management and working interest owners of Hartzog Draw Unit for permission to publish this paper. Rod Tillman and Ken Harpole provided helpful suggestions.

REFERENCES

Chase, C. A., and Todd, M. R. (1984). "Numerical Simulation of CO_2 Flood Performance", Soc. Pet. Eng. J., 597-605.

Claridge, E. L. (1982). "CO_2 Flooding Strategy in a Communicating Layered Reservoir", J. Pet. Tech., 2746-2756.

Craig, F. F., Jr. (1971). The Reservoir Engineering Aspects of Waterflooding, SPE Monograph Vol. 3, Society of Petroleum Engineers of AIME, Dallas, TX, 63-65.

Davis, J. C. (1973). Statistics and Data Analysis in Geology, John Wiley and Sons, New York.

Haldorsen, H. H., and Lake, L. W. (1984). "A New Approach to Shale Management in Field-Scale Models", Soc. Pet. Eng. J, 447-457.

Hearn, C. L., Ebanks, W. J., Jr., Tye, R. S., and Ranganathan, V. (1984). "Geological Factors Influencing Reservoir Performance of the Hartzog Draw Field", J. Pet. Tech., 1335-1344.

Hobson, J. P., Jr., Fowler, M. L., and Beaumont, E. A. (1982). "Depositional and Statistical Exploration Models, Upper Cretaceous Offshore Sand Complex, Sussex Member, House Creek Field, Wyoming", Amer. Assoc. Petrol. Geol. Bull., 689-707.

Hunt, R. D., and Hearn, C. L. (1982). "Reservoir Management of the Hartzog Draw Field", J. Pet. Tech., 1575-82.

Kortekaas, T. F. M. (1983). "Water/Oil Displacement Characteristics in Cross-Bedded Reservoir Zones", Paper SPE 12112 presented at the 1983 Society of Petroleum Engineers Annual Technical Conference and Exhibition, San Francisco, CA, Oct. 5-8.

Martinsen, R. S., and Tillman, R. W. (1978). "Hartzog Draw, New Giant Oil Field", (abst.), Amer. Assoc. Petrol. Geol. Bull., 540.

Martinsen, R. S., and Tillman, R. W., (1979). "Facies and Reservoir Characteristics of a Shelf Sandstone, Hartzog Draw Field, Powder River Basin, Wyoming", (abst.), Amer. Assoc. Petrol. Geol. Bull., 491.

Rester, S., and Todd, M. R. (1984). "An Automated Method for Predicting Full-Scale CO_2 Flood Performance Based on Detailed Pattern Flood Simulations", paper SPE/DOE 12663, Proceedings of the SPE/DOE Fourth Symposium on Enhanced Oil Recovery, Tulsa, OK, April 15-18.

Tillman, R. W., and Martinsen, R. S. (1979). "Hartzog Draw Field, Powder River Basin, Wyoming", in Flory, R. W. (ed.), Rocky Mountain High, Wyoming Geol. Assoc., 28th Ann. Meeting, Core Seminar Core Book, 1-38.

Tillman, R. W., and Martinsen, R. S. (1984). "The Shannon Shelf-Ridge Sandstone Complex, Salt Creek Anticline Area, Powder River Basin, Wyoming", in Tillman, R. W., and Siemers, C. T., (eds), Siliciclastic Shelf Sediments, Soc. of Econ. Paleont. and Mineral., Spec. Public. 34, 85-142.

Tillman, R. W., and Martinsen, R. S. (in prep.). "Sedimentologic Model and Production Characteristics of Hartzog Draw Field, Wyoming, a Shannon Shelf-Ridge Sandstone, in Tillman, R. W., and Weber, K. (eds.), Reservoir Sedimentology and Energy, Soc. of Am. Paleont. and Mineral., Spec. Public.

Todd, M. R., Cobb, W. M., and McCarter, E. D. (1982). "CO_2 Flood Performance Evaluation for the Cornell Unit, Wasson San Andres Field", J. Pet. Tech., 2271-2282.

$$\frac{1}{P_c^2} = (\text{Constant})\ S_w^* \tag{3}$$

$$K_{rw} = S_w^{*4} \tag{4A}$$

$$K_{rn} = \left[1-S_w^*\right]^2 \left[1-S_w^{*\,2}\right] \tag{4B}$$

where,

$$S_w^* = \frac{S_w - S_{wr}}{1 - S_{wr}} \tag{5}$$

Subsequent studies by Naar, et al.[5,6,7] investigated imbibition (increase in the wetting phase saturation) relative permeabilities and noted their differences from the drainage (decrease in wetting phase saturation) values. Since then, the literature[8-17] evaluates the accuracy of the above equations and their various forms/modifications by comparing their results with laboratory data. Land[10-11], formulated the trapping of the non- wetting phase in the imbibition cycle and derived a more general form of Corey's equations by incorporating a pore size distribution index (Equation 6).

$$S_w^* = \left[\frac{\text{Constant}}{P_c}\right]^{\lambda} \tag{6}$$

λ is calculated from the shape of the capillary pressure curve of the Leverett J-function and is related to the ratio of capillary pressure to capillary entry pressure and normalized wetting phase saturation. If the logarithm of both sides of Equation 6 are taken, then a plot of Log P_c and Log S_w^* will result in $-1/\lambda$. Substituting Equation 6 for the P_c term in Equations 2A and 2B and integrating, Land obtained the following equations for a gas-water (non-wetting-wetting) system.

$$K_{rn} = S_{nF}^{*\,2}\left[1 - (1 - S_{nF}^*)^{\epsilon-2}\right] \tag{7A}$$

$$K_{rw} = \left[S_w^*\right]^{\epsilon} \tag{7B}$$

where

$$\epsilon = \frac{2 + 3\lambda}{\lambda} \tag{8}$$

Equation 7A has been widely used and verified for the non-wetting phase with minor adjustments for both drainage and imbibition cycles[14,16]. Standing[13] found that Equation 7A, in its complete form, can be written as:

$$K_{rn} = K_{rn}^{\circ}\left[\frac{(S_m - 1)}{(S_m - S_{wr})} + S_{nF}^*\frac{(1 - S_{wr})}{(S_m - S_{wr})}\right]^2\left[1 - (1 - S_{nF}^*)^{\epsilon-2}\right] \tag{9}$$

K^o_{rn} is the maximum value of the non-wetting phase relative permeability in a complete drainage cycle. S_m is a factor representing both the critical non-wetting phase saturation and the anisotropy. S_m can be taken as unity for formations with no anisotropy. Formations that show anisotropy in the dip normal direction, should have z-directional relative permeabilities calculated separately. S^*_{nF} is the normalized, non-wetting phase saturation free from the trapped portion and related to Land's trapping constant C^*. Where C^* is:

$$C^* = \frac{1}{\dfrac{S_{nr}}{1 - S_{wr}}} - \frac{1}{\dfrac{S_{ni}}{1 - S_{wr}}} \quad (10)$$

The K_{rw} value from Equation 7B does not always provide satisfactory results. Land stated that there are minor differences in the drainage and imbibition curves for the wetting phase when $\epsilon = 4$ (Corey's equation). He also found that the wetting phase imbibition and drainage curves differ at higher values of ϵ. Based on his findings, Standing[13] stated that ϵ 's are more likely to be within the range of 5 to 6.3.

Killough[14], recognizing the hysteretic effect (changing saturation history), treated the problem of the wetting phase by interpolation between the drainage curve and experimental end points.

This paper presents an alternative approach to represent the hysteretic behavior of the wetting phase. The proposed empirical equations permit the relative permeabilities of different lithologies to be represented by a pore size distribution index. The initial wetting phase saturations are also accounted for in the hysteresis cycle.

WETTING PHASE HYSTERESIS

The following equation was taken from the imbibition relative permeability (gas-water) studies performed by Land[10].

$$K_{rw} = S_w^{*2} \left[\frac{\int_0^{S_w^* + S_{nt}^*} \frac{dS^*}{P_c^2}}{\int_0^1 \frac{dS^*}{P_c^2}} - (S_{nr}^*)^2 \frac{\int_{S_{wi}^*}^{S_w^* + S_{nt}^*} \frac{dS^*}{[1 - (1 - S_{nr}^*) S^*]^2 P_c^2}}{\int_0^1 \frac{dS^*}{P_c^2}} \right] \qquad (11)$$

Substituting Equation 6 into Equation 11 and integrating, we obtain:

$$K_{rw} = \frac{2+\lambda}{\lambda} \left[S_w^{*2}\right] \left[\frac{(S_w^* + S_{nt}^*)^{\frac{2+\lambda}{\lambda}}}{\frac{2+\lambda}{\lambda}} - (S_{nr}^*)^2 \int_{S_{wi}^*}^{S_w^* + S_{nt}^*} \frac{(S^*)^{2/\lambda}\, dS^*}{[1 - (1 - S_{nr}^*) S^*]^2} \right] \qquad (12)$$

If the portion of Equation 12 containing the integral function is defined as "I.F.", it is possible to rearrange Equation 12 to:

$$K_{rw} = (S_w^*)^2 (S_w^* + S_{nt}^*)^{\frac{2+\lambda}{\lambda}} - \frac{2+\lambda}{\lambda} (S_w^*)^2 (S_{nr}^*)^2 (\text{I.F.}) \qquad (13)$$

This is the general form of the equation for a wetting phase relative permeability. For the drainage cycle, since there is no trapping of the non-wetting phase, S_{nt}^* becomes zero. Thus, Equation 13 will reduce to the more familiar form of:

$$K_{rw} = (S_w^*)^2 (S_w^*)^{\frac{2+\lambda}{\lambda}} \qquad (14)$$

or

$$K_{rw} = (S_w^*)^{\frac{2+3\lambda}{\lambda}} \qquad (15)$$

For the imbibition cycle, S^*_{nt} is not zero and the "I.F." term cannot be neglected for high values of ϵ .

In order to solve Equation 13, the wetting phase relative permeability, several parameters must be known. These variables include: irreducible and initial wetting phase saturations, maximum residual non-wetting phase saturation or Land's trapping constant, and pore size distribution index. While all of these unknowns are important, the accuracy in irreducible wetting phase saturation governs the latter two variables, since they are calculated by using normalized saturations. Although irreducible water saturation can be measured for individual cores in the laboratory, it is difficult to obtain a representative "average" value for a heterogeneous or stratified lithology. As an example, in the Prudhoe Bay Field where vast quantities of irreducible saturation data are available, the value can vary by as much as 10 saturation units within the same lithology. To account for this uncertainty, a modified version of Equation 12 was used where the saturations are not normalized. This resulted in Equation 16. Land's trapping constant (C) and pore size distribution index (λ^o) were calculated on the same non-normalized basis. Figure 1 compares the two methods, wetting phase relative permeabilities calculated by Equation 12 and the other by the pragmatic approach of Equation 16. A minimal difference is observed in the shape and departure between the two curves.

$$K_{rw} = \frac{2+\lambda^\circ}{\lambda^\circ}\left[S_w^{\,2}\right]\left[\frac{(S_w + S_{nt})^{\frac{2+\lambda^\circ}{\lambda^\circ}}}{\frac{2+\lambda^\circ}{\lambda^\circ}} - (S_{nr})^2 \int_{S_{wi}}^{S_w + S_{nt}} \frac{(S)^{2/\lambda^\circ}\, dS}{[1-(1-S_{nr})\,S\,]^2}\right] \quad (16)$$

THE PROPOSED RELATIONSHIPS

The hysteretic relative permeabilities calculated from Equation 16 are impractical for use in numerical simulation. For ease of use, it is desirable to modify the equation so that it will be in a Corey-type form. Equation

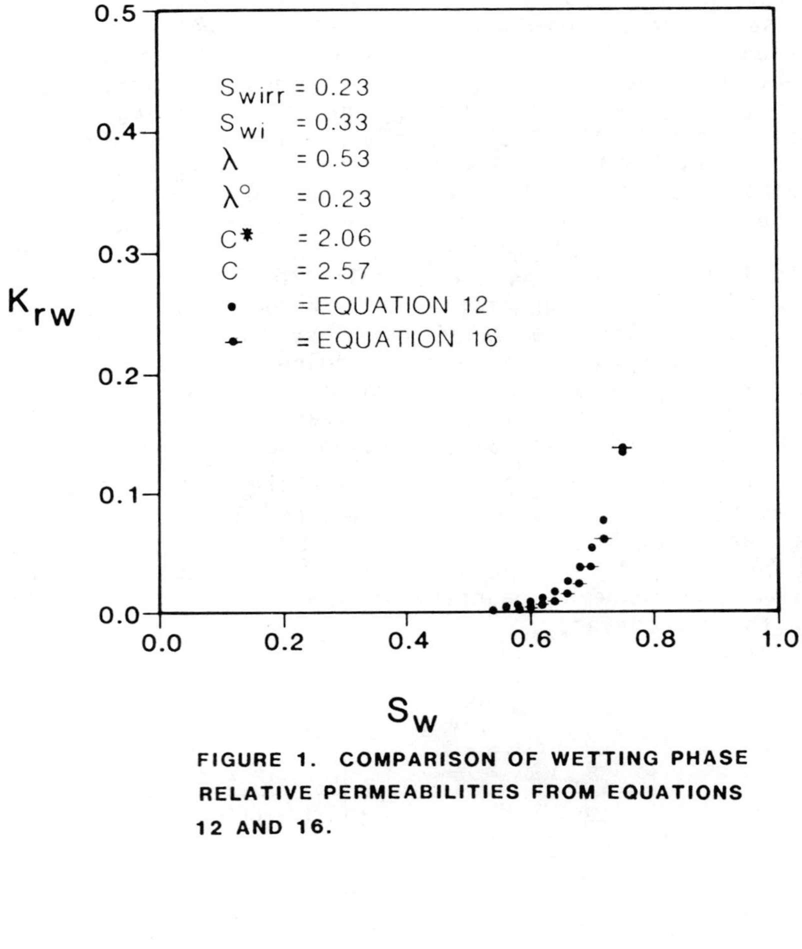

FIGURE 1. COMPARISON OF WETTING PHASE RELATIVE PERMEABILITIES FROM EQUATIONS 12 AND 16.

7B can be redefined by another exponent ϵ', which is a function of a non-normalized pore size distribution index, λ^o, Land's trapping constant, C, and initial wetting phase saturation. The irreducible saturations in S_w* are replaced by the initial saturation. The resulting equation is empirical in form and its validity is presented in this paper.

$$K_{rw} = \left[\frac{S_w - S_{wi}}{1 - S_{wi}}\right]^{\epsilon'} \quad (17)$$

where,

$$\epsilon' = \epsilon'\left[\lambda^o, C, S_{wi}\right]$$

The trapping constant was fixed at a typical Prudhoe Bay value of 2.6. The relative permeabilities were calculated using Equation 16 for a range of initial saturations and pore size distribution indices, λ^o. The exponent for Equation 17 was obtained through a linear regression of the logarithm of the relative permeability values and of the logarithm of the S_w* terms. When the new exponents ϵ' were plotted on a semi-log paper, a series of straight lines with constant slope was obtained (Figure 2) and followed the form of:

$$\epsilon' = 0.85\ \epsilon^o * 10^{-1.06\ S_{wi}} \quad (18)$$

Next, the pore size distribution index was fixed at an average value of 0.35 (non-normalized) and relative permeabilities were calculated for a range of initial saturations and trapping constants. Once again, a linear regression fit was utilized and the new exponents were plotted. As can be seen in Figure 3, the sensitivity of ϵ' is minimal for different values of C. Therefore, Equation 18 can be maintained in the above form since it is C-independent. This exponent is used in a Corey-type equation to calculate wetting phase relative permeability. Figure 4 shows a comparison of Equation 12 and Equation 18 values with the same parameters as in Figure 1. The

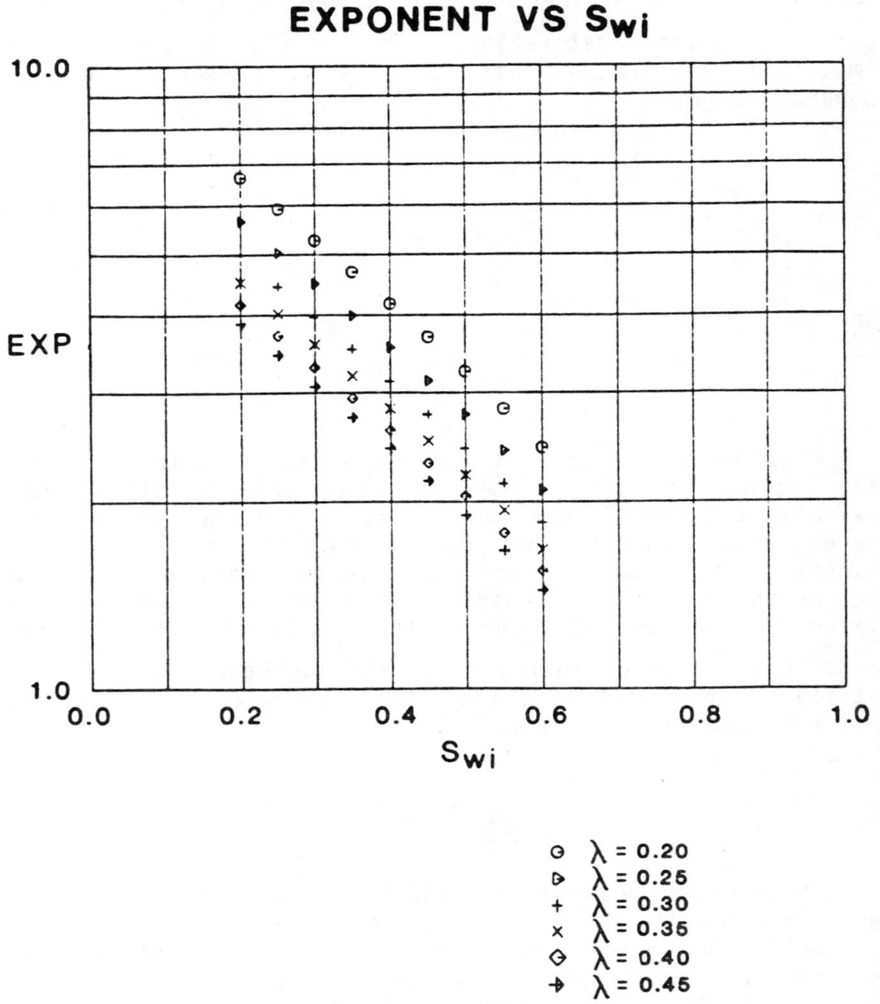

FIGURE 2. WETTING PHASE EXPONENT (ϵ') VERSUS INITIAL SATURATION (SWI) FOR DIFFERENT PORE SIZE DISTRIBUTION INDICES.

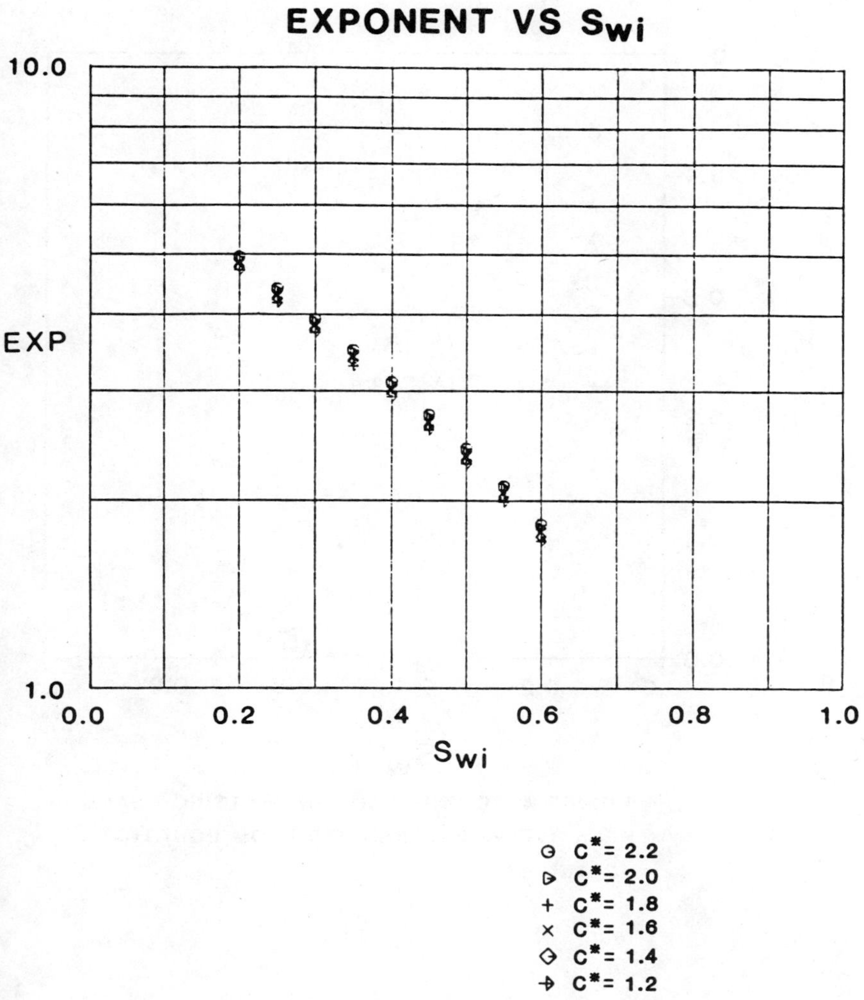

FIGURE 3. WETTING PHASE EXPONENT (ϵ') VERSUS INITIAL SATURATION (SWI) FOR DIFFERENT TRAPPING CONSTANTS.

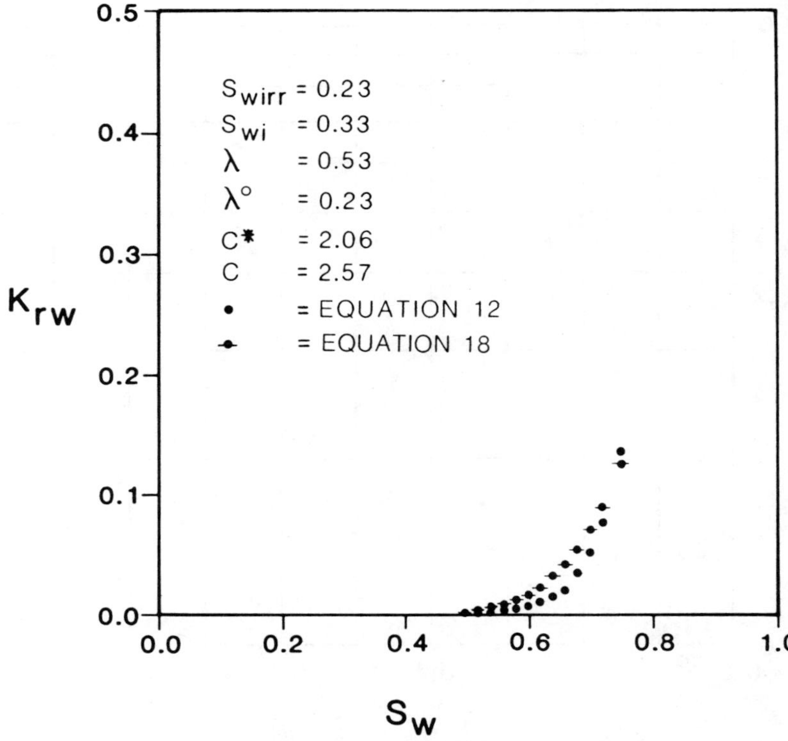

FIGURE 4. COMPARISON OF WETTING PHASE RELATIVE PERMEABILITIES FROM EQUATIONS 12 AND 18.

non-wetting phase relative permeabilities are calculated from Equation 14.

VERIFICATION

The complex depositional environment at Prudhoe Bay resulted in lateral and vertical variations in reservoir rock properties.[18-20] For geological and engineering use the main producing horizon, the Ivishak Sandstone of the Sadlerochit group, has been subdivided into several zones based upon petrophysical and shale continuity criteria. The Prudhoe Bay zonation shown in Figure 5 differentiates between the numerical and lettered terminology associated with the zones.[19] Zone 1 consists of a very fine to fine-grained sandstone and abundant interbedded shales. Zone 2 is predominantly a fine grained sandstone with increasing amounts of conglomerate towards the top of he zone. It contains an abundance of local shales. Zone 3 consists primarily of a sandy conglomerate with very few shales. Zone 4 is a more homogeneous fine-grained sandstone with local non-correlatable shales.

This complex stratification almost certainly leads to different capillary pressure characteristics and thus different distributions of oil, water and gas within the pore structure. Therefore, there is the need to account for the influences that the different lithologies have upon relative permeability values.

In addition, there have been numerous oil-water relative permeability core displacement tests from Prudhoe Bay. The tests have included both unsteady state and steady state methods; composite cores and short cores; and preserved and restored cores. This wide range of inputs creates a difficult task for the practicing engineer to obtain "averaged" relative permeability curves by lithology or those for the entire reservoir. The proposed method accounts for the effects of stratification upon relative permeability and also avoids the heuristic methods of averaging groups of relative permeability data.

Several-fold more oil-water capillary pressure tests compared to relative permeability tests have been run at Prudhoe Bay. They have been normalized through a Leverett J- function and categorized by area and zone (lithology). The pore size distribution indices were calculated for each zone and the appropriate equations were applied to calculate oil-water relative permeabilities. The values were compared to steady-state core tests. Figure 6 shows that the two methods agree very well.

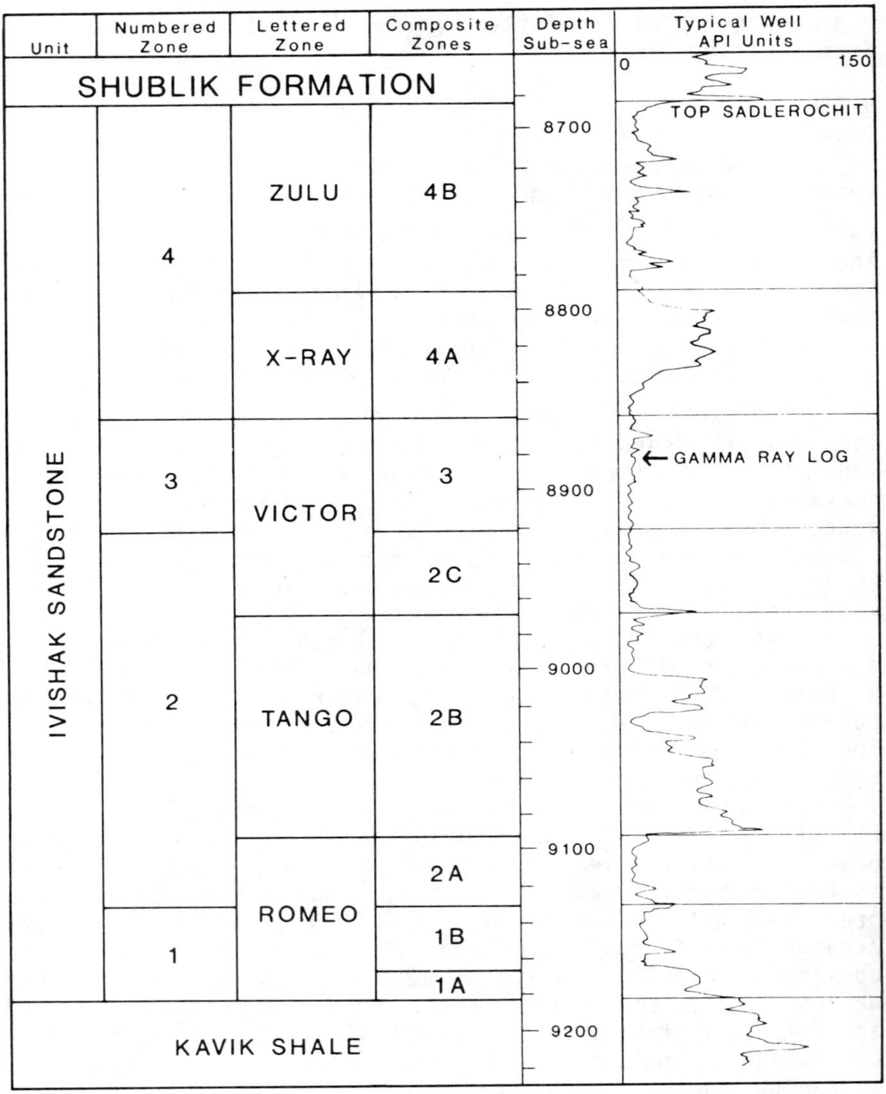

FIGURE 5. PRUDHOE BAY ZONATION

PE-I-NS-S-1897

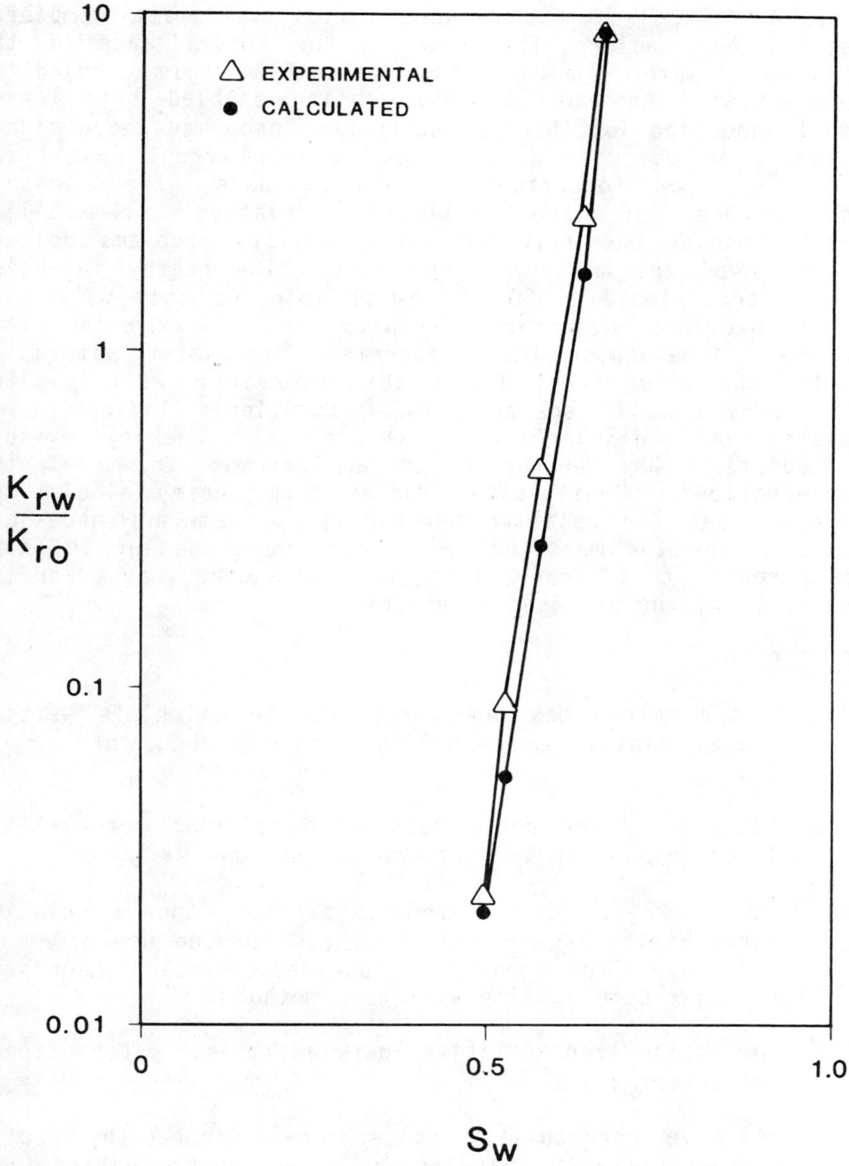

FIGURE 6: COMPARISON OF EXPERIMENTAL Krw/Kro VALUES TO CALCULATED Krw/Kro VALUES.

A fiberglass cased observation well was completed across the Sadlerochit formation for surveillance of the planned water alternating miscible gas injection project.[21] The non-conductive casing enabled time lapsed dual induction log (DIL) measurements to be made to evaluate changes in water saturation as the flood front passed the well.[22] An opportunity to evaluate the in-situ performance of the oil-water relative permeability relationships was available when facility problems delayed startup of the miscible injections. The nearby injection well (approximately 500 ft. away) injected only water and this provided waterflood results in a compressed time frame. Time-lapsed DIL measurements of water saturation after the water front passed the observation well resulted in very small changes in saturations indicating a piston-like displacement. A radial Buckley-Leverett calculation was performed in an attempt to model the observations. Sensitivity studies from these calculations showed that the relative permeability parameters necessary to describe the measured saturations were similar to those proposed in this paper. This provided another check for the validity of the proposed equations.

CONCLUSIONS

1). A new method has been presented to calculate wetting phase relative permeability with knowledge of λ , C and S_{wi}.

2). The non-wetting phase saturation relative permeability is calculated in the same manner as Land[10].

3). The effects of stratification upon relative permeabilities were accounted for through use of λ , C and S_{wi}. This procedure avoided invoking heuristic relative permeability averaging methods.

4). The calculated relative permeabilities match steady state core tests.

5). Relative permeability values are inferred by radial Buckley-Leverett calculations from an observation well in a water-alternating-miscible gas injection pattern.

NOMENCLATURE

C^*	=	Land's trapping constant
C	=	Land's trapping constant non-normalized
f_w	=	Fractional flow of Water
K	=	Absolute permeability
K_{rn}	=	Relative permeability of non-wetting phase
K_{rn}^o	=	maximum value of the non-wetting phase relative permeability in a complete drainage cycle
K_{rw}	=	Relative permeability of wetting phase
K_z	=	Kozeny constant
P_c	=	Capillary pressure
S_m	=	Wetting phase saturation at which non-wetting phase tortuosity is infinite
$S_n{}^*$	=	Normalized non-wetting phase saturation
$S_{ni}{}^*$	=	Normalized Initial non-wetting phase saturation
$S_{nF}{}^*$	=	Normalized free non-wetting phase saturation
$S_{nr}{}^*$	=	Normalized residual non-wetting phase saturation
$S_{nt}{}^*$	=	Trapped, non-wetting phase saturation

Sp	=	internal surface area per unit pore volume
S_w	=	Wetting phase saturation
S_{wi}	=	Initial wetting phase saturation
S_{wi}^*	=	Normalized initial wetting phase saturation
S_{wr}^*	=	Normalized irreducible wetting phase saturation
λ	=	Pore distribution index
λ^o	=	Non-normalized Pore distribution index
ϵ	=	Corey exponent
ϵ^o	=	Non-normalized Corey Exponent
ϵ'	=	Modified Corey Exponent
ϕ	=	Porosity

ACKNOWLEDGEMENT

The authors would like to thank the management of SOHIO Petroleum Company for permission to publish this paper. It should be noted that the other Prudhoe Bay owners may not agree with the methods and results presented. A special thanks to F. J. Fayers for his guidance and to Angie Lingad and L. A. Fujikawa for typing this manuscript and, to the Drafting Department for the illustrations.

REFERENCES

1). Carman, P. C.: "Flow of Gases Through Porous Media", Butterworths, London (1956).

2). Purcell, W. R.: "Capillary Pressures - Their Measurement Using Mercury and Calculation of Permeability Therefrom", Trans., AIME, February 1949, 39-48.

3). Burdine, N. T.: "Relative Permeability Calculations from Pore Size Distribution Data", Trans., AIME, 198, 71-78.

4). Corey, A. T.: "The Interrelation Between Gas and Oil Relative Permeabilities", Prod. Monthly, 1954, V19, 38-41.

5). Naar, J. and Henderson, J. H.: "An Imbibition Model - Its Application to Flow Behavior and the Prediction of Oil Recovery", SPEJ, June 1961, 61-70.

6). Naar, J. and Wygal, R. J.: "Three Phase Imbibition Relative Permeability", SPEJ, December 1961, 254-258.

7). Naar, J., Wygal, R. J., and Henderson, J. H.: "Imbibition Relative Permeability in Unconsolidated Porous Media", SPEJ, March 1962, 13-17.

8). Torcaso, M. A. and Wyllie, M. R. J.: "A Comparison of Calculated Kro/Krg Ratios with a Correlation of Field Data", Trans., AIME, 213, 436-437.

9). Pirson, S. J., Boatman, E. M., and Nettle, R. L.: "Prediction of Relative Permeability Characteristics of Intergranular Reservoir Rocks from Electrical Resistivity Measurements", JPT, May 1964, 564-570.

10). Land, C. S.: "Calculation of Imbibition Relative Permeability for Two- and Three-Phase from Rock Properties", SPEJ, June 1968, 149-156.

11). Land, C. S.: Comparison of Calculated with Experimental Imbibition Relative Permeability", SPEJ, December 1971, 419-425.

12). LeFebvre du Prey, E. J.: "Factors Affecting Liquid-Liquid Relative Permeabilities of a Consolidated Porous Medium", SPEJ, February 1973, 39-47.

13). Standing, M. B.: "Notes on Relative Permeability Relationships", Trondheim, Norway, 1974.

14). Killough, J. E.: "Reservoir Simulation with History Dependent Saturation Functions", SPEJ, February 1976, 37-48.

15). Molina, N. N.: "A Systematic Approach to the Relative Permeability Problem in Reservoir Simulation", SPE Paper 9234, 1980.

16). Meads, R. M. and Bassiouni, Z.: "Combining History and Petrophysical Correlations to Obtain Representative Relative Permeability Data", SPE Paper 12113, 1983.

17). Heaviside, J. and Black, C. J. J.: "Fundamentals of Relative Permeability: Experimental and Theoretical Considerations", SPE Paper 12173, 1983.

18). Wadman, D. H., Lamprecht, D. E. and Mrosovsky, I.: "Joint Geological/Engineering Analysis of the Sadlerochit Reservoir, Prudhoe Bay Field", JPT, July 1979, pp. 933-940.

19). Chang, D. M., Haldorsen, H. H. and Kirwan, P. A.: "History Matching and Predicting GOR Behavior with Analytical Functions", SPE 12763, 1984.

20). Inman, K.: Personal Communication, Sohio, 1984.

21). Rupp, K. A., Nelson, W. C., Christian, L. D., Zimmerman, K. A., Metz, B. E., and Styler, J. W.: design and Implementation of a Miscible Water-Alternating-Gas flood at Prudhoe Bay", SPE Paper 13272, 1984.

22). H. J. Mayson: Personal Communication, Sohio, 1984.

correctly the saturation profiles in the reduced dimension. Whenever the chosen profiles correspond closely to the actual dip-normal distribution, a good match is obtained between the two and three dimensional calculations.

A review of previous work suggests that two dimensional models incorporating pseudo functions can be divided into three categories depending upon the assumptions made regarding dip-normal saturation distributions. The simplest assumption made is that the capillary transition zone is infinitesimally thin. At any location, one fluid completely saturates the rock and a sharp transition zone forms between the different fluids. A more refined assumption, as shown by Martin[1], is to assume thick capillary transition zones with each phase in dip-normal hydrostatic equilibrium. This assumption is frequently referred to as the vertical equilibrium, VE, assumption. By assuming hydrostatic equilibrium, capillary pressures can be readily calculated and dip-normal saturation profiles determined. The third approach, commonly termed the Dynamic Pseudo Function Method, makes no physical assumptions about the nature of fluid flow.

Examples of the first category are the works of Coats[3] and Hearn[8]. These examples differ from each other in the vertical distribution of the wetting and non-wetting fluids. Coats assumes vertical segregation while Hearn assumes the fluids fully dispersed in the vertical. Although easy to use, this approach is limited to rock types having low capillary pressures.

An example of the VE method is the work of Coats[2]. This method appears useful and physically justifiable in many practical cases. In addition, by evaluating dimensionless criteria, it is possible to determine when the VE assumption is valid. For each rock curve, an analogous pseudo curve can be easily developed and used in conjunction with a two dimensional numerical calculation. Thus, under proper conditions the VE method has been shown to be cost effective and accurate.

In an attempt to remove the restrictions imposed by the VE assumption, Jacks et. al.[4] introduced the dynamic pseudo function approach. Additional extensions and modifications were presented by Kyte and Berry[5] and

Thomas[6]. These methods make no assumptions about the state of flow inside the reservoir and, in principle, are valid under all conditions including those of VE. In practice, however these methods are inherently difficult to apply and have several drawbacks. Like all the above methods, one needs to know dip-normal saturation distributions in order to evaluate integrals defining the pseudo functions. To ensure accuracy, it is first necessary to carry out a three dimensional calculation and determine saturation profiles for each grid block. However, the motivation for using pseudo functions is to avoid making a three dimensional calculation in the first place. It is sometimes suggested that a cross-sectional calculation might approximate the correct saturation profiles, but it is not always clear which cross-section is "representative" of the reservoir. Additional complications arise because pseudo curves under dynamic conditions vary with time and location. Thus, it is necessary to develop many sets of curves.

The philosophical approach of this work is that under certain limiting conditions, one can use two dimensional models to describe three dimensional reservoir phenomena. However, one can never completely account for three dimensional flow using a two dimensional model. From a practical point of view, the VE model is most useful since it is easy to implement and it applies to many types of reservoirs. To that effect, this work extends the VE model to three phase flow. New pseudo capillary curves, that are always positive, are introduced. Dimensionless criteria, one for each phase, are derived and shown useful in determining whether a given reservoir satisfies the VE conditions. These criteria differ from that proposed by Coats[2] and involve fewer simplifying assumptions in their derivation. Using a fully implicit black oil simulator, a parameter study is carried out in a quarter of a five-spot pattern to determine the numerical values and validity of the proposed dimensionless criteria.

In what follows, theoretical concepts are introduced in Section II. Comparative two and three dimensional parametric simulations are carried out and discussed in Section III. Based on the parametric simulations, refined VE criteria are proposed in Section IV. Finally, conclusions are presented in Section V.

II. THEORY

Derivation of the governing two dimensional equations which utilize pseudo functions is presented in Appendix A. These equations are summarized below as follows:

$$\bar{\bar{\nabla}} \cdot \left[\frac{k^p_{rw}\,\bar{k}}{\mu_w B_w}\left(\bar{\bar{\nabla}}\,\bar{P}_w - \gamma_w\,\bar{\bar{\nabla}} D\right)\right] = \frac{\partial}{\partial t}\left(\frac{\bar{\phi}\,S^p_w}{B_w}\right), \tag{1}$$

$$\bar{\bar{\nabla}} \cdot \left[\frac{k^p_{ro}\,\bar{k}}{\mu_o\,B_o}\left(\bar{\bar{\nabla}}\,\bar{P}_o - \gamma_o\,\bar{\bar{\nabla}} D\right)\right] = \frac{\partial}{\partial t}\left(\frac{\bar{\phi}\,S^p_o}{B_o}\right), \tag{2}$$

$$\bar{\bar{\nabla}} \cdot \left[\frac{k^p_{ro}}{\mu_o\,B_o} R_{so}\bar{k} \cdot \left(\bar{\bar{\nabla}}\,\bar{P}_o - \gamma_o\,\bar{\bar{\nabla}} D\right)\right]$$

$$+ \bar{\bar{\nabla}} \cdot \left[\frac{k^p_{rg}}{\mu_g\,B_g}\,\bar{k} \cdot \left(\bar{\bar{\nabla}}\,\bar{P}_g - \gamma_g\,\bar{\bar{\nabla}} D\right)\right]$$

$$= \frac{\partial}{\partial t}\left[\bar{\phi}\left(\frac{R_{so}}{B_o}\,S^p_o + \frac{S^p_g}{B_g}\right)\right], \tag{3}$$

$$\bar{P}_{cow} = \bar{P}_o - \bar{P}_w, \tag{4}$$

$$\bar{P}_{cgo} = \bar{P}_g - \bar{P}_o, \tag{5}$$

$$S^p_w + S^p_o + S^p_g = 1. \tag{6}$$

In the above equations the Laplacian operator is two dimensional, that is:

$$\bar{\bar{\nabla}} \equiv \left(\frac{\partial}{\partial x} + \frac{\partial}{\partial y}\right) \tag{7}$$

where the coordinate directions are displayed in Figures 1 and 2. An analogy between two dimensional black oil equations and those above, simulating three dimensional flow, is apparent. However, the analogous terms have a different physical interpretation. In the above equations, pseudo saturations replace block saturations, pseudo relative permeabilities replace rock relative permeabilities and dip-normal averages of porosity, permeability and pressure replace terms containing rock porosity, permeability and pressure; respectively. All variables representing averages in the dip-normal direction are designated with an overbar (i.e., $\bar{k}$, $\bar{\phi}$, $\bar{P}_o$, $\bar{P}_w$, $\bar{P}_g$, $\bar{P}_{cow}$ and $\bar{P}_{cgo}$) and consistently defined as:

$$\bar{f} \equiv \frac{1}{h} \int_0^h f \, dz \; . \tag{8}$$

Pseudo relative permeability and pseudo saturation for the water phase (and similarly for oil and gas) are defined respectively as:

$$k^p_{rw} \equiv \frac{1}{\bar{k}h} \int_0^h k_{rw} \, k \, dz \; ,$$

$$S^p_w \equiv \frac{1}{\bar{\phi}h} \int_0^h \phi \, S_w \, dz \; . \tag{9}$$

With the aid of mean capillary and pseudo relative permeability curves, Equations (1) through (6) are solved for the dip-normal average phase pressures and pseudo phase saturations (i.e., $\bar{P}_w$, $\bar{P}_o$, $\bar{P}_g$, S^p_w, S^p_o and S^p_g).

Equations (1) through (6) are similar to those derived by Martin[1]. The pseudo relative permeabilities and pseudo saturations are interpreted identically. However, the pressure terms in Martin's equations refer to

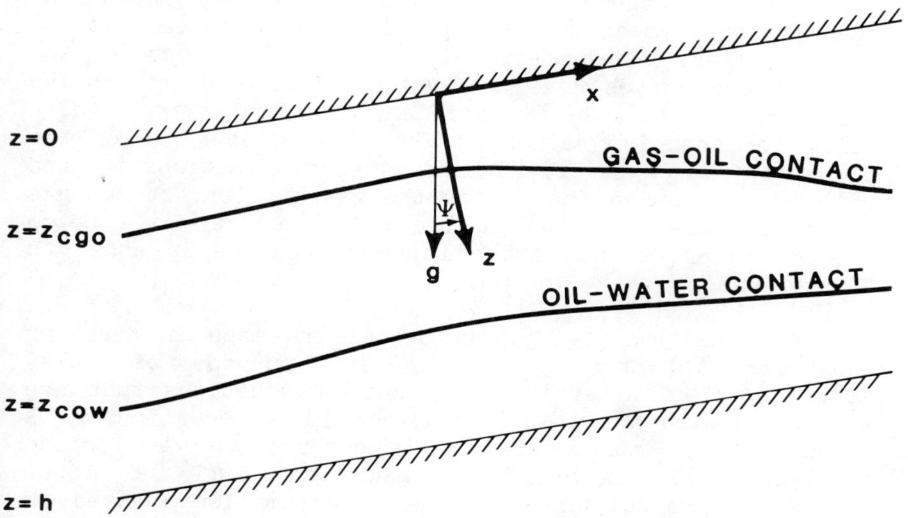

Figure 1: Formation cross section containing three flowing phases under VE conditions.

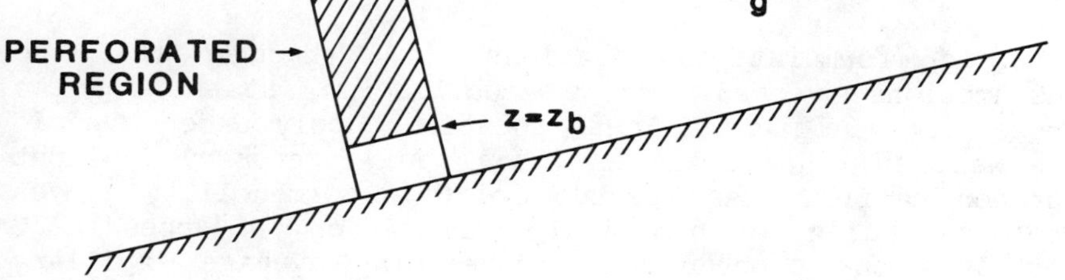

Figure 2: Illustration of the perforated region for a partially completed well.

the top of the sand while here they are interpreted as dip-normal averages. It is advantageous to employ the mean pressure as an unknown, particularly when calculating mean capillary pressure curves, given by Equations (4) and (5). These curves have a clear physical meaning (unlike the conventional pseudo capillary pressure curves which allow for negative values). When taking averages of the pressure, (as on the right hand sides of Equations (4) and (5)) it is assumed that below the gas-oil contact the gas phase pressure is equal to the oil pressure while below the oil-water contact the oil phase pressure is equal to that of water.

Several simplifying assumptions are made in deriving Equations (1) through (6). The purpose of these assumptions is to arrive at a set of equations that are analogous with the two dimensional flow equations. A standard black oil simulator, in conjunction with pseudo functions, can be used to make the two and three dimensional calculations. A major assumption invoked is that the reservoir is at sufficient depth and/or sufficiently thin that the pressure deviation in the dip-normal direction is small when compared with the dip-normal average pressure. This assumption allows one to calculate all fluid properties in terms of the average dip-normal pressure. Additional assumptions made are: no flow across the top and bottom boundaries of the formation and permeability in the x, y coordinate plane is isotropic. As shown in Appendix A, if one relaxes the no flow assumption, an additional term must be added to Equations (1) through (3). If one also includes permeability anisotropy in the x, y plane, it is necessary to introduce directionally dependent pseudo relative permeabilities. However, it is not necessary to make the latter two assumptions if the simulator is modified to account for the additional computations.

In formulating Equations (1) through (6), no assumptions regarding the equilibrium state of the reservoir are made. These equations apply under dynamic as well as VE conditions. It is only necessary to input proper capillary and pseudo relative permeability curves before making a numerical calculation. Appendix C outlines the procedure for developing average capillary and pseudo relative permeability curves when making use of the VE assumption. Under this assumption, it is shown that the capillary pressure and the pseudo relative permeability are functions of the pseudo saturation only. Using braces to indicate functional dependence, these relations can be summarized as:

$$\bar{P}_{cow}\ \{S_w^p\},\ \bar{P}_{cgo}\ \{S_g^p\},\ k_{rw}^p\ \{S_w^p\},\ k_{rg}^p\ \{S_g^p\}$$

and $k_{ro}^p\ \{S_w^p,\ S_g^p\}$.

When employing the VE assumption, criteria are necessary for assessing its validity. By performing a dip-normal force balance and assuming friction forces are small compared to gravity and pressure forces, a dimensionless criterion is developed for each phase in Appendix B. Specializing these criteria for a five-spot pattern and rewriting in terms of field units yields:

$$N_w \equiv \frac{|q_{wp} - q_{wi}| B_w \mu_w^*}{L^2\, k_{rw}^*\, \gamma_w^* \cos\psi} \left(\overline{\frac{1}{k_z}}\right)_w \ll 2.4 \times 10^{-7} ,$$

$$N_o \equiv \frac{|\,(q_{wp} - q_{wi})\, B_w + q_{op}\, B_o| \mu_o^*}{L^2\, k_{ro}^*\, \gamma_o^* \cos\psi} \left(\overline{\frac{1}{k_z}}\right)_o \qquad (10)$$

$$\ll 2.4 \times 10^{-7} ,$$

$$N_g \equiv \frac{|\,(q_{wp} - q_{wi})\, B_w + q_{op}\, B_o| \mu_g^*}{L^2\, k_{rg}^*\, \gamma_g^* \cos\psi} \left(\overline{\frac{1}{k_z}}\right)_g$$

$$\ll 2.4 \times 10^{-7} ,$$

where:

$$\left(\overline{\frac{1}{k_z}}\right)_w \equiv \frac{1}{h} \int_0^h \frac{1}{k_z}\, dz ,$$

$$\overline{\left(\frac{1}{k_z}\right)}_o \equiv \frac{1}{z_{cow}} \int_{z_{cow}}^{h} \frac{1}{k_z}\, dz \ , \qquad (11)$$

and:

$$\overline{\left(\frac{1}{k_z}\right)}_g \equiv \frac{1}{z_{cgo}} \int_{z_{cgo}}^{h} \frac{1}{k_z}\, dz \ .$$

Henceforth, N_w, N_o, and N_g will be called Equilibrium Numbers for the water, oil, and gas phases, respectively. Variables with superscript asterisks indicate volumetric averages in the region where the given phase is present. The above criteria must be satisfied simultaneously to insure that the VE assumption is valid. The right hand sides in Relations (10) are theoretical upper limits and provide a rough measure of comparison. With the aid of comparative studies, in the next Section, more refined values are obtained. In general, the left hand sides vary with time, (i.e., flow rates and relative permeabilities change) thus, it is conceivable that a given reservoir may be in the VE state only part of the total production time.

Inspection of Relations (10) shows which formation parameters are significant in determining the equilibrium state of the reservoir. It appears that large spacing between wells for a five spot pattern is most influential in ensuring VE. In addition, Relations (10) also show that large absolute vertical and relative permeabilities, large specific gravities, small production rates, and small fluid viscosities enhance the state of VE, as intuitively expected. The absolute vertical permeability is harmonically averaged. Thus, for a layered reservoir the layer having the smallest permeability exerts the greatest influence on the harmonic average. Surprisingly, the formation thickness does not appear explicitly in the above criteria and has no bearing on the dynamic state of the reservoir. This result is true so long as the thickness is sufficiently small that the fluid properties can be specified in terms of the mean dip-normal pressure.

III. COMPARATIVE STUDIES

In this Section, comparative calculations, three dimensional versus two dimensional, are carried out to determine when the pseudo function approach is advantageous. Simple hypothetical cases are chosen and the formation parameters varied in order to validate and refine the criteria established in Relations (10). When varying parameters, the reservoir description becomes physically implausible, at times. Nevertheless, these comparisons are extremely useful in illustrating parameter sensitivity and the conditions necessary to establish VE. By making use of hypothetical problems instead of actual field cases, one eliminates uncertainties with respect to the formation parameters, reservoir geometry, fluid properties, and the quality of wellbore pressure and production data.

All examples considered in this work assume a five-spot well pattern. For calculation purposes, a quarter of the five spot pattern is used. Due to symmetry, it is only necessary to simulate an eigth of a five-spot pattern. However, a quarter of the pattern is chosen because the specification of boundary conditions for square geometry is easily defined. The grid size for all three dimensional calculations is 5 by 5 by 10 (10 grid blocks in the dip-normal direction). The corresponding two dimensional calculations uses a 5 by 5 grid system. Originally, a 10 by 10 by 10 mesh was used, but sample calculations showed no loss in accuracy with the smaller mesh size. Numerical calculations are carried out with the aid of a fully implicit black oil simulator using a direct solver.

Example 1

In this example, a five-spot pattern having a 10 acre well spacing is undergoing a waterflood. The formation is assumed horizontal. Comparative calculations are made to study the effects of vertical permeability changes while holding all the other parameters fixed. Simulation is carried out over a period of 1500 days. At the end of this period, the reservoir becomes substantially depleted in oil and gas reserves and produces water at a rate nearly equal to that injected. Both production and injection wells are fully completed in a formation having uniform horizontal permeability, porosity and thickness of 100 md, 20%, and 46 ft, respectively. The production well is set at a bottom hole flowing pressure of 2000 psia. The water injection rate is set at 200 barrels per day throughout the simulation period.

Initial distribution of fluids in the three dimensional calculation is determined by specifying the location of the water-oil and gas-oil contacts. The water-oil contact is set at the bottom of the formation while the gas-oil contact is located 15.5 ft below the top of the formation. Below the water-oil contact, an initial residual oil saturation of 20% is obtained from the capillary pressure curves. Saturation distributions are calculated with the aid of rock capillary pressure curves (Figures 3 and 4) for hydrostatic equilibrium conditions. Figures 3 and 4 also show the corresponding pseudo capillary pressure curves. These curves were calculated in accordance with procedures outlined in Appendix C.

Fluid properties used in the following calculations are listed in Table 1. The water phase is assumed incompressible and free of solution gas. Bubble point pressure for the oil is 2580 psia. This also corresponds to the oil phase pressure at the gas-oil contact.

Initialization of the analogous two dimensional calculation is accomplished by specifying the pseudo saturation for each phase. These are calculated using initial saturation distributions from the three dimensional calculation and integrating according to Equation (9).

Three phase relative permeability curves for water and gas are plotted in Figures 5 and 6. Table 2 lists the relative oil permeability as a function of water and gas saturations. Along with the rock curves, these Figures and Table show the corresponding pseudo relative permeabilities for all three phases.

Making use of the aforementioned input conditions, the results of a comparative calculation are shown in Figure 7. Production rates for oil, water, and gas phases are plotted with respect to time. The three dimensional

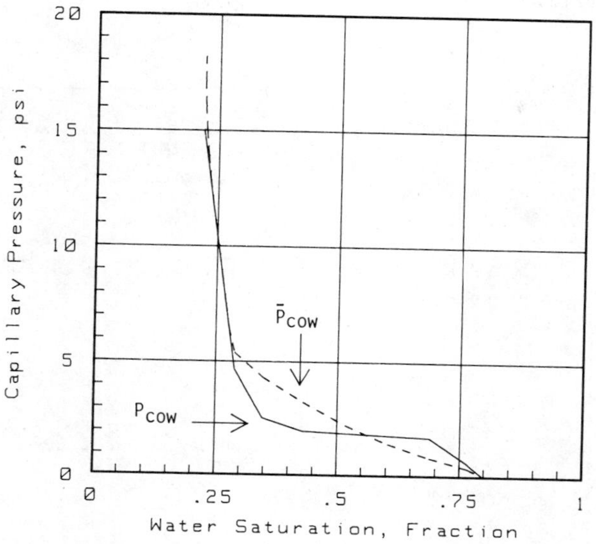

Figure 3: Rock and mean oil-water capillary pressure curves for Examples 1 through 4.

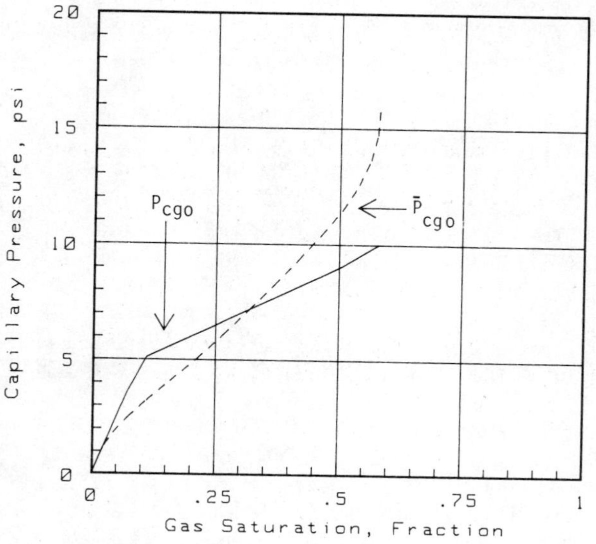

Figure 4: Rock and mean gas-oil capillary pressure curves for Examples 1 through 4.

Table 1: Fluid Properties for Example 1 through Example 4

P (PSIA)	R_{so} (SCF/STB)	B_o (RB/STB)	B_g (RB/MSCF)	ρ_o (lbm/ft^3)	ρ_g (lbm/ft^3)	μ_o (CP)	μ_g (CP)
100.0	11.37	1.0698	32.6348	50.2047	.2855	2.36	0.0133
200.0	25.61	1.0756	16.1608	50.0569	.5765	2.29	0.0133
300.0	43.22	1.0828	10.6715	49.8732	.8729	2.20	0.0134
400.0	64.52	1.0917	7.9286	49.6500	1.1750	2.09	0.0135
500.0	89.74	1.1023	6.2844	49.3847	1.4824	1.98	0.0136
600.0	119.07	1.1149	5.1897	49.0753	1.7951	1.86	0.0138
700.0	152.67	1.1294	4.4092	48.7205	2.1129	1.73	0.0139
800.0	190.67	1.1461	3.8252	48.3198	2.4354	1.60	0.0141
900.0	207.28	1.1535	3.3720	48.1451	2.7627	1.54	0.0142
1000.0	224.28	1.1611	3.0108	47.9666	3.0942	1.49	0.0144
1100.0	241.63	1.1689	2.7164	47.7849	3.4295	1.44	0.0146
1200.0	259.31	1.1769	2.4723	47.6004	3.7682	1.39	0.0148
1300.0	277.29	1.1850	2.2669	47.4133	4.1096	1.34	0.0150
1400.0	295.56	1.1934	2.0919	47.2239	4.4533	1.30	0.0153
1500.0	314.10	1.2019	1.9414	47.0326	4.7986	1.25	0.0155
1600.0	332.88	1.2106	1.8108	46.8395	5.1448	1.21	0.0157
1700.0	351.91	1.2195	1.6965	46.6450	5.4912	1.17	0.0160
1800.0	371.17	1.2285	1.5960	46.4492	5.8730	1.13	0.0163
1900.0	390.64	1.2376	1.5070	46.2522	6.1817	1.09	0.0165
2000.0	410.32	1.2469	1.4279	46.0543	6.5244	1.05	0.0168
2100.0	430.21	1.2564	1.3571	45.8556	6.8644	1.02	0.0171
2200.0	450.28	1.2660	1.2937	45.6563	7.2012	0.99	0.0174
2300.0	470.54	1.2757	1.2365	45.4565	7.5341	0.95	0.0177
2400.0	490.97	1.2855	1.1849	45.2563	7.8626	0.92	0.0181
2500.0	511.58	1.2955	1.1380	45.0560	8.1862	0.90	0.0184
2580.0	528.19	1.3036	1.1036	44.8955	8.4413	0.87	0.0186

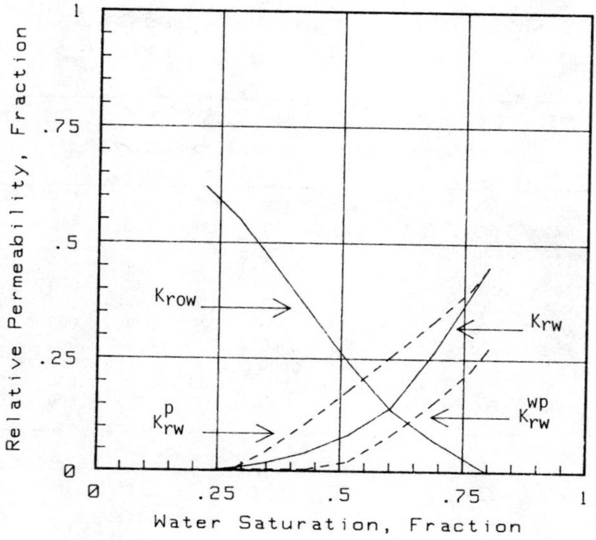

Figure 5: Rock and pseudo relative permeabilities for water in Examples 1 through 3.

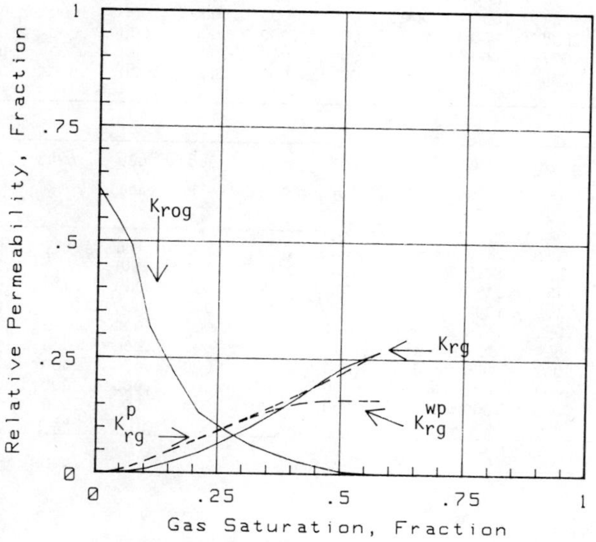

Figure 6: Rock and pseudo relative permeabilities for gas in Examples 1 through 3.

Table 2: Rock and Pseudo Relative Oil Permeabilities for Example 1 through Example 3

S_g \ S_w	.220	.278	.336	.394	.452	.510	.568	.626	.684	.742	.800
	Reservoir Oil Rock Relative Permeability, K_{ro}										
.000	.620	.562	.490	.408	.327	.248	.178	.119	.073	.035	.000
.058	.521	.471	.409	.339	.269	.199	.137	.081	.034	.000	.000
.116	.304	.272	.233	.188	.141	.093	.045	.000	.000	.000	.000
.174	.191	.168	.139	.107	.071	.033	.000	.000	.000	.000	.000
.232	.118	.100	.078	.052	.022	.000	.000	.000	.000	.000	.000
.290	.077	.061	.041	.019	.000	.000	.000	.000	.000	.000	.000
.348	.051	.035	.016	.000	.000	.000	.000	.000	.000	.000	.000
.406	.029	.013	.000	.000	.000	.000	.000	.000	.000	.000	.000
.464	.014	.000	.000	.000	.000	.000	.000	.000	.000	.000	.000
.522	.004	.000	.000	.000	.000	.000	.000	.000	.000	.000	.000
.580	.000	.000	.000	.000	.000	.000	.000	.000	.000	.000	.000

S_g^p \ S_w^p	.220	.278	.336	.394	.452	.510	.568	.626	.684	.742	.800
	Reservoir Oil Pseudo Relative Permeability, K_{ro}^p										
.000	.620	.562	.493	.427	.364	.301	.239	.176	.114	.052	.000
.058	.496	.446	.379	.314	.252	.192	.133	.048	.000	.000	.000
.116	.409	.364	.299	.235	.175	.118	.000	.000	.000	.000	.000
.174	.342	.302	.238	.177	.119	.055	.000	.000	.000	.000	.000
.232	.280	.245	.184	.125	.054	.000	.000	.000	.000	.000	.000
.290	.218	.190	.132	.052	.000	.000	.000	.000	.000	.000	.000
.348	.158	.136	.044	.000	.000	.000	.000	.000	.000	.000	.000
.406	.100	.084	.000	.000	.000	.000	.000	.000	.000	.000	.000
.464	.048	.038	.000	.000	.000	.000	.000	.000	.000	.000	.000
.522	.014	.007	.000	.000	.000	.000	.000	.000	.000	.000	.000
.580	.000	.000	.000	.000	.000	.000	.000	.000	.000	.000	.000

S_g^p \ S_w^p	.220	.278	.336	.394	.452	.510	.568	.626	.684	.742	.800
	Production Well Oil Pseudo Relative Permeability, K_{ro}^{wp}										
.000	.377	.347	.340	.335	.321	.291	.236	.176	.113	.052	.000
.058	.254	.232	.226	.223	.212	.181	.131	.048	.000	.000	.000
.116	.171	.154	.150	.148	.138	.110	.000	.000	.000	.000	.000
.174	.113	.100	.097	.094	.086	.050	.000	.000	.000	.000	.000
.232	.065	.056	.054	.053	.037	.000	.000	.000	.000	.000	.000
.290	.029	.024	.023	.017	.000	.000	.000	.000	.000	.000	.000
.348	.010	.007	.005	.000	.000	.000	.000	.000	.000	.000	.000
.406	.003	.001	.000	.000	.000	.000	.000	.000	.000	.000	.000
.464	.000	.000	.000	.000	.000	.000	.000	.000	.000	.000	.000

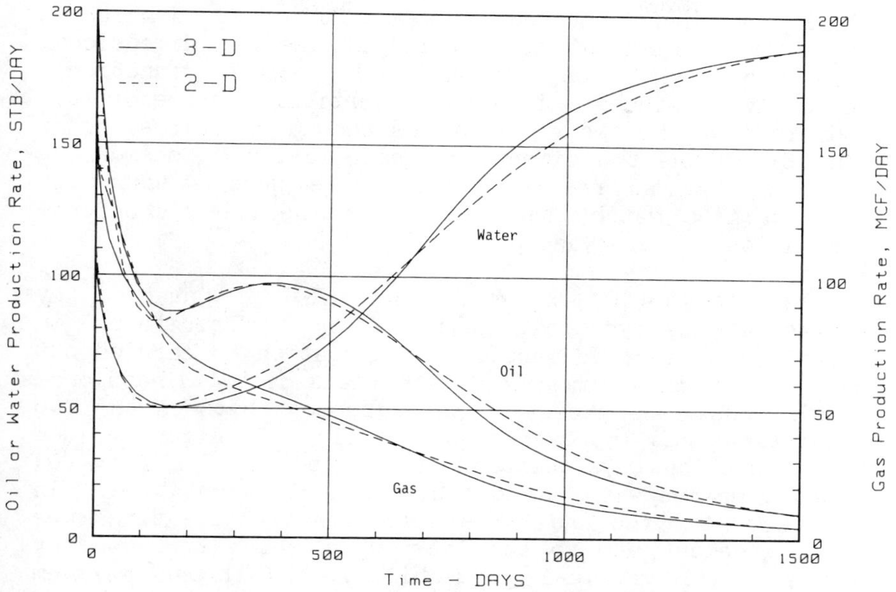

Figure 7: Comparative calculations for Example 1 where k_z = 10 md.

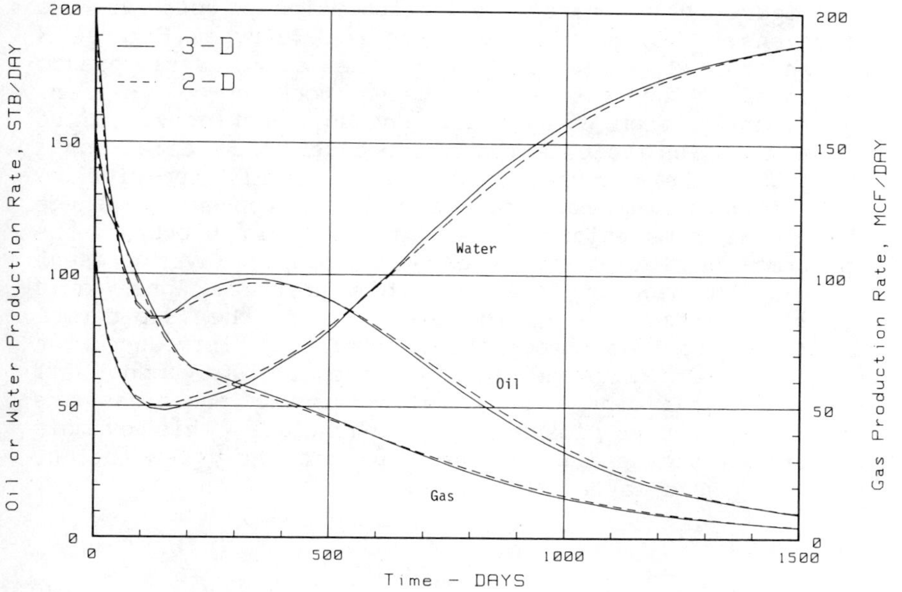

Figure 8: Comparative calculations for Example 1 where k_z = 100 md.

calculation uses a vertical permeability of 10 md which is 10% of the horizontal permeability. The two dimensional calculation, in conjunction with pseudo functions, implicitly assumes infinite vertical permeability. Comparison of the two calculations show a maximum error of about 8% in the two dimensional calculation. Considering the fact that no reservoir parameters were adjusted to obtain a good match, the results of the two dimensional calculation is quite acceptable.

Relations (10) suggest that VE is enhanced by increasing the vertical permeability. Therefore, in Figure 8, the three dimensional calculation is carried out for k_z = 100 md and compared with the same two dimensional calculations. As expected, the agreement between the two calculations has improved significantly. In Figure 9, the three dimensional calculation is carried out for k_z = .01 md and compared with the two dimensional calculation. In this case, the two and three dimensional calculations are in disagreement due to the lack of VE which is caused by such a small vertical permeability. Although maximum errors are about 20%, the general trends are in agreement and, therefore, the two dimensional calculation may still be acceptable.

Figure 10 compares two calculations, both in two dimensions. One calculation uses rock curves (Figures 3 through 6 and Table 2) while the other uses pseudo functions. The calculation using rock curves differs significantly from that using pseudo functions. Since pseudo function calculations reproduce three dimensional results, in this example, they are clearly superior to calculations using rock curves. It is typical in three dimensional simulations to assign one grid block in the dip-normal direction for each permeability layer present in the formation. Based on this practice, one would normally assign one grid block, (in the dip-normal direction) to the hypothetical problems introduced in Example 1. As illustrated above, such an assignment will lead to significant errors unless pseudo functions are used. For three dimensional calculations, this example illustrates the need to refine dip-normal grid resolution, even for single layer formations.

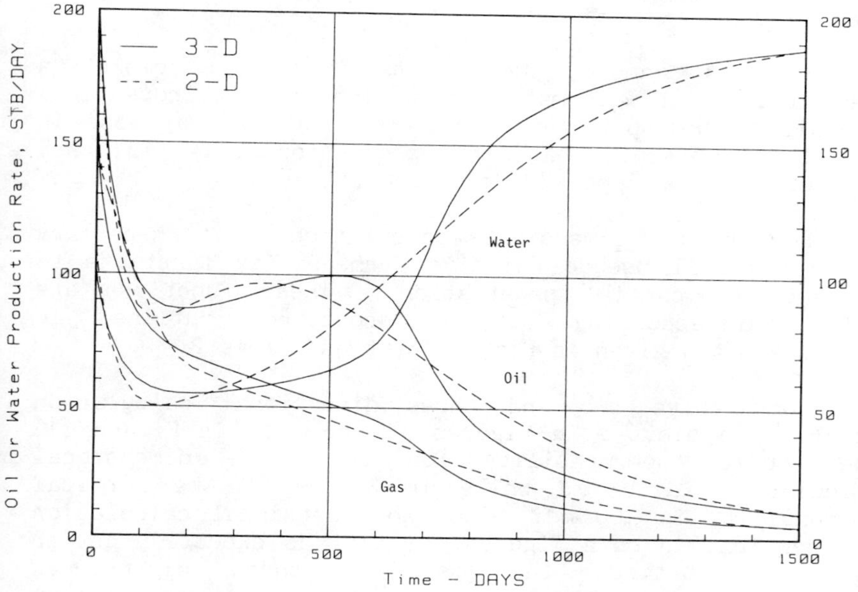

Figure 9: Comparative calculations for Example 1 where k_z = 0.01 md.

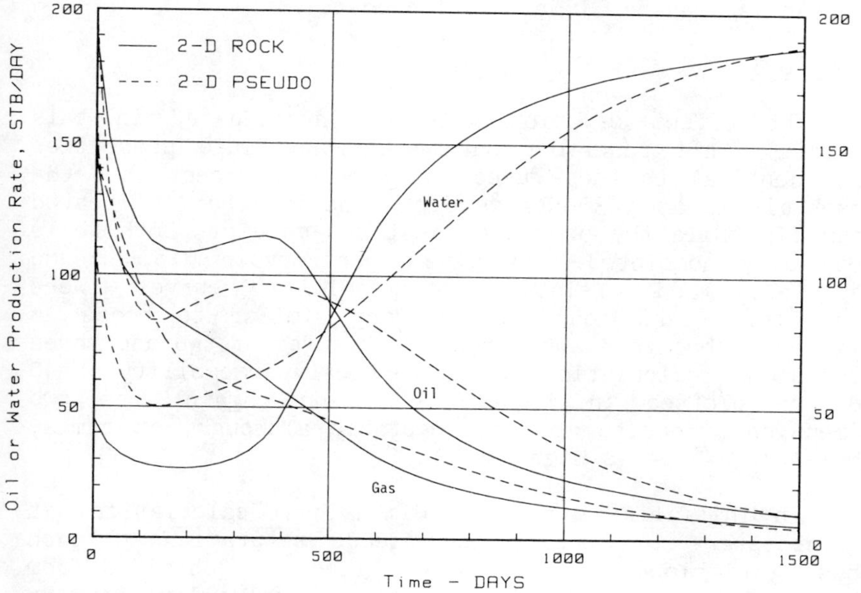

Figure 10: Comparative two dimensional calculations for Example 1 using rock and pseudo functions.

Example 2

In this example, the effects of partial completion are studied. The formation conditions, fluid properties, and all other input parameters are identical to those in Example 1 except that the production well is only completed in the upper 30 ft of the formation.

For the two dimensional calculation, it is necessary to employ well pseudo relative permeability functions to account for partial penetration. These functions are calculated according to the procedures outlined in Appendix D and given in Figures 5, 6 and Table 2.

Comparative two and three dimensional calculation results are plotted in Figures 11 and 12. In Figure 11, the vertical permeability for the three dimensional calculation is 10 md while in Figure 12 the vertical permeability is 100 md. The two dimensional calculation is identical in both figures. Again, as expected, better agreement, between the two and three dimensional calculations, results when the larger vertical permeability is used. From an economical viewpoint, the two dimensional calculation can be seen to reproduce the three dimensional calculation, in this example, at substantial savings in simulation cost.

Example 3

The effect of dip angle is investigated in this example. All reservoir parameters and fluid properties are identical to those used in Example 1 except that the reservoir is at a 30 degree dip. The injector is located down dip while the producer is at the up dip. Both wells are fully completed. Because of reservoir dip, pseudo capillary and relative permeability curves were recalculated and found to be very similar to those in Figures 3 through 6 and Table 2. Results of two and three dimensional calculations for a vertical permeability of 10 md are plotted in Figure 13. Once again, the two dimensional results appear acceptable although, at times, the error may be as high as 20%.

In initializing the two dimensional calculation, it is necessary to specify the pseudo saturation of each phase grid block by grid block. These are obtained from the initialization of the three dimensional calculation by

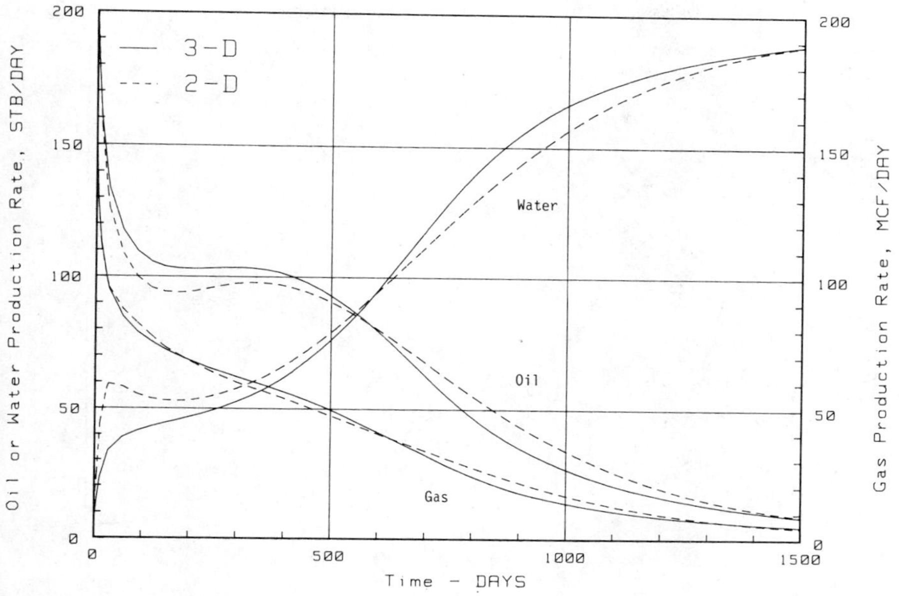

Figure 11: Comparative calculation for Example 2 assuming k_z = 10 md and the producer is partially completed.

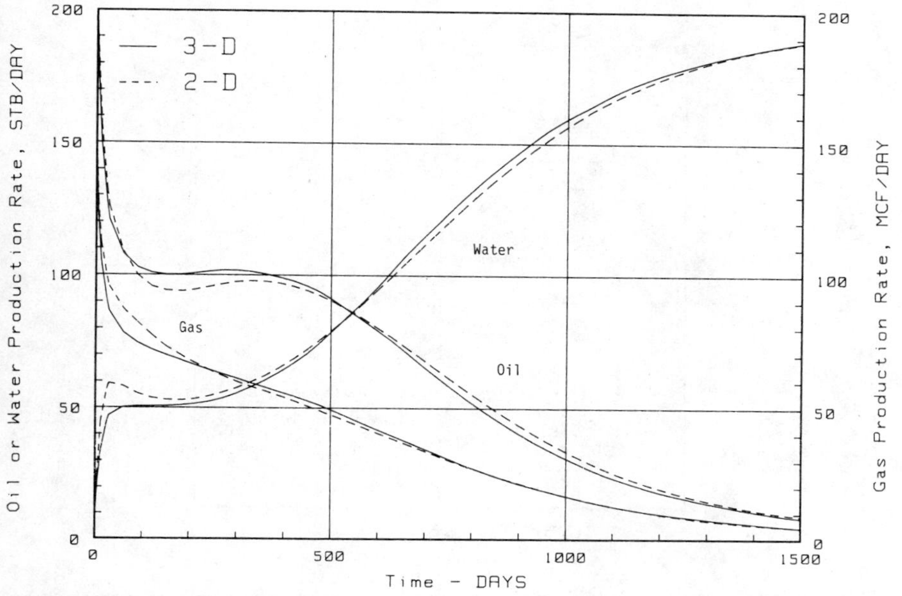

Figure 12: Comparative calculations for Example 2 assuming k_z = 100 md and the producer is partially completed.

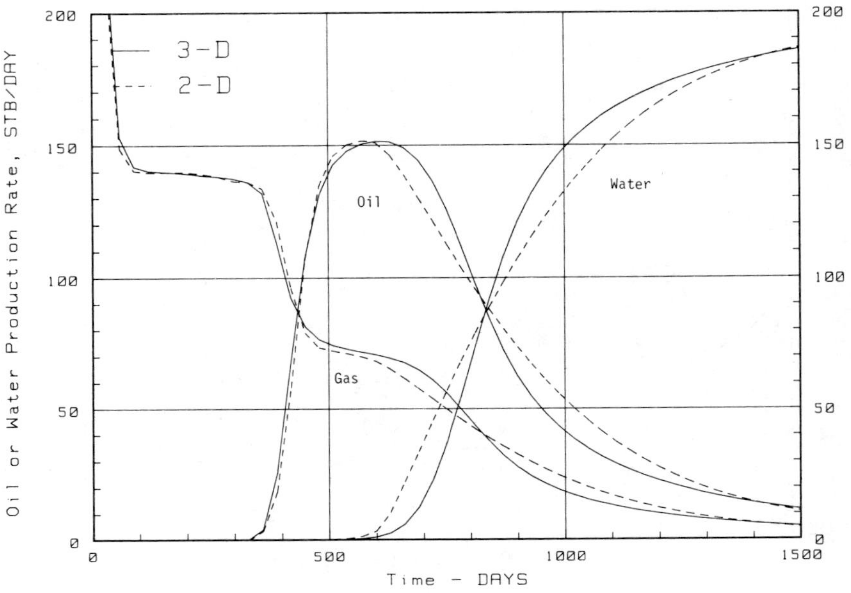

Figure 13: Comparative calculation for Example 3 assuming k_z = 10 md and the reservoir is at a 30 degree dip.

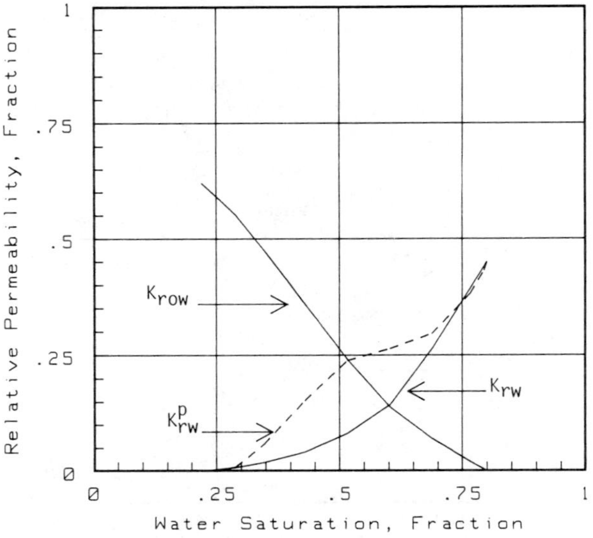

Figure 14: Rock and pseudo relative permeabilities for water in Example 4.

appropriate dip-normal averaging. Since the initial water-oil and gas-oil contacts are horizontal, but the reservoir is dipping, the dip-normal pseudo saturations vary along the length of the reservoir with large water pseudo saturations down dip and large gas pseudo saturations up dip.

Example 4

In this example, the concept of pseudo functions, under VE assumption, is extended to stratified reservoirs with low permeability layers. The reservoir is assumed horizontal and composed of three layers. The top layer is 13 ft thick having a horizontal permeability of 100 md. The middle and bottom layers are 15 and 18 ft thick, respectively, with horizontal permeabilities of 1 and 100 md, respectively. Like previous Examples, water is injected at the rate of 200 barrels per day while the producer is set at a bottom hole pressure of 2000 psia. Both injection and production wells are fully completed. Fluid properties, porosity, rock capillary and relative permeability data are the same as in Example 1. However, the pseudo relative permeability data differs and is shown in Figures 14 and 15 and Table 3.

Results of a comparative calculation where the vertical permeability of each layer is set equal to the horizontal are plotted in Figure 16. It is quite apparent that in this situation the two dimensional calculation is inadequate with errors close to 40%. By setting the vertical permeability equal to 100 md in all three layers (but leaving the horizontal permeability fixed as before), a better match is obtained as can be seen in Figure 17. Relations (10) suggest that by increasing the interwell spacing one can enhance the establishment of VE. The example in Figure 16 is repeated (i.e., vertical permeability is equal to the horizontal in each layer) except that a 160 acre well spacing five-spot pattern is assumed. Because the drainage area is much larger, simulation is carried out to 20,000 days. The comparative results are plotted in Figure 18. Clearly, the three and two dimensional calculations are in closer agreement when compared to the results obtained in Figure 16. This calculation validates the importance of interwell spacing when making use of the VE assumption.

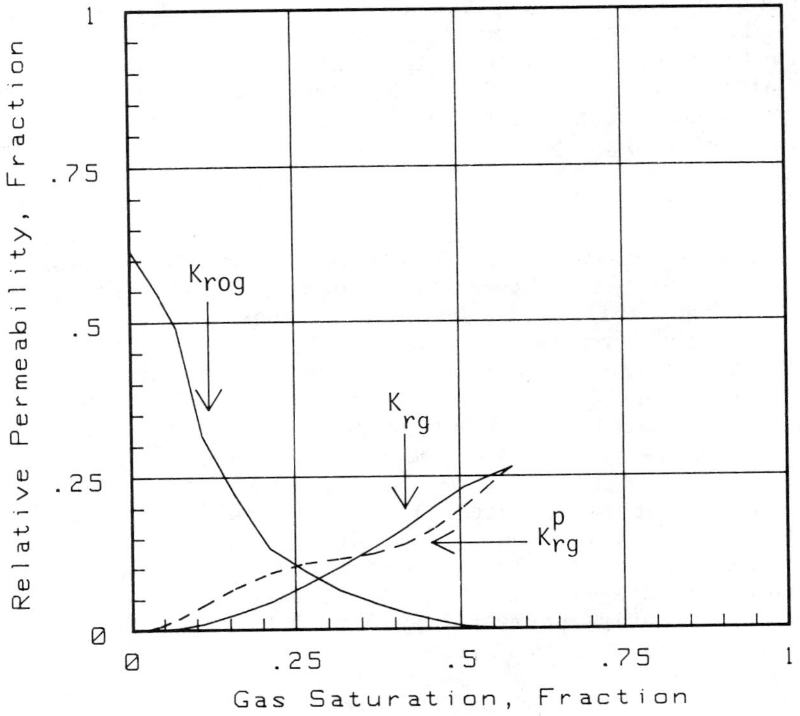

Figure 15: Rock and pseudo relative permeabilities of gas for Example 4.

Table 3: Oil Pseudo Relative Permeability for Example 4

S_g^p \ S_w^p	.220	.278	.336	.394	.452	.510	.568	.626	.684	.742	.800
	Reservoir Oil Pseudo Relative Permeability, K_{ro}^p										
.000	.621	.561	.465	.372	.296	.245	.230	.204	.157	.074	.000
.058	.469	.419	.325	.233	.157	.111	.096	.051	.000	.000	.000
.116	.389	.342	.250	.158	.083	.040	.000	.000	.000	.000	.000
.174	.352	.308	.218	.130	.058	.013	.000	.000	.000	.000	.000
.232	.321	.282	.194	.109	.027	.000	.000	.000	.000	.000	.000
.290	.279	.245	.162	.052	.000	.000	.000	.000	.000	.000	.000
.348	.218	.190	.058	.000	.000	.000	.000	.000	.000	.000	.000
.406	.143	.123	.000	.000	.000	.000	.000	.000	.000	.000	.000
.464	.070	.056	.000	.000	.000	.000	.000	.000	.000	.000	.000
.522	.021	.011	.000	.000	.000	.000	.000	.000	.000	.000	.000
.580	.000	.000	.000	.000	.000	.000	.000	.000	.000	.000	.000

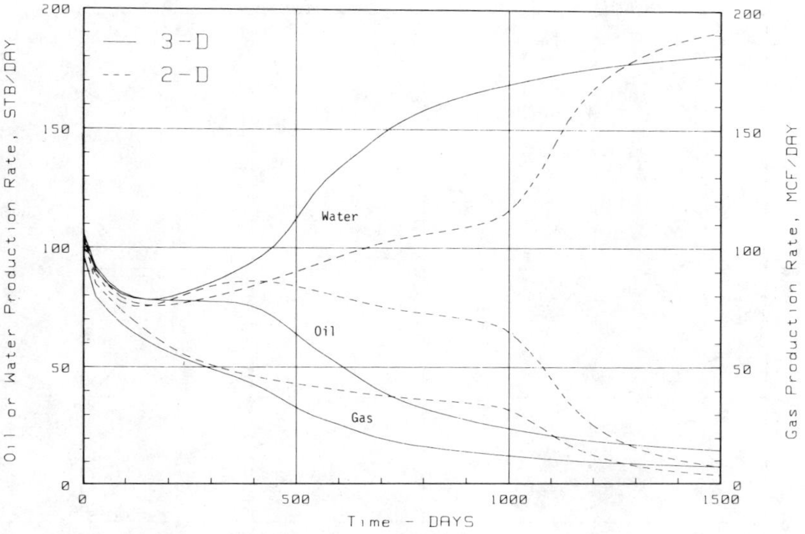

Figure 16: Comparative calculation for Example 4 where the vertical permeability for each layer is set equal to the horizontal permeability.

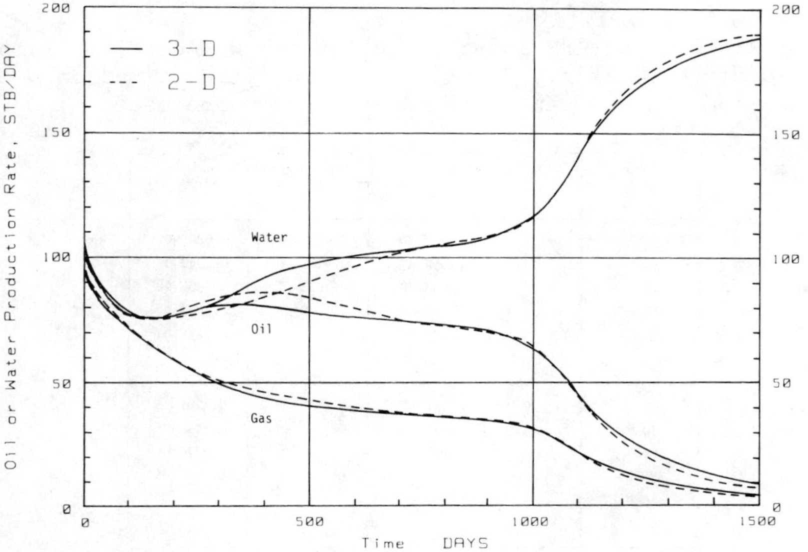

Figure 17: Comparative calculation for Example 4 where the vertical permeability for each layer is set equal to k_z = 100 md.

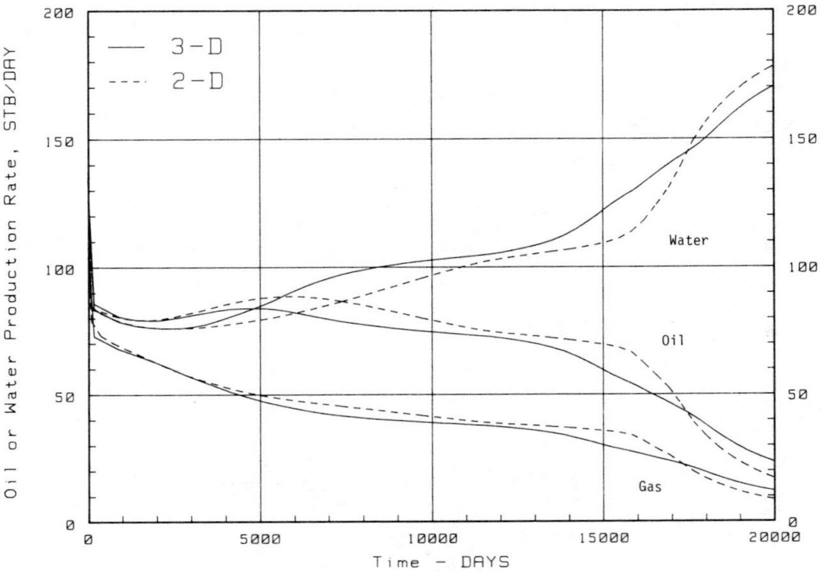

Figure 18: Comparative calculation for Example 4 where the well spacing is increased to a 160 acre five spot pattern.

Table 4: Oil, water, and gas Equilibrium Numbers for Example 1 at different times. Relative permeabilities based on pseudo saturations at the center of the reservoir are used.

Time Days	k_z MD	At Center of Reservoir N_o	N_w	N_g
0.2	100.0	0.139E-07	0.444E-08	0.524E-07
30.0		0.228E-08	0.948E-08	0.104E-07
60.0		0.362E-09	0.105E-07	0.138E-08
330.0		0.103E-08	0.571E-08	0.263E-07
660.0		0.302E-09	0.242E-08	0.623E-07
1260.0		0.165E-09	0.477E-09	
0.2	10.0	0.124E-06	0.462E-07	0.128E-05
30.0		0.241E-07	0.953E-07	0.111E-06
60.0		0.116E-08	0.104E-06	0.417E-08
330.0		0.123E-07	0.572E-07	0.299E-06
660.0		0.230E-08	0.247E-07	
1260.0		0.209E-08	0.461E-08	
0.2	0.01	0.123E-03	0.500E-04	0.128E-02
30.0		0.255E-04	0.918E-04	0.117E-03
60.0		0.336E-05	0.100E-03	0.118E-04
330.0		0.151E-04	0.558E-04	0.415E-03
660.0		0.407E-05	0.276E-04	
1260.0		0.125E-05	0.417E-05	

IV. ANALYSIS OF EQUILIBRIUM NUMBERS

In the previous Section parametric studies were carried out for four examples. Example 1 illustrates the effects of changes in the vertical permeability on VE for a uniform single layer formation. Equilibrium Numbers, defined in Relations (10) and derived in Appendix B, are calculated and listed in Table 4 for different vertical permeabilities and at various times. Calculations for high vertical permeability yield small Equilibrium Numbers and satisfy the theoretically derived criteria in Relations (10) while the calculation using k_z = 0.01 md yield large numbers and fail the test criteria (see Figures 8 and 9). The three dimensional calculation for k_z = 0.01 md shows the least agreement with the two dimensional calculation. This is due to the infinite vertical permeability assumption in the two dimensional calculation. The Gas Equilibrium Number is usually the largest among the three numbers but still satisfies the VE criteria in Relations (10) in two cases. At large simulation times, the Gas Number is not computed because the gas phase is absent from the reservoir. In general, if VE criteria in Relations (10) are satisfied at early times (i.e., 30 days) they continue to be satisfied at later times.

Calculation of Equilibrium Numbers is made difficult by the need to obtain mean relative permeabilities that are representative of each phase. Values calculated in Table 4 make use of the following procedure. Referring to the three dimensional calculation, a dip-normal average saturation (pseudo saturation) for each phase is calculated at the center of the reservoir. It is assumed that the pseudo saturation at the center of a horizontal reservoir approximately equals a volumetric average saturation. The water-oil and gas-oil contact locations are calculated with the aid of pseudo capillary curves and procedures outlined in Appendix C. Pseudo saturations for oil and gas are corrected by dividing by the formation thickness fraction above the water-oil and gas-oil contacts, respectively. Hence, one obtains a better estimate of the volumetric average saturation in the volume occupied by the given phase. Based on the corrected average saturations, the pseudo relative permeabilities are determined and used to calculate the Equilibrium Numbers. While evaluating gas and water saturations to determine pseudo relative oil permeability, the pseudo gas and water saturations are corrected to approximate a volume average above the water-oil contact.

As a practical matter, the calculation of Equilibrium Numbers becomes less accurate at later times. Implicit in the aforementioned corrections for the saturations is the existence of clearly defined contacts under conditions of gravity segregation. These conditions only exist at early times when most reservoirs are near hydrostatic equilibrium.

To ensure accuracy, pseudo saturations, used to obtain pseudo relative permeabilities and Equilibrium Numbers, were determined from the three dimensional calculations. In practice, however, it is suggested that one calculate Equilibrium Numbers, for early times, based on saturations determined from a two dimensional calculation. The VE assumption can then be tested aposteriori. From trial calculations, it appears that Equilibrium Numbers based on two and three dimensional calculations are found to be nearly equal at early times.

Equilibrium Numbers for Example 2 are listed in Table 5. Two cases are presented having different vertical permeability values. The Water Numbers are larger in this case when compared to Example 1 because the completion interval excludes most of the water zone. Consequently, vertical velocity and frictional effects increase due to the injection of water and little corresponding production. In both cases the VE criteria in Relations (10) is satisfied and comparative calculations show good agreement as depicted in Figures 11 and 12.

For the dipping reservoir, Example 3, Equilibrium Numbers are listed in Table 6. In this case, the pseudo saturations used to determine relative permeability are taken at the production well, center of the reservoir, and at the injection well. Fluids in place are initially segregated in the vertical direction. Therefore, the dip-normal pseudo saturation at the injector is representative of the water phase. Water and Gas Equilibrium Numbers are extremely large at the center since these phases are present in trace amounts and the consequent relative permeabilities are extremely small. Considering only the numbers taken at locations representative of the given phase, the VE criteria are satisfied and the match between

Table 5: Oil, water, and gas Equilibrium Numbers for Example 2 at different times. Relative permeabilities based on pseudo saturations at the center of the reservoir are used.

Time Days	k_z MD	At Center of Reservoir		
		N_o	N_w	N_g
0.2	100.0	0.633E-08	0.143E-07	0.657E-07
30.0		0.562E-10	0.115E-07	0.432E-09
60.0		0.623E-09	0.111E-07	0.457E-08
330.0		0.956E-09	0.623E-08	0.261E-07
660.0		0.341E-09	0.237E-08	0.203E-07
1260.0		0.624E-09	0.453E-09	0.966E-08
0.2	10.0	0.600E-07	0.143E-06	0.622E-06
30.0		0.296E-08	0.125E-06	0.223E-07
60.0		0.490E-08	0.119E-06	0.341E-07
330.0		0.901E-08	0.577E-07	0.342E-06
660.0		0.303E-08	0.235E-07	
1260.0		0.319E-08	0.414E-08	

Table 6: Oil, water, and gas Equilibrium Numbers for Example 3 at different times. Relative permeabilities based on pseudo saturation at the producer, center, and injector are used.

Time Days	k_z MD	At Production Well		
		N_o	N_w	N_g
0.2	10.0	0.294E+11	0.619E-03	0.178E-07
30.0		0.360E+11	0.244E-03	0.161E-07
60.0		0.360E+11	0.251E-03	0.161E-07
600.0		0.190E-08	0.650E-05	0.975E-07
870.0		0.163E-09	0.415E-07	
1260.0		0.109E-08	0.727E-08	

Time Days	k_z MD	At Center of Reservoir		
		N_o	N_w	N_g
0.2	10.0	0.585E-07	0.145E-03	0.773E-06
30.0		0.871E-07	0.550E-04	0.178E-06
60.0		0.840E-07	0.391E-04	0.212E-06
600.0		0.509E-08	0.671E-07	
870.0		0.411E-09	0.259E-07	
1260.0		0.361E-08	0.606E-08	

Time Days	k_z MD	At Injection Well		
		N_o	N_w	N_g
0.2	10.0	0.106E-06	0.105E-06	
30.0		0.341E-06	0.564E-07	0.657E-05
60.0		0.626E-06	0.517E-07	0.175E-04
600.0		0.121E+08	0.475E-07	
870.0		0.521E+06	0.222E-07	
1260.0		0.200E+07	0.574E-08	

two and three dimensional calculations demonstrates acceptable results as shown in Figure 13.

Equilibrium Numbers for layered reservoirs, studied in Example 4, are listed in Table 7. All the numbers appear to satisfy VE criteria in Relations (10) even though some of the comparative calculations show poor agreement (see Figure 16). In a relative sense, cases showing poor agreement correspond to Equilibrium Numbers that are relatively large. Therefore, the VE criteria in Relations (10) can be refined for layered reservoirs.

Examples studied in this work can be broadly classified into two categories: those concerning single layer and those concerning multi-layered formations. The VE criteria, developed from theoretical considerations in Appendix B, can be refined based on the parametric studies carried out in this work. It is suggested that for single layer formations:

$$N_w, N_o, N_g \leq 1.1 \times 10^{-7} \qquad (12)$$

where the Equilibrium Numbers should be calculated at an early time (i.e., after the flow rates have stabilized, such as 30 days). The right hand side of (12) comes from Table 4 and corresponds to the largest Equilibrium Number found for which the overall match between two and three dimensional calculations is acceptable. It is further suggested that for layered reservoirs the proper criteria are:

$$N_w, N_o, N_g \leq 1.3 \times 10^{-8} \qquad (13)$$

where the right hand side of (13) is chosen from Table 7 using the same method as above.

V. CONCLUSIONS

To the best of the authors knowledge, this is the first published work to make use of pseudo functions for three phase flow. The theory used is an extension of previous work by Coats[2], which assumes dip-normal hydrostatic equilibrium when calculating the pseudo functions. A clear and simple derivation of the pseudo functions and their associated two dimensional equations is presented. These are derived by integrating the governing three dimensional equations in the dip-normal

Table 7: Oil, water, and gas Equilibrium Numbers for Example 4 at different times. Relative permeabilities based on pseudo saturations at the center of the reservoir are used.

Time Days	k_z MD	At Center of Reservoir N_o	N_w	N_g	Spacing Acres
0.2	100.0	0.720E-08	0.317E-08	0.435E-07	10.0
30.0		0.300E-09	0.594E-08	0.140E-08	
60.0		0.656E-09	0.625E-08	0.309E-08	
330.0		0.530E-09	0.449E-08	0.235E-07	
660.0		0.112E-09	0.280E-08	0.110E-07	
1050.0		0.172E-09	0.166E-08		
1260.0		0.359E-09	0.605E-09		
1500.0		0.111E-08	0.252E-09		
0.2	0.1k	0.235E-06	0.112E-06	0.737E-06	10.0
30.0		0.178E-07	0.197E-06	0.628E-07	
60.0		0.132E-07	0.205E-06	0.456E-07	
330.0		0.344E-07	0.138E-06	0.193E-07	
660.0		0.487E-08	0.645E-07		
1050.00		0.453E-08	0.318E-07		
1260.00		0.107E-08	0.235E-07		
1500.00		0.141E-08	0.185E-07		
0.2	0.1k	0.115E-07	0.827E-08	0.360E-07	160.0
180.0		0.637E-09	0.131E-07	0.194E-08	
720.0		0.170E-08	0.134E-07	0.499E-08	
2520.0		0.230E-08	0.124E-07	0.111E-08	
3600.0		0.183E-08	0.109E-07	0.160E-08	
7200.0		0.155E-09	0.767E-08	0.374E-09	
10080.0		0.180E-09	0.597E-08	0.398E-09	
14580.0		0.133E-09	0.424E-08		
16560.0		0.893E-11	0.316E-08		
18180.0		0.933E-11	0.230E-08		
20000.0		0.310E-12	0.140E-08		

direction. Although the resulting equations are similar to those previously derived by Martin[1], the method of derivation and interpretation of flow variables is unique. Conceptually, the current presentation is more consistent and easier to interpret physically.

As a result of the above analysis pseudo capillary pressure curves are introduced which are always positive. In this work, the pseudo capillary pressures are defined as dip-normal averages of the local rock capillary pressure. Since rock capillary pressures are positive, their average value is also positive. From a physical point of view, this approach appears superior to the traditional pseudo capillary pressure definition. When initializing a two dimensional calculation, it is necessary to specify the pseudo saturation of each phase. However, when using traditional pseudo capillary pressure curves, it is possible to initialize by specifying contact locations. To some investigators, this may be seen as a major advantage of the traditional pseudo capillary curves over those introduced here. In most practical applications, the results are insensitive to the choice of capillary curves except when initializing a reservoir for fluids in place. This insensitivity is due to the relatively small values of capillary pressure (compared to the pressure drop at the well) and numerical dispersion which results when large grid blocks are used. However, under certain special conditions the capillary effects at the well may become significant in which case the use of positive capillary pressure curves is expected to yield more accurate results.

It is clearly illustrated, through several examples, that a two dimensional calculation (using pseudo functions) accurately simulates reservoir behavior when the reservoir operates under VE conditions. This calculation is particularly useful for water flooding where the overall reservoir pressure is maintained.

To help identify conditions under which the VE assumption is valid, new dimensionless criteria are introduced, one for each phase present. These Equilibrium Numbers are derived from a vertical force balance. With the aid of parametric studies, these criteria are refined for single and multi-layered formations. For practical applications, one carries out a two dimensional calculation and based on the results determines the

Equilibrium Numbers. The validity of the VE assumption is assessed aposteriori.

Acknowledgement

We acknowledge the management of Production Research Division, Gulf Research & Development Company for support and permission to publish this paper. We are also grateful to our colleagues S. H. Leventhal and W. H. Chen for helpful discussions and technical support. Manuscript review was carried out by J. N. Crane and H. Weinstein. This manuscript was skillfully typed by Giulia Peterson and Valerie Redd.

Nomenclature

B_w, B_o, B_g	=	Formation volume factors for the water, oil and gas phases, respectively
h	=	Formation thickness
$\tilde{k}$	=	Absolute permeability tensor
k	=	Horizontal absolute permeability
k_x, k_y, k_z	=	Absolute permeability in the x, y and z coordinate directions, respectively
k_{rw}, k_{ro}, k_{rg}	=	Three phase relative permeabilities for water, oil, and gas, respectively
k_{row}, k_{rog}	=	Two phase relative permeabilities for oil in the water and gas phases, respectively
k_{ro}^{o}	=	Oil relative permeability at irreducible water saturation and zero gas saturation
L	=	Distance between wells in a five spot pattern
N_w, N_o, N_g	=	Water, oil, and gas Equilibrium

Numbers, respectively, (defined in (10)) used to determine whether a given reservoir is in VE

P_w, P_o, P_g = Water, oil and gas pressures, respectively

P_{cow}, P_{cgo} = Capillary pressures for the oil-water and gas-oil systems, respectively

q_{wp}, q_{op}, q_{gp}, = Production flow rates at surface conditions for the water, oil and gas phases respectively

q_{wi} = Water injection flow rate at surface conditions

R_{so}, R_{sw} = Solution gas-oil and gas-water ratios, respectively

S_w, S_o, S_g = Saturations for the water, oil and gas phases, respectively

S_{gr} = Residual gas saturation

u_{wz}, u_{oz}, u_{gz} = Velocity in the dip-normal direction of the water, oil and gas phases, respectively

x, y, z = Cartesian coordinates with the z coordinate parallel to the dip-normal direction

z_t, z_b = Cartesian coordinates corresponding to the top and bottom of the perforation interval

z_{cgo}, z_{cow} = The dip-normal distances from the upper formation boundary to the gas-oil and water-oil contacts

γ_w, γ_o, γ_g = The product of density and gravitational acceleration for the water, oil and gas phases, respectively

θ = Cylindrical coordinate

μ_w, μ_o, μ_g = Viscosity of water, oil, and gas, respectively

ν_{owc}, ν_{goc} = Dip-normal velocity of the oil-water

		and gas-oil contacts
ρ	=	Fluid density
ϕ	=	Porosity
φ	=	Formation dip angle
∇	=	Two dimensional Laplacian operator $\left(\frac{\partial}{\partial x} + \frac{\partial}{\partial y}\right)$

Subscripts

w, o, g	=	Water, oil and gas, respectively
wp, op, gp	=	Water, oil and gas production respectively
wi	=	Water injection

Superscripts

~	=	Tensor
_	=	Dip-normal average
*	=	Volume average in regions where the given phase is present
p	=	Pseudo function
wp	=	Well pseudo function

REFERENCES

1. Martin, J. C.: "Partial Integration of Equations of Multiphase Flow," Soc. Pet. Eng. J. (Dec., 1968) 370-380.

2. Coats, K. H., Nielsen, R. L., Terhune, M. H. and Weber, A. G.: "Simulation of Three-Dimensional, Two-Phase Flow in Oil and Gas Reservoirs," Soc. Pet. Eng. J. (Dec., 1967) 377-388.

3. Coats, K. H., Dempsey, J. R. and Henderson, J. H.: "The Use of Vertical Equilibrium in Two-Dimensional Simulation of Three-Dimensional Reservoir Performance," Soc. Pet. Eng. J. (March, 1971) 63-71.

4. Jacks, H. H., Smith, J. E. O., and Mattax, C. C.: "The Modeling of a Three-Dimensional Reservoir with a

Two-Dimensional Reservoir Simulator - The Use of Dynamic Pseudo Functions," Soc. Pet. Eng. J. (June, 1973) 175-185.

5. Kyte, J. R. and Berry D. W.: "New Pseudo Functions to Control Numerical Dispersion," Soc. Pet. Eng. J. (August, 1975) 269-276.

6. Thomas, G. W.: "An Extension of Pseudo Function Concepts," SPE Paper No. 12274, presented at the Reservoir Simulation Symposium held in San Francisco, CA, November 15-18, 1983.

7. Dietrich, J. K.: "Relative Permeability During Cyclic Steam Stimulation of Heavy Oil Reservoirs," SPE Paper No. 7968, presented at the California Regional Meeting of the Society of Petroleum Engineers of AIME held in Ventura, California, April 18-20, 1979.

8. Hearn, C. L.: "Simulation of Stratified Waterflooding by Pseudo Relative Permeability Curves," J. Pet. Tech. (July, 1971) 805-813.

APPENDIX A

In this Section, an integration of the governing three dimensional Equations is carried out in the dip-normal direction thereby reducing the dimensionality of the problem at hand. Although the resulting equations are mathematically two dimensional, they describe three dimensional flow in an average sense. In general, the averaged equations differ from those describing real two dimensional flow. However, under certain conditions the two become analogous (i.e., no flow across the upper and lower formation boundaries and isotropic permeability in the horizontal direction). This derivation compliments the previous work of Martin[1] but differs on certain key points. Integration is carried out on three dimensional equations yielding a reduction in dimensionality as a consequence. Capillary pressures and pressures, in general, are interpreted as averages of the local quantities in the dip-normal directions.

A system of equations describing the simultaneous flow of oil, water, and gas phases can be written as follows:

$$S_w + S_o + S_g = 1 \ , \qquad \text{(A-1)}$$

$$P_{cow}\{S_w\} = P_o - P_w \ , \qquad \text{(A-2)}$$

$$P_{cgo}\{S_g\} = P_g - P_o \ , \qquad \text{(A-3)}$$

where the braces indicate a functional relationship between dependent and independent variables.

Mass Conservation for Water

$$\nabla\cdot\left[\frac{k_{rw}}{\mu_w B_w}\tilde{k}\cdot(\nabla P_w - \gamma_w \nabla D)\right] = \frac{\partial}{\partial t}\left(\frac{\phi\, S_w}{B_w}\right) \ , \qquad \text{(A-4)}$$

Mass Conservation for Oil

$$\nabla\cdot\left[\frac{k_{ro}}{\mu_o B_o}\tilde{k}\cdot(\nabla P_o - \gamma_o \nabla D)\right] = \frac{\partial}{\partial t}\left(\frac{\phi S_o}{B_o}\right) \ , \qquad \text{(A-5)}$$

Mass Conservation for Gas

$$\nabla\cdot\left[\frac{k_{rw}}{\mu_w B_w} R_{sw}\tilde{k}\cdot(\nabla P_w - \gamma_w \nabla D)\right] + \nabla\cdot\left[\frac{k_{ro}}{\mu_o B_o} R_{so}\tilde{k}\cdot(\nabla P_o - \gamma_o \nabla D)\right]$$

$$+ \nabla\cdot\left[\frac{k_{rg}}{\mu_g B_g}\tilde{k}\cdot(\nabla P_g - \gamma_g \nabla D)\right] = \frac{\partial}{\partial t}\left[\phi\left(\frac{R_{sw}}{B_w} S_w + \frac{R_{so}}{B_o} S_o + \frac{S_g}{B_g}\right)\right] \ ,$$

(A-6)

where $\tilde{k}$ is the permeability tensor while the unknown variables are P_w, P_o, P_g, S_w, S_o, and S_g.

In what follows, Equations (A-1) through (A-4) are integrated in the dip normal direction (refer to Figure 1). For the purpose of brevity, integration of (A-5) and (A-6) is omitted since it is analogous to that of (A-4).

Multiplying (A-1) by $\phi\{x,y,z\}$ and integrating term by term yields:

$$\int_0^h \phi(S_w + S_o + S_g)\, dz = \int_0^h \phi\, dz \,. \tag{A-7}$$

Introducing definitions for pseudo saturation and mean porosity:

$$S_w^p \equiv \frac{1}{\bar{\phi}h} \int_0^h S_w\, \phi\{x,y,z\}\, dz \,,$$

$$\bar{\phi} \equiv \frac{1}{h} \int_0^h \phi\, dz \,, \tag{A-8}$$

and substituting into (A-7) yields the constraint Equation for pseudo saturations as:

$$S_w^p + S_o^p + S_g^p = 1 \,. \tag{A-9}$$

Equations (A-2) and (A-3) are integrated and expressed in terms of averages as:

$$\bar{P}_{cow} = \bar{P}_o - \bar{P}_w \,, \tag{A-10}$$

$$\bar{P}_{cgo} = \bar{P}_g - \bar{P}_o \,. \tag{A-11}$$

Since capillary pressures for water wet reservoirs are greater than zero, the dip normal averages must also be positive. In evaluating the right hand sides of (A-10) and (A-11), it is assumed that the oil and water pressures are equal below the oil-water contact while the gas and oil pressures are equal below the gas-oil contact.

Rewriting (A-4) in cartesian coordinates, aligning the coordinate axes along principal permeability directions, integrating and exchanging the order of differentiation and integration yields:

$$\frac{\partial}{\partial x} \int_0^h \left[\frac{k_{rw}}{\mu_w B_w}\, k_x\left(\frac{\partial}{\partial x}P_w - \gamma_w \frac{\partial}{\partial x}D\right)\right] dz$$

$$+ \frac{\partial}{\partial y} \int_0^h \left[\frac{k_{rw}}{\mu_w B_w} k_y \left(\frac{\partial}{\partial y} P_w - \gamma_w \frac{\partial}{\partial y} D\right)\right] dz$$

$$+ \left[\frac{k_{rw} k_z}{\mu_w B_w} \left(\frac{\partial}{\partial z} P_w - \gamma_w \frac{\partial}{\partial z} D\right)\right]\Big|_{z=h} - \left[\frac{k_{rw} k_z}{\mu_w B_w} \left(\frac{\partial}{\partial z} P_w - \gamma_w \frac{\partial}{\partial z} D\right)\right]\Big|_{z=0}$$

$$= \frac{\partial}{\partial t} \int_0^h \left(\frac{\phi S_w}{B_w}\right) dz \ . \qquad \text{(A-12)}$$

Note that the third and fourth terms of (A-12) are the vertical Darcy velocities $(u_{wz}/B_w)|_{z=h}$, $(u_{wz}/B_w)|_{z=0}$ divided by the formation volume factor. As a step towards simplifying (A-12), the pressure is defined in terms of its dip-normal mean plus a deviation about the mean:

$$P_w\{x,y,z,t\} \equiv \bar{P}_w\{x,y,t\} + P'_w\{x,y,z,t\} \ . \qquad \text{(A-13)}$$

It is further assumed that:

$$\bar{P}_w >> P'_w \ ,$$

$$\frac{\partial}{\partial x} \bar{P}_w >> \frac{\partial}{\partial x} P'_w \ , \qquad \text{(A-14)}$$

$$\frac{\partial}{\partial y} \bar{P}_w >> \frac{\partial}{\partial y} P'_w \ .$$

The above assumptions imply that the ratio of the dip normal to the lateral dimension is small and that the reservoir is at sufficient depth such that the mean reservoir pressure is much greater than any local deviation from the mean.

As a consequence of (A-13) and (A-14) the following approximations result:

$$P_w\{x,y,z,t\} \simeq \overline{P}_w\{x,y,t\} ,$$

$$\frac{\partial}{\partial x} P_w \simeq \frac{\partial}{\partial x} \overline{P}_w ,$$

$$\frac{\partial}{\partial y} P_w \simeq \frac{\partial}{\partial y} \overline{P}_w , \qquad \text{(A-15)}$$

$$\frac{\partial}{\partial z} P_w = \frac{\partial}{\partial z} P'_w ,$$

$$B_w\{P_w\{x,y,z,t\}\} \simeq B_w\{\overline{P}_w\{x,y,t\}\} ,$$

$$\mu_w\{P_w\} \simeq \mu_w\{\overline{P}_w\} ,$$

$$\rho_w\{P_w\} \simeq \rho_w\{\overline{P}_w\} .$$

Substituting (A-15) into (A-12) and moving functions independent of the z coordinate outside the integrals yields:

$$\frac{\partial}{\partial x} \left[\frac{k^p_{rwx}}{\mu_w B_w} \overline{k}_x \left(\frac{\partial}{\partial x} \overline{P}_w - \gamma_w \frac{\partial}{\partial x} D\right)\right]$$

$$+ \frac{\partial}{\partial y} \left[\frac{k^p_{rwy}}{\mu_w B_w} \overline{k}_y \left(\frac{\partial}{\partial y} \overline{P}_w - \gamma_w \frac{\partial}{\partial y} D\right)\right] + F\{x,y,t\}$$

$$= \frac{\partial}{\partial t} \left(\frac{\overline{\phi}\, S^P_w}{B_w}\right) , \qquad \text{(A-16)}$$

where the pseudo relative permeabilities (k^p_{rwx}, k^p_{rwy}) and the function $F\{x,y,t\}$ are defined as:

$$k^p_{rwx}\ \bar{k}_x\ h \equiv \int_0^h k_{rw}\ k_x\ dz\ ,$$

$$k^p_{rwy}\ \bar{k}_y\ h \equiv \int_0^h k_{rw}\ k_y\ dz\ , \tag{A-17}$$

$$F\{x,y,t\} \equiv \frac{1}{hB_w}\left(u_{wz}|_{z=h} - u_{wz}|_{z=0}\right)\ .$$

Equation (A-16) differs from the water mass conservation equation for two dimensional flow in two respects: 1) the pseudo relative permeabilities exhibit an anisotropic character. 2) the presence of an additional term $F\{x,y,t\}$ accounting for net flux across the upper and lower formation boundaries. Equation (A-16) can be reduced to a form analogous with that representing two dimensional flow by assuming impermeable upper and lower formation boundaries ($F\{x,y,t\} = 0$) and a horizontally isotropic permeability tensor (i.e., $k_x = k_y = k$ thus, $k^p_{rwx} = k^p_{rwy} = k^p_{rw}$).

Although the averaged equations were derived assuming constant formation thickness and straight upper and lower formation boundaries, these conditions can be relaxed. The formation thickness may vary and the upper and lower formation boundaries may bend, assuming that the spatial gradients are small.

APPENDIX B

In this Section, a set of dimensionless criteria is developed for the purpose of determining if a given reservoir is being produced under conditions of VE. These criteria are derived from the momentum equations for each phase. Since the procedures involved are similar, a detailed derivation is presented for the water phase only.

The flow of fluids through a porous medium is assumed to be governed by a balance between friction, gravity, and pressure forces (Darcy's law). A momentum balance in the dip-normal direction (the z direction refering to Figure 1) can be written as:

$$u_{wz} \frac{\mu_w}{k_z k_{rw}} = - \left(\frac{\partial}{\partial z} P_w - \gamma_w \cos \psi\right) . \qquad \text{(B-1)}$$

For gravity-pressure equilibrium in the z direction, it is necessary that the friction forces be of much smaller magnitude than the other two forces. Mathematically this can be stated as:

$$|u_{wz}| \frac{\mu_w}{k_z k_{rw}} \ll \gamma_w \cos \psi . \qquad \text{(B-2)}$$

The above criterion is only valid locally at each point. It is conceivable that VE may exist in only a small portion of the reservoir. Therefore, a global criterion defining the overall state of the reservoir is necessary. With this in mind, both sides of (B-2) are integrated over the reservoir volume yielding:

$$\int_0^h \int_0^{L_2} \int_0^{L_1} |u_{wz}| \frac{\mu_w}{k_z k_{rw}} \, dx \, dy \, dz$$

$$\ll \cos \psi \int_0^h \int_0^{L_2} \int_0^{L_1} \gamma_w \, dx \, dy \, dz , \qquad \text{(B-3)}$$

where L_1 and L_2 are the lateral dimensions of the reservoir. By invoking the mean value theorem, (B-3) simplifies to:

$$\left(|u_{wz}| \frac{\mu_w}{k_{rw}}\right)' \int_0^h \int_0^{L_2} \int_0^{L_1} \frac{1}{k_z} \, dx \, dy \, dz$$

$$\ll \gamma_w' \, L_1 \, L_2 \, h \cos \psi , \qquad \text{(B-4)}$$

where the primed quantities refer to a representative or

average value of the variable at some point inside the reservoir. It is further assumed that the primed variables are of the same order of magnitude as:

$$\left(|u_{wz}| \frac{\mu_w}{k_{rw}}\right)' \sim \frac{|u_{wz}|^* \mu_w\{P_w^*\}}{k_{rw}\{S_w^*\}} ,$$

$$\gamma_w' \sim \gamma_w\{P_w^*\} \qquad \text{(B-5)}$$

where the variables with superscript asterisks are volumetrically averaged quantities over the region where that phase is present while the braces indicate a functional relationship between dependent and independent variables. Substituting (B-5) into (B-4), dividing both sides of (B-4) by the right hand side of (B-4) and assuming that $k_z = f\{z\}$ leads to the dimensionless criterion:

$$\frac{|u_{wz}|^* \mu_w\{P_w^*\}}{k_{rw}\{S_w^*\}\ \gamma_w\{P_w^*\}\ h \cos\psi} \int_0^h \frac{1}{k_z}\,dz \ll 1 . \qquad \text{(B-6)}$$

Similarly, the analogous criteria for the oil and gas phases respectively are:

$$\frac{|u_{oz}|^* \mu_o\{P_o^*\}}{k_{ro}\{S_o^*\}\ \gamma_o\{P_o^*\}\ h \cos\psi} \int_{z_{cow}}^h \frac{1}{k_z}\,dz \ll 1 , \qquad \text{(B-7)}$$

$$\frac{|u_{gz}|^* \mu_g\{P_g^*\}}{k_{rg}\{S_g^*\}\ \gamma_g\{P_g^*\}\ h \cos\psi} \int_{z_{cgo}}^h \frac{1}{k_z}\,dz \ll 1 . \qquad \text{(B-8)}$$

In this case, the volume integrals are carried out only in the region where each phase is present (i.e., above the water-oil contact for the oil phase and above the gas-oil contact for the gas phase). Thus, variables with superscript asterisks imply volumetric averages in the region where that phase is present.

Before calculating the left hand sides of (B-6) through (B-8), it is necessary to estimate the mean vertical velocity of each phase (i.e., $|u_{wz}|^*$, $|u_{oz}|^*$ and $|u_{gz}|^*$). In general, such an estimate depends on the reservoir geometry and well spacing patterns. As an example, a five-spot pattern with an interwell spacing of L is assumed. The vertical velocities of the oil-water contact (ν_{owc}) and gas-oil contact (ν_{goc}) (neglecting effects due to gas coming out of solution) can be estimated from a global material balance as:

$$\nu_{owc} \simeq \frac{1}{L^2} (q_{wp} - q_{wi}) B_w , \qquad \text{(B-9)}$$

$$(\nu_{goc} - \nu_{owc}) \simeq \frac{1}{L^2} q_{op} B_o . \qquad \text{(B-10)}$$

Assuming that the mean vertical velocity of the water phase is approximately equal to the velocity of the oil-water contact while the mean vertical velocities of the oil and gas phases is equal to the velocity of the gas oil contact (excluding solution gas effects), substituting into (B-9) and (B-10) and solving for the mean velocities yields:

$$|u_{wz}|^* \simeq |\nu_{owc}| \simeq \frac{1}{L^2} | q_{wp} - q_{wi} | B_w ,$$

$$|u_{oz}|^* \simeq |u_{gz}|^* \simeq |\nu_{goc}| \qquad \text{(B-11)}$$

$$\simeq \frac{1}{L^2} | (q_{wp} - q_{wi}) B_w + q_{op} B_o | .$$

Thus, the Water Equilibrium Number N_w, for a five-spot pattern (B-6) can be written using field units as:

$$N_w \equiv \frac{|q_{wp} - q_{wi}| \, B_w \, \mu_w^*}{L^2 \, k_{rw}^* \, \gamma_w^* \, \cos\psi} \left(\overline{\frac{1}{k_z}}\right)_w \qquad \text{(B-12)}$$

$$<< \; 2.4 \times 10^{-7}.$$

where:

$$\left(\overline{\frac{1}{k_z}}\right)_w \equiv \frac{1}{h} \int_0^h \frac{1}{k_z} \, dz \; .$$

Similarly, the Equilibrium Numbers for the oil and gas phases can be derived.

APPENDIX C

This Section outlines the procedure for calculating relevant pseudo functions for three phase flow under conditions of dip-normal gravity pressure equilibrium (VE). Under these conditions, it is established that the pseudo curves are functionally analogous to the rock curves (e.g., pseudo relative permeability for water is a function of pseudo water saturation only).

The procedure outlined below is an extension of Coat's[2] two phase treatment. Unlike previous analysis, however, a dip-normal mean capillary pressure which is always positive (for a water wet rock) is introduced. An interesting consequence of extending to a three phase system is the need to integrate a modified Stone's[7] formula in the z direction (assuming three phase relative permeability data is unavailable).

Most three phase black oil model formulations make simplifying functional assumptions regarding rock curves due to the scarcity of three phase data. For this study, the model used makes the following functional assumptions:

$$P_{cow} = P_{cow}\{S_w\} \quad ,$$

$$P_{cgo} = P_{cgo}\{S_g\} \quad ,$$

$$k_{rw} = k_{rw}\{S_w\} \quad , \qquad \text{(C-1)}$$

$$k_{rg} = k_{rg}\{S_g\} \quad ,$$

$$k_{ro} = k_{ro}\{S_g, S_w\} \ ,$$

where the braces indicate a functional relationship between the dependent and independent variables. If dip-normal gravity pressure equilibrium is assumed in the z direction (refer to Figure 1), one can express the capillary pressures as:

$$P_{cow}\{S_w\} = (\gamma_w - \gamma_o)\cos\psi\,(z_{cow} - z)$$

for $z < z_{cow}$,

$$P_{cow} = 0$$

for $z > z_{cow}$, (C-2)

$$P_{cgo}\{S_g\} = (\gamma_o - \gamma_g)\cos\psi\,(z_{cgo} - z)$$

for $z < z_{cgo}$,

$$P_{cgo} = 0$$

for $z > z_{cgo}$.

With the aid of rock capillary curves and (C-2), one can establish the saturation distributions in the z direction for fixed locations of the water-oil and gas-oil contacts. From physical considerations, the contacts must always be located inside the formation thickness boundaries. However, one can mathematically move the contacts below the lower formation boundary in order to calculate a variety of three phase saturation distributions under conditions where the contacts do not exist.

Having established that the capillary pressure is a linear function of the z coordinate, one can integrate both sides of (C-2) in the z direction to arrive at a relation between the dip-normal mean capillary pressure and the contact location as follows:

$$\bar{P}_{cow} = \frac{(\gamma_w - \gamma_o)\cos\psi}{2h}\, z_{cow}^2 \ , \qquad \text{(C-3)}$$

$$\overline{P}_{cgo} = \frac{(\gamma_o - \gamma_g)\cos\psi}{2h} z_{cgo}^2 .$$

Thus, the mean water oil capillary pressure is only a function of the oil-water contact location while the mean gas-oil capillary pressure depends on the location of the gas-oil contact.

A porosity weighted average of the saturation profile in the z direction is termed as the pseudo saturation in Appendix A. Rewriting the definition of the water pseudo saturation and splitting the integral into two parts yields:

$$S_w^p \equiv \frac{1}{\overline{\phi}h}\left[\int_0^h \phi\, S_w\, dz\right] = \frac{1}{\overline{\phi}h}\left[\int_0^{z_{cow}\{\overline{P}_{cow}\}} \phi\{z\}\, S_w\{P_{cow}\{z_{cow},z\}\}\, dz + (1-S_{or})\int_{z_{cow}\{\overline{P}_{cow}\}}^{h} \phi\{z\}\, dz\right] . \tag{C-4}$$

For the case where the contact is set below the formation boundary (i.e., $z_{cow} > h$), the pseudo saturation is determined from:

$$S_w^p = \int_0^h \phi\{z\}\, S_w\{P_{cow}\{z_{cow}, z\}\}\, dz . \tag{C-5}$$

By inspection of (C-4) and (C-5), it follows that the water pseudo saturation is a function of the dip-normal mean oil-water capillary pressure. For a given value of the mean capillary pressure, one value of the pseudo saturation is determined by a numerical integration of (C-4) or (C-5) with the aid of (C-2) and (C-3). Thus, a table of the dip-normal mean capillary pressure versus pseudo saturation can be generated by choosing a set of mean capillary pressures, calculating the corresponding set of contact locations, and evaluating the integrals above for a set of integration limits. In a similar manner, the mean capillary pressure versus pseudo saturation for the gas phase can be generated with the aid of the following integrals:

$$S_g^p \equiv \frac{1}{\bar{\phi}h}\int_0^h \phi\, S_g\, dz = \frac{1}{\bar{\phi}h}\left[\int_0^{z_{cgo}\{\bar{P}_{cgo}\}} \phi\{z\}\, S_g\{P_{cgo}\{z_{cgo}, z\}\}\, dz + S_{gr}\int_{z_{cgo}\{\bar{P}_{cgo}\}}^{h} \phi\{z\}\, dz\right], \quad \text{(C-6)}$$

while for $z_{cgo} > h$

$$S_g^P = \frac{1}{\bar{\phi}h}\int_0^h \phi\{z\}\, S_g\{P_{cgo}\{z_{cgo}, z\}\}\, dz .$$

With the aid of tables for mean capillary pressure versus pseudo saturation, one proceeds to calculate the pseudo relative permeabilities for the water and gas phases (assuming isotropic permeability in the x,y directions) by carrying out a numerical integration of:

$$k_{rw}^p = \frac{1}{\bar{k}h}\left[\int_0^{z_{cow}\{\bar{P}_{cow}\{S_w^p\}\}} k\{z\}\, k_{rw}\{S_w\{P_{cow}\{z_{cow}, z\}\}\}\, dz + k'_{rw}\int_{z_{cow}\{\bar{P}_{cow}\{S_w^P\}\}}^{h} k\{z\}\, dz\right], \quad \text{(C-7)}$$

where k'_{rw} is the maximum value of the water permeability.

and,

$$k_{rg}^p = \frac{1}{\bar{k}h}\int_0^{z_{cgo}\{S_g^p\}} k\{z\}\, k_{rg}\{S_g\{Z_{cgo}, z\}\}\, dz . \quad \text{(C-8)}$$

For Z_{cow}, $Z_{cgo} > h$, the limits of integration must be adjusted as in (C-5) and (C-6). An inspection of the above integrals shows that the pseudo relative permeability for water and gas are only functions of the pseudo saturations for water and gas, respectively. Through repeated integration of (C-7) and (C-8) (due to a change in integration limits), a table of pseudo relative permeability versus pseudo saturation is developed.

For this study, the oil relative permeability in three phase flow was determined from two phase data with the aid of a modified Stone's formula. The following modified version of Stone's[7] formula was used:

$$k_{ro} = k_{ro}^{o} \left[\left(\frac{k_{row}}{k_{ro}^{o}} + k_{rw} \right) \left(\frac{k_{rog}}{k_{ro}^{o}} + k_{rg} \right) - k_{rg} - k_{rw} \right] . \qquad \text{(C-9)}$$

Assuming that the locations of the gas-oil and water-oil contacts are inside the formation thickness, one can write the definition of the oil pseudo relative permeability as the sum of three integrals

$$k_{ro}^{p} \equiv \frac{1}{\bar{k}h} \int_{o}^{h} k \, k_{ro} \, dz$$

$$= \frac{1}{\bar{k}h} \left[\int_{0}^{z_{cgo}\{S_g^p\}} k\{z\} \, k_{ro}\{S_g\{z_{cgo},z\} \, , \, S_w\{z_{cow},z\}\} \, dz \right.$$

$$\left. + \int_{z_{cgo}\{S_g^p\}}^{z_{cow}\{S_w^p\}} k \, k_{ro} dz + \int_{z_{cow}\{S_w^p\}}^{h} k \, k_{ro} dz \right] . \qquad \text{(C-10)}$$

In cases where the contacts are set outside the formation thickness, one cannot split the integral into three parts as above. An inspection of (C-9) shows that for $z > z_{cow}$, $k_{ro} = 0$ while for $z_{cow} > z > z_{cgo}$, $k_{ro} = k_{row}$. Consequently, (C-10) can be rewritten as:

$$k_{ro}^{p} = \frac{1}{\bar{k}h} \left[\int_{0}^{z_{cgo}} k \, k_{ro} \, dz + \int_{z_{cgo}}^{z_{cow}} k \, k_{row} \, dz \right] . \qquad \text{(C-11)}$$

Equation (C-11), with the aid of (C-9), shows that the oil pseudo relative permeability is a function of pseudo gas and water saturations. As in previous cases, repeated numerical integration of (C-11) yields a table for pseudo relative oil permeabilities as a function of the pseudo water and gas saturations.

APPENDIX D

In this Section, the well pseudo relative permeabilities for partially penetrating wells are defined. It is assumed that the flow near a well is in dip-normal gravity pressure equilibrium (i.e., the flow does not have any large velocity components in the dip-normal direction as might result from coning). For cases where the production well is perforated over the entire interval, the well pseudo relative permeability is identical to the reservoir pseudo relative permeability.

Net fluid production into the wellbore is due to the velocity component normal to the well surface. For the water phase (see Figure 2), the normal velocity component can be written as:

$$u_{wn} = - \frac{k_{rw} k}{\mu_w} \left(\frac{\partial}{\partial r} P_w \Big|_{r=r_w} - \gamma_w \cos \psi \right) . \tag{D-1}$$

By integrating the normal velocity over the perforated surface (see Figure 2), one can calculate the total flow rate as:

$$q_w = \int_0^{2\pi} \left[\int_{z_t}^{z_b} u_{wn} \, dz \right] r_w \, d\theta . \tag{D-2}$$

Assuming that:

$$\frac{\partial}{\partial r} P_w\{r,\theta,z,t\} \simeq \frac{\partial}{\partial r} \overline{P}_w\{r,\theta,t\} ,$$

$$\mu_w\{P_w\} \simeq \mu_w\{\overline{P}_w\} ,$$

$$\gamma_w\{P_w\} \simeq \gamma_w\{\bar{P}_w\} ,$$

where the braces indicate a functional relationship between the dependent and independent variables.

Substituting (D-1) into (D-2) results in:

$$q_w = - \int_0^{2\pi} [\frac{1}{\mu_w} (\frac{\partial}{\partial_r} \bar{P}_w \Big|_{r=r_w} - \gamma_w \cos \psi) \times \tag{D-3}$$

$$\int_{z_t}^{z_b} k_{rw}\ k\ dz]\ r_w\ d\theta .$$

Defining the well pseudo relative permeability for water as:

$$k_{rw}^{wp} \equiv \frac{1}{\bar{k}h} \int_{z_t}^{z_b} k_{rw}\ k\ dz , \tag{D-4}$$

and substituting into (D-3) yields:

$$q_w = - \int_0^{2\pi} \frac{k_{rw}^{wp}\ \bar{k}}{\mu_w} (\frac{\partial}{\partial r} \bar{P}_w \Big|_{r=r_w} - \gamma_w \cos \psi)\ h\ r_w\ d\theta . \tag{D-5}$$

Thus, (D-5) describes a well flow rate condition for three dimensional flow that is mathematically analogous to the condition used in describing two dimensional flow. By similar means, the well pseudo relative permeabilities for the oil and gas phases can be written as:

$$k_{ro}^{wp} \equiv \frac{1}{\bar{k}h} \int_{z_t}^{z_b} k_{ro}\ k\ dz ,$$

$$k_{rg}^{wp} \equiv \frac{1}{\bar{k}h} \int_{z_t}^{z_b} k_{rg}\ k\ dz . \tag{D-6}$$

NOTES ON STOCHASTIC SHALES; FROM OUTCROP TO SIMULATION MODEL

HELGE H. HALDORSEN
DAVID M. CHANG

SOHIO PETROLEUM COMPANY
SAN FRANCISCO, CALIFORNIA

I. ABSTRACT

This article deals with the role of discontinuous vertical flow barriers (e.g. shales with lateral extent less than the well spacing) in reservoir engineering. Clearly, when such vertical flow barriers are present within the sand, the "homogeneous rock condition" as referred to by Aronofsky et al.[1-3] is severely violated. In particular, depending on the number, size and spatial disposition of the vertical flow barriers, the vertical permeability of the formation (and of the large-scale grid blocks in a simulation model) may be drastically reduced in comparison to those of homogeneous sands.

First, an overview of reservoir engineering problems associated with continuous and discontinuous shales is provided. Second, statistical procedures for synthetic shale descriptions are outlined. Third, the procedures are illustrated and verified utilizing an outcrop study of the Assakao sandstone as control, and finally, effective single-phase vertical permeabilities of large-scale grid blocks containing sand and shales are calculated utilizing the latest statistical-analytical techniques available. These techniques for estimating effective vertical permeability are shown to be quick, reliable and inexpensive, requiring a minimum of input data. Nevertheless, correct barrier length/width/thickness statistics and interdependencies among them are of fundamental importance, and, hence, the new procedures rely even more heavily on geological considerations and input than in the past.

II. INTRODUCTION

The major gross inhomogenity in many sandstone reservoirs may well be discontinuous shales embedded in, and continuous shales completely separating, the sandstone units. These shales either divide the reservoir (or parts of it) into separate hydraulic units,(4) or, if the shales are discontinuous, set a tortuous environment in which fluid transport occurs.(5) In some reservoirs, the most important element of a good reservoir description might well be a reliable shale description since this information literally defines the internal geometry of the geological domain.(6)

While continuous correlatable shales are handled with ease in reservoir simulation models (by reducing vertical transmissibilities between adjacent layers or by imposing a no-flow condition), discontinuous shales historically have represented a difficult reservoir engineering problem both in terms of their description and their effects on fluid flow.

Listed below are a number of factors which illustrate the importance of adequate shale modelling in numerical studies.

A. GAS UNDERRUNNING SHALES (7, 8)

In order to achieve the gas underrunning (cusping) phenomenon in simulation studies (Figure 1), it is necessary to have the shales located properly with correct lateral dimensions. When the producing rate exceeds the critical rate, the rate where the viscous forces override the gravity forces, gas cap displacements become unstable. For a flank gas cap drive, such as in Figure 1, the gas-oil contact will tilt until it is parallel to the formation dip, thus allowing the gas to overrun the oil or underrun the shales. Similarly, shales aid in initiating water tonguing from the aquifer below.

B. CONING PROTECTION FROM SHALES (9)

The presence of shales (continuous and/or discontinuous) above/below the perforations provides some, moderate, or complete protection from gas and water coning (Figure 2). The critical coning rate is a function of the radius of the shale as indicated in Figure 2.

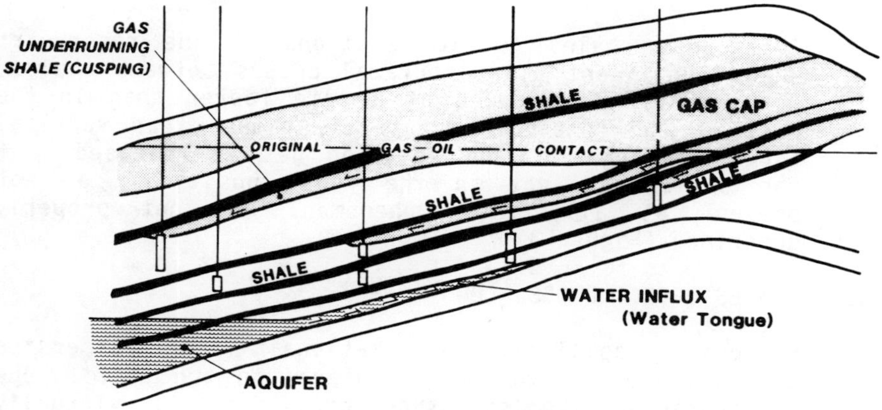

FIGURE 1: GAS UNDERRUNNING SHALES.

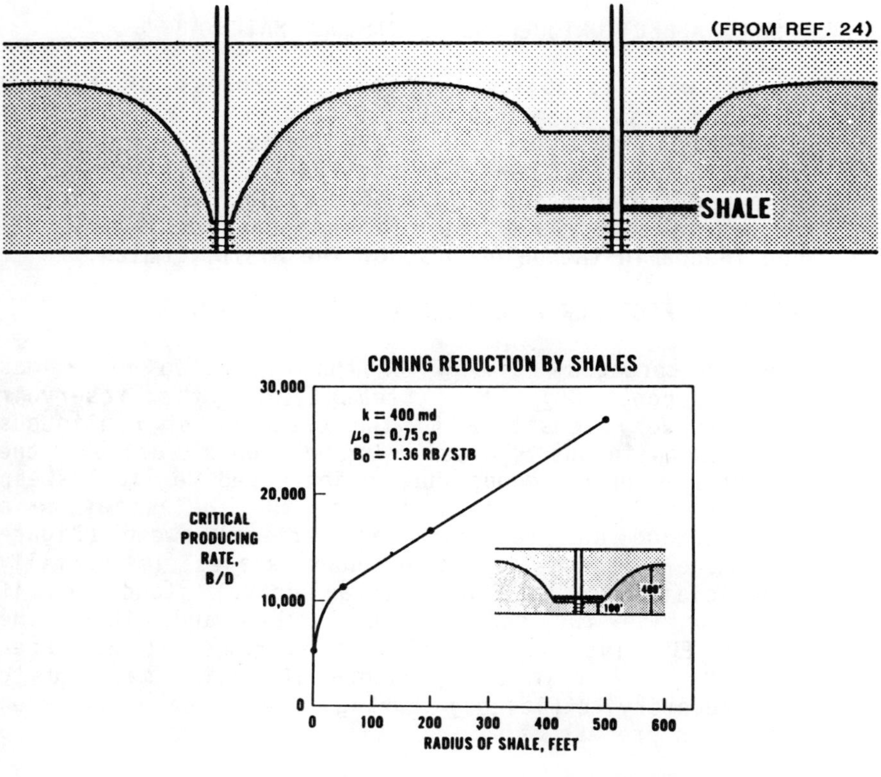

FIGURE 2: CONING PROTECTION FROM SHALES.

C. SHALES AFFECT PRIMARY PRODUCTION [10,11]

Continuous shales isolate sections of the reservoir. When significant oil withdrawal occurs between shales, the pressure is reduced at a rate faster than in the rest of the reservoir. Lateral, rather than vertical flow is triggered, and solution gas is released. If the shales are not in the right positions, or not present at all, these phenomena are not properly described (Figure 3).

D. SHALES AFFECT CROSS-FLOW [12,25]

Viscous, capillary, gravitational and dispersive cross-flow are commonly encountered in the field. Due to continuous shales, such cross-flow is virtually eliminated (depending on the shale's vertical permeability and areal extent), and due to scattered shales, such cross-flow may be increased, reduced, or essentially eliminated (Figure 4).

E. SHALES AFFECT UNIQUENESS OF HISTORY MATCH [13]

It is well established that the process of history matching is a non-unique exercise. In numerical reservoir simulation there are quite a few physical parameters which can be adjusted to yield the same response in pressure, gas-oil ratio, and water-oil ratio. A good shale description increases the overall confidence in the uniqueness of the history match.

F. SHALES AFFECT EOR PROCESSES [6,14,15,17,39]

Any EOR process relying on the injection of a gas (Hydrocarbon, CO_2, N_2, Steam) into the reservoir will be very sensitive to the presence of continuous and discontinuous shales. In horizontal drives, the shales are advantageous due to increased vertical sweep (Figure 5). In vertical drives, the shales may be a disadvantage due to decreased vertical sweep (Figure 6). The presence of continuous shales is normally beneficial from an EOR (lateral drive) standpoint if they confine the sand to be flooded and, thus, the costly EOR fluids. In dipping systems, it has been shown [17,39] that the presence of shales may add to the recovery efficiency during a secondary nitrogen injection process.

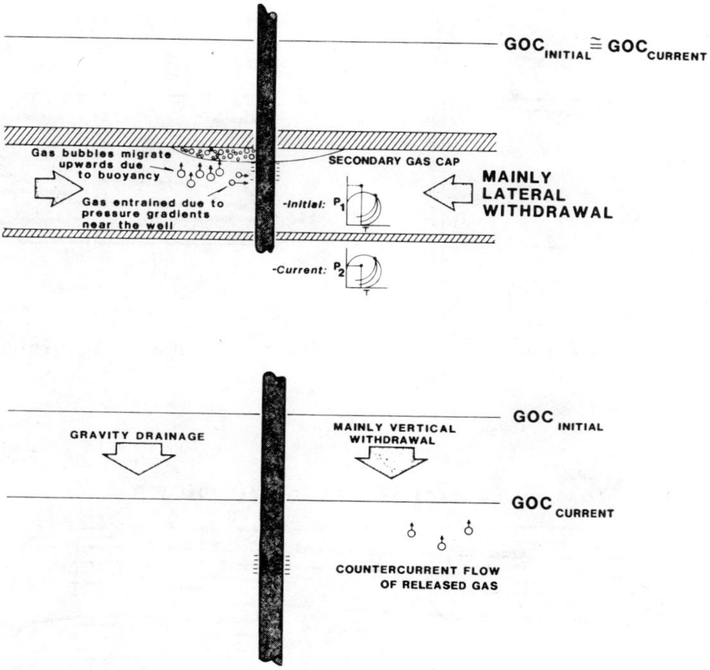

FIGURE 3: SHALES AFFECT PRIMARY PRODUCTION.

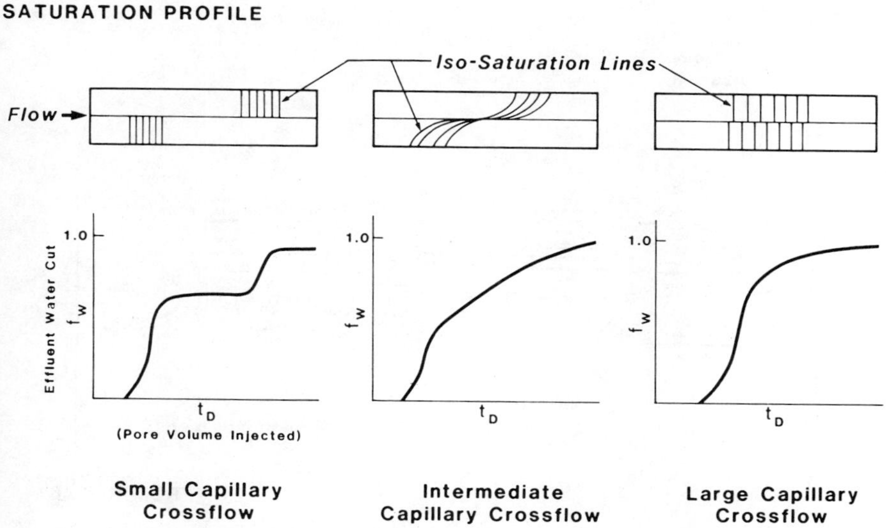

FIGURE 4: SHALES AFFECT CROSS-FLOW. (FROM REF. 25)

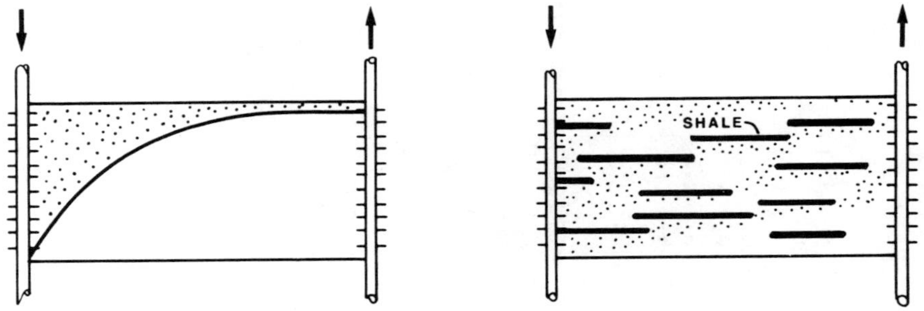

FIGURE 5: IN HORIZONTAL DRIVES SHALES ARE ADVANTAGEOUS DUE TO INCREASED VERTICAL SWEEP.

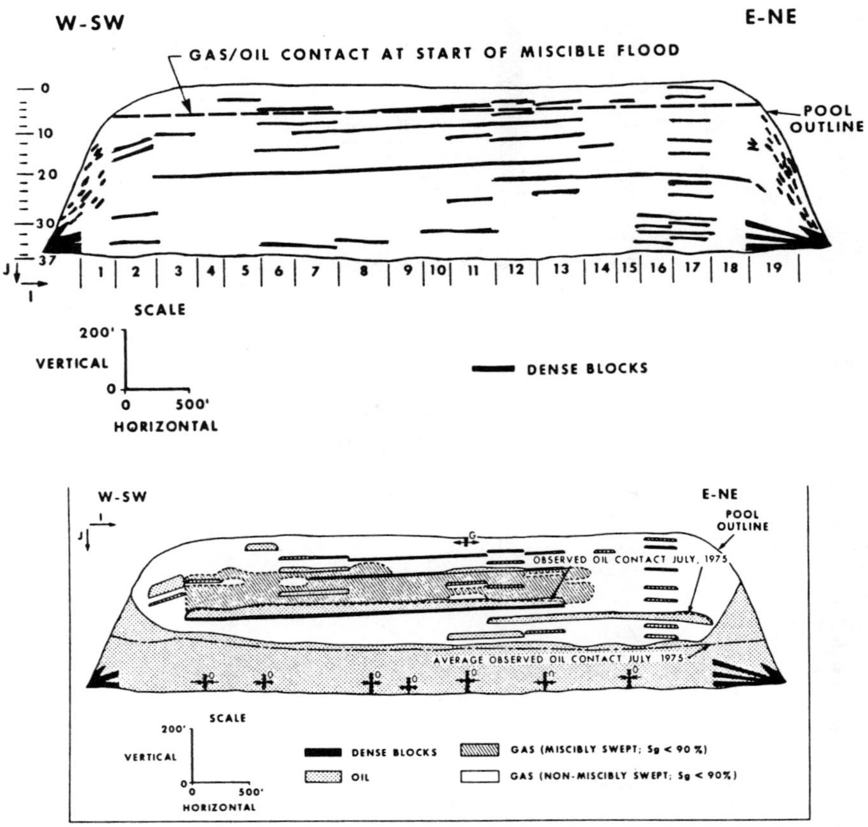

(FROM REF. 6)

FIGURE 6: IN VERTICAL DRIVES SHALES MAY BE A DISADVANTAGE DUE TO DECREASED VERTICAL SWEEP.

G. SHALES AFFECT GRAVITY DRAINAGE (16-18)

Shales affect the gravity drainage process in a variety of ways (Figure 7):

1). Uneven advancement of the gas-oil contact.
2). Hold-up of oil on top of shales - reduced volumetric sweep.
3). Barriers for upward segregation of liberated solution gas.
4). Barriers for downward inter-zonal flow of oil.

H. SHALES AFFECT WATERFLOODING (4,19)

Shale continuity is a key to waterflooding in many fields.(19) They may form effective barriers to vertical flow and thus aid in confining injected water to reduce slumping and improve the vertical sweep efficiency. In other cases, shales may confine high-permeability layers, and "cycling" of water is a major problem. To model a waterflooding operation correctly, it is essential to have a good shale description. Continuous major shales as well as scattered, random shales may play a key role.

In summary, shales or other vertical flow barriers affect most oil and gas recovery schemes that we are trying to model with numerical simulators.

III. STATEMENT OF THE PROBLEM

The following three questions are addressed in the remainder of this article: 1) What are the lateral dimensions of discontinuous shales, 2) How are the shales distributed within the reservoir, and 3) What is the effective vertical permeability of large grid blocks containing a number of discontinuous shales? The descriptive techniques rely on shale occurrence statistics from the wells, shale length statistics from outcrops of the appropriate depositional environment (or other sources), and a number of assumptions, some of which are reinforced in a statistical analysis of the shales exposed in an actual outcrop (fluvial sandstone cliff in the Assakao (Tassili) Region, Central Sahara (22)).

The effective vertical permeability of a given shale configuration is estimated through a statistical streamline technique recently developed by Begg and King (23).

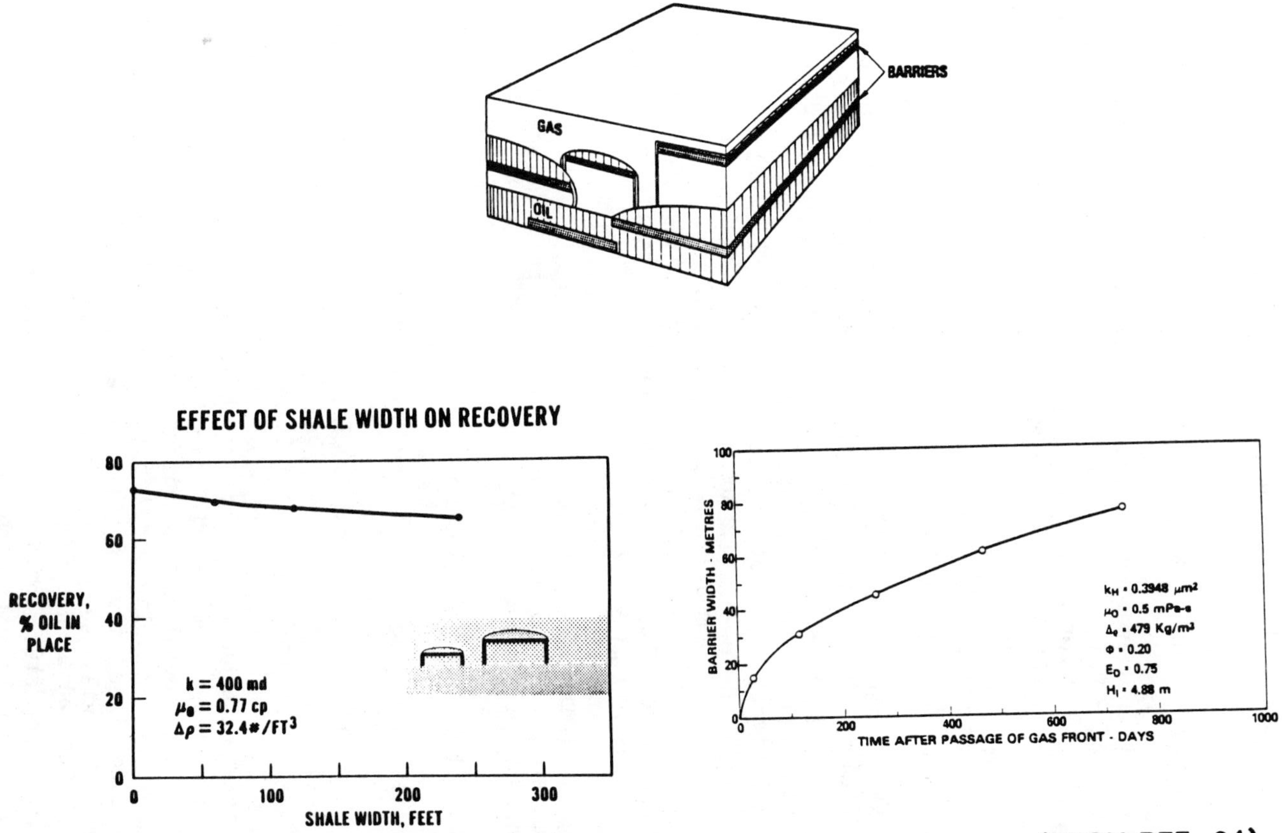

(FROM REF. 24)

FIGURE 7: SHALES AFFECT GRAVITY DRAINAGE.

IV. PROCEDURES FOR SHALE DESCRIPTION [20,21]

Sand-shale sequences in wells may be inferred from either logs (particularly gamma ray logs) or from cores (Figure 8). An empirical distribution of shale thicknesses can be determined from the observed sand/shale sequence.[20] Also, the number of shale occurrences, N_{SH}, over the gross pay (or over a particular hydraulic unit), H, may be recorded. This "shale density" (number of shales/unit vertical distance) at the observation point (the well with areal coordinates (x,y)) is then given by:

$$s(x,y) = \left(\frac{N_{SH}}{H}\right) \qquad \ldots(1)$$

With enough point values of s(x,y), a shale density map may be constructed for the whole reservoir thickness or for individual zones (Figure 9).

Unless the well spacing is extremely dense, nothing is revealed about the shales' lateral dimensions. In some cases a few shales correlate between wells while the majority do not. Hence, two different types of shales are often distinguished (Figure 10). These are 1) deterministic shales that are continuous and correlatable between wells (Figure 10B). and 2) stochastic shales which are uncorrelatable and appears to be randomly distributed in space (Figure 10A). When building the simulation model, deterministic shales are forced to coincide with the boundary between two layers in the model. Stochastic shales must be handled differently. They will be generated, i.e., a synthetic shale description is to be constructed.

The following assumptions apply to the 2-D shale generation:

1). The shale density, s, at a wellbore (which may or may not be) intersecting the 2-D cross-sectional model which a reservoir engineer is setting up is a representative measure of s for the whole cross section.
2). The location of stochastic shales is random, uniform and independent of other shales. Note that if any well(s) actually intersects the 2-D cross section under study, the known succession(s) of sands and shales is maintained in the synthetic description through a conditioning step.
3). A distribution of shale lengths for the reservoir studied can be obtained. The primary source of such information is visual studies on shale lengths in outcrops of ancient sediments.[27-32]

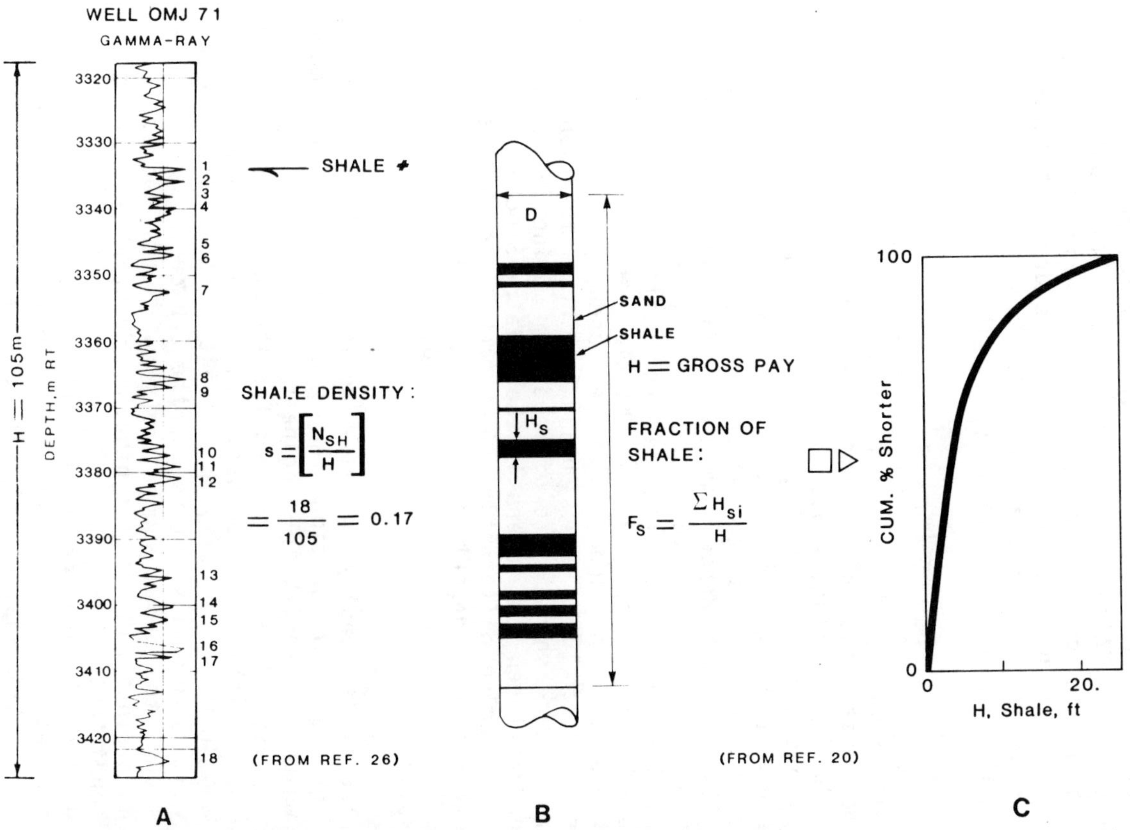

FIGURE 8: (A) SHALES MAY BE INFERRED FROM THE GAMMA-RAY LOG, (B) SHALES MAY BE OBSERVED IN CORES, (C) EMPIRICAL CUMULATIVE DISTRIBUTION FUNCTION OF SHALE THICKNESSES.

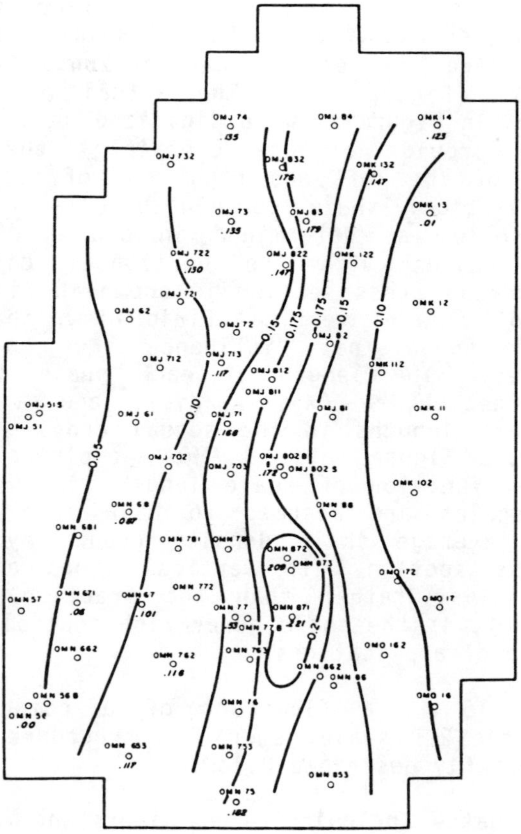

FIGURE 9: SHALE DENSITY (s(x,y)) MAP. (FROM REF. 26)

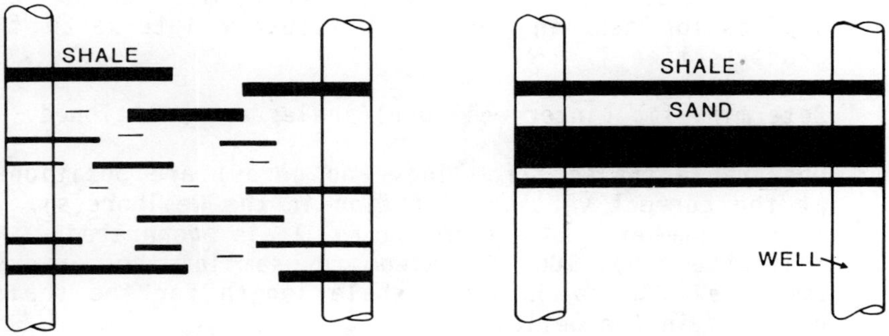

(A) (B)

FIGURE 10: (A) STOCHASTIC SHALES, (B) DETERMINISTIC SHALES. (FROM REF. 20)

The shale statistics used in the examples are from an outcrop study[22] of the Assakao (fluvial) sandstone which contains a large number of near-horizontal non-continuous shale breaks (Figure 11). The actual positions of the shales shown in Figure 11 were digitized on a VAX/Intergraph station to provide a tape containing the exact (X,Z) coordinates of the left and right end of each shale. The "average" vertical shale coordinate $(Z_{1i} + Z_{2i})/2$ of a digitized shale was determined in order to avoid having shales at angles with the horizontal direction. The corrected actual cross section is shown in Figure 12. From the point of view of vertical fluid flow, this geometrical adjustment should not influence K_{VE} estimations significantly. The length of each shale is now easily determined as: $L_i = (X_{2i} - X_{1i})$, and a histogram of all the exact lengths in the actual cross section can be constructed, Figure 13. The empirical cumulative distribution function of shale lengths is shown in Figure 14. The shales are assumed to have zero thickness and hence, the average shale density (found by sampling the actual cross section with vertical lines at 1000 random X-locations), s, rather than the fraction of shale, F_s (see Ref. 20), is the shale generation controlling parameter in this type of applications.

Figure 15 is a flowchart of a computer algorithm describing the 2-D shale generation procedure. The various steps are briefly described below:

1) Input data including the dimensions of the cross section, X-location of the well, Z-positions of shales at the wellbore, deterministic shales, and the cumulative distribution function (CDF) of shale lengths are loaded. [The technique shown is for a 2-D application assuming one well actually intersects the cross section.]

2) Deterministic (inter-wellbore) shales are positioned.

3) Observable shales (from logs and cores) are positioned at the correct vertical location in the wellbore(s). A random number, RN, from 0 to 1 is generated. RN multiplied by 100 is used to sample the inverse empirical CDF to obtain a shale length for the shales observed in the well(s).

4) Left and right extensions of shales in the well(s) are randomly selected (Figure 16).

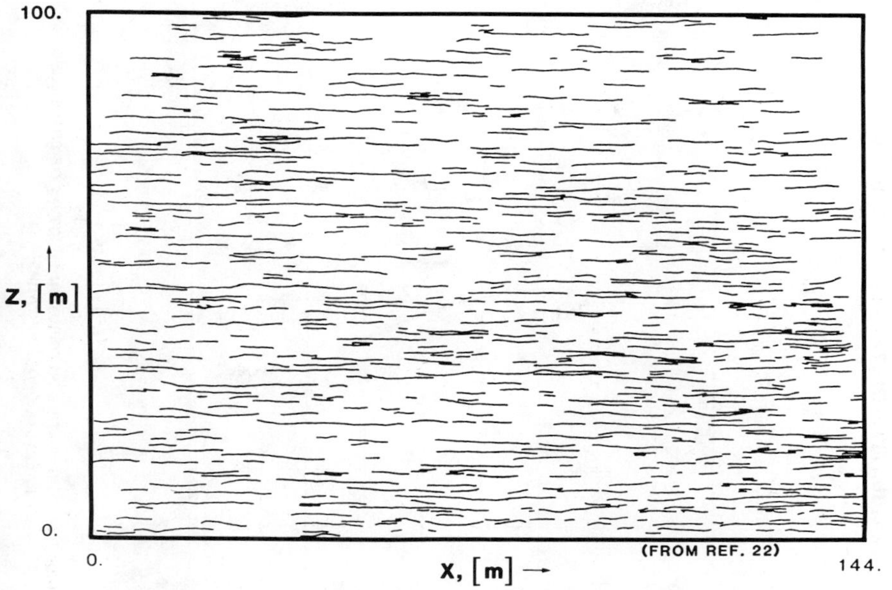

FIGURE 11: ASSAKAO SANDSTONE OUTCROP WITH NON-CONTINUOUS SHALE BREAKS (ACTUAL SHALE DISTRIBUTION).

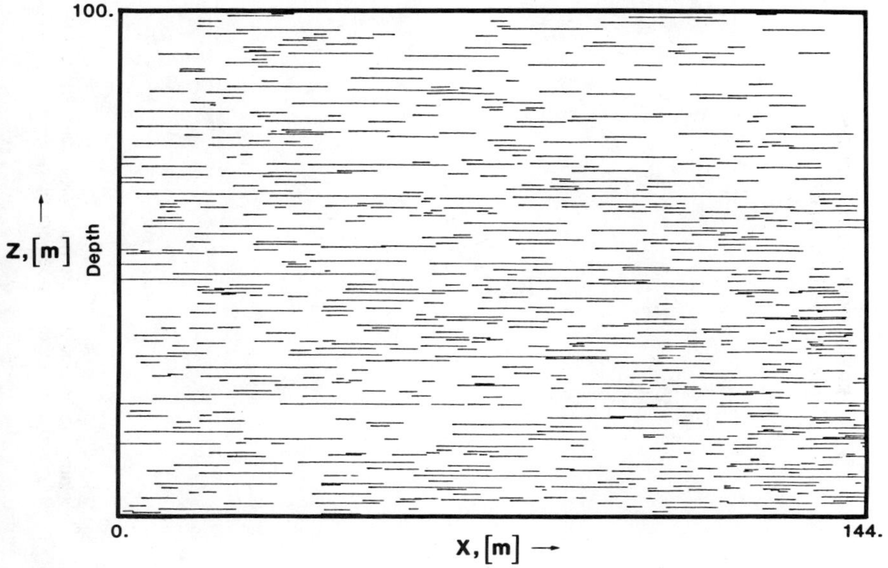

FIGURE 12: CORRECTED ACTUAL CROSS SECTION.

FREQUENCY DISTRIBUTION OF SHALE LENGTH

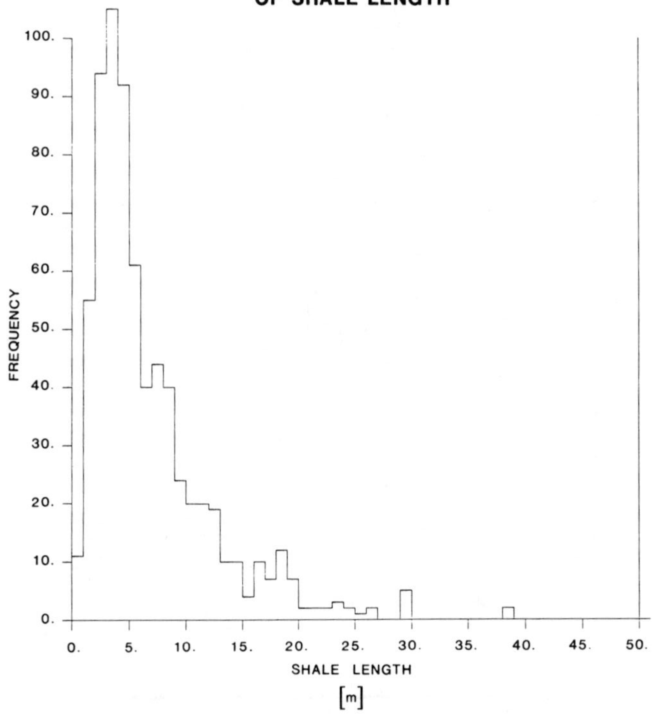

FIGURE 13: FREQUENCY DISTRIBUTION OF SHALE LENGTHS IN ACTUAL CROSS SECTION.

CUMULATIVE DISTRIBUTION FUNCTION

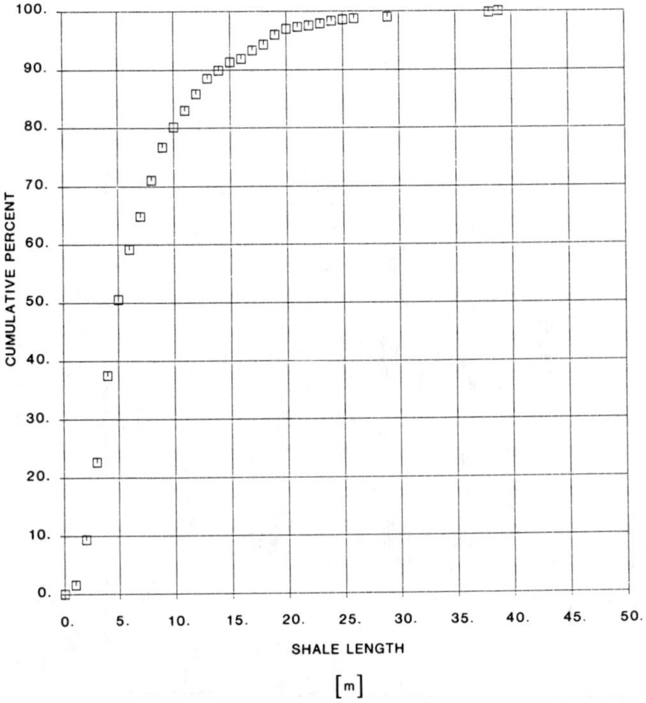

FIGURE 14: CUMULATIVE DISTRIBUTION FUNCTION OF SHALE LENGTHS IN ACTUAL CROSS SECTION.

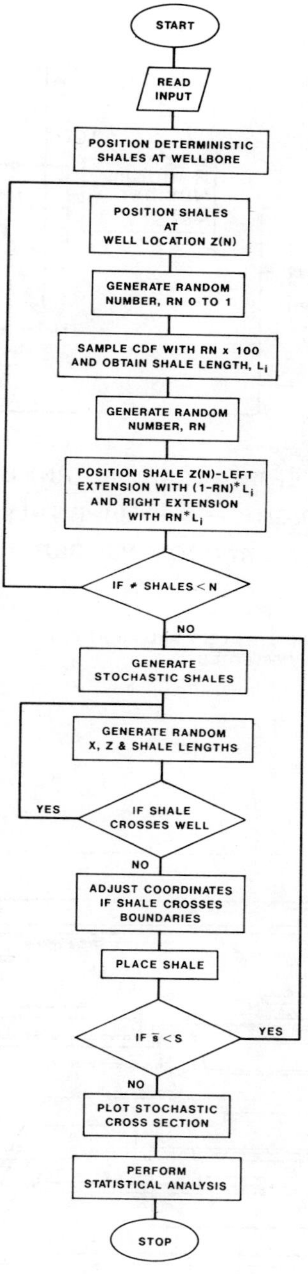

FIGURE 15: FLOWCHART FOR SHALE GENERATION.

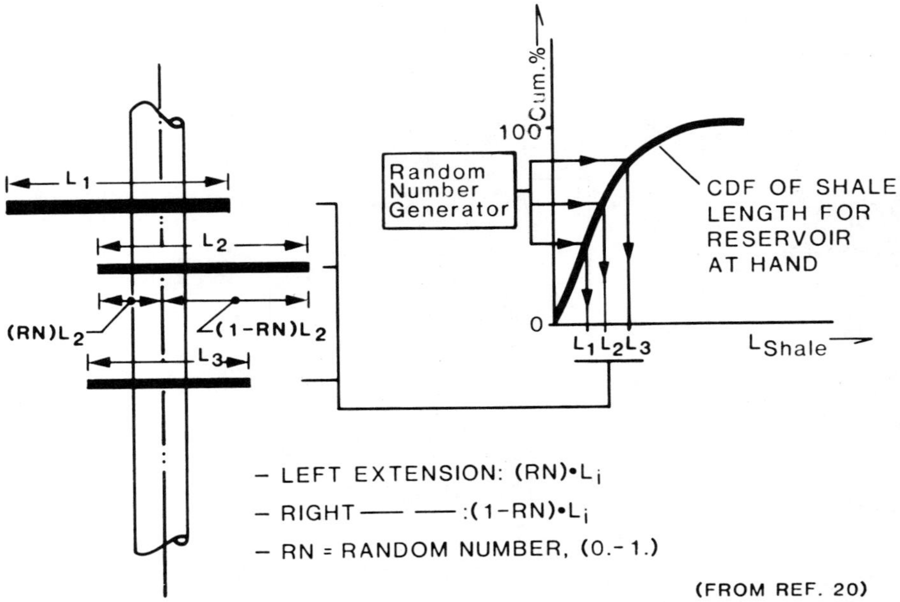

FIGURE 16: LENGTH, AS WELL AS LEFT AND RIGHT EXTENSIONS OF SHALES IN THE WELL(S) ARE RANDOMLY SELECTED.

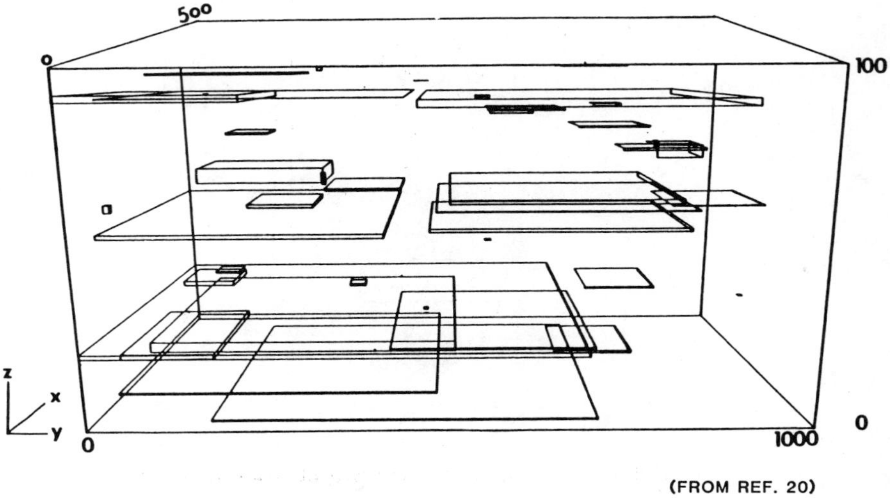

FIGURE 17: GRID BLOCK WITH INTERNAL STOCHASTIC SHALES.

5) Stochastic shales are generated one by one to fill the cross section. Random (X,Z) locations and shale length values are generated. If generated shales cross the well(s) or the system boundaries they are rejected or adjusted respectively.

6) The overall average cross section shale density, $\bar{s}$, is continuously calculated to see if it is equivalent to the wellbore shale density, s. If $\bar{s}$ is less than s, then step 5) is repeated until $\bar{s}$ = s.

7) A plot of the resulting synthetic cross section is generated.

8) Relevant shale statistics for the synthetic cross section are tabulated. An anistropy factor, α defined as the ratio of effective single-phase vertical permeability of a system with sand and internal shales, K_{VE}, to that of homogeneous sand, K, is determined.

V. SHALE THICKNESS INCLUSION AND 3-D APPLICATION

Instead of representing shales as zero-thickness lines, the empirical shale thickness statistics obtained from wells may be utilized to also generate a thickness for each shale. In this case shales are represented as rectangular areas in 2-D. No [20], or partial[33] or complete[34] correlation between shale thickness and length may be assumed.

In 3-D applications the shale must be given a length, a width, a thickness and a random spatial (X,Y,Z) location of the center. Hence, they are represented as rhombohedral bodies. The shale generation scheme in principle is identical to the one presented for 2-D applications. Length, width and thickness statistics must be user-supplied, and any desired interdependence among the three parameters must be built in (Figure 17).

VI. STATISTICAL ANALYSIS OF CROSS SECTIONS

Without statistical comparisons of the actual and synthetic cross sections it is not possible to know whether the stochastic cross section is an accurate representation of the actual observations. And, furthermore, it is not possible to know if the shale description methodology presented is viable.

The synthetic cross section generated from the parameters in Table 1 and the shale length statistics in Figure 14 are shown in Figure 18. (Note the presence of a hypothetical well.)

It is thought that if the overall shale density and to a lesser extent, the average vertical distance between shales between the actual and synthetic cross sections are equivalent, then the two cross sections are equivalent from the point of view of fluid flow.

The cross sections were subdivided into one meter columns spanning the entire cross section. The shale density and the average distance between shales for each column were determined. From visual observation of the actual cross sections (Figures 11,12), an increasing shale density from left to right is apparent. This is confirmed in Figures 19A and 20A, where the shale density is increasing and the average distance between shales is decreasing from left to right. The shale density of the synthetic cross section (Figure 19B) matches that of the actual quite well. The average vertical distance between shales in the synthetic cross section has no trend (Figure 20B). This indicates a more "homogeneously heterogeneous" shale distribution. Column statistics from the actual and synthetic cross sections are presented in Table 2. The numbers suggest that the proposed technique is indeed applicable for generating "realistic" synthetic cross sections.

Desbarats[40] performed a geostatistical study on the same outcrop data. He first generated a grid of shale indicator values making it amenable to variogram analysis. A 146 x 101 grid of sample nodes was constructed on a 1x1 m spacing. The indicator function[41] was:

$$I(x) = \begin{cases} 1 & \text{if point } x \text{ is shale} \\ 0 & \text{if point } x \text{ is sand} \end{cases}$$

and a value of the indicator function was assigned to each sample node on the basis of its proximity to a shale. More precisely, an indicator value of 1 was assigned to a node if the distance from the node to the straight line representing a shale was less than half the grid spacing, i.e., 0.5 m. This procedure implicitly gives every shale a thickness of 1m and a length within 0.5m of its measured value. Desbarats found cross section parameters and column statistics as shown in Tables 1 and 2. In addition, Desbarats found that:

TABLE 1

CROSS SECTION PARAMETERS

	ACTUAL CROSS SECTION	SYNTHETIC CROSS SECTION
o Cross Section Length	144.42 Meters	Same
o Cross Section Thickness	100 Meters	Same
o Hypothetical Well Location	-	89 Meters (X-Location)
o Shales at Well	34	Same
o Shale Density	.34 Shales/ Meter	Same
o Fraction of Shale	0.05	Same
o Number of Stochastic Shales	706	751
o Assumed Shale Thickness	0.5 ft.	Same

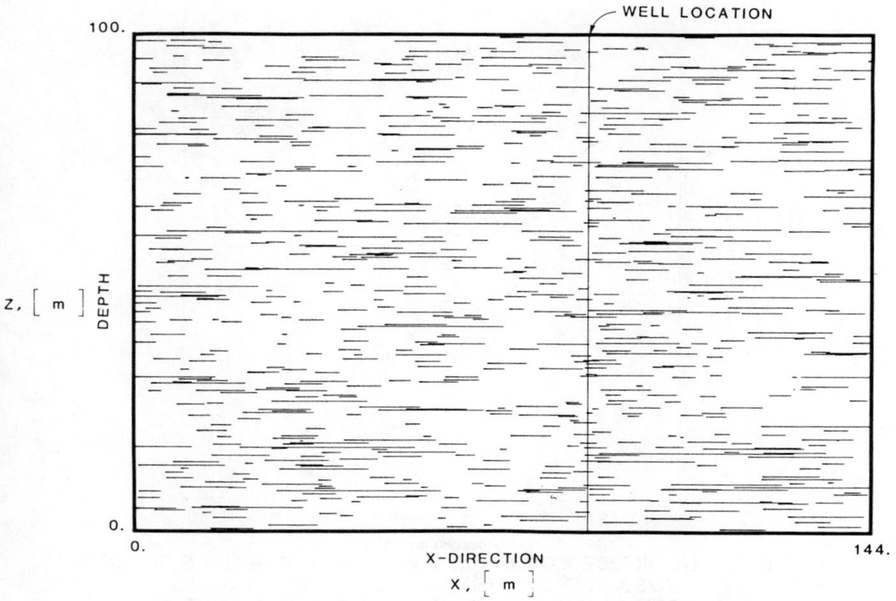

FIGURE 18: SYNTHETIC CROSS SECTION.

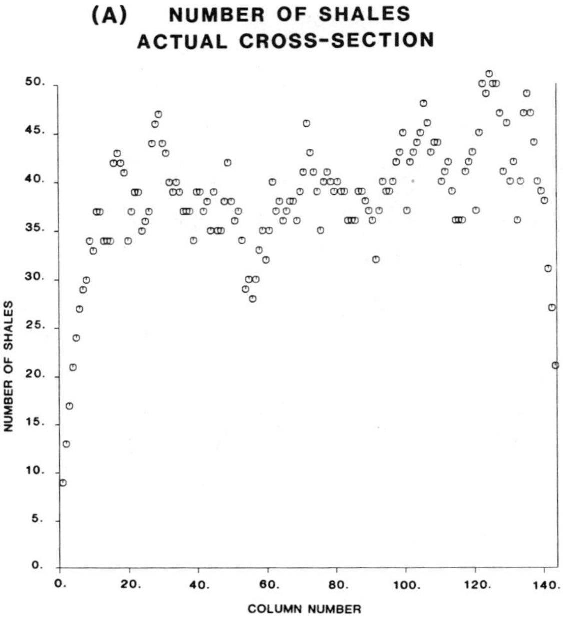

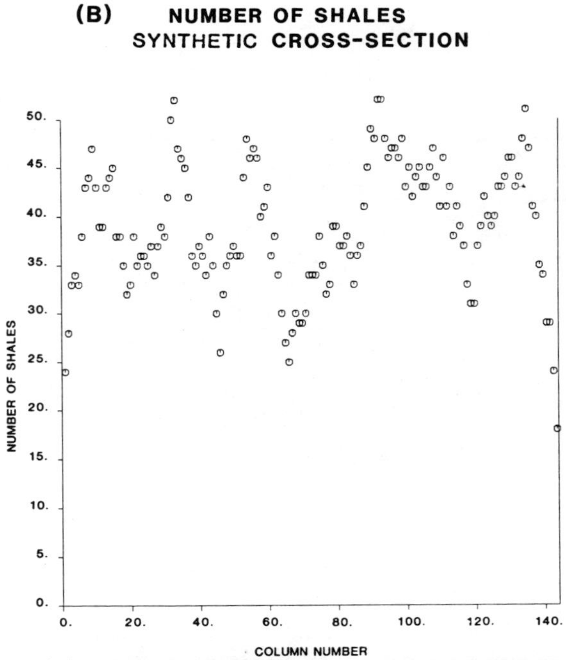

FIGURE 19: (A) NUMBER OF SHALES vs. COLUMN NUMBER IN ACTUAL CROSS SECTION, (B) NUMBER OF SHALES vs. COLUMN NUMBER IN SYNTHETIC CROSS SECTION.

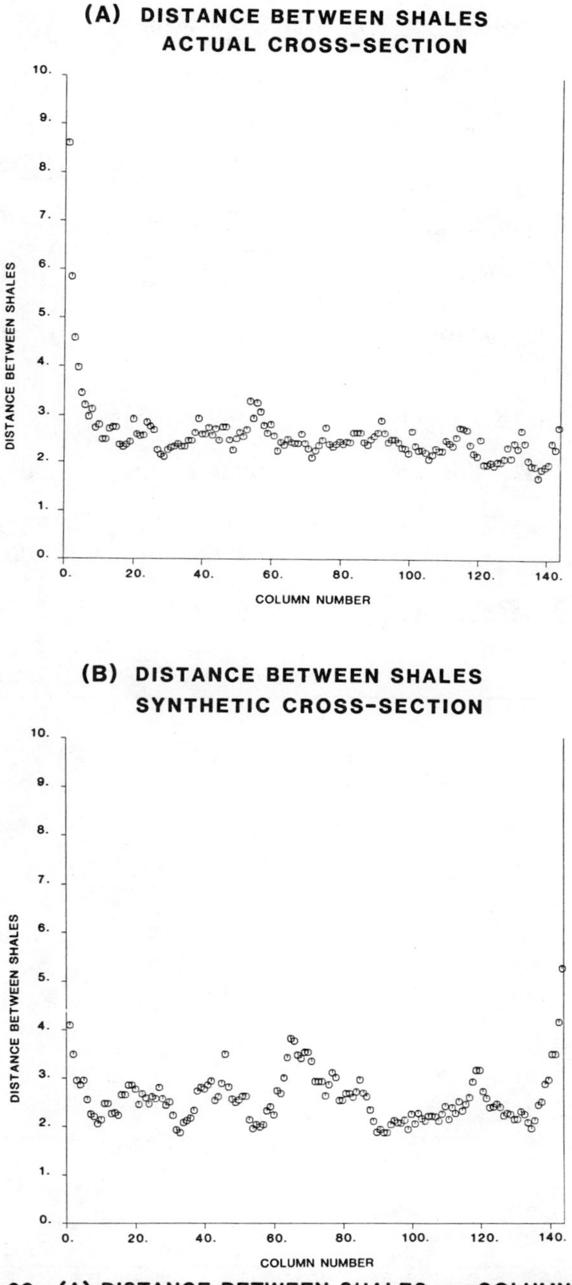

FIGURE 20: (A) DISTANCE BETWEEN SHALES vs. COLUMN NUMBER IN ACTUAL CROSS SECTION, (B) DISTANCE BETWEEN SHALES vs. COLUMN NUMBER IN SYNTHETIC CROSS SECTION.

TABLE 2

COLUMN STATISTICS

	ACTUAL CROSS SECTION	SYNTHETIC CROSS SECTION
o Average Shale Length	6.83 Meters	6.49
o Fraction of Shale	0.05	0.05
o Standard Deviation of Shale Length	5.50 Meters	5.07
o Maximum Shale Length	38.89 Meters	37.90
o Minimum Shale Length	.002 Meters	.045 Meters
o Average Shale Density (Columns)	38.15 Shales/ 100 Meters	38.84 Shales/ 100 Meters
o Average Distance Between Shales	2.57 Meters	2.58 Meters

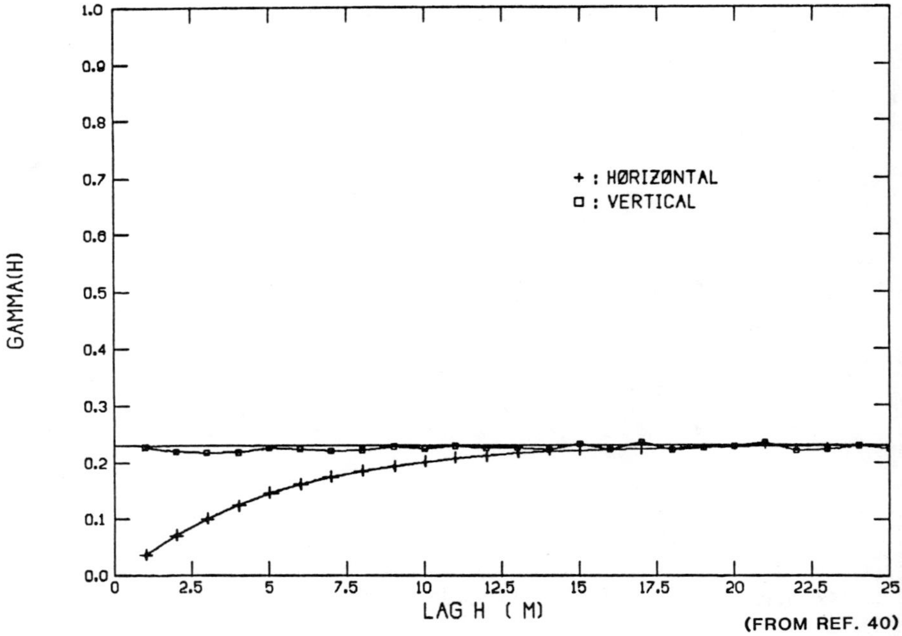

(FROM REF. 40)

FIGURE 21: EXPERIMENTAL SHALE INDICATOR VARIOGRAM (FOR SHALES IN FIGURE 11).

1) The average horizontal distance between adjacent shales is 14.2 m.

2) The experimental shale variogram in the vertical direction presents a total absence of spatial correlation between shales or a pure nugget effect (Figure 21). A small hole effect may be observable, however, at a lag of 2.5 m, corresponding to the mean vertical distance between shales. In the horizontal direction the experimental variogram exhibits a very smooth transition structure with a range of about 15 m. No nugget effect is apparent in this direction.

3) A plot of the center coordinates of the 706 shales (Figure 22) appears at first glance random, but careful examination reveals many local alignments.

4) A relationship between shale lengths and shale positions was revealed for the horizontal direction while the vertical variogram does not exhibit any structure (Figure 23).

The horizontal shale length variogram shows a transition structure with a range of about 10 m (Figure 23). Desbarats notes that it may be significant that the range of the shale length variogram plus the average shale length gives the approximate range of the shale indicator variogram which in turn is close to the mean lateral distance between shales.

Desbarats[40] concludes that while the disposition of shale centers may or may not be random, shale lengths are certainly correlated in space. This is an important conclusion with bearing on the adequacy of the shale generation procedure. In general, the utility of geostatistics in reservoir description and particularly its usefulness in heterogeneity quantification and scale transformations seems promising and should be further pursued.

VII. DETERMINING AN ANISOTROPY FACTOR

The resulting synthetic shale description arrived at in the previous section is too complex for practical simulation usage. When a user-defined grid system is superimposed on the synthetic cross section, large elemental units containing both sand and shales are defined. The goal, therefore, is to convert such heterogeneous grid blocks and cross sections into apparently homogeneous ones with effective properties (Figure 24).

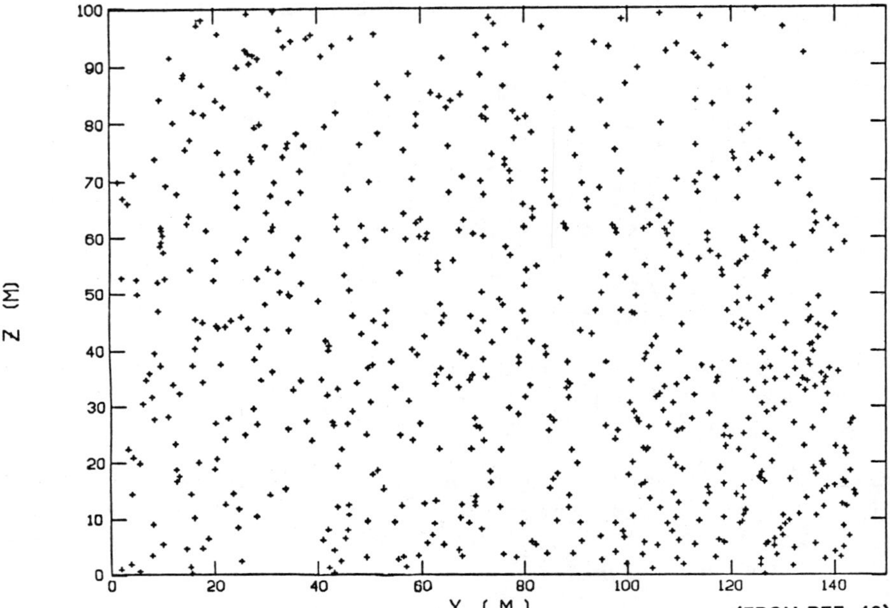

FIGURE 22: SHALE CENTERS FOR SHALES IN FIGURE 11.

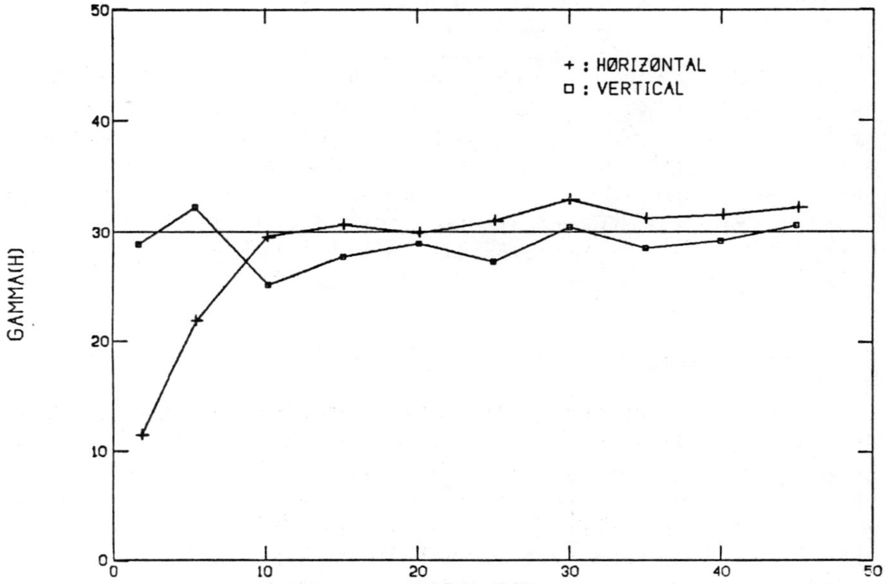

FIGURE 23: SHALE LENGTH VARIOGRAM (FOR SHALES IN FIGURE 11).

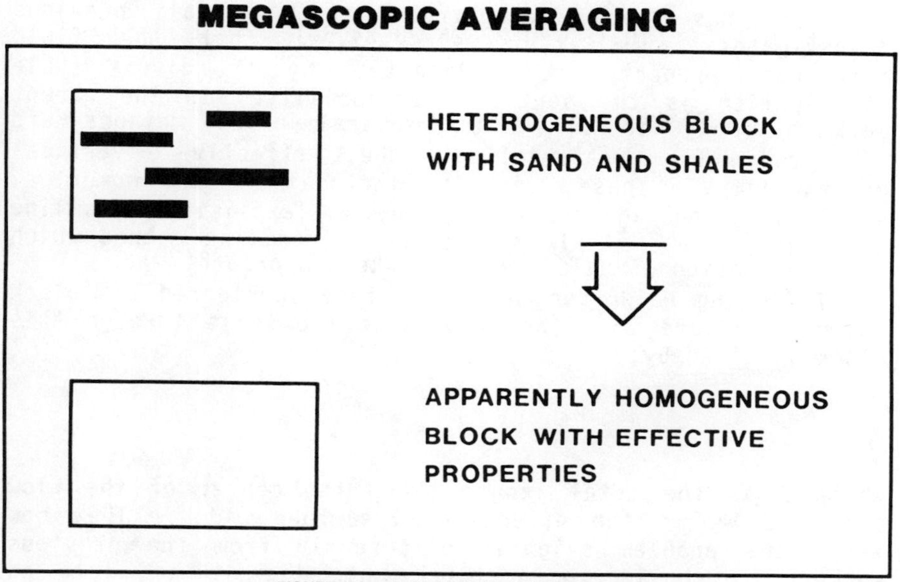

FIGURE 24: MEGASCOPIC AVERAGING.

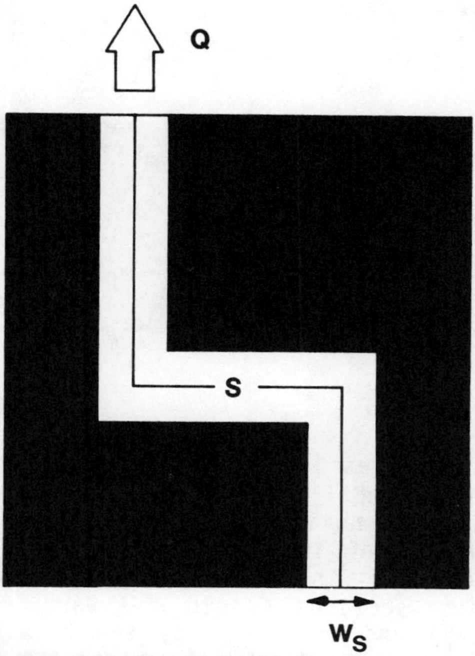

FIGURE 25: BLOCK WITH SHALE AND CONCEPTUAL TORTUOUS FLOW TUBE. (FROM REF. 23)

It has been demonstrated by several previous investigators (10,16, 20,22,26-28,34,36) that the fluid transport property most impacted by a given shale distribution is the vertical permeability. In the recent work by Begg et al. (23), several novel methods are suggested for calculating the effective vertical permeability. These methods include direct numerical simulation, an analytical method, an explicit streamline technique, and a simple statistical streamline method which is an extension of Haldorsen and Lake's approach(20).

Following Haldorsen and Lake, they considered the block shown in Figure 25. The volumetric flow rate through this block is given by:

$$Q = \left[\frac{K}{\mu} \right] W_S \left[\frac{\Delta P}{S} \right] \qquad \ldots(2)$$

where S is the total length along the center of the flow path and W_S is the effective streamtube width. They now treat the problem slightly differently from the previous workers by considering a block which is divided into an arbitrary sample of N_S equally spaced streamtubes, such as shown in Figure 26. In this case, Equation (2) can be rewritten as:

$$Q = \left[\frac{K}{\mu} \right] \sum_{i=1}^{N_S} \left[\frac{\Delta P}{S_i} \right] W_{Si} \qquad \ldots(3)$$

Now suppose that this block is replaced by one with a homogeneous effective vertical permeability so that the flow rate is:

$$Q = \left[\frac{K_{VE}}{\mu} \right] L \left[\frac{\Delta P}{H} \right] \qquad \ldots(4)$$

By equating Equation (3) and (4) we obtain:

$$\left[\frac{K_{VE}}{K} \right] = \left[\frac{H}{L} \right] \sum_{i=1}^{N_S} \left[\frac{W_{Si}}{S_i} \right] \qquad \ldots(5)$$

For constant pressure (top and bottom) and no-flow (sides) boundary conditions, each streamtube starts and finishes at the same horizontal position. This is combined with the assumption of incompressibility, and they further assume that the areas (in 2-D) of the streamtubes are equal (though the cross-sectional width may vary along a stream tube (Figure 26)) and are given by the total area of sand divided by the number of streamtubes so that:

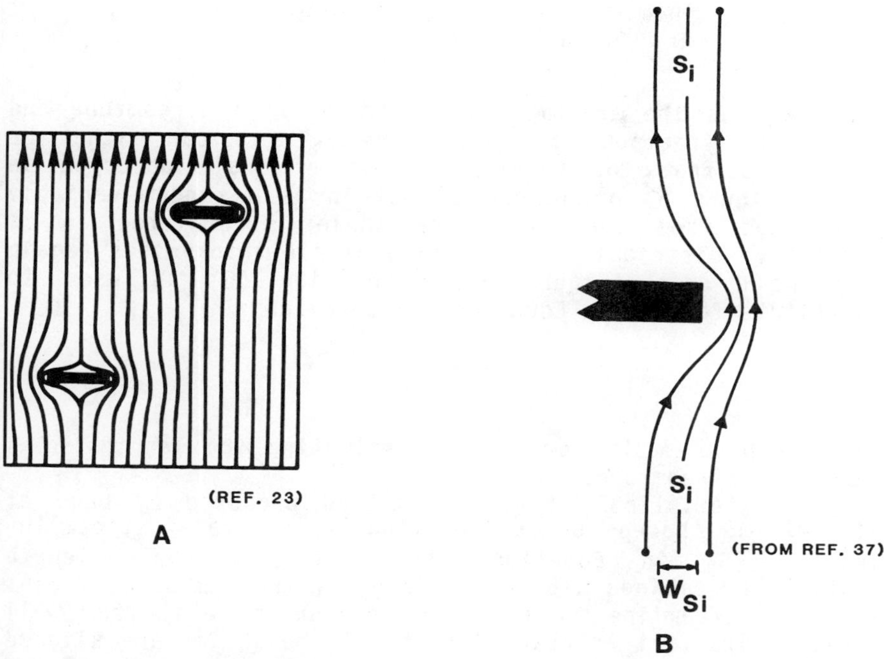

FIGURE 26: (A) STREAMLINES IN BLOCK WITH SHALES (B) FLOW TUBE.

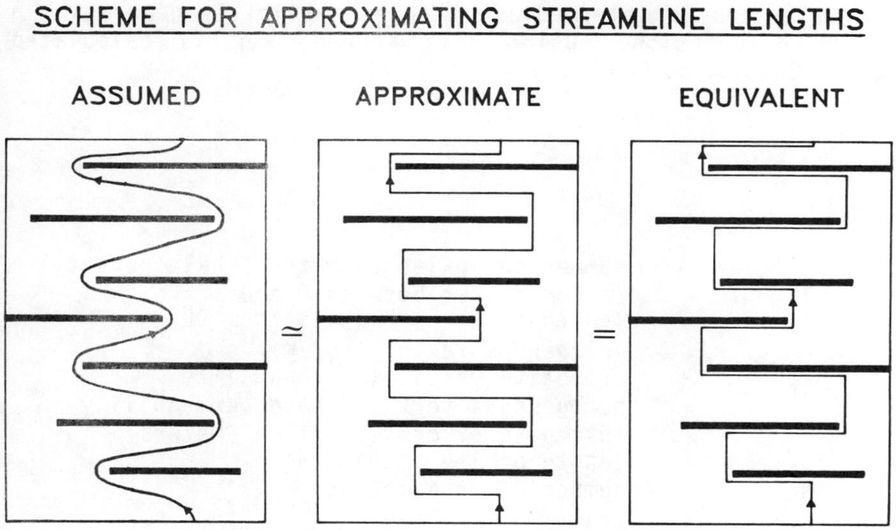

FIGURE 27. SCHEME FOR APPROXIMATING STREAMLINE LENGTHS.

$$A_i = \overline{W}_{Si} \; S_i = \left[\frac{(1 - F_s)\; HL}{N_S}\right] \qquad \ldots(6)$$

where $\overline{W}_{Si}$ is the arithmetic mean width of a streamtube and F_S is the fraction of shale. The F_S term can either be determined from the total thickness of shale observed in wells (Figure 8) or it can be put in as a random variable calculated from the number of shales encountered and a cumulative distribution function of shale thicknesses (again from well data (Figure 8)). Equation (6) is used to substitute for W_{Si} in Equation (5) to give:

$$\alpha = \left[\frac{K_{VE}}{K}\right] = \left[\frac{(1 - F_s)\; H^2}{N_S}\right] \sum_{i=1}^{N_S} \left[\frac{1}{S_1^2}\right] \qquad \ldots(7)$$

and the problem is now one of estimating the lengths of a sample of streamtubes.

The statistical streamline method proposed by Begg et al.[23] is based on calculating approximate streamline lengths (S_i in Equation (7)) from the shale length statistics combined with a knowledge of the number of shales which a streamline is likely to encounter (e.g. from well data). First, it is assumed that all the shales are aligned normal to the flow and that a streamline travels a random fraction of the length of each shale encountered plus a length equal to the height of the model H (Figure 27). Secondly, the shale lengths are taken directly from the cumulative distribution function of shale lengths without recourse to the intermediate step of explicitly generating a shale description. Using this scheme, K_{VE} is calculated using Equation (7) where:

$$S_i = H + \sum_{j=1}^{(Hs + 1)} (RN) \; l_j \qquad \ldots(8)$$

and,

s	=	number of shales permeter (shale density)
RN	=	random number between 0 and 1
l_j	=	length of the jth shale from CDF
α	=	anisotropy factor (K_{VE}/K)
K_{VE}	=	effective vertical permeability
K	=	homogeneous vertical sand permeability
H	=	thickness of cross section (meters)
S_i	=	length of ith streamline
N_S	=	number of streamtubes
F_S	=	fraction of shale

An average actual shale thickness of one half foot was assumed and a fraction of shale, F_S, of .05 was determined for the actual cross section shown in Figure 11. Iterating Equations (7) and (8) 10,000 times, an anisotropy factor of $\alpha = 0.20$ was found.

If the formation is sufficiently thick Begg et al.(23) showed that the tortuosity for all arbitrary streamtubes is approximately equal. Therefore, Equation (7) becomes:

$$\alpha = \left[\frac{K_{VE}}{K}\right] = \left[\frac{\left(1 - F_s\right)}{\left(1 + s\,\frac{\bar{l}}{C}\right)^2}\right] \qquad \ldots(9)$$

where,

$\bar{l}$ = average shale length
C = geometry constant; C = 2 for 2-D, and C = 3 for 3-D applications

This simplified approach gave a 2-D anisotropy factor of $\alpha = 0.18$. (See input data in Table 2) and a 3-D anisotropy factor of $\alpha = 0.27$. Note that 3-D is more "forgiving" than 2-D in terms of vertical permeability reduction. (Note also that we have implicitly assumed that $K_H = K_V = K$ for the sand and that shales are squares in 3-D.)

Dupuy and LeFebvre[22] provide a unique opportunity for validating the α - values computed above. In their original work they determined K_{VE}/K values for the cross section in Figure 11 by several independent means. In order to simulate single phase vertical flow they reproduced Figure 11 on conducting paper by simply cutting a slit along the shales to represent the impermeable barriers. Then they measured the resistances to electrical flow in the vertical and horizontal directions by imposing equipotentials (conducting paint) both on the horizontal and vertical sides of the cross section. From the measured resistances they derived the effective resistivities in the two directions of flow and the coefficient of anisotropy. A 2-D α - value of 0.203 was found (Figure 28). Based on an analytic approach, Dupuy and LeFebvre[22] found a 3-D anisotropy factor of 0.32. Hence, we conclude that the calculated α - values utilizing the statistical streamline technique proposed by Begg et al.(23) are in excellent agreement with the results reported by Dupuy and LeFebvre[22] (Table 3.)

Therefore, we have demonstrated that the simple method suggested by Begg et al. [23] is a quick, inexpensive and accurate technique for estimating the effective vertical single-phase permeability of rock units containing scattered, discontinuous shales. As pointed out in their

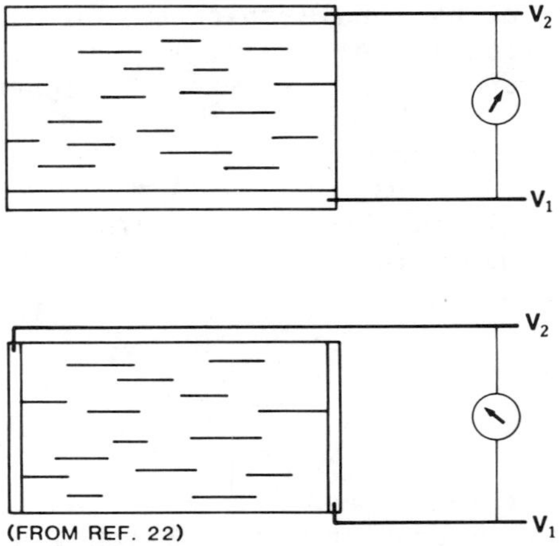

(FROM REF. 22)

- VERTICAL RESISTANCE: 9,870 Ω:
- HORIZONTAL RESISTANCE: 4,150 Ω:
- HEIGHT: 50 cm:
- WIDTH: 72 cm:
- VERTICAL RESISTIVITY: 9,870 x 72/50 = 14,200 Ω
- HORIZONTAL RESISTIVITY: 4,150 x 50/72 = 2,880 Ω
- ▶ **ANISOTROPY FACTOR: 0.203.**

FIGURE 28: ANISOTROPY DETERMINATION (FOR CROSS SECTION IN FIGURE 11) BY MEANS OF CONDUCTING PAPER.

TABLE 3

COMPARISON OF ANISOTROPY ESTIMATES

	$\left(\frac{K_{VE}}{K}\right)$
o Dupuy et al. [22] - Conducting Paper (2-D)	0.203
o Statistical Method (Begg [23]) with 10000 Streamlines (2-D)	0.205
o Simplified Statistical Method (Begg [23]) (2-D)	0.18
o Dupuy et al. [22] - Analytical Approach (3-D)	0.32
o Simplified Statistical Method (Begg [23]) (3-D)	0.27

original paper, the effective vertical permeability is strongly dependent on the shale dimensions, the shale density and the system thickness.

VIII. SHALE LENGTH AND WIDTH STATISTICS

The most important information to be specified by the geologist is shale length and shale width statistics and some guidance as to their interdependence. Some information already has been obtained from outcrops (and subsurface correlations) on the geometrical characteristics (length, width, thickness and their interdependence) of vertical flow barriers.[10,16,22,27-32,34] More data is, however, needed.

In order for the geologist to be prepared to aid the practicing engineer with the shale description in numerical models of, for example, point bar, meandering stream, fan, or deltaic systems, more emphasis should be placed on the quantification of geometrical characteristics of sands and shales. Classical geological information would be that shales are wide, extensive or lenticular. This type of information is of little use for the modeller. It would be a major step forward if the geologist for a given reservoir could supply information on min/max/most likely barrier length and width (Figure 29). Along with the more easily determined H, s and $F_s = 1 - N/G$ (Figure 30) the 3-D effective single phase vertical permeability is then determined using Equation (7) where S_i now is given by:

$$S_i = H + \sum_{j=1}^{(Hs + 1)} \min \left[RN_1 \, l_j, \, RN_2 w_j \right] \qquad \ldots(10)$$

A length (l_j) and a width (w_j) is randomly chosen for each shale encountered (from the same or different distribution functions). The streamline will take the shortest path as illustrated in Figure 30.

Since barrier thickness may be determined in wells, a relationship between barrier length and thickness would be of great aid. Delhomme et al.[26] concluded that in the deltaic system studied, "the thickness of a shale break is related only to the local topography prevailing at the time of deposition, and thus independent of the length of the shale." For the alluvium of the lower Mississippi Valley, Wu et al.[34] found that for pervious sands embedded in silt or fine sand a strong relationship between length and thickness existed. From regression analysis they found (correlation coefficient = 0.72):

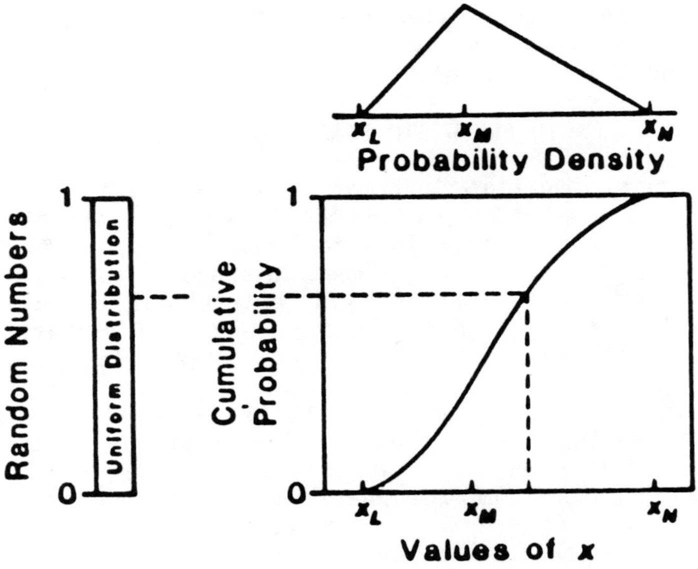

X = SHALE LENGTH OR WIDTH

Given: x_L = lowest value expected

x_M = most probable value expected

x_H = highest value expected

r = random number between 0 and 1

if, $r \geq (x_M - x_L) / (x_H - x_L)$

$$x = x_H - [(1 - r)(x_H - x_L)(x_H - x_M)]^{1/2}$$

if, $r \leq (x_M - x_L) / (x_H - x_L)$

$$x = x_L + [r\,(x_H - x_L)(x_M - x_L)]^{1/2}$$

(FROM REF. 38)

FIGURE 29: IF INFORMATION ON MIN/MAX/MOST LIKELY BARRIER LENGTH/WIDTH COULD BE SUPPLIED, RANDOM VALUES CAN EASILY BE GENERATED.

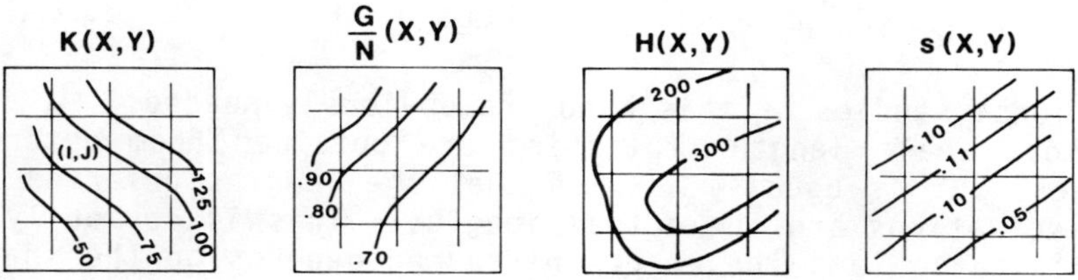

-  = simulation grid
- For Block (I,J) interpolate on grids to find K, $\frac{N}{G}$, H, and s at that point
- Calculate the effective vertical permeability from:
- $$K_{VE}(I,J) = K(I,J)\left[\frac{\left(\frac{N}{G}(I,J)\right)\left(H(I,J)\right)^2}{N_s}\right]\left[\sum_{i=1}^{N_s}\frac{1}{S_i^2}\right]$$

and,

- $$S_i = H(I,J) + \sum_{j=1}^{H(I,J)s(I,J)+1} \min\left[(RN)_1\,\ell_j,\ (RN)_2\,w_j\right]$$
- Distributions of barrier lengths and widths (ℓ_j,w_j) for a given reservoir must be supplied by the geologist
- Shale geometry statistics could be made a function of position

FIGURE 30: GRID OPERATIONS AND SIMPLE SUBROUTINE TO CALCULATE K_{VE} (I,J).

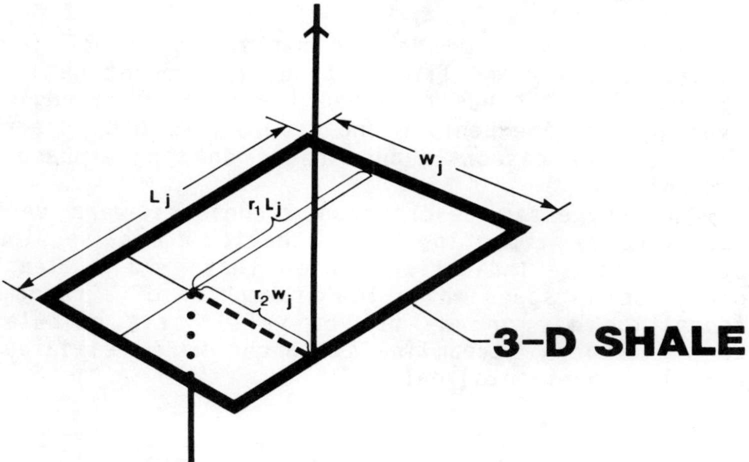

FIGURE 31: CONCEPTUAL ILLUSTRATION OF STREAMLINE PATH IN 3-D.

$$\text{Log}(L) = 1.5604 + 0.7684\ \text{Log}(H) \quad \ldots(11)$$

Additional studies of this kind are obviously needed.

Most shale length statistics are obtained from outcrop studies. It is usually assumed that the lengths observed in the excavations are the actual lengths. This is obviously a simplification and the error introduced depends on the shape of the shales and their orientation with respect to the excavation's face. If the direction of deposition (paleocurrent) is known as well as the angle between the outcrop face and the depositional direction, some form of "correction factor" might be devised for the shale length statistics obtained (Figure 31). Wu et al.(34) studied this problem and found that errors were negligible if the shales were assumed to be of circular shape. Additional studies of this problem are recommended.

Three component systems, i.e., vertical flow barriers and sand lenses both embedded in a silty low-permeability matrix, should receive more attention in terms of a stochastic geological description. Shale "overlapping," erosion (formation of "holes") of more continuous shales and the permeability of shales themselves (to non-wetting and wetting fluids) also should be studied in more detail.

The tremendous importance of depositional environment control on shale continuity and how log data may aid in shale facies identification is outlined in detail by Geehan et al.(42) Obviously, the geologist and the engineer must work closely together.

IX. SUMMARY AND CONCLUSIONS

In this work, we have shown the state of the art in stochastic shale modelling. Also, the recent publication by Begg et al. (23) has provided the practicing engineer with a very quick, inexpensive and accurate method to account for the effect of discontinuous shales on single-phase vertical permeability.

The suggested description techniques were verified by statistically comparing a synthetic cross section to an actual one. Indications are that shale lengths are correlated in space while the assumption of a random spatial disposition may or may not be valid. K_{VE} calculations by the statistical streamline technique were validated against reliable experimental data.

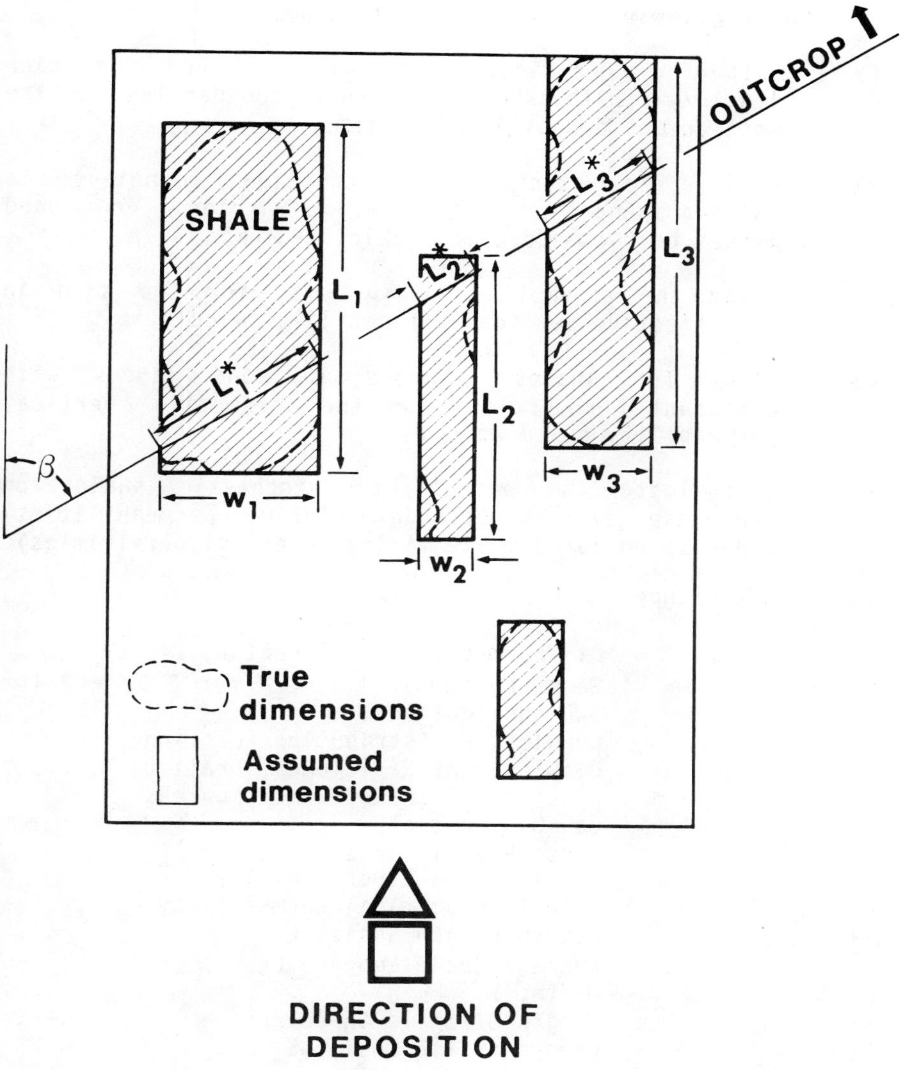

FIGURE 32: SHALE LENGTHS OBSERVED IN OUTCROPS MAY NOT BE L OR W.

Challenges for the future are to:

1) Gather quantitative data on barrier length/width/thickness and interdependencies in the most common depositional systems.

2) Apply the stochastic description technique to multi-component systems (e.g. barriers and sand lenses both embedded in a third medium).

3) Relax the assumption of $K_H = K_V$ for the sand in the effective vertical permeability calculation.

4) Allow for shales embedded in sand layers with different properties in the effective vertical permeability calculation.

5) Investigate the effects of stochastic shales on two-phase systems (pseudo relative permeabilities) and also on large-scale mixing rates (dispersivities).

X. NOMENCLATURE

B_o	=	Oil Formation Volume Factor, L^3/L^3
C	=	Geometry Constant (= 2 for 2-D, = 3 for 3-D applications)
CDF	=	Cumulative Distribution Function
E_D	=	Displacement Efficiency, Fraction
F_s	=	Fraction of Shale, Dimensionless
GOC	=	Gas-Oil Contact
H	=	Gross Pay, Thickness, L
K	=	Isotropic Sand Permeability, L^2
K_{VE}	=	Effective Vertical Permeability, L^2
l_j	=	Length of jth Shale, L
$\bar{l}$	=	Average Shale Length, L
L	=	Width, L
L*	=	Length of Shale in Outcrop, L
L_i	=	Length of ith Shale, L
N_S	=	Number of Streamlines
N_{SH}	=	Number of Shales
OWC	=	Oil-Water Contact
P	=	Pressure, $MT^{-2}L^{-1}$
RN	=	Random Number Between 0 and 1
s	=	Well Average Shale Density, L^{-1}
s	=	Cross Section Average Shale Density, L^{-1}
S_i	=	Length of ith Streamline, L
X	=	Coordinate of Well, L
X_{1i}, X_{2i}	=	Horizontal Coordinates of Shale Endpoints, L

w_j	=	Width of jth Shale, L
Y	=	Coordinate of Well, L
Z_{1i}, Z_{2i}	=	Vertical Coordinates of Shale Endpoints, L
ρ	=	Density, ML^{-3}
ϕ	=	Porosity, Fraction
α	=	Anisotropy Ratio (K_{VE}/K), Dimensionless
μ_o	=	Oil Viscosity, $ML^{-1}T^{-1}$

XI. ACKNOWLEDGEMENTS

The authors wish to express their appreciation to the management of Sohio Petroleum Company for permission to publish this paper. Special thanks are extended to A. J. Desbarats for allowing us to present some of his result, to S. H. Begg for stimulating discussions and some good figures, and to Angie A. Lingad and Lorry P. Navellier for typing the manuscript.

XII. REFERENCES

1) Aronofsky, J. S., Cull, G. W. L., Cox, T. F., and Gaffney, P. D.: "Use and Abuse of Reservoir Simulation - I," Oil & Gas Journal, November 5, 1984, pp. 79-82.

2) Aronofsky, J. S., Cull, G. W. L., Cox, T. F., and Gaffney, P. D.: "Use and Abuse of Reservoir Simulation - II," Oil & Gas Journal, November 12, 1984, pp. 81-85.

3) Aronofsky, J. S., Cull, G. W. L., Cox, T. F., and Gaffney, P. D.: "Use and Abuse of Reservoir Simulation - III," Oil & Gas Journal, December, 1984, pp. 116-118.

4) Harpole, K. J.: "Improved Reservoir Characterization - A Key to Future Reservoir Management for The West Seminole San Andreas Unit," SPE Paper No. 8274 presented at the 54th Annual Technical Conference and Exhibition of the SPE in Las Vegas, Nevada, September 23-26, 1979.

5) Bishlawi, M.: "Pressure Studies Optimize Montrose Field Production," Oil & Gas Journal, January 26, 1981, pp. 177-187.

6) Reitzel, G. A. and Callow, G. O.: "Pool Description and Performance Analysis Leads to Understanding Golden Spike's Miscible Flood," SPE Paper No. 6140, presented at the 51st Annual Technical Conference and Exhibition held in New Orleans, La., October 3-6, 1976.

7) Chang, D. M., Haldorsen, H. H. and Kirwan, P. A.: "History Matching and Predicting GOR Behavior with Analytical Functions," SPE 12763, 1984 SPE California Regional Meeting.

8) Killough, J. E., Pavlas, E. J. Jr., Martin, C. and Doughty, R. K.: "The Prudhoe Bay Field: Simulation of a Complex Reservoir," SPE Paper No. 10023, presented at the International Exhibition and Technical Symposium of the SPE in Beijing, China, March 18-26, 1982.

9) Addington, D. V.: "An Approach to Gas Coning Correlations for a Large Grid Cell Reservoir Simulation," J. Pet. Tech., November, 1981, pp. 2267-2274.

10) Wadman, D. H., Lamprecht, D. E., and Mrosovsky, I.: "Joint Geologic/Engineering Analysis of the Sadlerochit Reservoir, Prudhoe Bay Field," Journal of Petroleum Technology, July 1979, pp. 933-940.

11) Nadir, F. T.: "Thistle Field Development," EUR Paper NO. 165, presented in London, England, 1980.

12) Zapata, V. J. and Lake, L. W.: "A Theoretical Analysis of Viscous Crossflow," SPE Paper No. 10111 presented at the 56th Annual Technical Conference and Exhibition of the SPE in San Antonio, TX, October 5-7, 1981.

13) O'Brien, D. G., Brown, M. E. and Lederer, M. C.: "Use of Reservoir Performance Analysis in the Simulation of Prudhoe Bay," SPE Paper No. 13217, presented at the 59th Annual Technical Conference and Exhibition in Houston, TX, September 16-19, 1984.

14) Katz, M. L. and Stalkup, F. I.: "Oil Recovery by Miscible Displacement," 11th World Petroleum Congress Paper, London 1983 (RTD2).

15) Oglesby, K. D., Blevins, T. R., and Johnson, W. M.: "Status of the Ten-Pattern Steamflood, Kern River Field, California," SPE Paper No. 8833 Presented at the SPE 1980 Symposium On Enhanced Oil Recovery, Tulsa, OK, April 1980.

16) Richardson, J. G., et al.: "The Effect of Small, Discontinuous Shales on Oil Recovery," J. Pet. Tech. (November 1978), pp. 1531-37.

17) Ypma, J. G. J: "Analytical and Numerical Modeling of Immiscible Gravity-Stable Gas Injection Into Stratified Reservoir," SPE Paper No. 12158, presented at the 58th Annual Technical Conference and Exhibition in San Francisco, CA, October 5-8, 1983.

18) Haldorsen, H. H., Rego, C. A., Chang, D. M., Mayson, H. J., and Creveling, D. M., Jr.: "An Evaluation of the Prudhoe Bay Gravity Drainage Mechanism by Complementary Techniques," SPE 13651, 1985 SPE California Regional Meeting.

19) Leonard, J.: "Big Waterflood to Boost Prudhoe Recovery," Oil & Gas Journal, April 18, 1982.

20) Haldorsen, H. H. and Lake, L. W.: "A New Approach to Shale Management in Field-Scale Models," SPEJ Agusut 1984, pp. 447-452.

21) Haldorsen, H. H.: "Reservoir Charactrization Procedures for Numerical Simulation, Ph.D dissertation, U of Texas, (1983).

22) Dupuy, M., and LeFebvre Du Prey, E.: L'ANISOTROPIE D'ECOULEMENT EN MILIEU POREUX PRESENTANT DES INTERCALATIONS HORIZONTALES DISCONTINUES, Communication No. 34 Troisieme Collogque De L'Association De Recherche Sur les Techniques De Forage et De Production, June 10-14, 1968, PAU, France.

23) Begg, S. H. et al.: "Modelling the Effects of Shales on Reservoir Performance: Calculation of Effective Vertical Permeability," SPE Paper No. 13529 presented at the 8th SPE Symposium on Reservoir Simulation held in Dallas, TX, February 10-13, 1985.

24) Richardson, J. G., Tinker, C. N., and Sneider, R. M.: "RESERVOIR GEOLOGY OF SANDSTONES, SPE Short Course Presented at the 58th Annual Technical Conference and Exhibition held in San Francisco, California, October 1983.

25) Yokoyama, Y. and Lake, L. W.: "The Effects of Capillary Pressure on Immiscible Displacements in Stratified Porous Media," SPE Paper No. 10109 Presented at the 56th Annual Technical Conference and Exhibition in San Antonio, Texas, October 5-7, 1981.

26) Delhomme, K. and Giannesini, F.: "Reservoir Description Techniques Improve Simulation Results in Hassi-Messaoud Field Algeria," SPE Paper No. 8435 Presented at the 54th Annual Technical Conference and Exhibition of the SPE in Las Vegas, Nevada, September 23-26, 1979.

27) Martin, J.H., and Cooper, J. A.: "An Integrated Approach to the Modelling of Permeability Barrier Distribution in a Sedimentologically Complex Reservoir," SPE Paper No. 13051 presented at the 59th Annual Technical Conference and Exhibition in Houston, TX, Septe. 16-19, 1984.

28) Weber, K.J.: "Influence of Common Sedimentary Structures on Fluid Flow in Reservoir Models," J. Pet. Tech. (March 1982), 665-672.

29) Zeito, G.A.: "Interbedding of Shale Breaks and Reservoir Heterogeneities," J. Pet. Tech, (Oct. 1965), 1223-1228.

30) Potter, P,, E., Maynard, J. B., and Pryor, W. A.: SEDIMENTOLOGY OF SHALE, Springer-Verlag, New York, 1980.

31) Jones, J.R. Jr., Scott, A.J., and Lake, L.W.: "Reservoir Characterization for Numerical Simulation of Mesaverde Meanderbelt Sandstone, Northwestern Colorado," SPE Paper No. 13052 Presented at the 59th Annual Technical Conference and Exhibition in Houston, TX, Sept. 16-19, 1984.

32) Knutson, C.F., Maxwell, E.L., and Millheim, K.: "Sandstone Continuity in the Mesaverde Formation, Rulison Field Area Colorado," SPE 4a (1975), Field Case Historics Oil and Gas Reservoirs, pp. 156-164.

33) Newendorp, P.D., Campbell, J.M. "A Method for Treating Dependencies Between Variables in Simulation-Risk Analysis," J. Pet. Tech., (Oct. 1976), 1145-1150.

34) Wu, T.H., Vyas, S.K., and Chang, N.Y.: "Probabilistic Analysis of Seepage," Journal of the Soil Mechanics and Foundation Mechanics Division, April 1973, pp. 323-340.

35) Prats, M.: "The Influence of Oriented Arrays of Thin Impermeable Shale Lenses or of Highly Conductive Natural Fractures on Apparent Permeability Anisotropy," J. Pet. Tech., (Oct. 1972), 1219-1221.

36) Brydges, W.T., Gulati, S.T., and Baum, G.: "Permeability of Glass Ribbon-Reinforced Composites," Journal of Material Science, Vol. 10, 1975, pp. 2044-2049.

37) Private Communication, S.H. Begg.

38) McCray, A.W.; PETROLEUM EVALUATIONS AND ECONOMIC DECISIONS, Prentice-Hall, 1975.

39) Bath, P.G.H., van der Burgh, and Ypma, J.G.J.: "Enhanced Oil Recovery in the North Sea," RTD 2(2) Preprint of the 11th World Petroleum Congress, London, 1983.

40) Desbarats, A.J.: Ph.D. Dissertation in Progress, Stanford University.

41) Journel, A.G.: "Non-Parametric Estimation of Spatial Distributions," Math., Geology, Vol. 15, No. 3, pp. 445-468.

42) Geehan, G.W.: "Geologic Prediction of Shale Continuity, Prudhoe Bay Field," This Volume.

HOW HETEROGENEITY AFFECTS OIL RECOVERY

K. J. Weber

Shell Internationale Petroleum
Maatschappij B.V.
The Hague, The Netherlands

I. INTRODUCTION

Reservoir heterogeneity is among the major reasons why enhanced oil recovery is so difficult. Projects undertaken without detailed reservoir evaluation often end in failures related to unexpected baffles to flow, permeability heterogeneity, or wrong appreciations of the residual oil distribution. On the other hand in cases where much effort was put in reservoir analysis the results have been very beneficial for the understanding of the influence of heterogeneities on fluid flow. The closely spaced drilling, extensive coring, special logging, tracer tests, production, and pressure monitoring provide detailed information far beyond that available during the early development phase of a field.

Another reason for an improved understanding of the influence of heterogeneities is the possibility of simulating flow in very detailed reservoir prototype models with the aid of the new generation of large capacity computers.

Knowledge gathered over the years through outcrop and reservoir studies can now be used to better advantage because various types of heterogeneity can be quantified. Also the general understanding of the association of certain heterogeneity types with depositional environments and tectonic features can be very useful for designing suitable data gathering schemes at an early stage.

In principle reservoir heterogeneities should be quantified to a level sufficient to design a reservoir model from which the optimal production process, drainage, and completion pattern can be derived. Some heterogeneities can be adequately quantified by direct measurements on cores or in boreholes. However, evaluation of the overall effects of heterogeneities can best be achieved with the aid of models. Measured data and statistical data based on analog outcrop or reservoir observations can be combined into prototype models. Much research is carried out on this subject because there is still uncertainty about many aspects of the influence of heterogeneities.

This paper is aimed at clarifying some major aspects of the quantification of reservoir heterogeneities. Although many case histories in which some type of heterogeneity plays a role have been published, it is very difficult to derive general conclusions and principles. This is partly because of vague and incomplete descriptions but also because certain heterogeneities can have quite different effects depending on mobility ratio, PVT properties, aquifer strength, oil column, dip, depth, well spacing, and completion. The number of geological and non-geological parameters is so large that a comprehensive treatise of the subject is impractical. On the positive side there is much similarity between reservoirs and outcrops deposited in similar environments. Furthermore, there are often a number of very similar fields in a given area. Thus, certain heterogeneity types are common to many reservoirs.

For each reservoir the heterogeneities have to be described and analyzed in a way reflecting their respective impact on fluid flow. This calls for cooperation of geologists, petrophysicists, and reservoir engineers. An overview of the principles of classification, description, and analysis of heterogeneities is given to set the stage for reservoir modeling. This subject is discussed with emphasis on methods to account for the effect of heterogeneities and the influence of data density and reliability. It is demonstrated that the available data will usually not allow very detailed modeling in heterogeneous reservoirs with the exception of enhanced oil recovery pilot areas. Simulation experiments based on prototype models can indicate the relative merits of a given development scheme for a possible range of heterogeneity properties.

The presence and likely distribution of several heterogeneity types is predictable on the basis of their close relationship with depositional environments, diagenesis patterns, and tectonic style. An example is given of the use of such relationships. For each oil field area a data bank should be set up with information on reservoir heterogeneities and related production behavior observations. The wealth of data gathered in connection with enhanced oil recovery projects can form a particularly valuable source of information.

An important subject is the problem of averaging heterogeneity data for modeling purposes which can easily lead to under estimation of their effect. It is shown that averaging should be done via modeling of the appropriate geological configuration.

II. CLASSIFICATION, IMPORTANCE, AND SIGNIFICANCE OF RESERVOIR HETEROGENEITIES

A. Classification of Heterogeneities

A classification system can be based on size, genetic origin, and influence on fluid flow. An hierarchical sequence of heterogeneity types along the lines proposed by Pettijohn, et al. (fig. 1) appears to be the most practical approach (Pettijohn, Porter, and Siever, 1973). In this way we also get a system that follows the order in which heterogeneities can be recognized and quantified during field appraisal and development. Thus, size plays a major role and large tectonic features come before sedimentological characteristics.

Boundaries of permeable bodies within an unfaulted reservoir usually coincide with changes in reservoir facies, i.e. with the boundaries of genetic units. Within certain types of genetic units we often find characteristic patterns of permeability zonation. Furthermore, we may have baffles within genetic units with characteristic shapes and distributions related to environments of deposition or diagenetical processes. Within a given permeability zone inside a genetic unit we will encounter sedimentary structures of a specific range of types. Stepping down in size we arrive at the microscopic heterogeneity of the pore types and pore interconnections.

A further type of heterogeneity, always difficult to quantify, is fracturing both open and tight. Variation in oil viscosity and tar mats can also be regarded as reservoir heterogeneities but these will not be considered in this paper. Thus, we arrive at the basic classification system shown in fig. 2. This simple system is generally adequate to discuss overall reservoir characteristics. When it comes to detailed modeling a further split-up into sub-types is often required, e.g. between stylolites and tight streaks as baffles in a carbonate genetic unit.

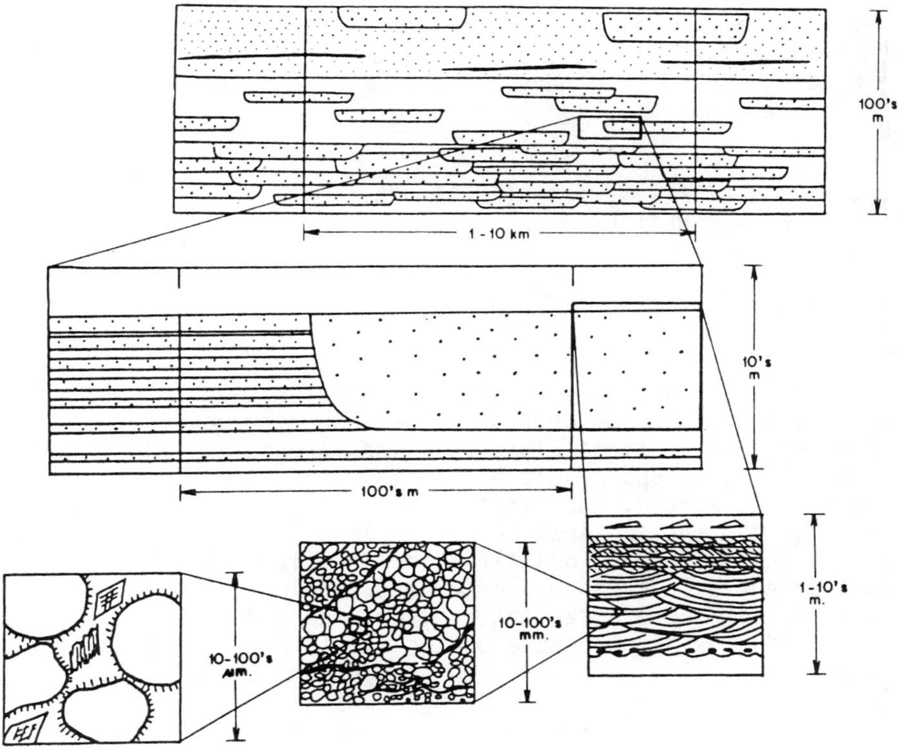

Figure 1. Levels of Reservoir Heterogeneity (Example in Fluviatile Reservoir)

B. Need to Account for Heterogeneities in Field Development

A review of recent literature showed that general and macroscopic geological reservoir descriptions are usually fairly clear and comprehensive. However, when it comes to detail there is much vagueness and undefined qualitative statements. Anybody who has attempted to make a table of geological characteristics of more or less comparable reservoirs on the basis of literature will agree that this is an almost hopeless undertaking. It is not surprising that many reservoir engineers turn to simplified models. Some even try to prove on the basis of very simplified prototypes that much detail can be neglected.

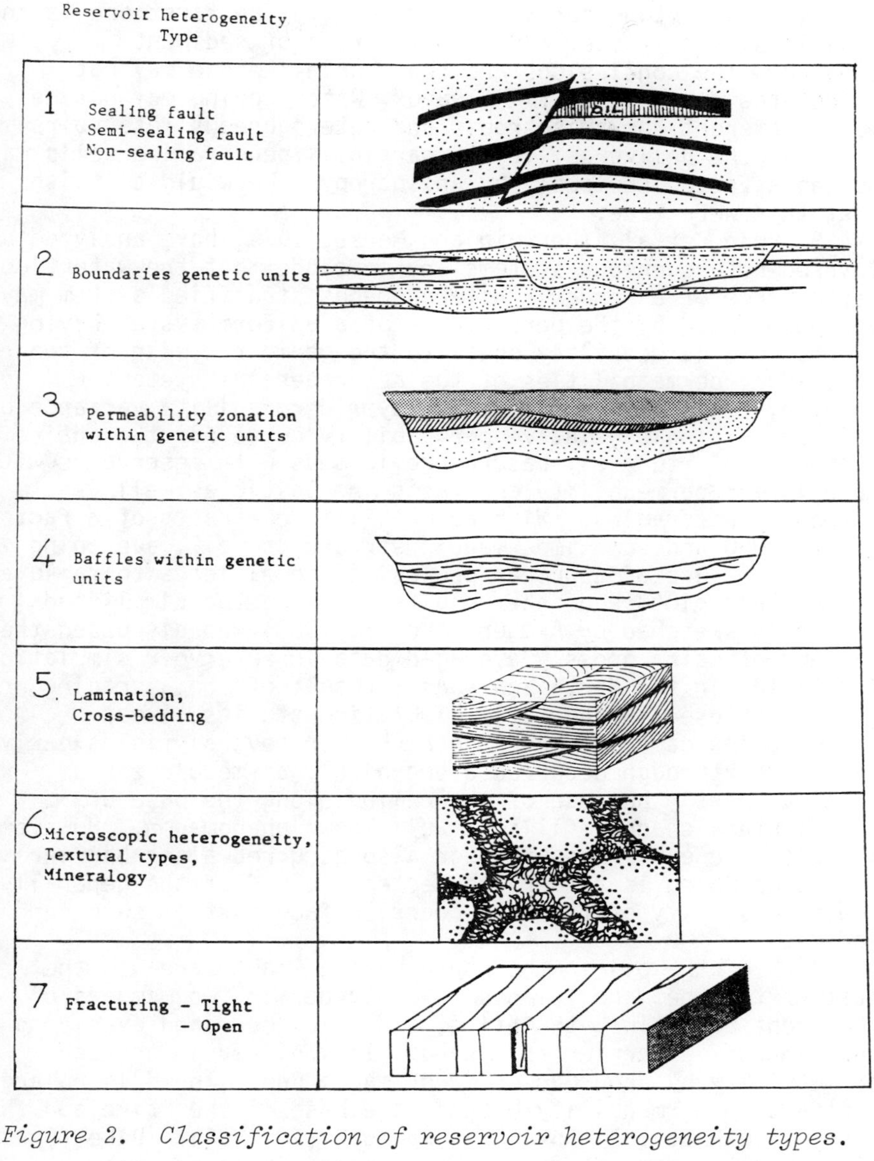

Figure 2. Classification of reservoir heterogeneity types.

Two cases stand out in this respect. The first one is an article by Rose (Rose, 1983) on the role of sediment bedding. The conclusions are that shale laminae may not affect the drainage of tight sands, water coning may not be much different for anisotropic and heterogeneous reservoirs, and that the deliverability of partially penetrating wells may not be affected seriously by anisotropy. We would all wish that this were true.

Berruin, et al. (Berruin and Morse, 1979) have analyzed layered heterogeneous systems and conclude that the waterflood performance of a randomly heterogeneous stratified system may be represented by the performance of a uniform system having an absolute permeability equal to the geometric mean of the individual permeabilities of the heterogeneous system. However, they only considered a layer permeability variance of 0.85 and a single relative permeability curve for the entire system. Unfortunately nature provides us with reservoirs with much larger permeability contrasts in clastic as well as carbonate reservoirs. With permeability contrasts of a factor 100 to 1000 and some impermeable streaks as well, averaging without the use of a proper geological model leads to failure.

A clear picture of the results of using too simplified models is sketched by Archer (Archer, 1983) who discussed the dangers of using grossly averaged data in reservoir simulation for fields in the North Sea. As a result of the averaging procedures used in the early simulation studies, sweep efficiencies came out much too high. In several fields early water breakthrough occurred along highly permeable zones. This was partly because of channeling along the base of distributary channel-fills, a well-known phenomenon. However, in this case early breakthrough also occurred along barrier bar sands which is rather unexpected because of the generally low permeability of the basel barrier face part of such sand bodies.

Much of the oil in the North Viking Graben area to the east of the Shetland Islands is contained in two members of the Brent formation, the Etive, a barrier bar sand overlying the Rannoch, a barrier face sand. They differ in average permeability by about an order of magnitude. There is a large contrast in permeability between the base of the Etive and the tight streak at the top of the Rannoch (fig. 3A). Dake (Dake, 1982) modeled the effect of this configuration on water-flooding (fig. 3B). The tight streak is not a seal as shown by RFT measurements. Most of the injected water enters into the Etive and will travel at high speed along the base of this member (up to 24 m/day) causing early water breakthrough. The transmissibility from the Etive to the Rannoch could be estimated with the aid of vertical pulse tests calibrated with

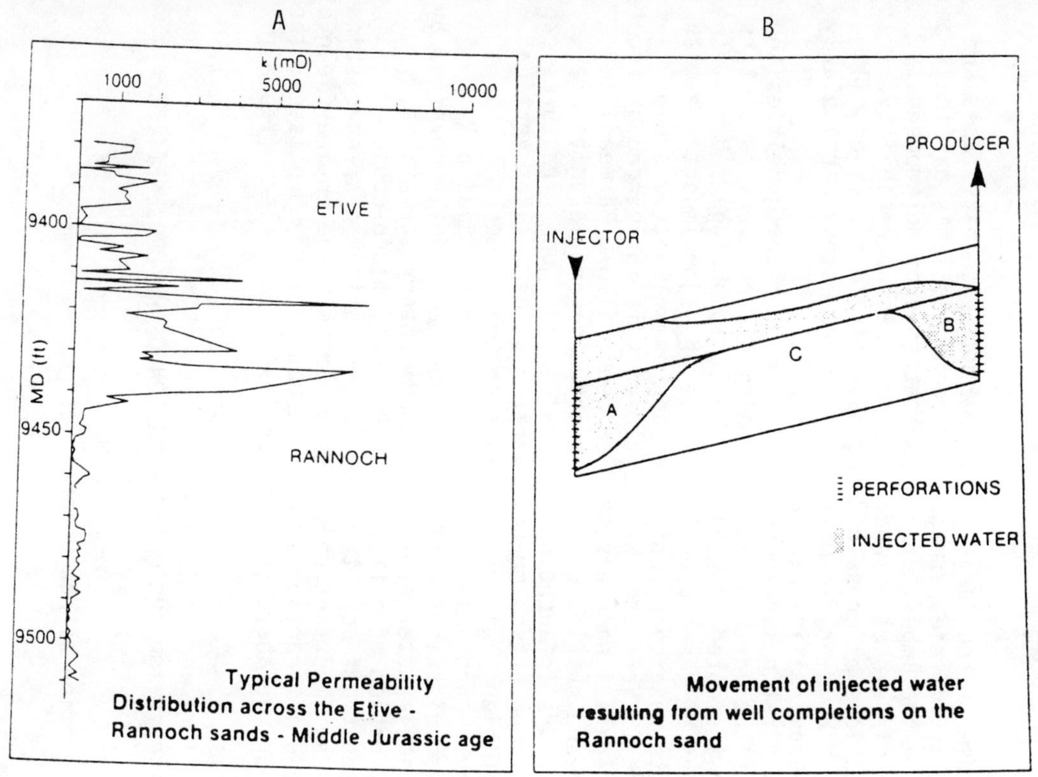

Figure 3. Permeability distribution and movement of injected water in the Etive and Rannoch Sands of the Brent Formation, North Sea.

a 3-D simulation model. Without a realistic value for this parameter it is not possible to design an optimal development plan.

C. Significance of Reservoir Heterogeneity for Oil Recovery

Because of the large variety of heterogeneity types and the severity of the permeability contrasts it is difficult to generalize the impact of a given category of heterogeneity. There are indeed cases where scattered low permeability intercalations or cross-bedding have little effect on oil recovery. However, in most recovery processes, particularly EOR, heterogeneities play a vital role.

As a first guide to the influence of heterogeneities the effect of the various types on major oil recovery parameters has been indicated in table I. Fault morphology and boundaries of genetic units will be of major importance with respect to reservoir continuity and on volumetric sweep efficiency. Vertical sweep efficiency is influenced by faults if bed parallel barriers to flow occur also. Permeability zonation and baffles within and between genetic units are major factors influencing vertical sweep. Open fractures can have a strong effect on both horizontal as well as vertical sweep efficiency with an emphasis on the former.

Recovery problems such as nonuniform residual oil saturation distribution in swept zones are mostly governed by smaller scale heterogeneities. Sedimentary structures, interlamination of different rock types, microscopic heterogeneity of pore types and mineralogical characteristics all play an important role. If we inspect the entire field of rock/fluid interaction in hydrocarbon recovery processes, the effects of the above rock properties becomes clear (table II).

III. DESCRIPTION AND QUANTIFICATION OF HETEROGENEITIES

A. Description of Heterogeneities

The description of geological features is a problem of size. Faults and thick reservoirs may be clearly visible on seismic sections. On the other hand, many important heterogeneities have an extent less than the average well spacing. If we take the schematic example shown in fig. 4, we

Table I. Significance of reservoir heterogeneity type for oil recovery.

Reservoir heterogeneity Type	Reservoir continuity	Sweep efficiency		ROS in swept zones	Rock/fluid interaction
		Horizontal	Vertical		
1. Sealing fault	X	X			
Semi-sealing fault	(X)	X	X		
Non-sealing fault	(X)	X	X		
2. Boundaries genetic units	X	X	X		
3. Permeability zonation within genetic units		(X)	X	(X)	
4. Baffles within genetic units		(X)	X	(X)	
5. Lamination, Cross-bedding		(X)	(X)	X	
6. Microscopic heterogeneity,				X	(X)
Textural types,				X	X
Mineralogy					X
7. Fracturing - Tight		(X)		X	
- Open		X	X	X	

X strong effect (X) moderate effect
ROS = remaining oil saturation

Table II. Rock Fluid Interaction

OPERATION	PHENOMENON	EFFECT	REMARKS
Drilling	Shale dehydration	Well-bore instability, drilling problems, solids build-up and subsequent formation damage	
	Filtrate interaction	Invasion and formation damage	
Acid stimulation	Solution of iron-rich minerals and formation of iron hydroxides	Formation damage and loss of permeability	Can be avoided by pH control and/or use of chelating agents
	HF solution of quartz, feld-spars and clays to form silicon and aluminium fluoride complexes and silicon hydroxide	Loss of permeability and productivity	Function of stimulation design.
	Ca^{++} minerals plus HF to form insoluble precipitates	Loss of permeability	Can be avoided by correct pre- and after-flush.
Hydraulic fraccing (tight gas reservoirs)	Interaction of frac fluid with formation	Impairment of matrix permeability and reduction in fracture conductivity. Water blockage	Dependent upon clay type.
Water injection	Clay swelling and/or fines movement due to incompatible fluids Precipitation of iron and sulphur compounds	Loss of injectivity	
CO_2 flooding	Carbonate solution	Increase in permeability Formation damage. Sand failure.	Causes of formation damage are not well understood. CO_2/crude interaction could be a factor.

Table II. Rock Fluid Interaction (continued)

OPERATION	PHENOMENON	EFFECT	REMARKS
Caustic flooding	Changes in wettability Ion exchange Dissolution and/or precipitation of silicate minerals, particularly clays	Increase in recovery. Caustic consumption, $Ca(OH)_2$ $Mg(OH)_2$ precipitation, changes in permeability/porosity, and caustic consumption	**Gypsum, anhydrite, siderite, dolomite promote caustic consumption.** **Na^+ clays preferable.**
Surfactant flooding	Cation/divalent ion exchange	Phase behaviour changes. Formation damage (clays). Surfactant precipitation and loss. Increased oil solubility	**Ca^{++} minerals ($CaSO_4$) and Ca^{++} clay minerals are particularly.**
Polymer flooding	Cation/divalent ion exchange. Adsorption	Crosslinking and increase in viscosity. Degradation of polymer. Formation damage.	
	Solution of iron minerals in the presence of O_2	Polymer degradation/precipitation.	
Steam injection	Clay swelling, fines movement	Formation impairment	
	Clay mineral transformation	Loss of porosity and permeability, and production of toxic solutions.	
	Silica/silicate dissolution and precipitation	Gravel pack and formation dissolution, reduction of formation porosity and permeability	**Particularly when pH > 11.**
In-situ combustion	Mineralogical transformations	? Changes in porosity and permeability	**Mineralogical transformations can be and are used as temperature indicators.**
	Coke deposition controlled by clay content	Increasing clay content increases combustion temperature	

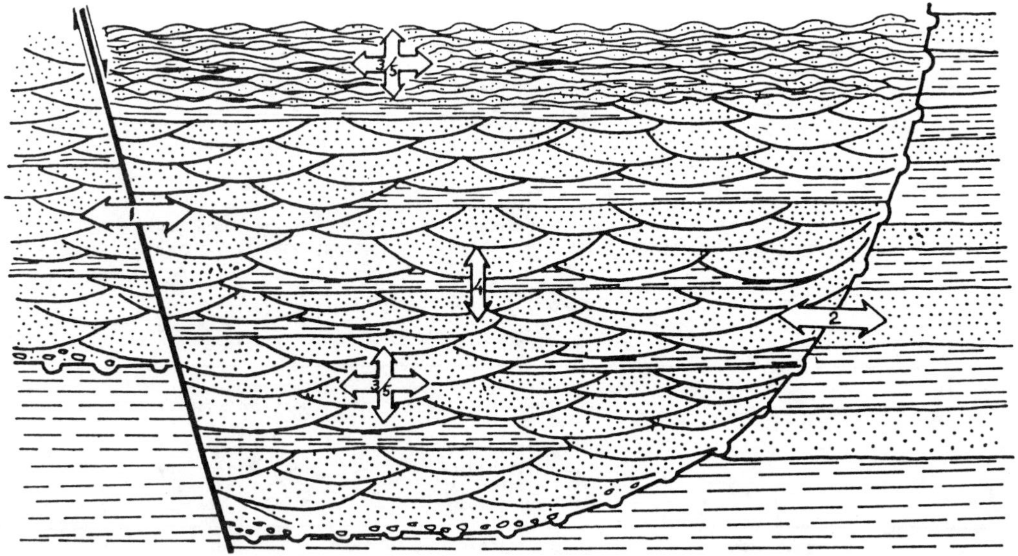

1. Faultplane which may be a poorly permeable baffle to flow.
2. Boundary between channel-fill and adjacent sand bed may be clay covered and of low permeability.
3. Lower part of channel-fill is usually of much higher permeability than the upper part.
4. Clay and silt intercalations often determine the overall vertical permeability of a sand body.
5. Permeability anisotropy on a meter scale mainly depends on the cross-bedding type and the permeability contrast of the composing laminae.

Figure 4. Schematic example of faulted channel-fill reservoir with multiple heterogeneity types.

observe a fault, a boundary between a channel-fill and adjacent beds, intervals with respectively medium-size and small-size cross-bedding and a number of shale intercalations. The fault may be seen on seismic or may intersect a well. The exact position of the channel-fill can only be determined when closely spaced wells have been drilled. The cross-bedding type identification still requires cores at present although novel developments of dipmeter logging tools show promise in this respect. The extent of the discontinuous shale intercalations will be difficult to determine from logs and correlations.

Reservoir heterogeneity exists at a number of scales (fig. 1). Large-scale heterogeneities, of well spacing size, can often be analyzed on the basis of detailed well log correlations and sedimentological models derived from core descriptions. Occasionally detailed seismic analysis may help. Medium-scale heterogeneities of a size range from 10 to 100 meters are much more difficult to evaluate realistically. Core data and well logs can reveal certain patterns but the reservoir models will have to be based on outcrop-analogue sedimentary prototypes. In some cases of enhanced oil recovery projects, well spacings of 100 meters or less are used and very detailed correlations are feasible. Small-scale heterogeneities in the meter-to-micrometer range are only clearly discernable in cores. Analysis of side wall samples, micrologs and the new dipmeter tools recently introduced can give useful indications.

In general, it can be said that except for the faulting and fracturing, the key to the analysis of heterogeneities is the correct identification of environments of deposition and diagenetic history. Cores are thus indispensable. Combination of modern logs can give much information in non-cored wells (fig. 5) but only after proper calibration via core studies. The use of log shapes for facies recognition is well-known in clastic reservoirs (fig. 6). Sneider, et al. give a good explanation of this method (Sneider, Tinker, and Meckel, 1978). Much use can be made of well-studied modern analogues as was demonstrated by Pryor, et al. for deltaic environment sediment types (Pryor and Fulton, 1978). Much research is being carried out to derive sedimentological information from logs. An overview of such methods is given by Wolf, et al. (Wolf and Pelissier-Combescure, 1982). The basis for these techniques is multi-variate analysis of the combined logging data such as cluster analysis and discriminant analysis. The former method is often used to identify natural groups. However, such a black box approach

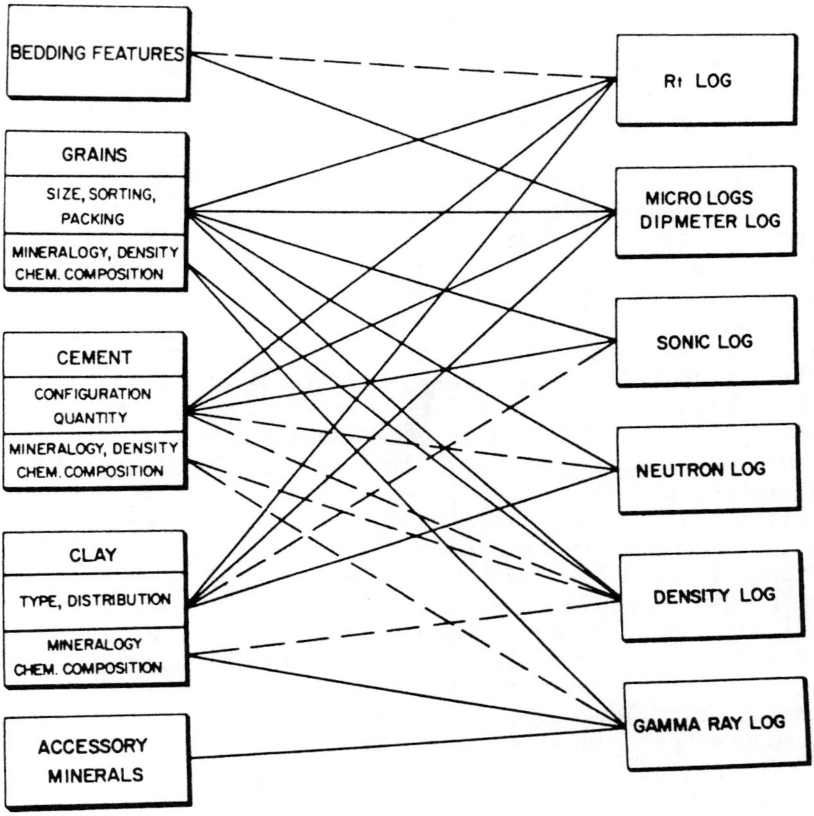

Figure 5. Log response to geological features in clastic sediments.

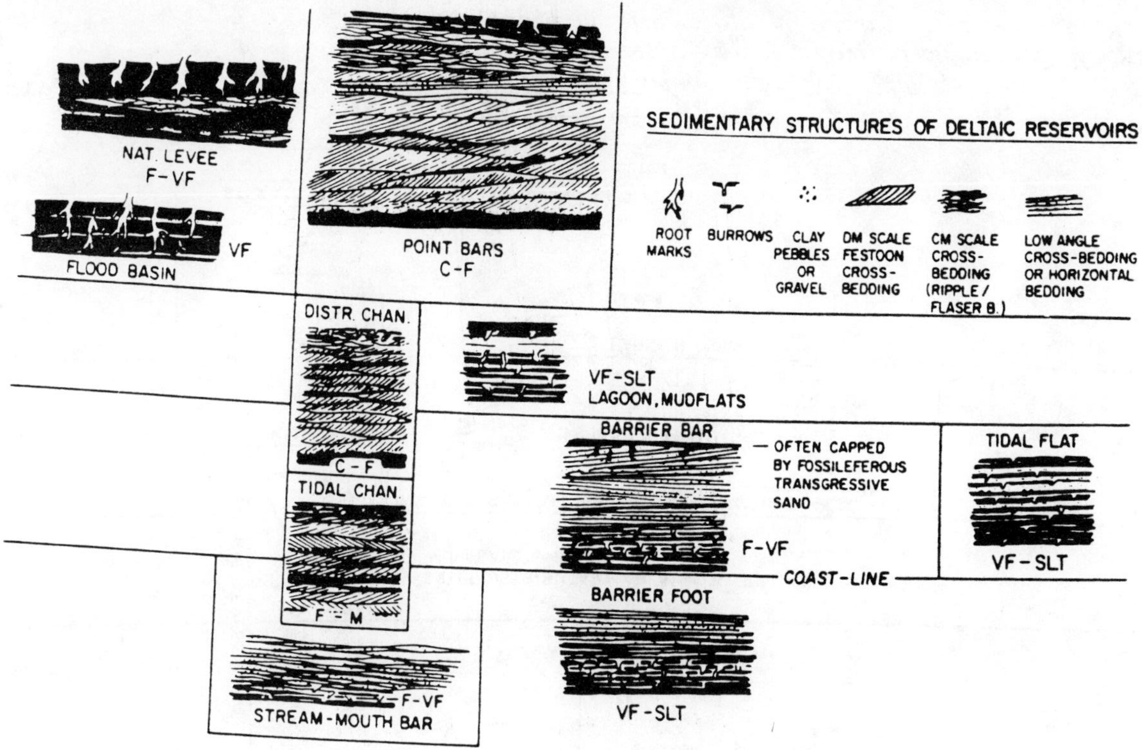

Figure 6. Sedimentary structures of wave dominated deltaic sediments and corresponding shapes of the natural gamma-ray log.

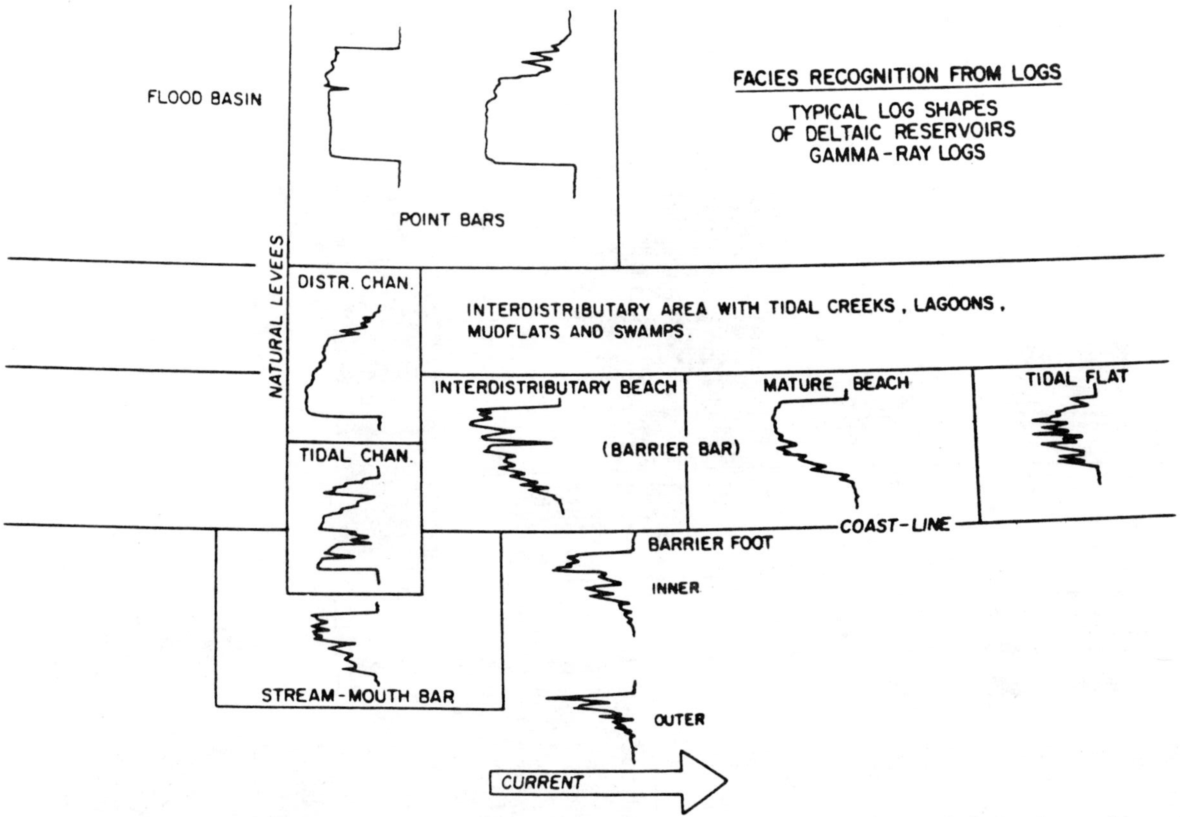

Figure 6. *Sedimentary structures of wave dominated deltaic sediments and corresponding shapes of the natural gamma-ray log (continued).*

is dangerous. Discriminant analysis of log response based on prior knowledge of facies characteristics derived from cores is more rewarding.

Besides the use of numerical log data one can also describe log shapes in some detail and integrate this additional parameter into the interpretation scheme. This approach was used successfully to distinguish between several rock types with overlapping porosity ranges but widely different permeabilities (Epting, 1980). By looking at the combination of porosity, lithology, and log character the rock types could be recognized.

When the reservoir is very heterogeneous and/or thin-bedded, many closely spaced cores are required. A good example of the effective use of intensive coring is described by Langston, et al. (Langston, Shirer, and Nelson, 1981) for a Smackover formation reservoir, a carbonate reservoir with 18 correlative zones with variable permeability and some intercalated tight streaks. No less than 119 wells were cored (some 8500 m of core). In addition, radioactive tracer tests, interference tests, and bottomhole pressure monitoring were carried out and special logs such as the TDT were run. With these data the heterogeneity pattern was analyzed. Discontinuous unswept zones were identified and infill wells and selective recompletions could be planned.

Microscopic heterogeneity is mostly related to the occurrence of a mixture of pore types, i.e. vugs in a finely crystalline matrix. With thin sections and scanning electron microscope observations good descriptions can be made. Todd, et al. (Todd and Tweedie, 1978) prepared a very complete scheme for a core characterization procedure (fig. 7).

The presence of open fractures may be apparent from mud loss and the analysis of production tests. Borehole televiewer logs are beginning to become quite useful. Again a thorough core study is required to unravel the fracture system and to estimate fracture porosity. The structural style will give much information on the type of fracture system which can be expected. Usually the fracture systems have a preferred orientation relative to the structure of the reservoir. Outcrop studies in the vicinity of fields have frequently been very helpful to understand the prevailing fracture systems (Kempthorne and Irish, 1981).

In summary, it is possible to identify and analyze almost every kind of heterogeneity provided one has suitable data and a proper data density. In table III an overview is given of the merits of a great variety of data for the identification and quantification of reservoir heterogeneities. A major conclusion is that by itself the data gathering for the analysis of heterogeneities has a large impact on reservoir

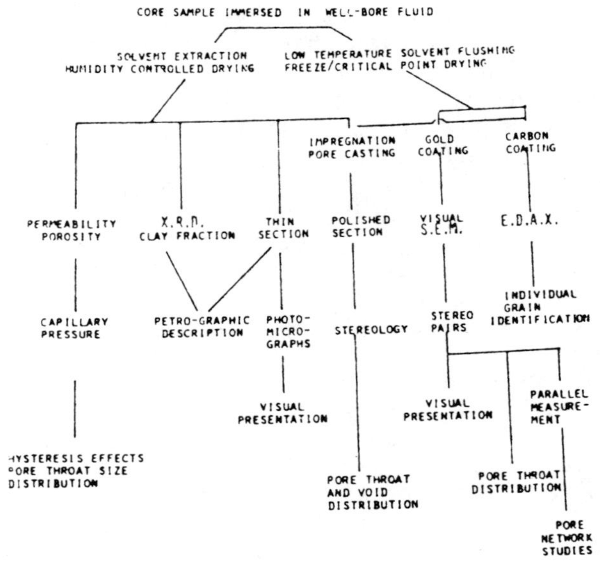

Figure 7. Core characterization procedure.

development planning. Data gathering schemes have to be carefully planned and the timing of the availability of certain data is crucial to the decision making process. Also the data gathering and analysis is a major cost item. There is always pressure to reduce expenditure on these activities and much management oriented explanation is usually required to ensure adequate support.

Of particular importance is the reservoir analysis phase prior to the start of a secondary or enhanced oil recovery pilot project, i.e. prior to the execution of a main data gathering programme. The art of making realistic reservoir models with the limited data available at this stage is important for three reasons. Firstly, we may discover unsuspected geological causes for oil recovery problems. Secondly, we may find that the distribution of the residual or target oil is different from that assumed in the planning of the intended recovery method. Thirdly, one may conclude that the pilot area is less representative of the main field area than previously thought. This implies that we have to retain much flexibility at this stage and not pin ourselves down too early to a rigid choice of pilot area or data gathering scheme.

A typical result of reservoir heterogeneity is a nonuniform distribution of residual oil. Quotations of

Table III. Value of data for identification and quantification of reservoir heterogeneities.

Reservoir Heterogeneity Type	Detailed seismic	Reservoir pressure		Production data/tests					Well logging			Rock samples		Outcrop/ other reservoir analogous
		Hor. distrib.	Vert. distrib.	Prod. tests	Pulse tests	Tracer tests	Prod. history	Prod. logs	Standard	Special	ROS	Cores	SWS cuttings	
1. Sealing fault	X	X	(X)	(X)	X	(X)	X	-	X	(X)	(X)	(X)	-	(X)
Semi-sealing fault	X	X	(X)	(X)	(X)	(X)	(X)	-	X	-	-	(X)	-	(X)
Non-sealing fault	X	(X)	(X)	-	-	-	-	-	X	-	-	(X)	-	(X)
2. Boundaries genetic units	(X)	X	X	(X)	(X)	(X)	(X)	(X)	X	(X)	(X)	X	(X)	X
3. Permeability zonation within genetic units	-	-	(X)	(X)	(X)	-	-	(X)	X	X	X	X	(X)	X
4. Baffles within genetic units	-	-	(X)	-	(X)	-	(X)	(X)	X	(X)	-	X	-	X

Table III. Value of data for identification and quantification of reservoir heterogeneities (continued).

Reservoir Heterogeneity Type	Detailed seismic	Reservoir pressure		Production data/tests					Well logging			Rock samples		Outcrop/ other reservoir analogous
		Hor. distrib.	Vert. distrib.	Prod. tests	Pulse tests	Tracer tests	Prod. history	Prod. logs	Stan-dard	Special	ROS	Cores	SWS cuttings	
5. Lamination Cross-bed-ding	–	–	–	–	–	–	–	–	–	(X)	–	X	–	X
6. Microscopic heteroge-neity, Textural types	–	–	–	–	–	–	–	–	–	(X)	(X)	X	(X)	(X)
7. Fracturing														
Tight	–	–	–	X	–	–	X	–	–	–	–	X	–	(X)
Open	(X)	X	(X)	X	X	(X)	X	X	(X)	–	–	X	(X)	(X)

X large value (X) occasionally of value

percentages of residual oil are quite meaningless unless clearly labeled and subdivided into their proper category, viz. 'unconnected,' 'by-passed' (unswept) and 'oil-trapped in pores in a rock volume swept by water or gas' (fig 8.). It is precisely this subdivision that is generally insufficiently known when enhanced oil recovery projects are being planned. Because this distribution is mostly governed by reservoir heterogeneities the study of residual oil and heterogeneities should be carried out as one combined project. Methods to analyze the distribution and nature of residual oil include logging and coring of infill wells, tracer tests, through casing logging of existing wells and modeling on the basis of reservoir prototypes. In all cases, a detailed reservoir model is indispensable to correctly interpret the results.

B. Quantification of Heterogeneities

The quantification of heterogeneities is partly possible by direct measurements as summarized in table III, but for many configurations we can only evaluate the effect through modeling. Furthermore, there is often a strong interdependence of the effects of different heterogeneities. It would require a book to describe all these possibilities in detail. Some major aspects of quantifying various important heterogeneities for reservoir modeling purposes will be reviewed.

1. Sealing Capacity of Faults

Faults may be clearly visible on seismic or from log correlations but their sealing capacity is often unknown except if we observe different fluid contacts or different pressures across the fault. Three major sealing mechanisms occur, clay smearing into the fault plane, cataclasis of fault gouge material, and diagenetic healing of faults.

Clay smearing is a process that especially takes place during the movement of synsedimentary faults in unconsolidated sand/clay sequences. In deltaic sediments growth faults are particulary prone to be sealing because of clay smearing. Weber, et al. have described this phenomenon for the Niger Delta oil fields (Weber, Mandl, Pilaar, Lehner, and Precious, 1978) and in outcrops provided by West German open-pit lignite mines where similar phenomena can be observed (fig. 9). Smith demonstrated this process in the Louisiana Gulf Coast fields

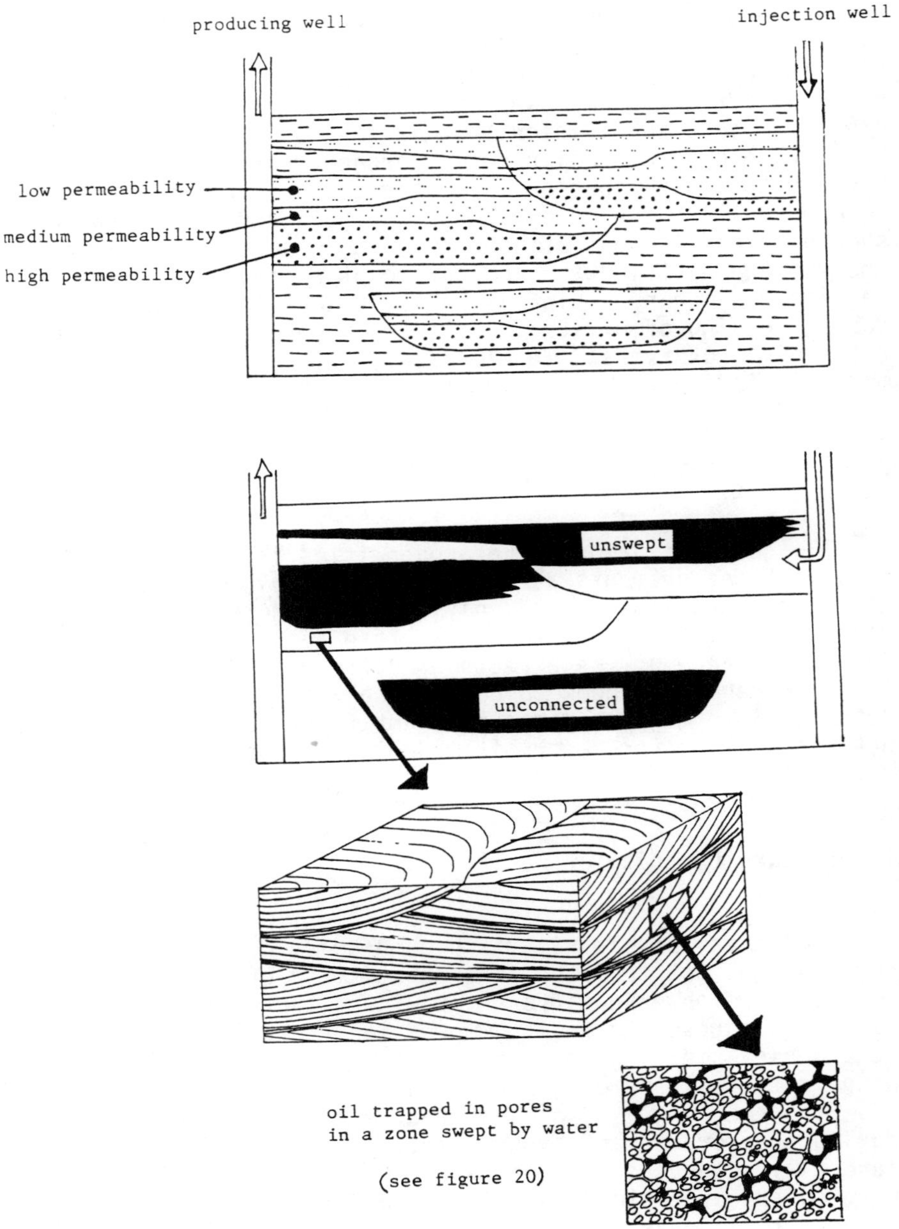

Figure 8. Terminology for the subdivision of different categories of residual oil.

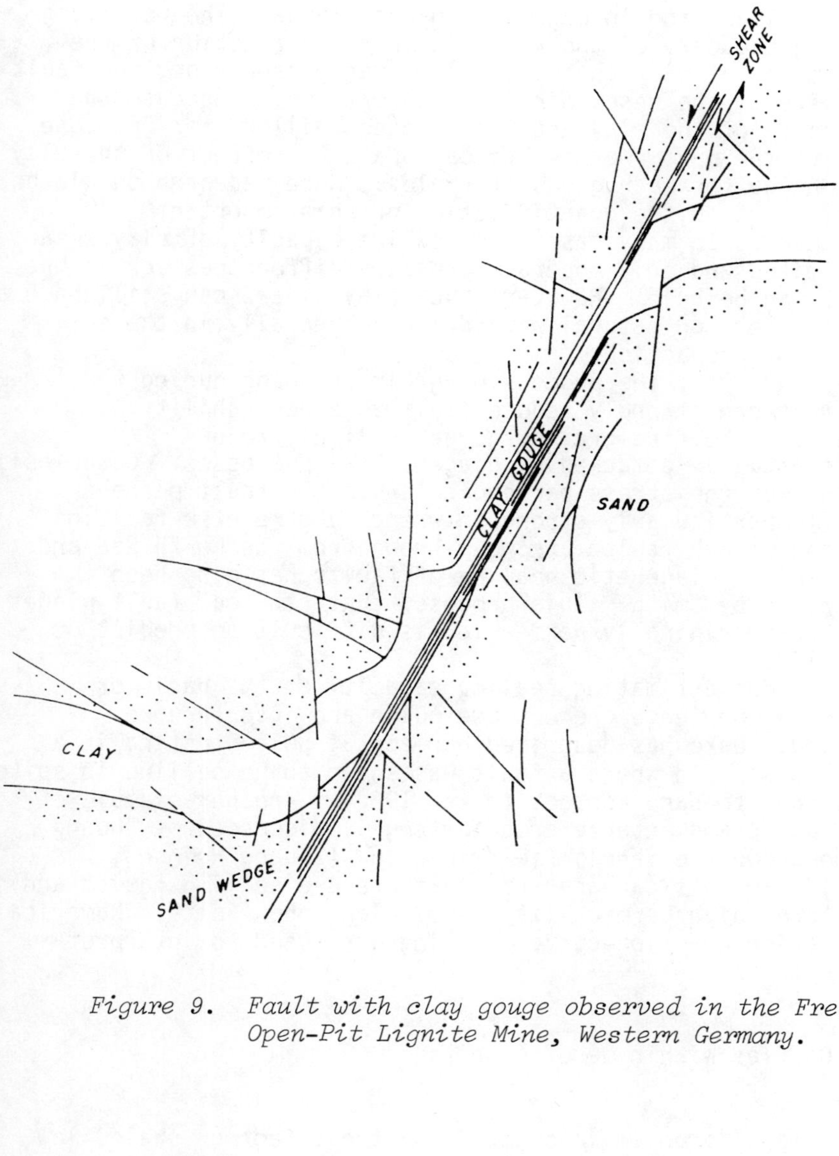

Figure 9. Fault with clay gouge observed in the Frechen Open-Pit Lignite Mine, Western Germany.

(Smith, 1980) and in outcrops in east Texas. The resulting sealing capacity of the fault is mainly a function of the percentage of clay in the section that passed along the fault opposite to the reservoir layer under study. Because the viscosity of the clay and the stresses will differ from one area to the next, we have to calibrate the effect of the clay percentage for each group of fields. More research should be carried out on the quantification of this important parameter. In many cases the sealing capacity of clay smears is limited and only moderate pressure differences across the fault can be held. However, such clay smears can still have a large effect on natural waterdrive, especially in the early stages of production.

Cataclasis, the process of grain crushing during fault movement can strongly reduce fault zone permeability. Moreover, the fine-grained gouge is likely to be preferentially attacked by diagenetical processes. Cataclasis depends on the stress perpendicular to the fault plane and thus is particularly strong in wrench and reverse faulting. Sealing wrench faults are well-known from the North Sea and Venezuela. Diagenetic healing of faults has also been described by Smith. This process, that can seal fault planes that were originally permeable, is difficult to predict or quantify.

Besides estimating sealing capacity on the basis of regional knowledge one can use pulse and interference testing. Dake has described horizontal pulse testing in a North Sea field where a fault was shown to be sealing in spite of a sand-to-sand contact (Dake, 1982). Another example of both pulse and interference testing is discussed by Zhuang Huinong for the Shenli field in China (Zhuang Huinong, 1984). In this carbonate field there are sealing faults and fractured high permeability zones along the faults. Numerical simulation and type-curve matching were used to interpret these tests.

2. Baffles Within Genetic Units

A major problem in quantifying the effect of shale, low permeable silt, and other baffles in clastic reservoirs is the fact that most thin baffles cannot be correlated from well to well. For marine shales continuity can be considerable but in fluviatile sands the average continuity is usually small (fig. 10). Efforts have been made to collect data on shale continuity as a function of environment of deposition (Weber, 1982). This survey resulted only in a very limited data set

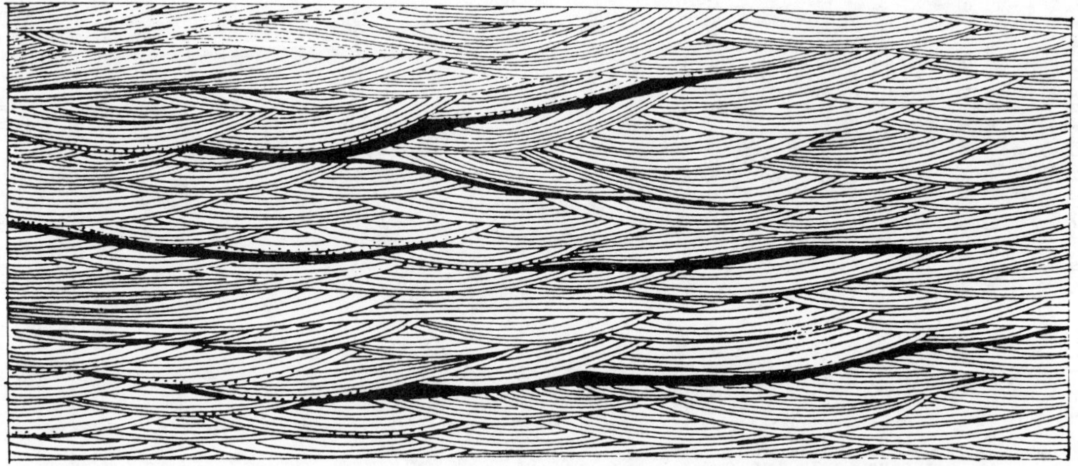

Figure 10. Discontinuous clay drapes in a Festoon cross-bedded distributary channel-fill.

from which nevertheless certain trends of continuity distribution could be derived (fig. 11).

Recently a method was proposed by Haldorsen and Lake to model the continuity and spatial disposition of shales with statistical techniques (Haldorsen and Lake, 1984). These techniques are based on the above shale continuity statistics. With the distribution histograms and the assumption of a random distribution in space models can be generated with a computer. These models can be conditioned to fit the well data. When a user-defined grid system is superimposed on the resulting description, effective permeabilities can be calculated for each gridblock.

A problem is still the quantification of the effect of baffles at an angle to the bedding such as in epsilon cross-bedding in point bars or the mud drapes in subtidal channel-fills (fig. 12). These last ones are particularly prevalent in such sand bodies (Yang and Nio, 1985). Both phenomena occur in important oil fields. Obviously their presence will result in large horizontal anisotropies. Research on permeability quantification in such sand bodies is required.

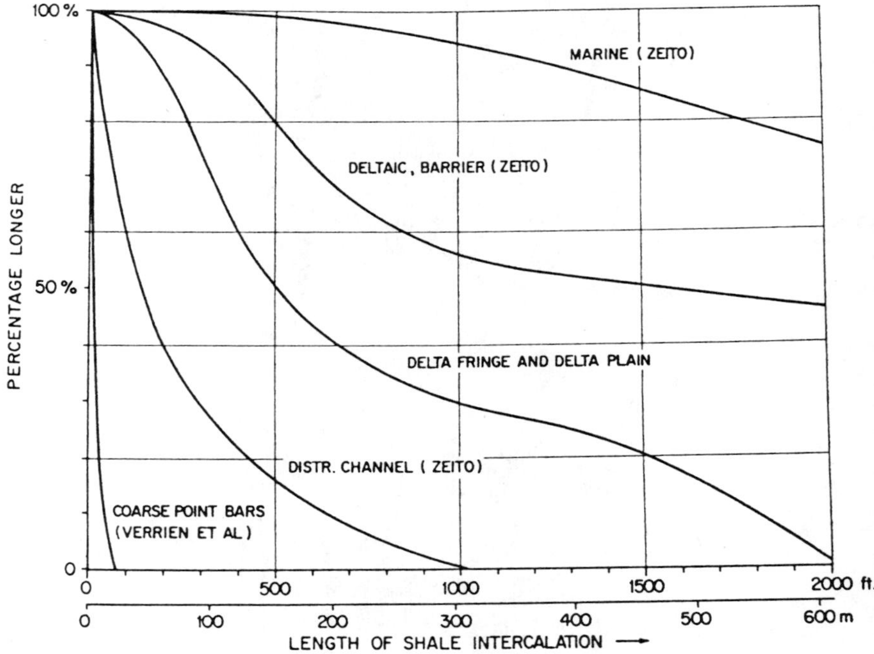

Figure 11. Continuity of shale (Silt) intercalations as a function of depositional environment.

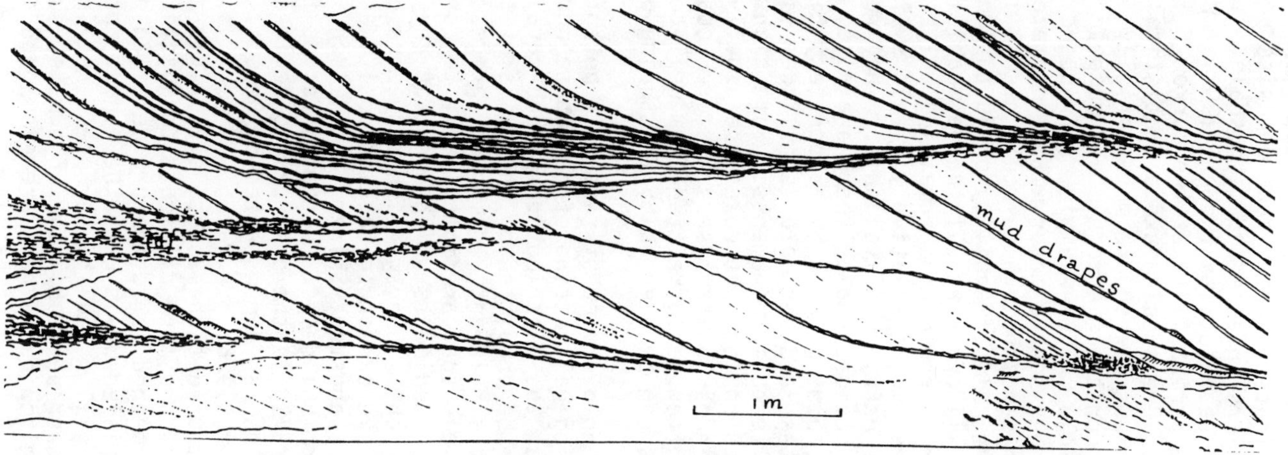

Figure 12. Large scale cross-bedding in a subtidal channel-fill with the characteristic double mud drapes along the foreset laminae.

Although fairly continuous, thin marine shale intercalations are often pierced by burrows. This is the reason that interlaminated barrier face deposits and shallow marine storm sand layers interbedded with shale often have an appreciable vertical permeability (Weber, 1982). With the mini-permeameter the permeability of the burrow-fills can be estimated (fig. 13). One should make a statistical estimate on the number of burrows per square meter of clay intercalation, then construct a model, and compute vertical permeability (fig. 14). As an example, a frequency of 25 to 50 piercing burrows per square meter with a diameter of 2 cm in a package of thin sand/shale beds would result in a ratio of horizontal to vertical permeability of 50 to 100. Ultra-violet photographs will often reveal the permeability of burrows by the fluorescence of the remaining oil saturation.

For carbonate reservoirs relatively little systematic data on the continuity of baffles is available. The problem is complicated because besides shales there are also, marly streaks, evaporites, beds plugged by evaporites, tight carbonates, stylolites, and bed parallel slicken-sided surfaces related to tectonic movements. Some tight zones can be recognized as such on logs but cores are usually required. Baffles in carbonate reservoirs are usually fairly continuous and can often be correlated from well-to-well. However, the baffles can be pierced by fracturing or leaching.

Lapré has given an overview of reservoir potential as a function of the geological setting of carbonate rocks (Lapré, 1980). He shows that the distribution of bed-parallel barriers to flow can be predicted on the basis of the overall characteristics of the relevant sedimentary process. Typically the intervals that are most likely to contain extensive baffles are sediments which occur at the interface of cycles of deposition. Thus, we find many more baffles in thin-bedded, layer-cake carbonate shoal deposits as described by Langston, et al. (Langston, Shirer, and Nelson, 1981) than in carbonate platform build-ups. Detailed facies mapping both vertically and laterally is the best way to delineate sedimentary baffels as was demonstrated by Bebout, et al. (Bebout and Pendexter, 1975).

Diagenetically caused tight zones, stylolitic zones, and slicken-sided baffles are more difficult to quantify. Generally, the best method is to measure detailed pressure profiles in newly drilled wells with the repeat formation tester (RFT). This will reveal higher pressures in vertically isolated zones of lower permeability. Flowmeter surveys and other production monitoring methods are also good indicators of the presence and position of baffles.

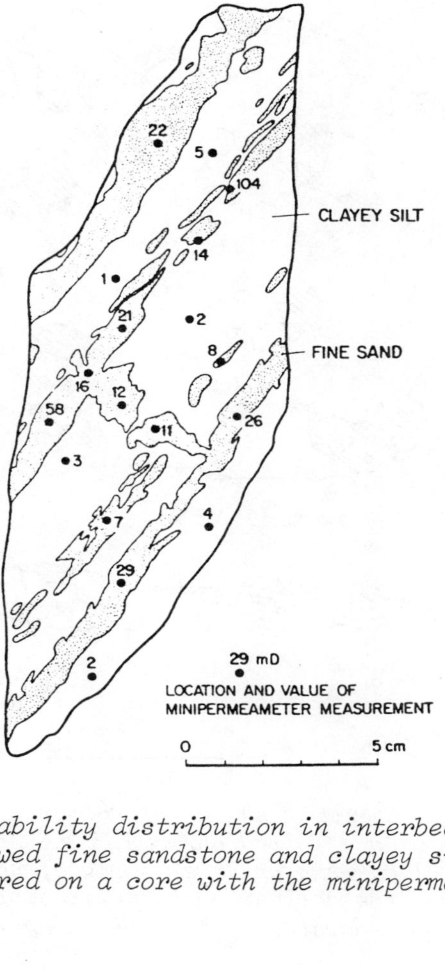

Figure 13. *Permeability distribution in interbedded burrowed fine sandstone and clayey silt measured on a core with the minipermeameter.*

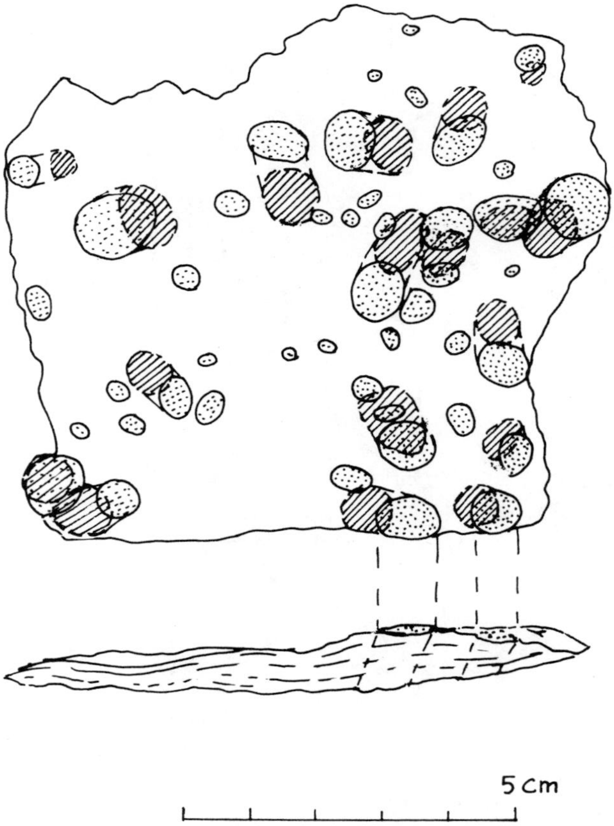

Figure 14. *Burrowed shale intercalation pierced by numerous sand filled burrows. Dotted areas indicated burrows on upper surface, hatched areas show the corresponding intersections with the underside of the shale layer.*

3. Lamination and Cross-Bedding

Thinly laminated deposits are very common. If strongly contrasting lithologies are intercalated it is often difficult to even recognize productive intervals. A good example of this configuration was presented by Bos (Bos, 1982). Sandstone layers with moderate amounts of kaolinite and permeabilities in the darcy range alternate with sandstones in which the pores are virtually plugged by kaolinite. This mixture results in average log-derived water saturations as high as 50-65 percent but water free production of up to 6,000 b/d. The kaolinite forms a micro-porosity system which is filled by immobile water while oil can flow through the clean macropore system of the adjacent permeable streak. Obviously such laminated rocks require coring for proper assessment. Recently, improved dipmeter logging tools have shown promise for the analysis of thin bedded reservoirs as demonstrated by Sallee, et al (Sallee and Wood, 1984).

Cross-bedding is a very important heterogeneity type in sandstones. There is controversy about the actual influence of cross-bedding on fluid flow. Much depends on the permeability contrasts of the cross-bed laminae and the associated bottom-set layers (fig. 15). The subject was

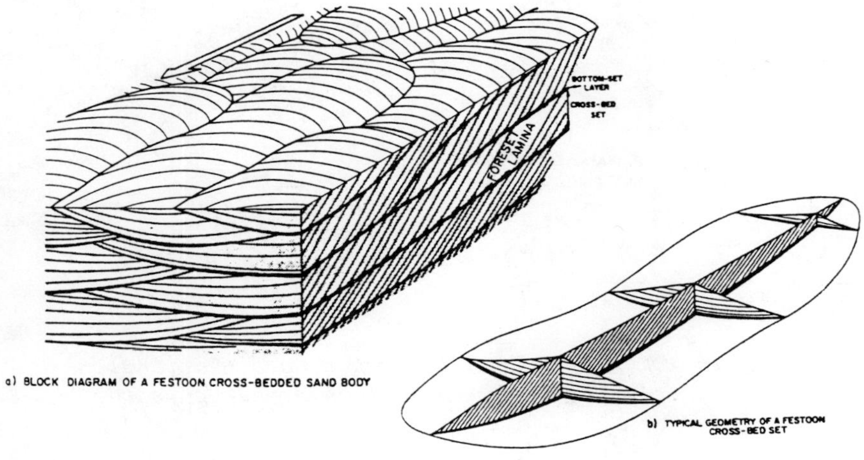

Figure 15. Festoon cross-bedding.

covered by Weber (Weber, 1982) who included the methods to measure and calculate realistic permeability anisotropy figures (fig. 16). It is also shown that anisotropy is strongly enhanced by the presence of discontinuous shale intercalations which force the fluid flow along a very tortuous path. Permeabilities of laminae and bottom-sets can be measured with a mini-permeameter, a simple and quick way to establish the degree of heterogeneity of cross-bed sets (Eype and Weber, 1971). For the many types of cross-bedding somewhat different anisotropy patterns can be expected.

FORMULAE :

$$\frac{1}{K_\alpha} = \frac{\cos^2\alpha}{K_{\parallel L}} + \frac{\sin^2\alpha}{K_{\perp L}}$$

$$\frac{1}{K_x} = \frac{d}{L K_B} + \frac{1}{K_\alpha}$$

$$\frac{1}{K_y} = \frac{d}{W K_B} + \frac{1}{K_{\parallel L}}$$

$$\frac{K_x}{K_y} = A_H$$

$$K_R = \sqrt{K_x K_y} = \text{RADIAL INFLOW PERMEABILITY INTO A WELL.}$$

$$\frac{H+d}{K_V} = \frac{H}{K(90-\alpha)} + \frac{d}{K_B}$$

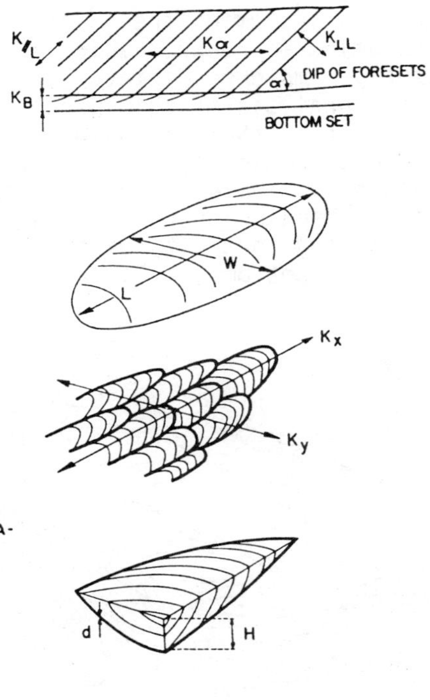

Figure 16. Formulae used in computing directional permeability in Festoon cross-bed sets.

In a recent distributary channel-fill in the Rhône delta permeability profiles were measured on cores with a mini-permeameter (fig. 17). The inset shows a detailed permeability profile of an interval of 60 cm comprising two cross-bed sets. In such recent deposits permeability contrasts are usually low (Weber, Eype, Leynse, and Moens, 1972). However, these contrasts can increase dramatically through diagenetical processes as shown by the measurements on a core from the Rotliegendes in the North Sea (fig. 18) where authigenic clays have preferentially plugged up the finer grained laminae. In this case, especially, the very low permeable bottom-set layers and other dune base sands are very effective in preventing upward flow of water. Because of the wedging out of the cross-bed sets the bottom-sets layers come together (fig. 19), and the lateral flow is also restricted (Weber, 1985).

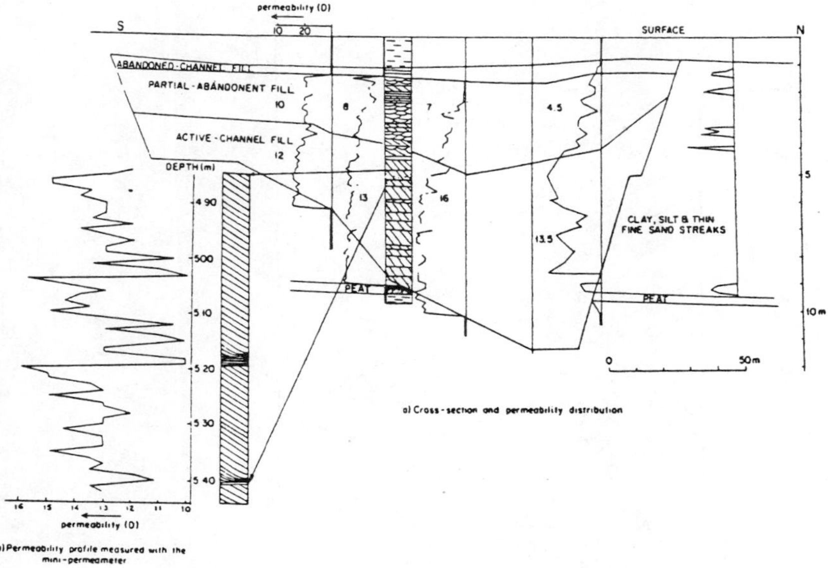

Figure 17. Distributary channel-fill in the Holocene Rhône Delta with the general permeability distribution measured on cores and the detailed permeability distribution within some Festoon cross-bed sets.

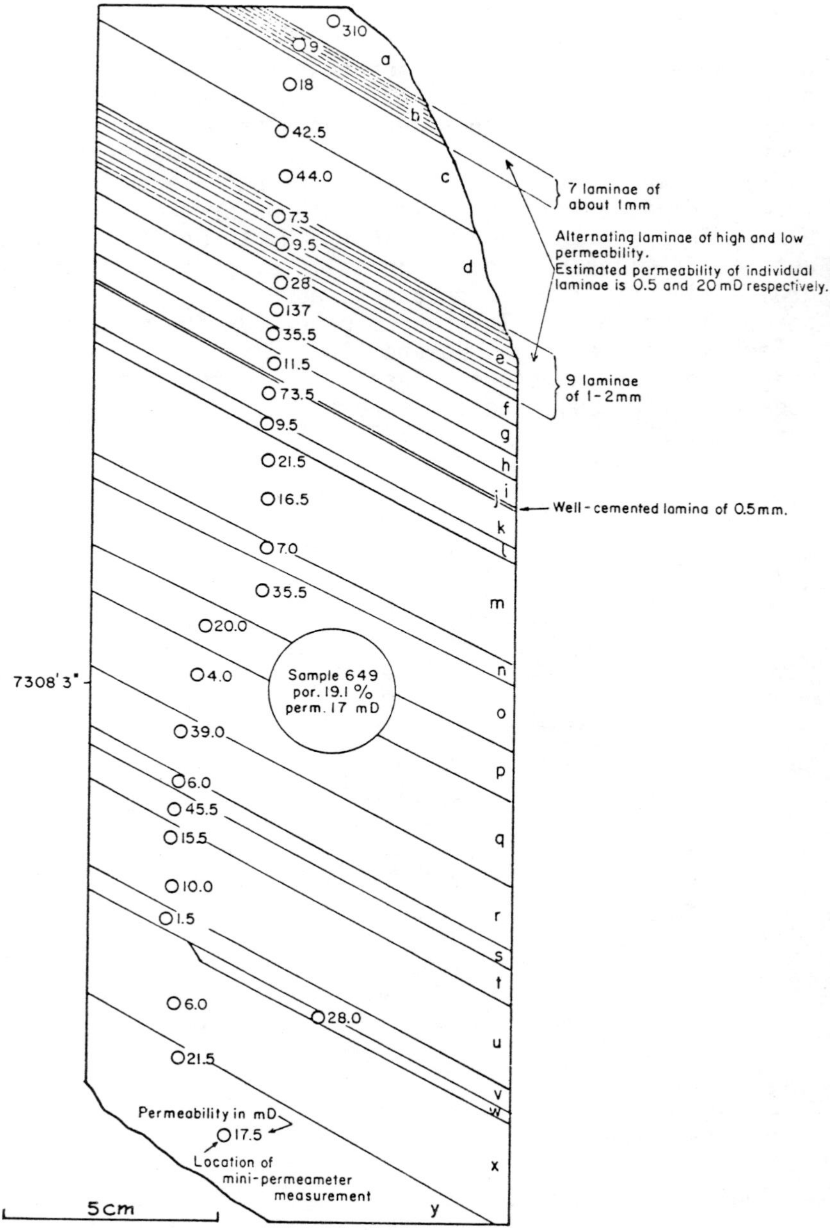

e 18. Permeability distribution in a cross-bedded aeolian interval of the Rotliegendes Formation in the North Sea, measured with a mini-permeameter.

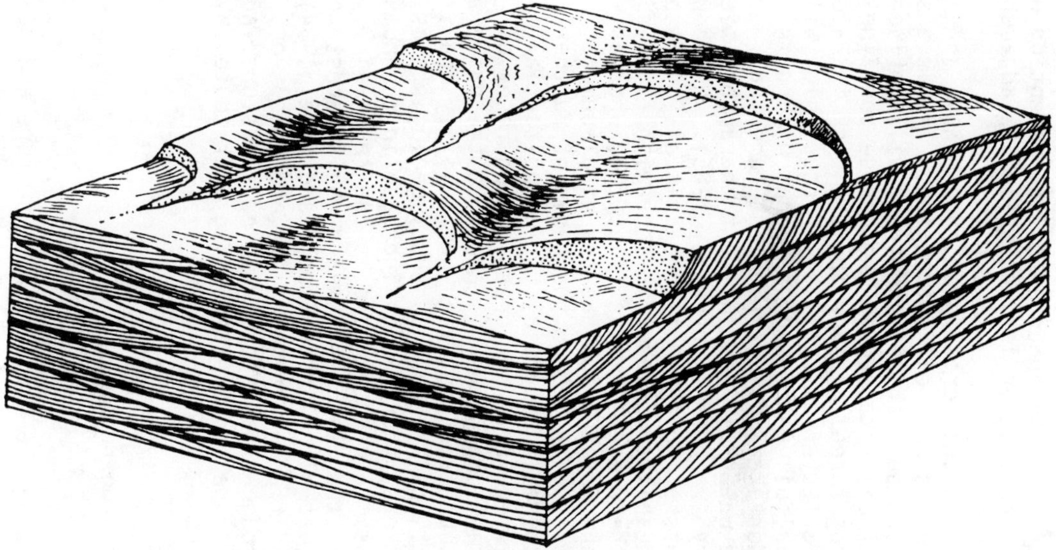

Figure 19. *Barchan shaped transverse dunes as interpreted for the Rotliegendes Formation in the Leman Field, U.K. North Sea. Average cross-bed thickness is about 4.5 m.*

Emmett, et al. describe the influence of the heterogeneity of the Wyoming Tensleep reservoir on infill drilling and secondary recovery (Emmett, Beaver, and McCaleb, 1971). In this case permeability parallel to the cross-bedding laminae is about four times higher than perpendicular to the laminae. Because of the high viscosity of the crude (42 cp) this anisotropy has a comparatively strong effect on the lateral flow. The effect is enhanced by the retention of asphaltic crude in the tighter laminae while the adjacent laminae can be flushed.

This example shows that cross-bedding can have an important effect on residual oil saturation. If viscous forces are strong the permeability contrast will influence the relative volumes of oil retained in the coarse and fine laminae. If viscous forces are low and the rock is water-wet one can expect an effect from capillary forces which will tend to keep the residual oil saturation somewhat higher in the coarse laminae. Kortekaas has carried out numerical simulations of the water/oil displacement characteristics in cross-bedded reservoir zones (Kortekaas, 1985). He shows that considerable amounts of movable oil are initially left behind in the higher permeability foreset laminae with fluid flow perpendicular to the preset laminae (fig. 20), whereas with flow parallel to the foreset laminae the displacement efficiency is good. He argues on the basis of further simulations that there is potential for a substantial improvement in oil recovery from cross-bedded reservoir zones if diluted surfactant or polymer is added to the drive water.

Although the subject of detailed permeability analysis requires much additional research both by geologists as well as reservoir engineers, there is an increase in the number of useful publications. A good example is the reservoir characterization of the Mesa Verde meanderbelt sandstone in NW Colorado by Jones, et al. (Jones, Scott, and Lake, 1984). They describe the different permeability and anisotropy levels associated with trough cross-beds, rippled sandstone, and troughs with clay drape and cross-bedding distributed by dewatering structures.

4. Microscopic Heterogeneity

Heterogeneities on the scale of rock texture and petrographic composition have to be described separately. In rare cases entire genetic units are composed of a single homogeneous rock type, but this is the exception rather than the rule. In most cases we have to distinguish between a

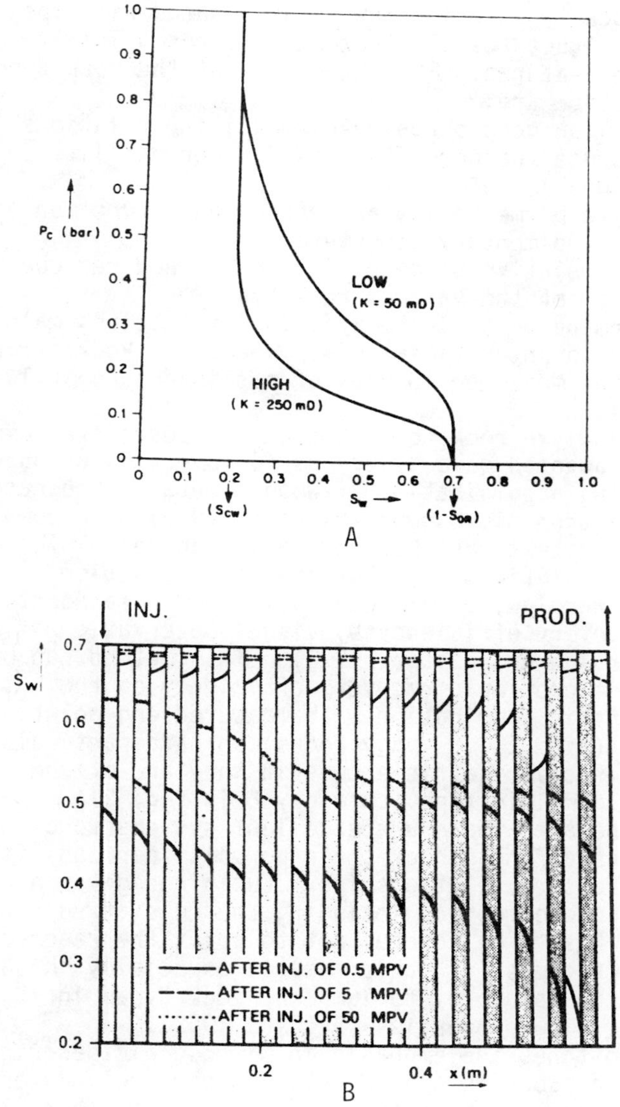

Figure 20. Capillary pressure curves of high and low permeability laminae of a model of a cross-bed set and the water saturation in the high permeability (grey) and low permeability laminae during a waterflood simulation.

number of rock types which may be associated with specific sedimentary structures or other more macroscopically identifiable features. The objectives of the rock type characterization are:

a. To establish good porosity-permeability relationships in order to use the porosity logs for permeability prediction.
b. To analyze permeability anisotropy as a function of rock type and sedimentary structure.
c. To study relative permeability curves and residual oil saturation of the various rock types.
d. To determine possible rock/fluid interactions mainly in relation to interstitial clay minerals. Rock strength has also to be measured in view of possible sand influx problems.

Methods to analyze rock for the above purposes are summarized in fig. 9. Swanson gave a very useful overview of pore type analysis using a combination of Wood's metal impregnation and scanning electron microscopy (Swanson, 1979). In complex carbonate rock textures this technique can be helpful to understand saturation distributions and production behaviour. Wardlaw, et al. have developed a method to estimate recovery efficiency by visual observation of pore systems (Wardlaw and Cassan, 1978). This method can be used to estimate microscopic recovery efficiency on the basis of thin sections in combination with measured end-point saturations (fig. 21). Vuggy rocks can have particularly low microscopic recovery efficiencies if they are of the unconnected oomoldic type (category E in fig. 21).

For carbonates there are excellent systems to describe rock types as a function of their porosity/permeability/capillary pressure relationships. Lucia has given a good overview of these methods including the role of vugs (Lucia, 1981). One has to be careful not to split the reservoir into too many rock types. In this respect especially the Archie classification is useful to lump the rock types into groups with similar flow properties (Archie, 1952).

For sandstones the emphasis in microscopic descriptions is on grain size, sorting and cement, and particularly on the type and configuration of interstitial clays. There is a vast literature on this subject with respect to porosity/permeability relationships, formation damage during drilling, and production and rock/fluid interaction during injection processes. Sandstone reservoirs can be damaged by migration of *in situ* particles (Lever and Dawe, 1984); expandable clays (Neasham, 1977); and sensitivity to acids (Walsh, Lake, and

Schechter, 1982). A useful formula for calculating permeability from average grain size, sorting and porosity has been proposed by Van Baaren (Van Baaren, 1979).

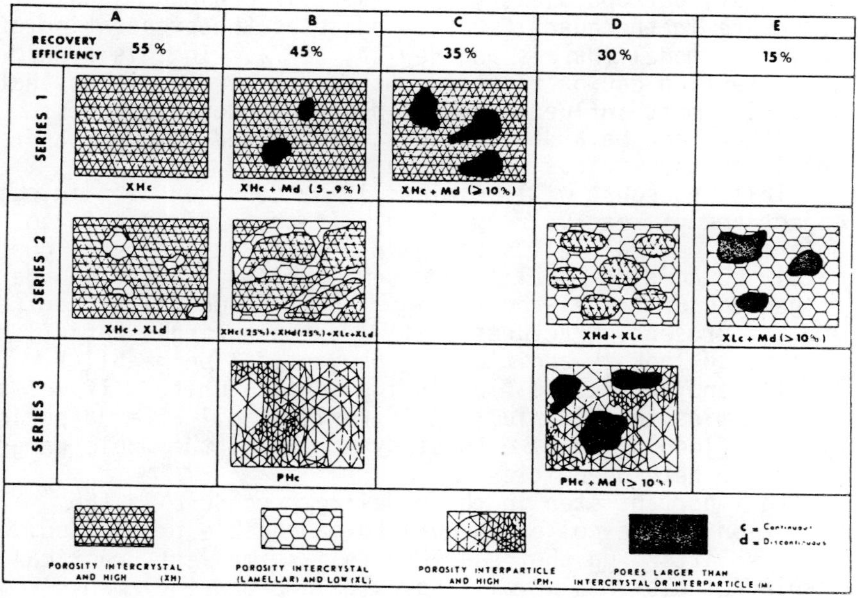

Figure 21. Classification of some important types of pore networks observed in carbonate reservoir rocks (with empirically assigned recovery efficiencies).

IV. RESERVOIR MODELING INCLUDING HETEROGENEITIES

A. Schemes for Reservoir Modeling

A general scheme for a reservoir study seen through the eyes of a reservoir engineer is shown in fig. 22. The box in the upper right hand corner looks quite different if we ask a geologist to design a scheme (fig. 23). Ideally the geologist, petrophysicist, and reservoir engineer should cooperate in the quantification phase of the preparation of a reservoir model (Harris and Hewitt, 1977). In this synergetic way attention can be focussed on those heterogeneities that will have most influence on the planned recovery process. Also there can be a dialogue on the required accuracy and detail of the various model parameters.

This procedure will be clear if we consider certain common objectives of reservoir modeling. If the objective is to determine optimal well spacing in primary development, one has mainly to consider sealing and semi-sealing faults, boundaries of genetic units related to significant permeability contrasts and, if present, fractures. If we are planning a waterflood we have to look in addition to the vertical permeability profile and baffles to flow inside genetic units. If we want to determine the distribution of residual oil after a period of waterflooding we have to study virtually the whole range of reservoir heterogeneities.

An important step in the modeling procedure is the reduction of the collected data to manageable proportion. After splitting up the reservoir to its smallest practically meaningful elements we will have to lump these elements together again into units we can handle in the quantification phase. Weber, et al. have described this procedure for a barrier bar reservoir in Nigeria (Weber, Klootwijk, Konieczek and Van der Vlugt, 1978). The cores showed a barrier bar deposit which could be subdivided into three intervals, a lower and upper barrier face and the clean barrier sand which in turn is overlain by a much lower permeability transgressive sand (fig. 24). The split up of the barrier face interval is subjective, but it is a practical way to obtain a layered system with reasonably constant layer properties particularly with respect to the vertical permeability. This in turn is very suitable for 3-D simulation (fig. 25). This simulation result incidentally shows the retardation of water break-through by the low horizontal and vertical permeability of the

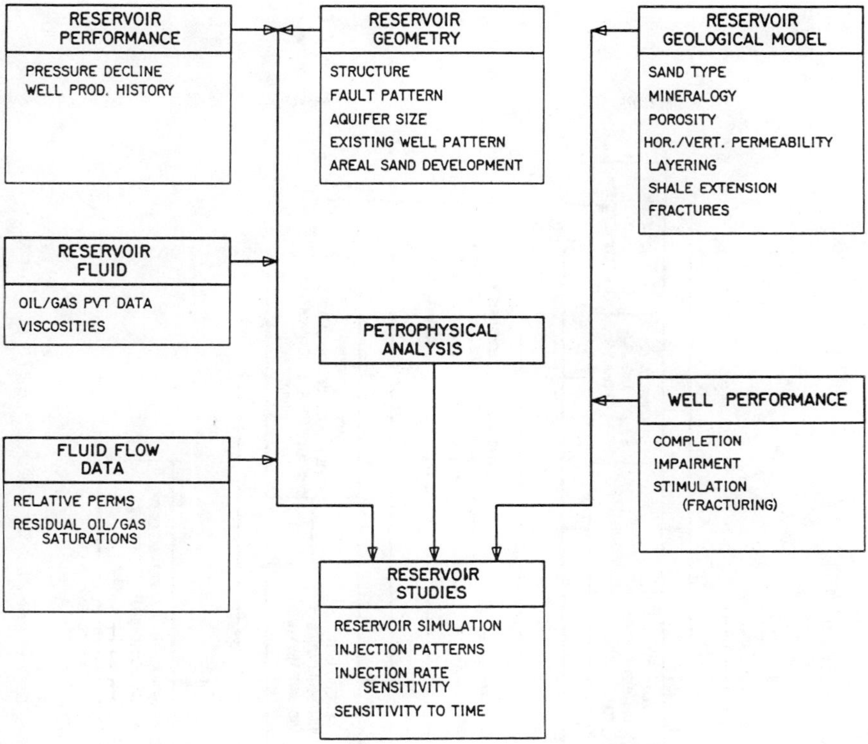

Figure 22. Scheme for a reservoir study.

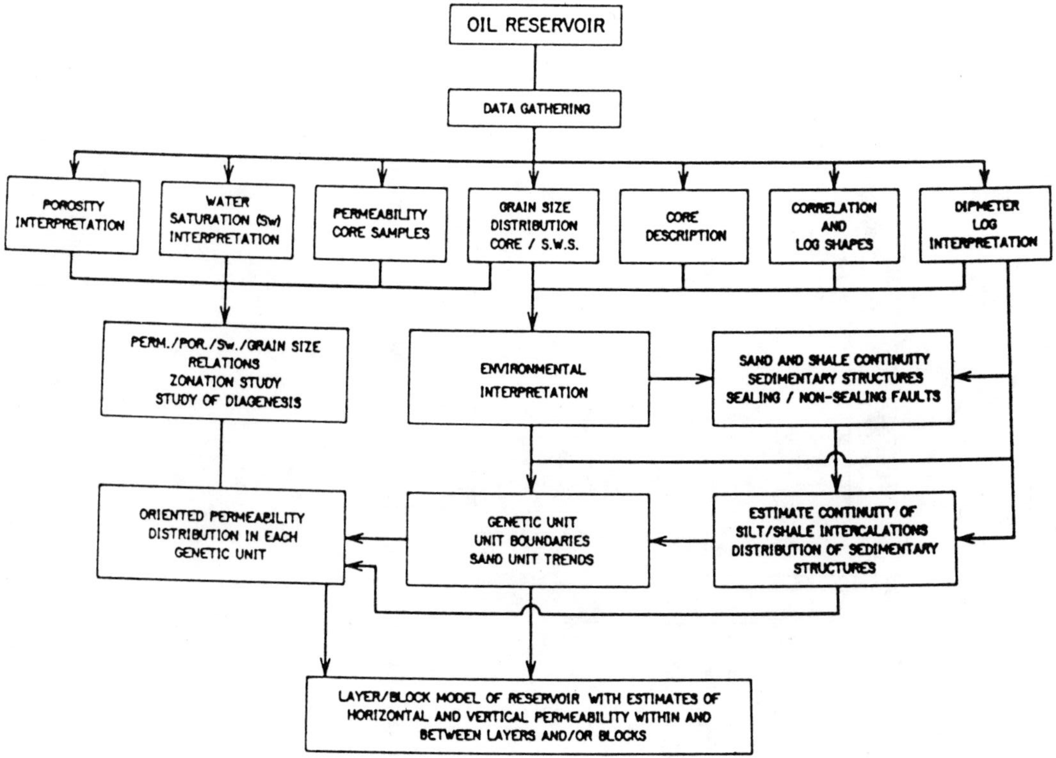

Figure 23. Scheme for a reservoir geological analysis.

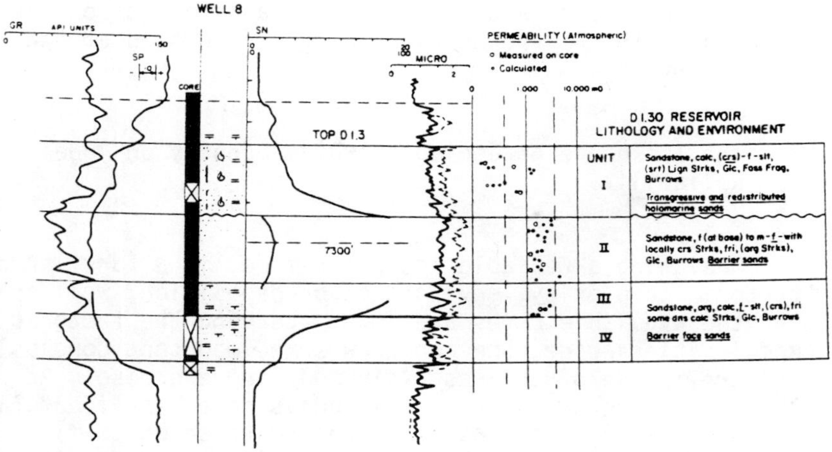

Figure 24. Core data and logs from the D 1.3 reservoir of the Obigbo North Field, Nigeria.

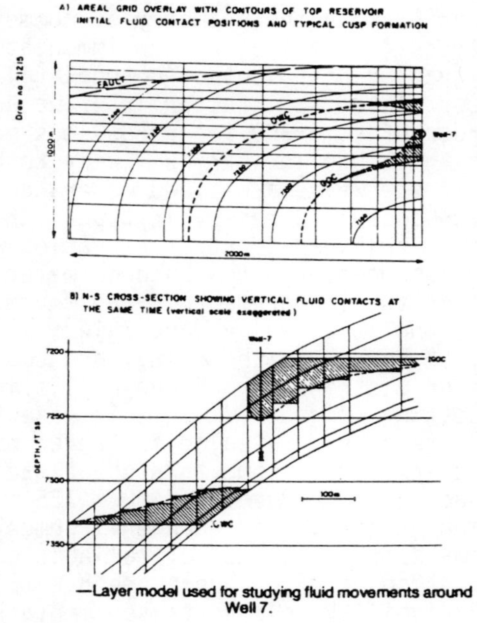

Figure 25. 3-D layer model for studying fluid movements around a well of the Obigbo North Field, D 1.3 reservoir.

barrier face sands and also the favorable effect on the gas cusping of the low permeable transgressive sand at the top of the reservoir.

B. The Influence of Geological Data Density on Model Reliability

Designing a reliable reservoir model is a function of tectonic complexity, environment of deposition, well spacing, and the available cores and logs. Leaving the tectonic complexities aside, one can draw some important conclusions at an early stage of any reservoir modeling exercise. As soon as the prevailing types of genetic units have been identified we can consider our chances to correlate these units satisfactorily across the existing well spacings. In this context one has to know the palaeogeographical setting of the reservoir, the general shape and order of deposition of the genetic units, and have a data bank of ranges of width/thickness/length relationships for each genetic unit type.

If we take as an example the geometry of common deltaic genetic sandbody types (table IV), we can immediately see that a number of non-migrating distributary channel-fills can probably only be correlated satisfactorily if we have well spacings of the order of some 300 meters or less. On the other hand, a 10-m thick barrier bar is likely to be correlatable even if the wells are 1,000 meters apart. However, even these widespread reservoir types can have large lateral variations in thickness, net/gross ratio, and permeability over distances of a few hundred meters (fig. 26).

A good example of the improvement of the correlation with decreasing well spacing is shown by Wang Zhiwu, et al. (Wang Zhiwu, Wang Qimin, Li Bohu, Lan Chengjing, and Luo Xiangzhong, 1982) for a series of distributary channel-fills in a fluviolacustrine setting (fig. 27). In this case the original 600 x 500 m well pattern only revealed scattered non-correlatable channel-fills and some thin sheet sands. Only in areas where the spacing was reduced to 300 x 125 m could a detailed correlation be made. For each environment of deposition an estimate of the chance of reliable correlation of the genetic units can be made. Next one has to consider internal lateral variability related to sedimentary structures, grain size distribution, and diagenetical changes. These variations are often too sudden to delineate with a normal well spacing (400-800 m) without much palaeogeographic and diagenetic history reconstruction. Thus

	Width (W) meters	Thickness (T) meters	W/T
Non-migrating distributary channel-fills	125- 1,200	3-45	15- 60
	larger ones (W > 500) tend to have W/T > 40		
Meandering channel-fills, point bars	1,200-20,000	9-90	100-800
	generally < 5000	generally 10-30	generally < 450
	larger ones (W > 3000) tend to have W/T > 300		
Barrier bars	1,400-12,000	8-30	150-500
	larger ones (W > 4000) tend to have W/T > 400		

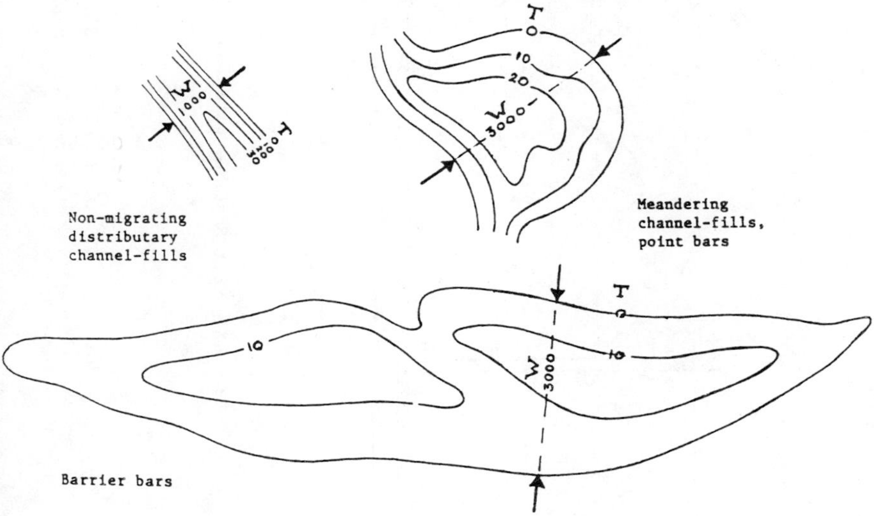

Table IV. Geometry of common deltaic genetic sandbody types.

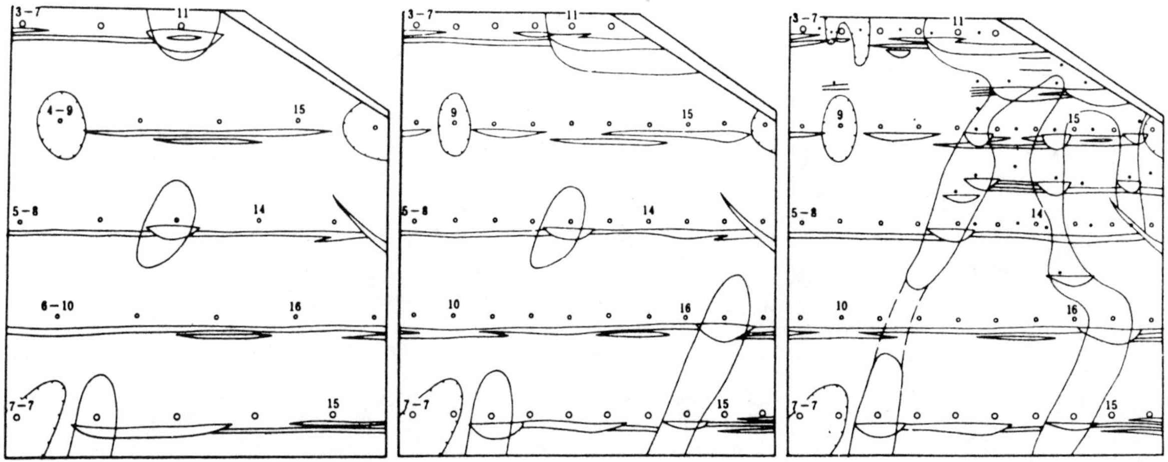

(1) WELL SPACING: 600 × 500m

(2) WELL SPACING: 600 × 250m

(3) IN SOME AREAS WELL SPACING HAS BEEN DECREASED TO: 300 × 125m

Figure 27. Effectiveness of reserve utilization under different well spacing of downstream distributary sand bodies.

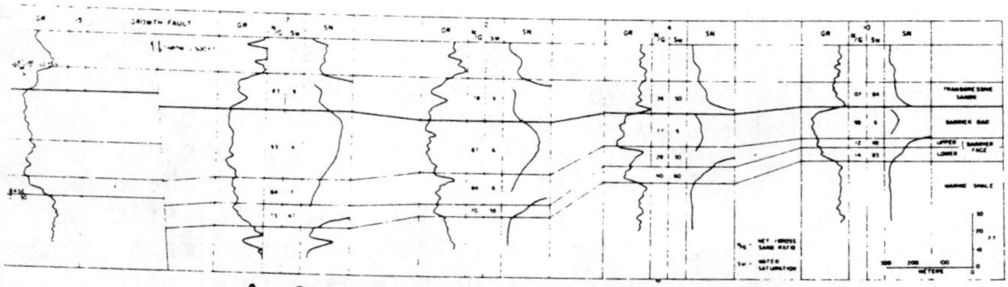

A —Stratigraphic cross-section about perpendicular to the growth fault.

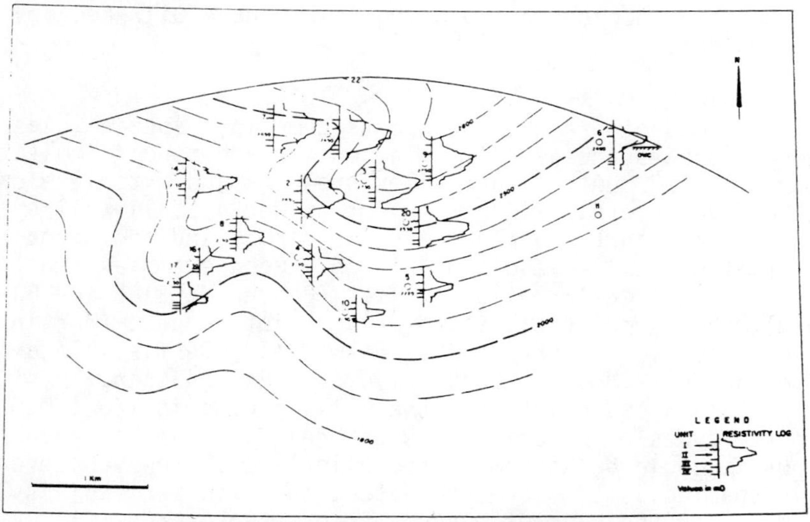

B —Resistivity logs and inferred permeability distribution of Unit 2.

Figure 26. Correlation of barrier bar and homogressive shallow-marine sands of the D 1.3 reservoir, Obigbo North Fields, Nigeria.

interpolation and conceptual modeling come into play, but, in consequence, relability goes down. As discussed in Section II the smaller scale heterogeneities are usually difficult to correlate and modeling on the basis of analogue prototypes and limited core and log data will have to be carried out.

With normal well spacing we will only rarely be able to construct reliable models that will provide realistic production forecasts without considerable adjustments via history matching and the input of non-geological data. However, even though inaccurate in detail, conceptual models can still be very useful for comparing the relative merits of different development schemes. Also the understanding of the typical configuration of a type of heterogeneity can help in predicting and interpreting production behaviour.

C. Detailed Geological Modeling in Enhanced Oil Recovery Projects

Enhanced oil recovery projects come in a special class. These projects are usually so expensive and so difficult to carry out and control that pilot projects and extensive data gathering pay off. In these pilot projects small well spacing between injectors and producers are normal and additional observation wells are often drilled. Good examples are the Sloss field micellar pilot project (Basan, McCaleb and Buxton, 1978), the Marmul field steam-drive pilot project (Martin and Cooper, 1984), and the Hartzog Draw field CO_2 miscible drive pilot project (Hearn, Ebanks, Tye, and Ranganathan, 1984).

In the Sloss pilot area the well spacing is 120-150 m and the two types of genetic units channel-fills and overbank deposits could be mapped in the pilot area. However, even this spacing is too large to detect specific permeability patterns in the discontinuous thin overbank sands (fig. 28). Tracer tests were required to unravel the well-to-well communication and permeability pattern.

The Marmul field case is more complicated with a large variety of fluvioglacial genetic units. The well spacing in the pilot area is in the 150-200 m range and major genetic units can be correlated. Here the problem is the occurrence of thin low permeable beds which are, with few exceptions, uncorrelatable between wells despite the small spacing. In this case probabilistic models were constructed on the basis of analogue outcrop data on the continuity of such baffles along the principles outlined by Haldorsen and Lake (Haldorsen and Lake, 1984). These models were used to test the relative merits of completing the steam injection wells high or low in the reservoir (fig. 29). Although the simulation result is

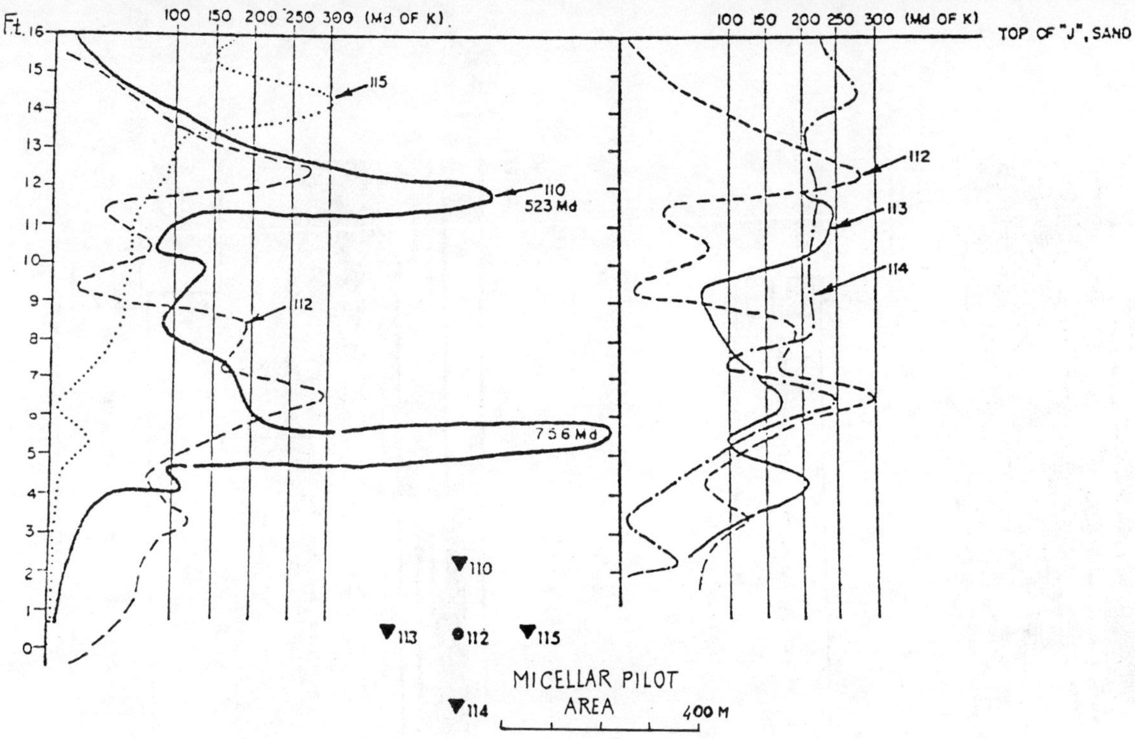

Figure 28. *Distribution of permeability among the pilot wells. Stratigraphic datum is the top of the Muddy J_1 interval.*

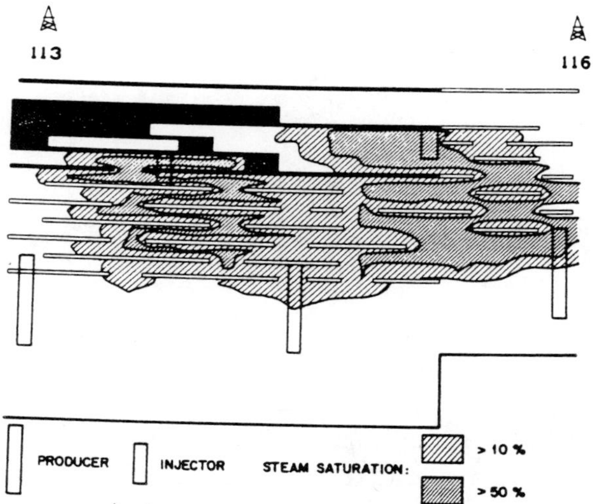

A —Example of predicted steam zone development: high-injector completions

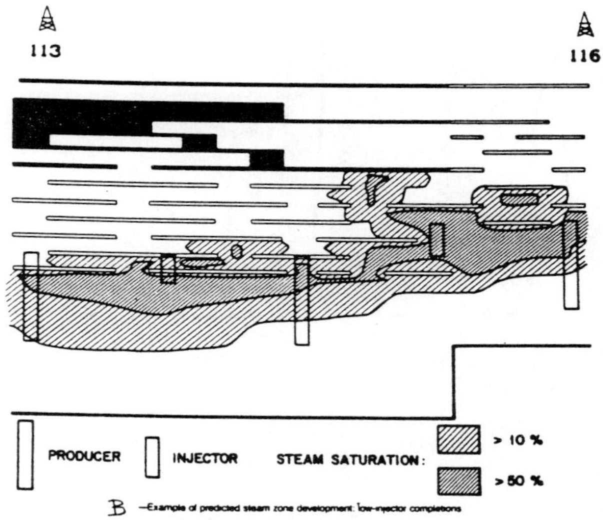

B —Example of predicted steam zone development: low-injector completions

Figure 29. Results of simulation of steam drive project in the Marmal Field with two different injection interval patterns.

strongly influenced by changes in the assumed geology of the pilot area, the results are sufficient to make a decision on how to start the project.

The Hartzog Draw field consists of a series of imbricated shallow marine sand ridges. Well spacing in the pilot area is 500 to 600 meters but for this environment of deposition fairly accurate delineation of the individual sand lenses and the intervening shales was possible. Hearn, et al. give a good description of the data gathering for the planning of closer spaced water injection and a CO_2 miscible flood project (Hearn, Ebanks, Tye, and Ranganathan, 1984). Nine wells were cored in the pilot area of which three cores were oriented. Another ten wells had already been cored nearby previously. The reservoir was subdivided into five flow units (fig. 30), and defined as a reservoir body that has similar porosity, permeability, and bedding characteristics throughout. They can be expected to have similar fluid flow properties. This approach can considerably simplify the preparation of reservoir simulation input.

Johnson and Krol carried out a detailed modeling of a very heterogeneous sandstone reservoir, the Statfjord formation in Brent field (Johnson and Krol, 1984). In this very large reservoir full downdip water injection is planned possibly augmented by miscible gas injection in some reservoir units. This study showed how complex, three-dimensional reservoir heterogeneity can be modeled with the latest large-capacity computers. The reservoir is composed of a 210-230 m thick package of fluvial sandstones and shales overlain by a blanket of transgressive shallow marine sand with a thickness of some 27 m.

The main problem in this case is the difficulty of correlating the fluvial sands. The well spacing is 1-6 km which necessitates the development of conceptual prototype models adapted to the well data. For fluvial channel sands theoretical distribution patterns have been constructed by Allen (Allen, 1978). He investigated how the net-to-gross ratio influenced the interconnectedness of the sandbodies. This method has also been used for the Statfjord formation. The resulting models (fig. 31) were used to test the performance of the reservoir under various alternative development schemes.

V. CONCLUSIONS

Heterogeneity interpretation and quantification is one of the major tasks of the geological engineer. The

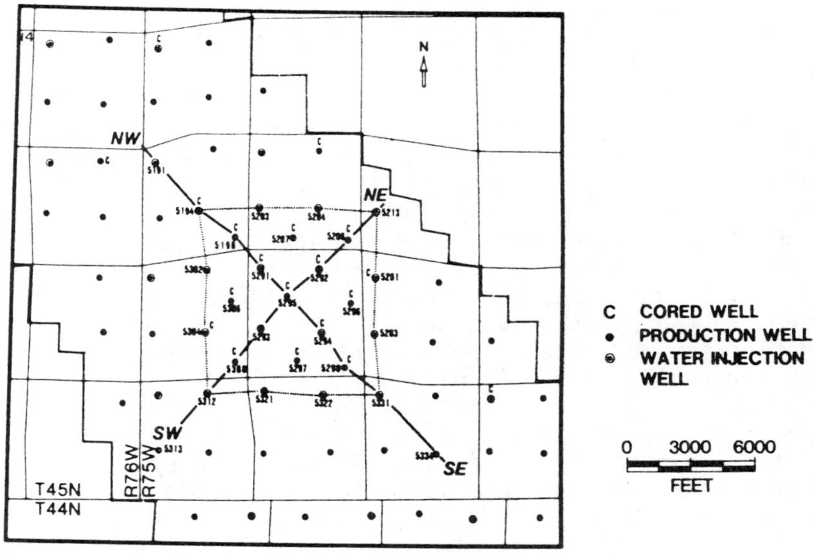

A —Pilot area map.

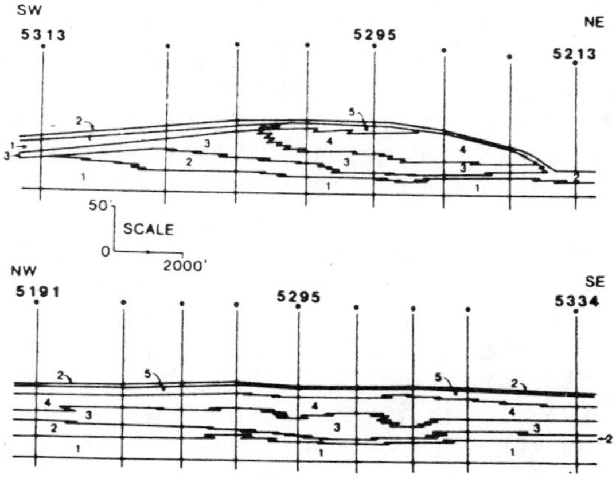

B —Flow unit cross sections (see Fig. A for locations).

Figure 30. Hartzog Draw Field pilot area and cross-sections through the Shannon sandstone reservoir showing correlation of flow units.

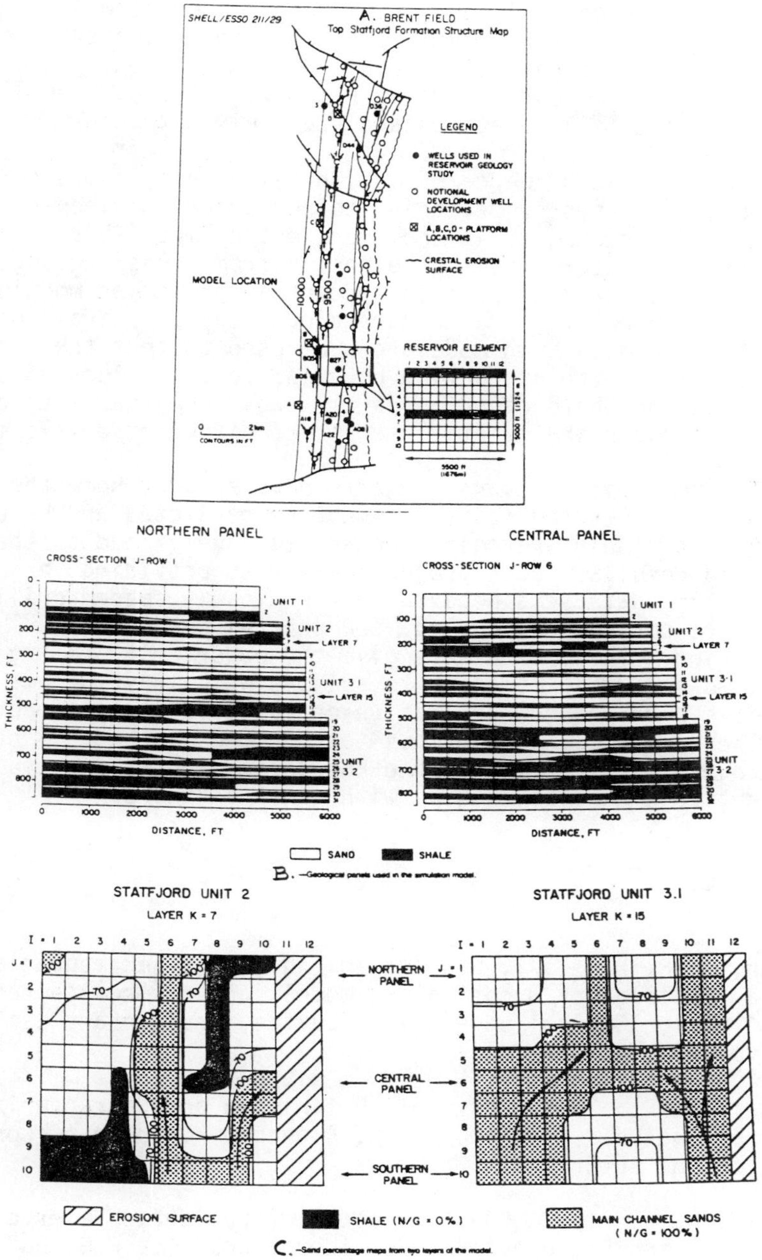

Figure 31. Model of the Statfjord formation in the Brent Field showing sand and shale distribution.

quantification can only be achieved in cooperation with petrophysicists and reservoir engineers in a synergetic manner with the aid of cores, logs, analogue models, conceptual modeling, and various reservoir production tests. Usually no single method is adequate to analyze a given type of heterogeneity.

The general influence of many heterogeneity types on recovery processes is beginning to be common knowledge. With normal well spacing it is, however, often impossible to quantify the heterogeneities directly from cores and logs. We will have to take recourse to conceptual prototype models based on the observed data and analogue examples from outcrops or other fields. Such models can be used to test the sensitivity of the production behavior to the values allotted to the various heterogeneity parameters. They can also be used to compare the relative merits of different development schemes.

Enhanced oil recovery projects have brought home the message that heterogeneities cannot be neglected and have shown that simple averaging procedures rarely lead to the expected results. Such projects are also providing very interesting data sets because of the closely spaced drilling, intensive coring and logging, and careful production monitoring. These data sets give the opportunity to try out detailed reservoir modeling techniques which have become possible with the new generation of large capacity computers.

There is still an enormous task for researchers to study heterogeneities to improve the knowledge about their typical properties and relationships with other features.

REFERENCES

Allen, J. R. L. 1978, Studies in Fluviatile Sedimentation; an Exploratory Quantitative Model for the Architecture of Avulsion-Controlled Alluvial Suites, Sedimentary Geology, Vol. 21, p. 129-147.

Archer, J. S. 1983, Reservoir Definition and Characterization for Analysis and Simulation, Eleventh World Petr. Cong., London, pD 6 (1).

Archie, G. E. 1952, Classification of Carbonate Reservoir Rocks and Petrophysical Considerations, A.A.P.G. Bull., Vol. 36, No. 6, p. 278-298.

Basan, P. B., McCaleb, J. A. and Buxton, T. S. 1978, Important Geological Factors Affecting the Sloss Field Micellar Pilot Project, Soc. Petr. Eng. paper 7047.

Bebout, D. G. and Pendexter, C. 1975, Secondary Carbonate Porosity as Related to Early Tertiary Depositional Facies, A.A.P.G. Bull., Vol. 59, No. 4, p. 665.

Berruin, N. A. and Morse, R. A. 1979, Waterflood Performance of Heterogeneous Systems, Jour. Petr. Techn., July 1979, p. 829-836.

Bos, M. R. E. 1982, Prolific Dry Oil Production from Sands with Water Saturations in Excess of 50%, A Study of a Dual Porosity System, SPWLA - Third Annual Logging Symposium, July 1982.

Chang-Shu Yang and Swie-Djin Nio 1985, The Estimation of Palaeo-Hydrodynamic Processes from Subtidal Deposits Using Time Series Analysis Methods, Sedimentology, Vol. 32, p. 41-57.

Dake, L. P. 1982, Application of the Repeat Formation Tester in Vertical and Horizontal Pulse Testing in the Middle Jurassic Brent Sands, paper EUR 270, Proc. Eur. Offshore Pet. Conf. 1982.

Emmett, W. R., Beaver, K. W. and McCaleb, J. A. 1971, Little Buffalo Basin Tensleep Heterogeneity - Its Influence on Infill Drilling and Secondard Recovery, Jour. Petr. Techn., February 1971, p. 161-168.

Epting, M. 1980, Sedimentology of Miocene Carbonate Build-ups, Central Luconia, Offshore Sarawak, Geol. Soc. Malaysia Bull. No. 12, p. 17-30.

Eype, R. and Weber, K. J. 1971, Mini-Permeameters for Consolidated and Unconsolidated Sands, A.A.P.G. Bull., Vol. 55, No. 2, p. 307-309.

Haldorsen, H. H. and Lake, L. W. 1984, A New Approach to Shale Management in Field-Scale Models, Jour. Petr. Techn., August 1984, p. 447-457.

Harris, D. G. and Hewitt, C. H. 1977, Synergism in Reservoir Management - The Geologic Perspective, Jour. Petr. Techn., July 1977, p. 761-770.

Hearn, C. L., Ebanks, W. J., Tye, R. S. and Ranganathan, V. 1984, Geological Factors Influencing Reservoir Performance of the Hartzog Draw Field, Wyoming, Jour. Petr. Techn., August 1984, p. 1335-1344.

Johnson, H. D. and Krol, D. E. 1984, Geological Modeling of a Heterogeneous Sandstone Reservoir: Lower Jurassic Statfjord Formation, Brent Field, Soc. Petr. Eng. paper 13050.

Jones, J. R., Scott, A. J. and Lake, L. W. 1985, Reservoir Characterization for Numerical Simulation of Mesa Verde Meanderbelt Sandstone, Northwestern Colorado, Soc. Petr. Eng. paper 13052.

Kempthorne, R. H. and Irish, J. P. R. 1981, Norman Wells - A New Look at One of Canada's Largest Oil Fields, Jour. Petr. Techn., June 1981, p. 985-991.

Kortekaas, Th. F. M. 1985, Water/Oil Displacement Characteristics in Cross-Bedded Reservoir Zones. to be published in the Soc. Petr. Eng. Jour. December 1985.

Langston, E. P., Shirer, J. A. and Nelson, D. E. 1981, Innovative Reservoir Management - Key to Highly Successful Jay/Lec Waterflood, Jour. Petr. Techn., May 1981, p. 783-791.

Lapré, J. F. 1980, Reservoir Potential as a Function of the Geological Setting of Carbonate Rocks, Soc. Petr. Eng., paper 9246.

Lever, A. and Dawe, R. A. 1984, Water-Sensitivity and Migration of Fines in the Hopeman Sandstone, Jour. of Petr. Geology, Vol. 7, No. 1, p. 97-108.

Lucia, F. J. 1981, Petrophysical Parameters Estimated From Visual Description of Carbonate Rocks: A Field Classification of Carbonate Pore Space, Soc. Petr. Eng. paper 10073.

Martin, J. H. and Cooper, J. A. 1984, An Integrated Approach To the Modeling of Permeability Barrier Distribution in a Sedimentologically Complex Reservoir, Soc. Petr. Eng. paper 13051.

Neasham, J. W. 1977, The Morphology of Dispersed Clay in Sandstone Reservoirs and Its Effect on Sandstone Shaliness, Pore Space, and Fluid Flow Properties, Soc. Petr. Eng. paper 6858.

Pettijohn, F. J., Potter, P. E. and Siever, R. 1973, Sand and Sandstone, Springer-Verlag, New York, Heidelberg, Berlin.

Pryor, W. A. and Fulton, K. 1978, Geometry of Reservoir Type Sand Bodies in the Holocene Rio Grande Delta and Comparison with Ancient Reservoir Analogues, Soc. Petr. Eng., paper 7045.

Rose, W. 1983, A Note on the Role Played by Sediment Bedding in Causing Permeability Anisotropy, Jour. Petr. Techn., February 1983, p. 330-332.

Salle, J. E. and Wood, B. R. 1984, Use of Micro-resistivity from The Dipmeter to Improve Formation Evaluation in Thin Sands, Northeast Kalimantan, Indonesia, Jour. Petr. Techn., September 1984, p. 1535-1544.

Smith, D. A. 1980, Sealing and Non-Sealing Faults In Louisiana Gulf Coast Salt Basin, A.A.P.G. Bull., Vol. 64, p. 145-172.

Sneider, R. M., Tinker, C. N. and Meckel, L. D. 1978, Deltaic Environment Reservoir Types and Their Characteristics, Jour. Petr. Techn., November 1978, p. 1538-1546.

Swanson, B. F. 1979, Visualizing Pores and Non-Wetting Phase in Porous Rock, Jour. Petr. Techn., January 1979, p. 10-18.

Todd, A. C. and Tweedie, J. 1978, Total Rock Characterization of North Sea Sandstones with Particular Reference to Interstitial Clays, paper EUR 93, Proc. Eur. Offshore Pet. Conf. 1978.

Van Baaren, J. P. 1979, Quick-Look Permeability Estimates Using Sidewall Samples and Porosity Logs, Sixth European Formation Evaluation symposium, London, 26-27 March 1979.

Walsh, M. P., Lake, L. W. and Schechter, R. S. 1982, A Description of Chemical Precipitation Mechanisms and Their Role in Formation Damage During Stimulation by Hydrofluoric Acid, SPE Fifth Symp. on Formation Damage Control, Proc., p. 7-27.

Wang Zhiwu, Wang Qimin, Li Bohu, Lan Chengjing and Luo Xiangzhong 1982, Ways to Improve Development Efficiency of Daging Oilfield by Waterflooding, International Meeting on Oilfield Development Techniques, Daging, China, September 7-14, 1982.

Wardlaw, N. C. and Cassan, J. P. 1978, Estimation of Recovery Efficiency by Visual Observation of Pore Systems in Reservoir Rocks, Bull. Can. Petr. Geol., December 1978.

Weber, K. J. 1985, Computation of Initial Well Productivities in Aeolian Sandstone on the Basis of a Geological Model, Leman Gas Field, U.K., In: Tillman R. W. and Weber K. J. (ed.) - Reservoir Sedimentology - Spec. Publ. Soc. Econ. Palaeont. Mineral (in press).

Weber, K. J. 1982, Influence of Common Sedimentary Structures on Fluid Flow in Reservoir Models, Jour. Petr. Techn., March 1982, p. 665-672.

Weber, K. J., Klootwijk, P. H., Konieczek, J. and Van der Vlugt, W. R. 1978, Simulation of Water Injection in a Barrier-Bar-Type, Oil-Rim Reservoir in Nigeria, Jour. Petr. Techn., November 1978, p. 1555-1565.

Weber, K. J., Mandl, G., Pilaar, W. F., Lehner, F. and Precious, R. G. 1978, The Role of Faults in Hydrocarbon Migration and Trapping in Nigerian Growth Fault Structures, Offshore Techn. Conf., Houston, paper OTC 3356.

Weber, K. J., Eype, R., Leynse, D. and Moens, C. 1972, Permeability Distribution in a Holocene Distributary Channel-fill Near Leerdam (The Netherlands) - Permeability Measurements and In-Situ Fluid-Flow Experiments, Geologie en Mijnbouw, Vol. 51, No. 1, p. 53-62.

Wolf, M. and Pelissier-Combescure, J. 1982, Faciolog-Automatic Electrofacies Determination, SPWLA Third Ann. Logging Symposium, July 1982.

Zhuang Huinong 1984, Interference Testing and Pulse Testing in the Kenli Carbonate Oil Pool - A Case History, Jour. Petr. Techn., June 1984, p. 1009-1017.

If reservoir models could be effectively described in terms of depositional units, the reservoir description process could be considerably simplified. Currently a reservoir model requires the specification of horizontal and vertical permeability and porosity for each grid block in the reservoir. For a large reservoir model, which might have 20,000 grid blocks, the number of unknowns make the problem ridiculously underdetermined and unwieldly to manage. Alternatively, if geological studies can establish the dimensions and frequency distribution of parameters for depositional units, the reservoir could be described in terms of this considerably reduced set of parameters. In history matching, it would be these depositional unit descriptors and not individual grid block properties which would be modified.

III. THE SIMULATION PROGRAM

The reservoir simulation program used in this study is a three-dimensional two-phase (water and oil) incompressible model using an IMPES solution approach. A variable-weighting scheme[2] which effectively eliminates numerical diffusion associated with saturation discontinuities is used. Boundary potential and flow conditions are treated implicitly in the pressure solution.

IV. THE SMALL SCALE

Heterogeneities on the scale of centimeters will be referred to as small-scale heterogeneities. Examples of mechanisms generating small-scale heterogeneities are clay and silt deposits in ripple marks, disturbances such as burrowing in well-sorted sediments, and uneven diagenetic processes in carbonates. Such small-scale heterogeneities are typically chaotic and are difficult to characterize quantitatively.

From a mechanistic point of view, we could also consider the small scale to be the viscous-capillary regime where gravity forces are generally unimportant.

Small-scale heterogeneities are important because they determine the distribution of residual oil and partially determine the distribution of bypassed oil which is more important at the medium scale.

Figure 1a shows a small-scale two-dimensional permeability distribution derived from actual minipermeameter measurements by Weber[3]. The average permeability in the sand (red) is roughly $30 \times 10^{-3} \mu m^2$, while the average permeability in the silt (blue) is roughly 3 x $10^{-3}\mu m^2$. If this chunk of rock were to be waterflooded, how would the permeability distribution influence the ultimate recovery of oil from this small section of the rock? The important factors are: the residual oil saturations in the two types of rock, the mobility contrast between the water and oil, the capillary pressure curves for the two rock types, and the mobility contrast between the water and oil. The results of three simulations shown in Figure 1 demonstrate qualitatively how these factors influence recovery.

The case shown in Figure 1b is for a viscosity ratio of unity. Water (shown in blue) is injected at the left displacing the oil (red). After 0.5 pore volumes of water injected, we see that even though there is significant channeling, at the left-hand boundary which is now behind the front the silty part is being swept almost as well as the sandy sections. At this time, we have recovered 48% of the initial oil-in-place.

Figure 1c shows the results at the same time for an unfavorable mobility in which the oil viscosity is 20 times that of water. Here the constrasts in permeability and mobility reduce the sweep efficiency drastically. A water-saturated sand section will have a hydraulic contrast 200 times that of an oil-saturated silty section. At this time, 27% of the initial oil-in-place has been recovered.

In Figure 1d, we have include capillarity as well as the adverse mobility. The silt is assumed to have a maximum capillary pressure of 10 kPa with no capillary pressure in the sand. The total pressure drop across the sample is rougly 10 kPa. Instead of sharp channels as in the previous case without capillarity, the water moves in broader fronts. At the time shown, small patches of oil are left behind in cul-de-sacs of high permeability material with high water saturation in the bordering low permeability (high capillary pressure) material. These cul-de-sacs, of course, are where viscous forces are less able to effectively sweep and the fluids are immobilized as a result of a balance between viscous and capillary forces. This behavior is analogous to the pore-scale case in which the non-wetting phase is trapped in the larger pores surrounded by small pore containing water. At this point, 37% of the oil-in-place has been recovered which is considerably more than the second case.

These mechanisms and observations occur at the small scale in what would otherwise appear to be "homogeneous" rock. Because the number of pore volumes swept through a small-scale volume should in general be large (if it is swept at all), the most important effect of small-scale heterogeneities is on the residual saturation. The residual saturation in certain types of small-scale heterogeneities might be increased by capillary effects as shown in Figure 1d. Other types of small-scale heterogeneities could mechanically trap a significant amount of oil in small cul-de-sacs, although this mechanism has not been observed to our knowledge.

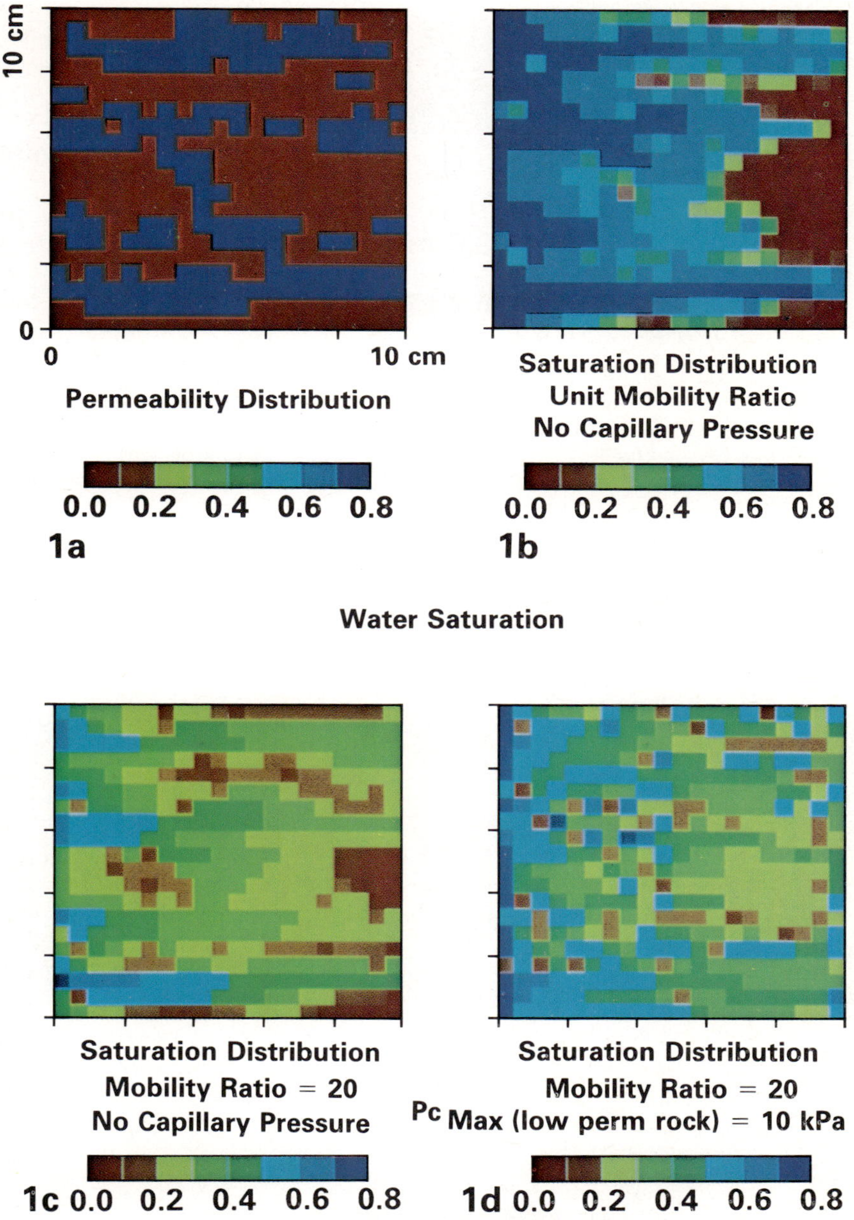

Figure 1. Small Scale Waterflood.
Time = 100,000 Seconds,
0.5 Pore Volume Water Injected.

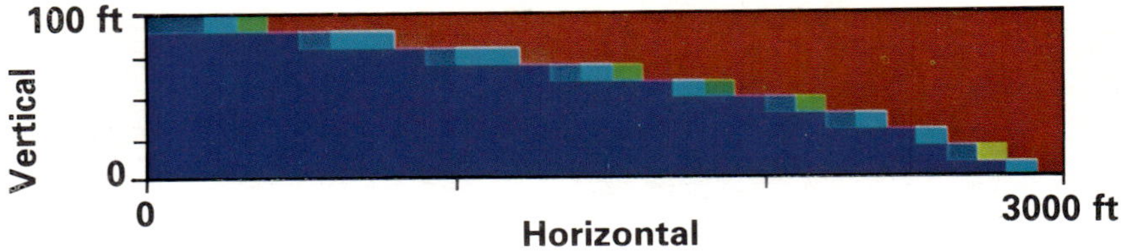

Channel Sand

Water Saturation

Permeability Distribution: Top = 100 md, Bottom = 1000 md

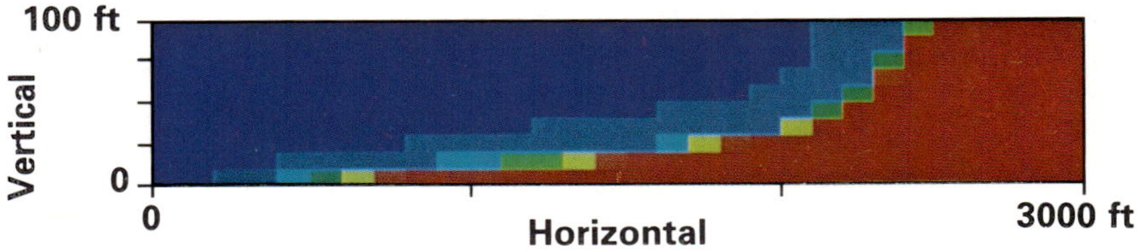

Bar Sand

Water Saturation

Permeability Distribution: Top = 1000 md, Bottom = 100 md

Figure 2. Medium Scale Waterflood - Graded Permeability

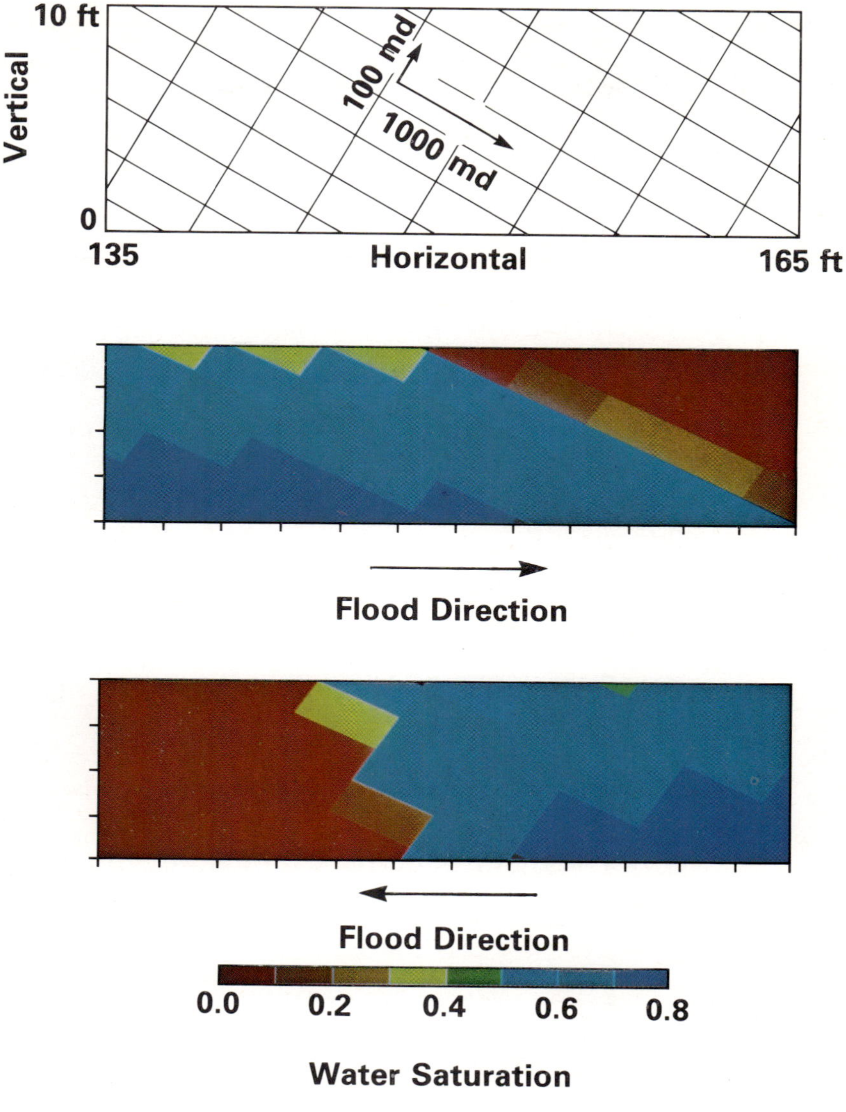

Figure 3. Medium Scale Waterflood - Cross-bedded Sand

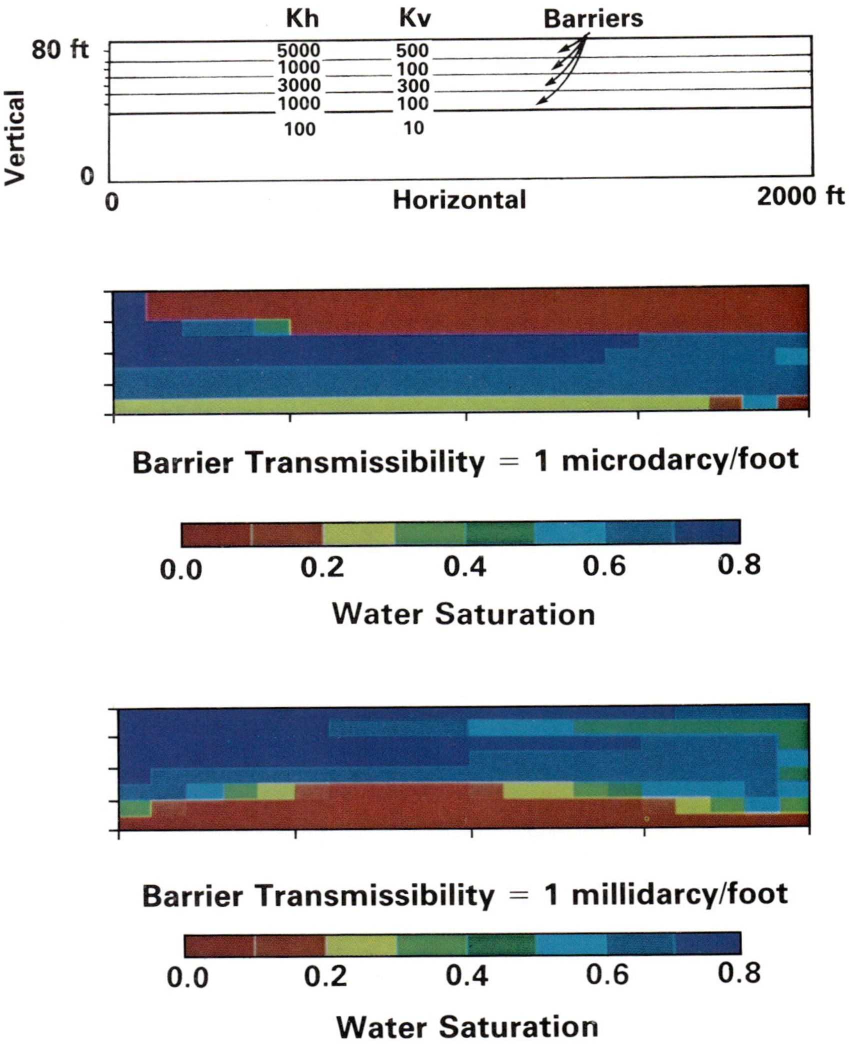

Figure 4. Large Scale Waterflood - Barriers
Time = 300 days 0.75 Pore Volumes Injected

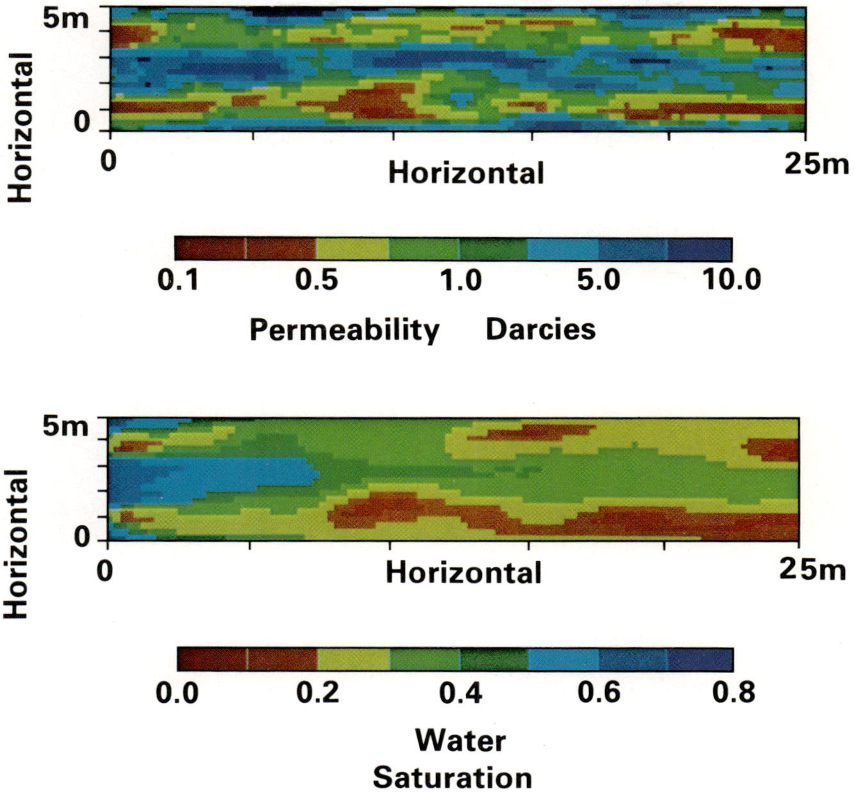

Figure 6. Scaling Up Process, Small Scale, Two-dimensional, Mobility Ratio = 20. Time = 10 days, 0.4 Pore Volumes Injected.

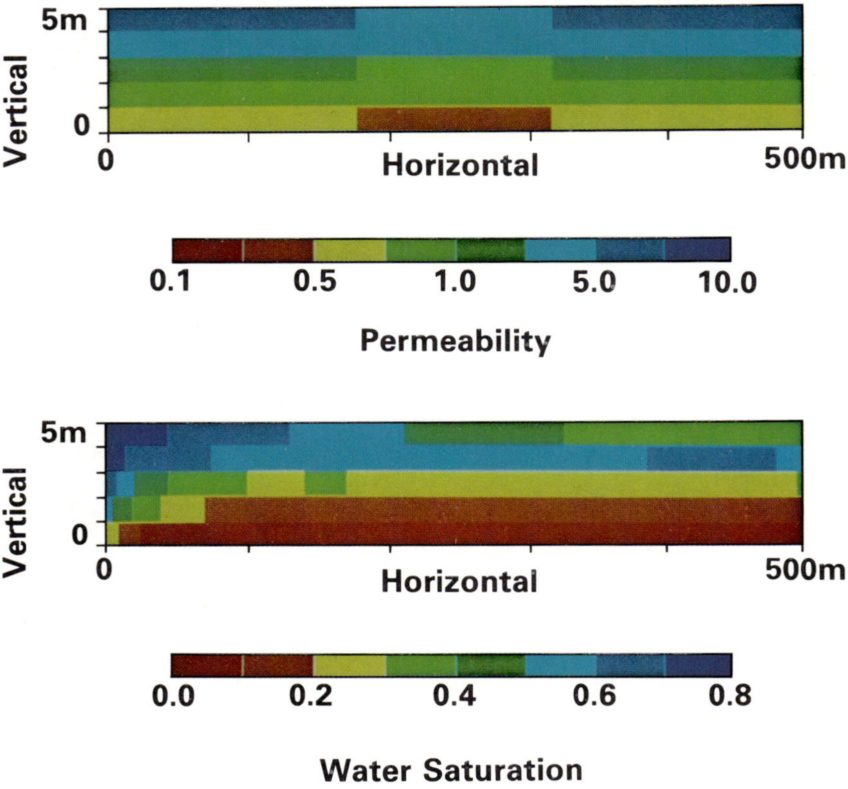

Figure 7. Scaling Up Process. Medium Scale, Three-dimensional, Mobility Ratio = 20. Time = 500 days, 1.0 Pore Volumes Injected.

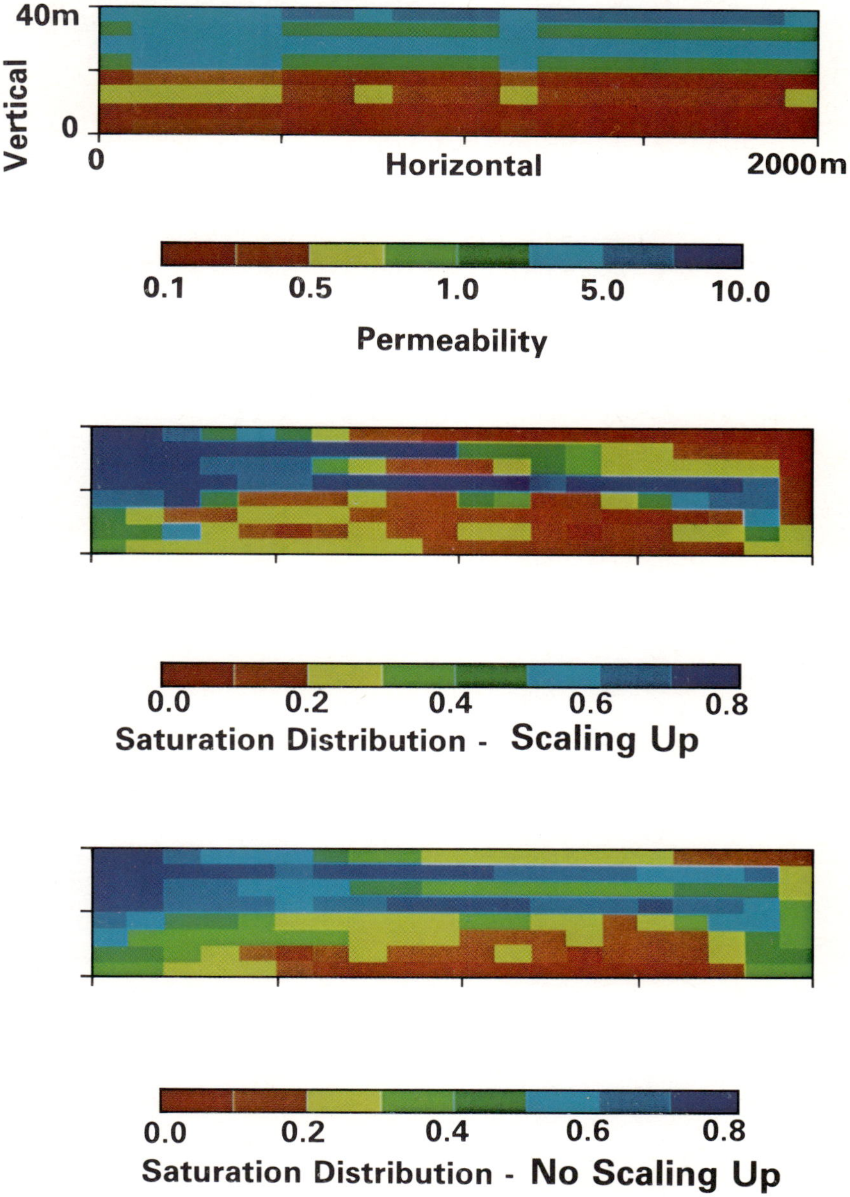

Figure 8. Scaling Up Process,
Three-dimensional, Mobility Ratio = 20.
Time = 5000 Days, 2.5 Pore Volumes Injected.

V. THE MEDIUM SCALE

The medium scale is the scale of depositional units. Because we have defined the depositional unit as being the result of the same single or repetitive depositional event, heterogeneities at this scale often appear more systematic than the more random heterogeneities at the small scale.

The medium scale could be considered to be the viscous-capillary-gravity regime, in which all three forces can play significant roles in determining the dynamic multiphase behavior.

While the description at the medium scale is typically quite complex, it should often be practical to model multiphase flow using somewhat simplified models in order to establish the influence of these heterogeneities on recovery. An example is the comparison in waterflood performance between a channel sand, in which the permeability generally increases downward, and a bar sand in which the permeability generally increases upwards as shown in Figure 2. In the channel sand, both gravity and the higher permeability on the bottom are pulling the water down. In a bar sand, the higher permeability pulls the water up, while gravity is pulling it down, resulting in better vertical sweep relative to the channel sand. While reality is considerably more complex than the idealized models shown here, such models should be able to capture the important features of systematic medium-scale heterogeneities.

Another interesting example of the effect of medium-scale heterogeneities on recovery is multiphase flow in cross-bedded sequences. Sweep efficiency parallel to the cross-bedding is obviously better than across the bedding. For flow across the bedding, it has also been conjectured that there may be a favorable direction of flow, either up-dip or down-dip with respect to the cross–bedding[4]. When the flow is in the down-dip direction with respect to the cross-bedding, the low-permeability streaks act as baffles, forcing the water down adding to the tendency of the water to move down because of gravity. When this happens, oil may be bypassed in the top of the set. When the flow is up-dip, the streaks tend to force the water up, counteracting gravity. This situation is analagous to the contrast in behavior between channel sands and bars sands previously described.

In order to model this problem, an inclined grid system was used in which the effective permeability normal to the cross-bedding is $100 \times 10^{-3} \mu m^2$ and the permeability parallel to the cross-bedding is $1000 \times 10^{-3} \mu m^2$ as shown in Figure 3. In Figure 3, we see that the shapes of the fronts are very different, but well behind the front, there is no sigificant difference in the sweep efficiency.

What might be the importance and role of these medium-scale heterogeneities? At the small-scale, the effect of the heterogeneities is principally on the residual saturation. At the medium scale, this effect is also important but the bypassing of oil becomes more important since the number of pore volumes swept through the volume is decreased. At the medium scale, the probability of mechanically trapping oil in cul-de-sacs formed by the juxtaposition of impermeable barriers could be an important effect in reducing ultimate recovery.

VI. THE LARGE SCALE

The large scale is the scale of many depositional units and perhaps several depositional environments. The large scale could also be defined as the viscous-gravity regime, in which capillary forces are less important than at smaller scales.

The essential features of the reservoir which control its large-scale behavior are lateral continuity and vertical communication. Physically, these aspects are determined by the dimensions of the depositional units, the contrast in permeability between high and low permeability regions, and post-depositional events such as faulting and fracturing. If we could accurately capture these essential features in a reservoir model, we will have come a long way in understanding reservoir behavior. The smaller-scale heterogeneities play a role in determining the amount of trapped and bypassed oil and so they must be given careful consideration, as well. But the dominant features in reservoir behavior are certainly the large-scale heterogeneities controlling lateral continuity and vertical communication.

Lateral continuity is particularly important in secondary and tertiary recovery processes since communication between injectors and producers is essential if efficient recovery is to be achieved. In addition, poor vertical communication as a result of extensive barriers often results in the differential flooding of layers, resulting in early breakthrough.

As an example of the importance of permeability barriers, consider the two cases shown in Figure 4 where we have compared saturation profiles for barriers whose vertical permeability divided by thickness is $10^{-6}\mu m^2$/foot and $10^{-3}\mu m^2$/foot. The pattern of the injection profile and the sweep efficiency changes drastically as can be seen. We can see that it is extremely important to know whether the vertical permeability of these barriers is $10^{-6}\mu m^2$ or $10^{-3}\mu m^2$: the sweep efficiency will be very different. Although we have no way of measuring effective vertical permeabilities over these large scales directly, the use of injection profiles and pulse testing in conjunction with a reservoir model could be very effective in determining these important reservoir parameters[5].

A key issue in reservoir simulation is how to scale up properly so that small-scale heterogeneities are effectively included in a large-scale reservoir model.

VII. THE SCALING UP PROCESS

The scaling up process is shown conceptually in Figure 5a. We start at the small scale and compute average properties and pseudo functions. The multiple grid block model at the small scale is replaced by a single grid block at the medium scale. Again effective properties are determined and we scale up to the next largest scale. By scaling up, we can effectively include the effects of smaller-scale heterogeneities, without an inordinately large number of grid blocks. For the example to be described, the large-scale model contains 800 grid blocks. The same model using small-scale grid blocks would contain 640 million grid blocks.

The pseudo function generation process which we will describe is simplified, but we feel it is realistic enough to investigate aspects of the scaling up process. We have assumed that the pseudo functions are the same everywhere at a given scale and that they are directionally independent. Spatial and directional dependence must be included in any realistic pseudo generation procedure. Where gravity and capillary forces are significant, rate dependence of the pseudos may also be a factor. These aspects should be investigated in future work.

The pseudo function generation process is outlined in Figure 5b. It is similar in some respects to the approach of Kyte and Berry[6], but differs in some important respects. The technique consists of first computing an effective directional permeability numerically by simply applying a constant potential in each coordinate direction across the model and computing the steady-state flux. For this calculation, we assume the fluid is single-phase with a viscosity of unity. The effective permeability is then computed from Darcy's equation:

$$k = \frac{Q \Delta X}{A \Delta \Phi} \tag{1}$$

where:

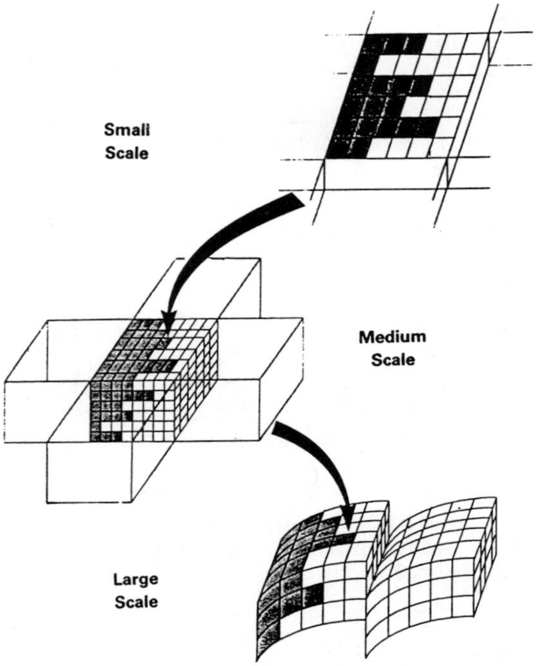

Figure 5a. The Scaling Up Process

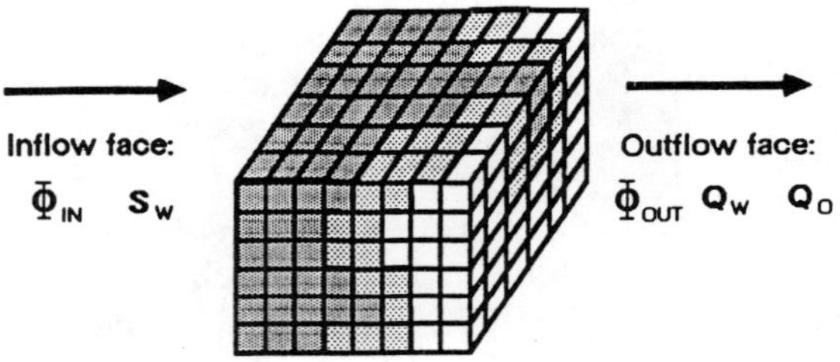

Compute effective permeability numerically

Run multiphase model:

- Compute pore—volume—averaged S_W versus time
- Compute outflow face fractional flow versus time
- Compute total transmissibility versus time

Derive pseudo relative permeabilities

Figure 5b. The Pseudo Generation Procedure

k	intrinsic permeability
Q	total volume flow rate
ΔX	distance between boundaries
$\Delta\Phi$	potential difference between boundaries
A	cross-sectional area

We then solve a multiphase flow problem, applying a constant potential (or constant flux) in one direction across the model and computing the outflow water and oil flow rates as a function of the pore-volume averaged saturation. The water and oil pseudo relative permeabilities are then computed from the equations:

$$k_{rw}(\tilde{S}_w) = \frac{Q_w \mu_w \Delta X}{kA\Delta\Phi_w} \tag{2}$$

$$k_{ro}(\tilde{S}_w) = \frac{Q_o \mu_o \Delta X}{kA\Delta\Phi_o} \tag{3}$$

where:

k_{rw}	water pseudo relative permeability
k_{ro}	oil pseudo relative permeability
$\tilde{S}_w$	pore-volume-averaged water saturation
Q_w	water volumetric flow rate
Q_o	oil volumetric flow rate
$\Delta\Phi_w$	water potential difference between boundaries
$\Delta\Phi_o$	oil potential difference between boundaries

In order to avoid boundary effects, an entrance section is necessary on the injection side so that the saturation profile is established prior to entering the volume over which the pseudos are being computed. If capillary pressures are included, an exit section is also required to avoid capillary end effects. Where entrance or exit regions are used, we have to compute average potentials across the entrance and exit surfaces. The appropriate averaging is assumed to be a flux-weighted average of a harmonic average of grid block potentials between neighboring grid blocks across the surface.

The pseudo functions developed in this manner include the effects of heterogeneities as well as largely eliminating numerical diffusion as shown by Kyte and Berry[6].

In the small-scale simulations, the variable weighting procedure[2] was used to effectively eliminate numerical diffusion effects associated with the use of the rock relative permeability functions. Having done this, upstream weighting with pseudo functions are used at the medium and large scales.

For our scaling up example, we will consider a water-oil problem in which the oil viscosity is 20 cp and the water is 1 cp. No capillary pressure will be included.

The process begins with the small-scale permeability distribution shown in Figure 6. The model is two-dimensional, 5 meters wide and 25 meters long with a total of 2000 grid blocks. The pseudo functions are computed using the last 5 meters at the right-hand end of the model, i.e., there is a 20 meter entrance region. The injection rate is such that the average superficial velocity is 1 meter/day with constant potential at the left boundary and a fixed potential at the right boundary. The permeability distribution is a log-normal distributed fractal surface[7]. The mean value is 1 μm^2. Because the distribution is log-normal, a standard deviation is a factor multiplying the mean value. For this example, a standard deviation is a factor of two: one standard deviation down is $500 \times 10^{-3} \mu m^2$ and one standard deviation up is 2 μm^2. One hundred pore volumes were injected and pseudo functions computed. In examining the figure, we see that the water moves principally along a high permeability streak down the center of the model.

The small-scale pseudos were then used in a medium-scale simulation shown in Figure 7. The model here is three-dimensional: 500 meters long, 100 meters wide, and 5 meters thick with a total of 2000 grid blocks. The pseudo functions are computed using the last 100 meters at the right-hand end of the model with a 400 meter entrance region. Each grid block is the size of the previous small-scale model. Each layer is the same normalized log-normal fractal permeability distribution, but the mean value decreases from $1 \mu m^2$ in the top layer to $10 \times 10^{-3} \mu m^2$ in the bottom layer. Fifty pore volumes were injected, and the scaling up process repeated. From the figure, we see that because of the low permeability in the bottom layers, little oil is being displaced from them.

The large-scale simulation is shown in Figure 8. The model is three-dimensional: 2000 meters long, 1000 meters wide, and 40 meters thick. Each grid block is the size of the previous medium-scale model. Again the permeability distributions for each layer is a log-normal fractal distribution. For expendiency, the same normalized distribution was used in each layer. The mean values are the same as those shown for the four layers in Figure 4. The standard deviations in all layers was a factor of two. The water injector is completed in the top four layers at the left and the producer in the bottom three layers at the right.

The saturation distribution after 5000 days is shown for two cases. The first is the results using the scaled-up pseudos. The second is the result without scaling up, that is, using the original rock functions. The saturation distribution using the scaled-up pseudo functions shows more distributed regions of bypassed oil than the corresponding case for the rock functions. There is certainly much more of a finger down to the producer in the pseudo function case.

Figure 9 shows the cumulative percent oil produced as a function of time for three cases:

1. scaling up from small to medium to large (rock functions input at the small scale);

2. scaling up from medium to large (rock functions input at the medium scale);

3. no scaling up (rock functions input at the large scale).

In comparing the three cases, we see that the two scaling up cases have almost identical recovery curves. Compared to the case without scaling up, the scaling up cases recover approximately 4% less of the initial oil-in-place and 10% less cumulative oil.

The production water cut for these three cases is shown in Figure 10. In the scaling up cases, the water breakthrough is considerably earlier than the case in which there is no scaling up. Surprisingly, the water cut at late time for the scaling up cases appears to have stabilized at lower water cut values than the case with no scaling up.

Figures 11 and 12 show the pseudo relative permeabilities in the process of scaling up from small to medium and medium to large scales. Despite the heterogeneities at the small scale, the small-scale pseudos are not significantly different from the rock functions. It follows that the small-scale heterogeneities do not have a significant effect on recovery. The medium-scale pseudos on the other hand are significantly different from the rock functions, suggesting that the systematic layering at the medium scale does influence recovery at the large scale.

VIII. CONCLUSIONS

We have developed a systematic scaling-up procedure which permits us to effectively include small-scale heterogeneities in large-scale models. The scaling-up process involves the computation of volumetrically-averaged pseudo functions which allow the modeling of complex heterogeneities as well as correcting for numerical diffusion effects. In the pseudo generation procedure, an entrance region is necessary to establish a stablized saturation front entering

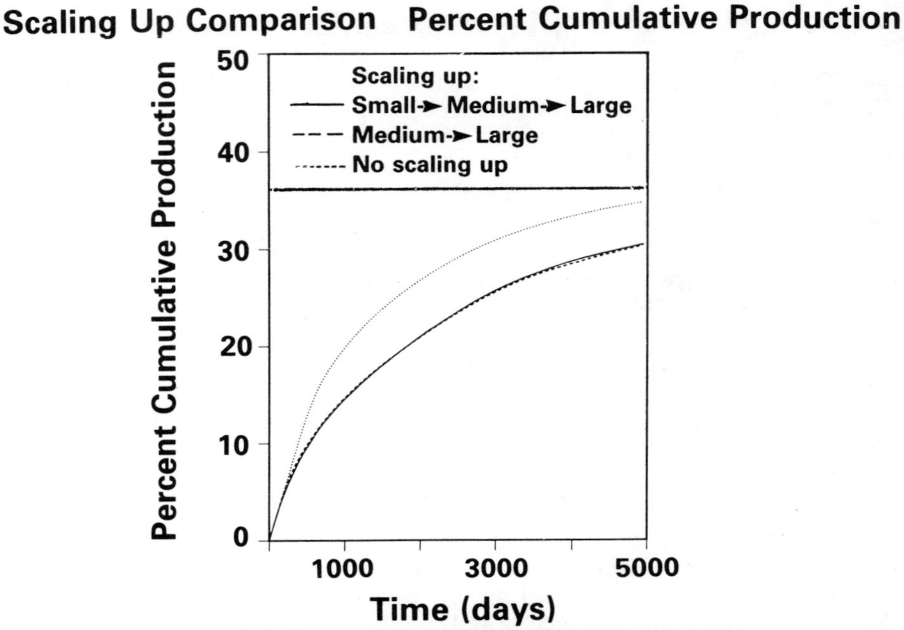

Figure 9. Cumulative Percent Oil Produced

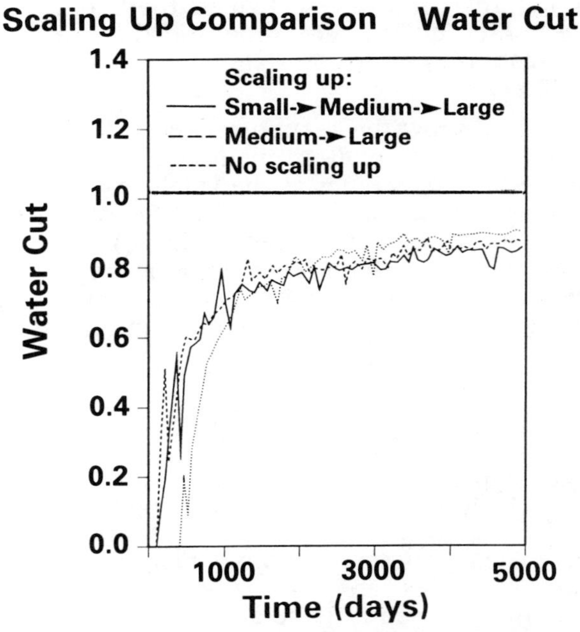

Figure 10. Production Fractional Flow

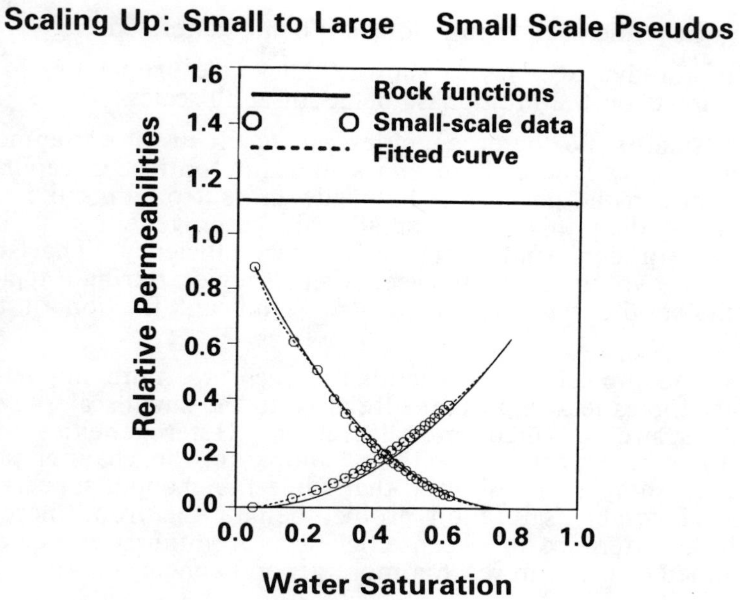

Figure 11. Small-scale Pseudo Functions - Relative Permeabilities

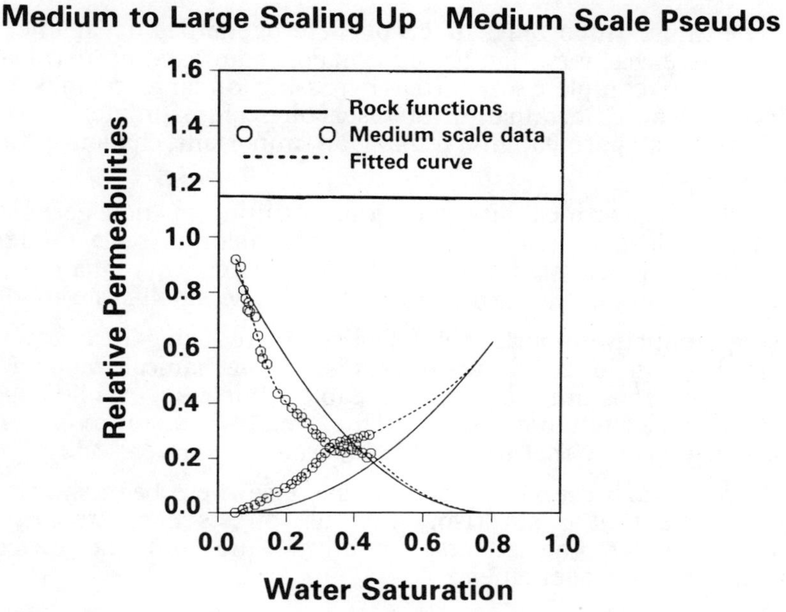

Figure 12. Medium-scale Pseudo Functions - Relative Permeabilities

the region where the pseudo functions are being generated.

In this study, we have shown that heterogeneities have significant effects on the multiphase behavior at all scales.

At the small scale, heterogeneities can result in oil entrapment in small cul-de-sacs where there is a significant contrast in capillary pressures between the porous media inside and surrounding the cul-de-sacs. Other than this effect, small-scale heterogeneities do not appear to have a significant effect on recovery efficiency. The large number of pore volumes which sweep a small region during a typical flood results in the rapid recovery of a significant fraction of the mobile oil.

At the medium scale, gravitational forces are more important and capillary forces less important. Relative to the small scale, fewer pore volumes sweep a medium-scale region. Heterogeneities tend therefore to have a greater effect. In comparing the channel sand and bar sand examples, we showed that while the channel appears to have a less favorable saturation profile, behind the front there is probably little difference in sweep efficiency. Additional cases need to be examined in order to reach a more general conclusion.

In our example, the medium-scale heterogeneities result in a much earlier water breakthrough. Medium-scale and even small-scale heterogeneities may be extremely important in highly adverse mobility situations such as gas and liquid.

The most important effect of heterogeneities is at the large scales where barriers, faults, and major contrasts in permeability result in the complete or partial bypassing of large regions of the reservoir. The determination of reservoir-scale estimates of vertical and horizontal permeability remain an important challenge for the industry.

Systematic permeability variations resulting in high permeability channels or layers such as at that for the medium-scale model, are undoubtedly more significant in reducing recovery than a more random distribution in permeability such as the small-scale model.

A potentially important mechanism in reducing recovery which has not been included in our models is the mechanical trapping of oil by the juxtaposition of impermeable barriers. While such a mechanism has not been seen at the small scale, outcrop maps of shale barriers show that it might be operative at larger scales[8].

Even though pseudo functions have elsewhere been shown to be very important, it is surprising that the effects here are not more substantial given the adverse mobility ratio and the degree of permeability heterogeneities.

To more realistically estimate reservoir performance, we must model reservoir heterogeneities at the various scales more effectively. This requires a more synergetic effort between geologists and reservoir engineers. We have concluded that reservoirs should be described and modeled using depositional units as the fundamental descriptors and building blocks. In addition, reservoir geology must develop a more quantitative approach in defining depositional units and the depositional environments in which they are formed. Finally, reservoir engineering must develop efficient means to use these more detailed, more quantitative, and more complex descriptions of reservoirs in reservoir simulation models.

REFERENCES

1. Sneider, R.M. et al, "Predicting Reservoir Geometry and Continuity in Pennsylvanian Reservoirs, Elk City Field, Oklahoma", Journal of Petroleum Technology, July 1977.

2. Lasseter, T.J., "The Extension of the Variable-flux-weighting Method to Multidimensional Problems Including Gravity and Capillarity", Eighth SPE Symposium on Reservoir Simulation, February 1985.

3. Weber, K.J., private communication.

4. Weber, K.J., "Influence of Common Sedimentary Structures on Fluid Flow in Reservoir Models", Journal of Petroleum Technology, March 1982.

5. Lasseter T.J., "The Interpretation of a Pressure Profile Pulse Test Using a Three-dimensional Numerical Model to Determine Reservoir-scale Properties", submitted to the Society of Petroleum Engineers.

6. Kyte, J.R. and Berry, D.W., "New Pseudo Functions To Control Numerical Dispersion", Society of Petroleum Engineers Journal, August 1975.

7. Voss, R. F., "Fourier Synthesis of Gaussian Fractals: 1/f noises, landscapes, and flakes", IBM Thomas J. Watson Research Center, unpublished report.

8. Jones, J.R., Scott, A., Lake, L.W., "Reservoir Characterization for Numerical Simulation of Mesaverde Meander Belt Sandstone, Northwestern Colorado", Society of Petroleum Engineers 1984 Fall Meeting, SPE 13052.

All methods used for calculation of the transfer functions are inaccurate because the reservoir conditions change with time. Moreover, the transfer functions depend on the production schemes. For example, fluid flow from matrix to fracture can be reversed during water or gas injection in the reservoir. Special sets of dynamic transfer functions may be required for different reservoir conditions.

A simplified approach with the use of a conventional single porosity simulator was presented by Bossie-Codreanu et al(9).

The next important step was the development of general purpose fractured reservoir simulators, with automatic calculation of the matrix contribution depending on changing reservoir conditions. It was done by extending Warren and Root dual porosity approach to a two-phase multidimensional case (Kasemi et al(10), 1976, Gilman, Kazemi(11), 1982). A three-phase, three-dimensional simulator for modeling naturally fractured reservoirs was developed by Thomas et al(12), 1980. Another three-phase, 3D fractured reservoir simulator, with the addition of matrix to matrix flow and multicomponent fluid representation, was published by Blaskovich et al(13), 1983.

Displacement of oil from the tight matrix blocks depends on capillary, gravity forces, and the pressure difference between matrices and fractures. The effects of capillary and gravity forces in fractured reservoirs require special consideration, different from a conventional single porosity system. It depends on the fluid surrounding a matrix block and the degree in which each matrix block is immersed in the gas or water. A detailed description of the fluid displacement processes for a single matrix block is given in the literature (Reiss(15), Van Golf-Racht(16)). This description includes the consideration of the specifics of capillary and gravity forces for matrix blocks in a given environment, but it does not allow calculation of the influx from matrix to fractures due to the pressure difference because the computation of pressure distribution requires a numerical simulator. At the same time, the fractured reservoir simulators(10,11,12,13) consider the matrix contribution due to pressure difference between matrices and fractures but the treatment of the capillary and gravity forces remains the same as in conventional reservoir simulators.

Thus, the fractured reservoir simulators presented in the literature can be divided into the following two groups:

1. Models for special applications which may consider the details of the gravity and capillary effects in the matrix blocks but can not be used universally for field studies of different fractured reservoirs.
2. General purpose three-dimensional, three-phase dual porosity reservoir simulators which are designed for field studies, but which don't consider all the details of gravity and capillary displacement in a single matrix block due to different conditions in the surrounding fractures.

The development presented in this paper incorporates the special treatment of capillary and gravity forces for the fracture-matrix media into a general purpose dual porosity three-phase, three-dimensional reservoir simulator which is designed for field applications.

II. FORMULATION OF FLOW EQUATIONS

The fractured reservoir is represented by the continuous system of fractures and discontinuous system of matrix blocks. Each grid block of the matrix grid may contain several equally spaced matrix blocks surrounded by fractures.

The equations describing a three-phase, three-dimensional flow in fractures and matrices in the finite-difference form can be written as follows.

Fracture flow equations

Oil

$$[\Delta(T_{of}\Delta p_f) + CG_{of}] + [T_{o_{mf}}(p_m - p_f) + CG_{o_{mf}}] - q_{of}$$

$$= \frac{V_b}{\Delta t} \delta \left(\frac{\emptyset S_o}{B_o} \right)_f \quad \ldots\ldots\ldots\ldots\ldots\ldots\ldots\ldots\ldots \text{(1)}$$

Water

$$[\Delta(T_{wf} \Delta p_f) + CG_{wf}] + [T_{w_{mf}} (p_m - p_f) + CG_{w_{mf}}] - q_{wf}$$

$$= \frac{V_b}{\Delta t} \delta \left(\frac{\emptyset S_w}{B_w} \right)_f \quad \ldots\ldots\ldots\ldots\ldots\ldots\ldots\ldots\ldots \text{(2)}$$

Gas

$$\Delta[(T_{gf} + R_{so} T_{of} + R_{sw} T_{wf}) \Delta p_f] + CG_{gf}$$

$$+ [(T_{g_{mf}} + R_{so} T_{o_{mf}} + R_{sw} T_{w_{mf}})(p_m - p_f)$$

$$+ CG_{g_{mf}}] - q_{gf} = \frac{V_b}{\Delta t} \delta \left[\emptyset \left(\frac{S_g}{B_g} + \frac{R_{so} S_o}{B_o} + \frac{R_{sw} S_w}{B_w} \right) \right]_f \quad \text{(3)}$$

Matrix flow equations

Oil

$$[\Delta(T_{om} \Delta p) + CG_{om}] - [T_{o_{mf}} (p_m - p_f) + CG_{o_{mf}}] - q_{om}$$

$$= \frac{V_b}{\Delta t} \delta \left(\frac{\emptyset S_o}{B_o} \right)_m \quad \ldots\ldots\ldots\ldots\ldots\ldots\ldots\ldots\ldots \text{(4)}$$

Water

$$[\Delta(T_{wm} \Delta p_m) + CG_{wm}] - [T_{wmf} (p_m - p_f) + CG_{wmf}] - q_{wm}$$

$$= \frac{V_b}{\Delta t} \delta \left(\frac{\emptyset S_w}{B_w} \right)_m \quad \ldots\ldots\ldots\ldots\ldots\ldots\ldots\ldots\ldots \text{(5)}$$

Gas

$$\{\Delta[(T_{gm} + R_{so}T_{om} + R_{sw}T_{wm})\Delta p_m] + CG_{gm}\}$$

$$- \{(T_{g_{mf}} + R_{so}T_{o_{mf}} + R_{sw}T_{w_{mf}})(p_m - p_f) + CG_{g_{mf}}\}$$

$$- q_{gm} = \frac{V_b}{\Delta t}\, \delta[\emptyset\ (\frac{S_g}{B_g} + \frac{R_{so}S_o}{B_o} + \frac{R_{sw}S_w}{B_w})]_m \quad \ldots\ldots\ldots(6)$$

Eqs. 1 through 3 describe the fluid flow in the fracture system with contribution from matrices. The first part of each equation (term $\Delta(T_{of}\Delta p) + CG_{of}$ in Eq. 1) represents the flow between fracture blocks, the second part (term $T_{omf}(p_m - p_f) + CG_{omf}$ in Eq. 1) describes the flow between matrices and fractures inside of each grid block. Similarly, the first parts of matrix Eqs. 4 through 6 depict the flow between matrix blocks. The second parts of these equations are the same matrix to fracture flow term as in the fracture equations but with a different sign.

The equations show separate contributions to the fluid flow from the three main mechanisms: pressure difference, capillary and gravity forces. Term $\Delta(T\Delta p)$ represents the influx due to pressure difference, term CG shows the capillary and gravity contributions. This formulation is more convenient for considering the capillary and gravity effects in the fractured reservoirs.

Capillary and gravity terms CG_f for flow between fractures in Eqs. 1 through 3 are calculated as follows:

$$CG_{of} = -\ \Delta[T_o\gamma_o\Delta D]_f$$

$$CG_{wf} = -\ \Delta[T_w(\gamma_w\Delta D + \Delta P_{c_{ow}})]_f$$

$$CG_{gf} = -\ \Delta[T_g(\gamma_g\Delta D - \Delta P_{c_{go}}) + R_{so}T_o(\gamma_o\Delta D)$$

$$+ R_{sw}T_w(\gamma_w\Delta D + \Delta P_{c_{ow}})]_f$$

Capillary and gravity terms CG_m for flow between matrices in Eqs. 4 through 6 are calculated similarly.

$$CG_{om} = - \Delta[T_o \gamma_o \Delta D]_m$$

$$CG_{wm} = - \Delta[T_w(\gamma_w \Delta D + \Delta P_{c_{ow}})]_m$$

$$CG_{gm} = - \Delta[T_g(\gamma_g \Delta D - \Delta P_{c_{go}}) + R_{so} T_o(\gamma_o \Delta D)$$

$$+ R_{sw} T_w(\gamma_w \Delta D + \Delta P_{c_{ow}})]_m$$

The calculation of the capillary and gravity terms CG_{mf} for flow between matrices and fractures will be discussed in the next section.

The given formulation was derived by substituting the capillary pressure relationships into the flow equations. The capillary pressure expressions for fractures and matrices are:

$$P_{c_{ow}} = p_o - p_w \qquad P_{c_{go}} = p_g - p_o$$

Saturation equations for fractures and matrices should also be added to the system of equations.

$$S_o + S_w + S_g = 1 \quad \ldots\ldots\ldots\ldots\ldots\ldots\ldots\ldots\ldots (7)$$

Altogether Eqs. 1 through 6, and the two Eqs. 7 represent eight equations with eight unknowns. The unknowns are the six saturations $(S_o, S_w, S_g)_f$, $(S_o, S_w, S_g)_m$, and the two pressures p_{of}, p_{om}.

The matrix Eqs. 4 through 6 will be much easier to solve if the matrix to matrix flow term is not included in the system. This term often represents insignificant portions of the flow compared to matrix/fracture term (see discussion "Transmissibilities" below). Eqs. 4 through 6 without the matrix to matrix flow term and without the production term can be written in the following form:

Oil

$$- T_{o_{mf}}(p_m - p_f) + CG_{o_{mf}} = \frac{V_b}{\Delta t} \delta \left(\frac{\emptyset S_o}{B_o} \right)_m \quad \ldots\ldots\ldots (8)$$

Water

$$- T_{w_{mf}} (p_m - p_f) + CG_{w_{mf}} = \frac{V_b}{\Delta t} \delta \left(\frac{\emptyset S_w}{B_w} \right)_m \quad \ldots\ldots\ldots(9)$$

Gas

$$- (T_{g_{mf}} + R_{so} T_{o_{mf}} + R_{sw} T_{w_{mf}})(p_m - p_f) + CG_{g_{mf}}$$

$$= \frac{V_b}{\Delta t} \delta \left[\emptyset \left(\frac{S_g}{B_g} + \frac{R_{so} S_o}{B_o} + \frac{R_{sw} S_w}{B_w} \right) \right] \quad \ldots\ldots\ldots\ldots(10)$$

The fracture system (described by Eqs. 1-3) in this case will provide the main path for the fluid flow in the reservoir with the matrices serving as a source. Matrix Eqs. 8 - 10 describe the fluid exchange between matrices and fractures.

Both formulations (Eqs. 4 - 6 and Eqs. 8 - 10) allow variations in the degree of fracturing throughout the reservoir by specifying different matrix block sizes and fracture porosities to different parts of the reservoir. Each grid block may have its own fracture and matrix properties. The highly fractured parts of the formation may have a small matrix block size. Non-fractured parts of the reservoir can be represented by assignment of the matrix rock property to Eqs. 1 - 3 and using these equations as a single porosity system. Eqs. 1 through 6 (or Eqs. 1 - 3, 8 - 10) will be reduced to the single porosity formulation if the matrix permeability is equal to zero.

The formulation of equations for the double porosity system requires not only a doubling of the number of equations compared to a conventional single porosity system, but also requires a different treatment of the displacement mechanism in the matrix rock and different presentation of the transmissibilities. These two questions will be discussed in the next two sections.

III. THE DESCRIPTION OF DISPLACEMENT MECHANISM IN THE MATRIX BLOCKS

A. Gravity Effect on the Matrix/Fracture Flow

The gravity forces in the conventional single porosity simulator depend on the difference in elevation between two neighboring grid blocks. Contrary to this the gravity effect on flow in a single matrix block surrounded by fractures is determined by the degree to which the matrix block is immersed in either water or gas.

For example, the gravity effect for two neighboring blocks with elevations D_1 and D_2 in a single porosity system will be

$$P_G = \gamma_\alpha (D_2 - D_1) \quad \text{..............................(11)}$$

where γ_α is a specific weight of fluid α, $\gamma_\alpha = \rho_\alpha/144$ psi/ft (kPa/m).

The gravity effect for a single matrix block surrounded by fractures in a water invaded zone will be

$$P_G = (\gamma_w - \gamma_o)(Z_m - Z_f) \quad \text{......................(12)}$$

where Z_m, Z_f represent the water levels in the fractures and in the matrix block.

The conventional single porosity approach (Eq. 11) gives a zero gravity effect for a horizontal one layer model since in this case $D_1 = D_2$ and $P_G = 0$. However, the actual gravity effect on the influx of oil from the matrices into fractures can be substantial if the matrix blocks are surrounded by water filled fractures.

Also, the values of the elevations D_i in the single porosity representation (Eq. 11) are constant for the whole run. On the other hand the water (or gas) levels in the fractures and the matrices used in Eq. 12 are changing dynamically during the run since the water (or gas) may move faster through the highly permeable fractures leaving behind matrix blocks which are totally or partially immersed in water (or gas).

Generally speaking, the conventional approach (Eq. 11) can be used for the double porosity system if the matrix block is divided into several layers. This approach can normally be used only for a single matrix block study because it requires an excessive number of layers. For example, consider modeling a 500 ft. thick horizontal fractured formation with matrix blocks 10 feet in height. The use of three layers for each matrix block would require a 150 layer model (150 layers = 3 layers/matrix block x 50 matrix blocks).

The effect of gravity on the matrix/fracture flow was not included in Thomas' formulation(12). Pseudo relative permeability and capillary pressure curves were used in this work to account for the gravity effect. Kazemi and Gillman(11) used different fracture and matrix(5) depth within a given grid block in order to represent the gravity effect. This approach still treats the gravity forces in the framework of the conventional formulation (Eq. 11). The procedure described below considers the gravity effect (Eq. 12) and does not increase the number of layers in the model.

B. Capillary Pressure Effect on the Matrix/Fracture Flow

Conventional single porosity formulations consider the capillary pressure effect as a function of the water saturation. The whole grid block is subject to imbibition in this case. However, in fractured reservoirs, imbibition takes place only in a portion of the grid block immersed in the water. This is explained by the segregated flow conditions in the fractures and the discontinuity of the matrix rock which makes the matrices above the water-oil level out of reach of the water in the fractures. The imbibition (or drainage) process in the fractured reservoir should be related to the water (or gas) levels in the fractures. Non zero capillary pressure can exist in the matrix blocks, but imbibition takes place only when water moves through the fractures and contacts the matrix block. In other words, the capillary pressure effect in this case is considered not just as a function of the water (or gas) saturations but also as a function of the change in the water (or gas) level in the fractures.

This approach is useful not only for the calculation of the capillary pressure effect during the run but also for specifying the initial equilibrium conditions for the matrix/fracture media.

The procedure for the calculation of the initial water or gas saturations in conventional reservoirs is based on the assumption that the water saturation in a transitional zone above the water-oil contact is gradually reduced, going up in the structure corresponding to the capillary pressure curve. A similar procedure is used for the gas-oil transitional zone.

The initial equilibrium conditions in a fractured reservoir can be completely different from that of a conventional reservoir. The water saturation in the matrix blocks is not related to the water-oil contact due to the matrix discontinuity. It is defined only by the properties of the matrix rock. Tighter matrices may have higher water saturation because of higher capillary pressure. As a result, a higher initial water saturation can be observed in zones above a low water saturation zone.

These initial conditions can be represented in a fractured simulator by specifying the term CG_{mf} initially equal to zero. In this case the system will be in equilibrium if the matrix and fracture pressures are equal for the same grid block. The described initial conditions mean that the capillary pressure is equal or higher than gravity force in a matrix block, which prevents water from moving out of the matrix block. At the same time, imbibition has not occurred because water is not available in the surrounding fractures. Imbibition takes place when water moves up in the fractures during production and thereby becomes available. The water-oil capillary force is calculated as a difference in capillary pressures between matrix and fracture blocks multiplied by coefficients C_w, C_g which represents the change in water (or gas) level in the fracture block.

C. Combined Effect of Gravity and Capillary Pressure on Matrix/Fracture Flow

Capillary imbibition will act in the same direction as the gravity forces for single matrix block immersed in the water (for water-oil system with water wet rock). Water will move in the matrix blocks under these forces and displace the oil.

The combined effect of gravity and capillary forces in this case will be

$$P_{GC} = P_G + P_C \qquad (13)$$

The capillary pressure P_c can be substantially larger in the dual porosity system (compared to single porosity model) since P_{cf} in the highly permeable fractures is close to zero and

$$P_c = P_{cm} - P_{cf} = P_{cm}.$$

The displacement mechanism in a gas invaded zone is different. Gravity forces will assist to the displacement of oil by gas in matrices but the capillary forces would act in the opposite direction. This is understandable since the liquid will be the wetting phase in this case and capillary pressure will resist the removal of oil from the matrix blocks. Besides, the gas will move into the matrix blocks only if the driving force (combined viscous and gravity contributions) exceeds the threshold capillary pressure. However, the gravity forces will be larger in the gas-oil system as compared to the water-oil system due to the substantial difference between the densities of the oil and gas. For large matrix blocks the gravity force can exceed the negative effect of the capillary pressure.

The combined effect of gravity and capillary pressure in the gas invaded zone will be

$$P_{GC} = -P_G + P_C \quad \ldots\ldots\ldots\ldots\ldots\ldots\ldots\ldots\ldots\ldots (14).$$

The described mechanism can be incorporated in the gravity and capillary pressure terms CG_{mf} of the flow equations. The procedure is the following.

1. Calculate the CG_{SMB} term for a single matrix block assuming that the matrix block is completely immersed in either water or gas.
2. Multiply the term CG_{SMB} by the number of matrix blocks contained in one grid block.
3. Calculate the portion of the whole grid block C_w (or C_g) immersed in the water or gas using the change of the water or gas levels in the fractures.
4. Calculate CG_{mf} term for a whole grid block taking into account the coefficient C_w (or C_g).

This procedure allows the formulation of the physical processes in a single matrix block surrounded by fractures by use of Eqs. 1 - 10, which deals with the whole grid blocks. In other words, this approach permits us to run fieldwide simulation studies and at the same time takes into account the specifics of the displacement mechanism in the matrix rock.

IV. TRANSMISSIBILITIES

The x-direction transmissibilities in the conventional single porosity formulation are calculated for each phase as follows

$$T_x = 0.001127(k_{r\alpha}/\mu_\alpha B_\alpha)(k\Delta y\Delta z/\Delta x) \quad \ldots\ldots\ldots\ldots\ldots(15)$$

Fracture to fracture and matrix to matrix transmis -sibilities in the double porosity simulator are calculated similar to Eq. 15.

$$T_{f_x} = 0.001127(k_{r\alpha}/\mu_\alpha B_\alpha)_f(k_f\Delta y\Delta z/\Delta x)\emptyset_f \quad \ldots\ldots\ldots(16)$$

$$T_{m_x} = 0.001127(k_{r\alpha}/\mu_\alpha B_\alpha)_m(k_m\Delta y\Delta z/\Delta x) \quad \ldots\ldots\ldots\ldots(17)$$

But the matrix/fracture transmissibilities are calculated differently, using so-called geometrical factor σ (see Kazemi et al(10), Thomas et al(12)).

$$T_{mf} = 0.001127(k_{r\alpha}/\mu_\alpha B_\alpha)_m(k_m\Delta x\Delta y\Delta z)\sigma \quad \ldots\ldots\ldots(18)$$

The geometrical factor(called also "matrix shape factor" or "fracture spacing parameter") was introduced first by Warren and Root for one dimensional flow equation in the form $\sigma = 4n(n+2)/L^2_m$, where n is normal set of fractures, n = 1,2,3; L_m is characteristic dimension of matrix.

For n = 1

$$\sigma = 12/L^2_m \quad \ldots\ldots\ldots\ldots\ldots\ldots\ldots\ldots\ldots\ldots\ldots\ldots(19)$$

Kazemi modified Eq. 19 for a more general case, when $L_x \neq L_y \neq L_z$

$$\sigma = 4\ (1/L^2_x + 1/L^2_y + 1/L^2_z) \quad \ldots\ldots\ldots\ldots\ldots\ldots\ldots(20)$$

where L_x, L_y, L_z are x, y and z dimensions of the matrix block.

The geometrical factor is a very important parameter effecting the results of the simulation. The derivation and physical meaning of this parameter will be considered below. Some useful conclusions will also be given based on this consideration.

Let's consider, for simplicity, the case in which a matrix block is a cube with side L_m. The transmissibility of a single matrix block surrounded along all six sides by fractures can be expressed in this case as

$$T_{SMB} = 0.001127(k_{r\alpha}/m_\alpha B_\alpha)_m \times 6(k_m L^2{}_m/0.5\ L_m)$$

The number of the matrix blocks in a given grid block is

$$N_m = \Delta x \Delta y \Delta z / L^3{}_m$$

Eq. 18 for matrix transmissibility of the whole grid block can be derived by multiplying the transmissibility of a single matrix block T_{SMB} by the number of matrix blocks in a given grid block.

The geometrical factor represents the ratio between the surface area, size and volume of the matrix block

$$\sigma = 12A_m/L_m V_m = 12/L^2{}_m$$

This expression exactly repeats the Eq. 19 given by Warren and Root.

A similar approach reduces the geometrical factor to Eq. 20 for matrix blocks with sides $L_x \neq L_y \neq L_z$.

Eqs. 19 and 20 assume that each matrix block has a cubic shape and is surrounded by fractures along all six sides of the cube. The equations will be different if a matrix block has a different form or fractures are located differently. For example, if there are fractures only along two sides of the cubic matrix block in the three-dimensional equation, the value of geometrical factor (Eq. 19) should be reduced by the factor of three.

The information about the shape of the matrix blocks and frequency of fractures is usually not available or not accurate. Because of that it is easy to use one equation like 19 or 20 in the program and modify only the value of the matrix block side in order to change the matrix/fracture transmissibility instead of using different expressions for each configuration of fractures.

Given derivation of the matrix/fracture transmissibility illustrates the main idea of representing many matrix blocks in one grid block of the simulator. The same approach (using the properties of the single matrix block and combining it for the description of the whole grid block) was implemented also for the description of the gravity and capillary effect in the matrix/ fracture interaction in the previous section.

The consideration of the transmissibilities also allows us to make some useful conclusions.

1. The matrix to matrix flow term can be omitted from the simulator equations if matrix block size is much smaller than grid block size.

   A substantial amount of CPU time can be saved if the matrix to matrix flow term would not be included in Eqs. 1 - 6. The term can be omitted when fluid migration from matrix to matrix blocks is much less than the migration between matrices and fractures. The consideration of Eqs. 17 and 18 shows that the matrix to matrix transmissibilities T_m are substantially smaller than the matrix/fracture transmissibilities T_{mf} if the size of the matrix block is much less than the grid block size. The ratio between T_{mf} and T_m using cubic matrix blocks is

   $$T_{mf}/T_m = \sigma\Delta x^2 = 12\Delta x^2/L^2_m \quad \text{.....................(21)}$$

   Let's consider the following example. Suppose the grid block size in a field simulation study was 660 ft. (10 acres per grid block). If the matrix block size is 10 ft., the ratio of T_{mf} to T_m will be 52,272. If the matrix block size is 1 ft., the matrix to matrix transmissibility will be 5,227,200 times smaller than the matrix/fracture transmissibility. The matrix to matrix flow in this case will be negligible compared to the matrix/fracture flow. Thus, in a field study with matrix block size substantially smaller than the grid block size it is not required to consider the matrix to matrix flow and Eqs. 1 through 6 can be reduced to Eqs. 1-3, 8-10.

   The matrix to matrix flow in case of large matrix blocks can be represented by the formulation 1 - 6. It can also be done in formulation 1 - 3, 8 - 10 by specifying certain grid blocks as large matrix blocks of single porosity without fractures. Fluid exchange between such blocks represents the matrix to matrix flow.

2. The comparison between matrix/fracture, fracture/fracture and conventional single porosity transmissibilities indicates when the performance of the double porosity system is approaching the single porosity case. This will happen when the value of matrix/fracture transmissibility is large and fracture transmissibility is equal to the transmissibility of the equivalent single porosity system. In this case the double porosity system is reduced to a single porosity formulation with total porosity $\emptyset_t = \emptyset_f + \emptyset_m$ and effective permeabilities $k_e = K_f \emptyset_f$, $k_{re} \doteq k_{rf}$ (see Eqs. 15, 16).

V. SOLUTION TECHNIQUES

The formulation of the double porosity system initially has eight unknown vectors and eight equations. The procedure similar to IMPES reduces the number of unknown vectors to two: matrix and fracture pressures. The system of pressure equations is larger than a single porosity formulation and has considerably different coefficients for matrix and fracture blocks. The solution of this system may require a large amount of CPU time. The pressure system can be solved much faster for formulation 1 - 3, 8 -10 with the use of the partial preconditioning described below.

The pressure equations in this case can be written as follow.

For fractures

$$A_f P_f + E_{mf} P_m = B_f \quad \ldots\ldots\ldots\ldots\ldots\ldots\ldots\ldots (22)$$

For matrices

$$E_{mf} P_f + E_m P_m = B_m \quad \ldots\ldots\ldots\ldots\ldots\ldots\ldots\ldots (23)$$

The pressure P_m can be eliminated from Eq. 22 with the help of block Gaussian elimination. Then, the equation will be

$$A_{f1} P_f = B_{f1} \quad \ldots\ldots\ldots\ldots\ldots\ldots\ldots\ldots\ldots (24)$$

where

$$A_{f1} = A_f - E_{mf}E_m^{-1}E_{mf}$$

$$B_{f1} = B_f - E_{mf}E_m^{-1}B_m$$

The solution of system 22, 23 in this case consists in solving Eq. 24 with regard to P_f, and calculating P_m from Eq. 23. Thus the fracture and matrix pressures are found separately which results in a substantial reduction in CPU time.

The program uses an IMPES formulation with implicit treatment of the well term and uses the described procedure for the solution of the pressure equations.

EXAMPLE PROBLEM

The model represents a two-well section of a reservoir. The simulated area is a 40 acre horizontal square with wells located in the diagonal corners of the grid. The reservoir is divided into three layers. The oil zone is represented by two upper layers, the aquifer is simulated by the bottom layer. The reservoir properties for the example are given in Table 1.

The results of the computer runs were compared for four different cases:

1. Dispersed flow in fractures.
   This case represents a conventional treatment of the capillary pressure and gravity. The capillary imbibition can occur in the matrix blocks located in the entire oil zone. No gravity effect can be represented for flow between matrices and fractures.
2. Segregated flow in fractures.
   This case uses the new treatment of capillary pressure described in this paper. The capillary imbibition depends on the water level in the fractures. Only matrix blocks contacted by the water in fractures can imbide water. The gravity effect on matrix/fracture flow is included in the consideration.
3. Gravity effect only.
   This case uses the same formulation as in case 2 but $P_{cm} = 0$.

4. No gravity, no capillary pressure.
   This case uses the same formulation as in case 1 with $P_{cm} = 0$.

The results of the runs are presented in Fig. 1 and Fig. 2. Producing WOR (Fig. 1) is zero for dispersed flow in fractures as all water in the fractures is imbided into the matrices. Segregated case shows the substantial amount of water in the producing fluid. The runs without capillary pressure and gravity give higher

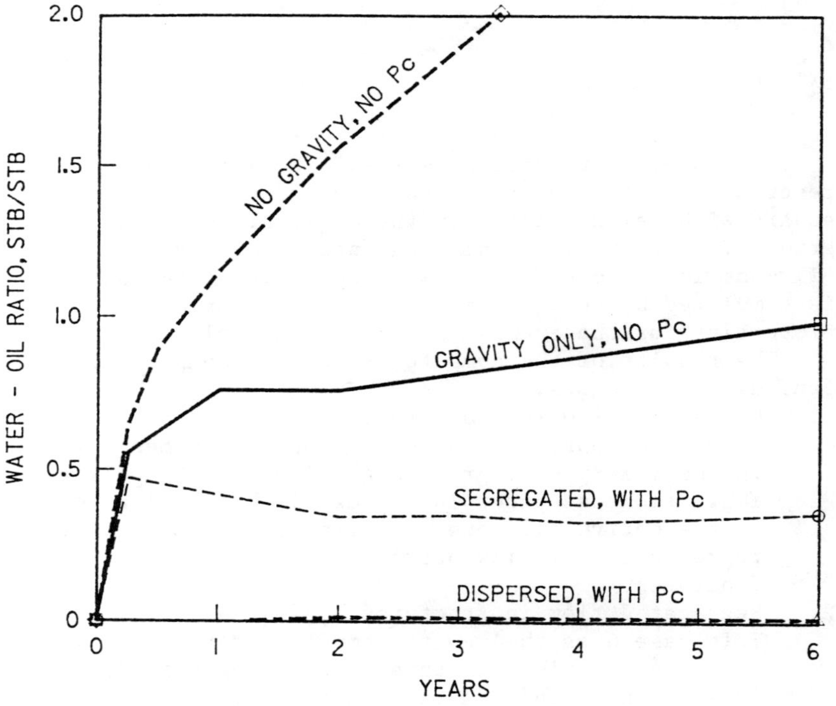

FIGURE 1 - WATER-OIL RATIO COMPARISON
EFFECT OF Pc AND GRAVITY

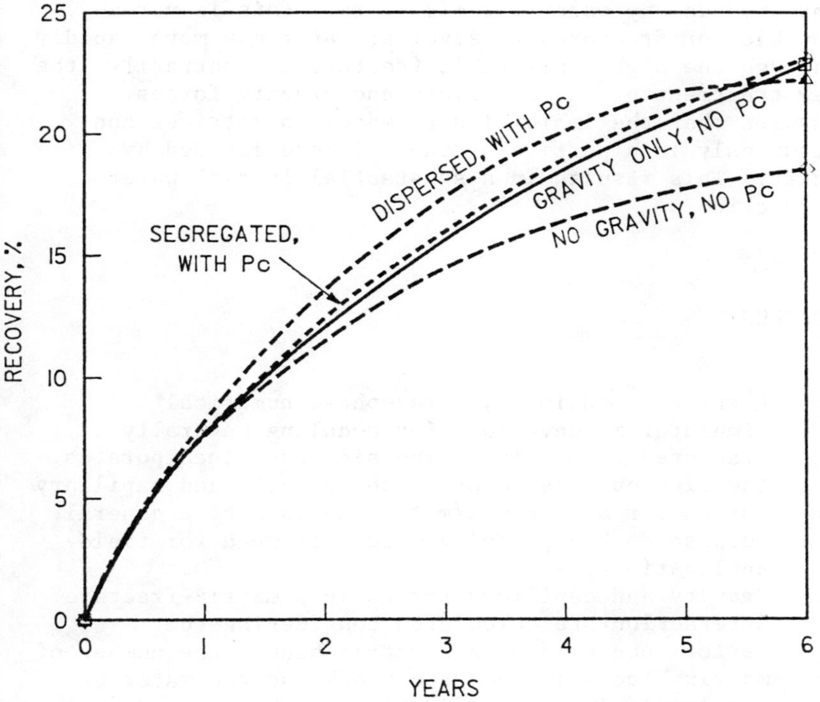

FIGURE 2 - RECOVERY COMPARISON
EFFECT OF Pc AND GRAVITY

WOR's. The oil recovery (Fig.2) is higher initially for the dispersed flow in fractures due to higher capillary effect. However, it approaches the recovery for segregated case when P_{cm} is declining in time (due to higher S_{wm}) and gravity effect in segregated case is increasing (due to high water level in fractures). The runs without capillary and gravity effects naturally have lower oil recoveries as compared to the corresponding runs with these effects.

This example demonstrates that the new treatment of capillary and gravity forces substantially changes the results of simulation for fractured reservoirs. The conventional treatment of capillary forces assumes that matrix blocks imbide the water in the entire 300 ft. thick oil zone in the layer above the aquifer. As a result, the well completed right above the aquifer did not produce any water for six years. This is unreasonable for fractured reservoirs; water can move rapidly through the highly permeable fractures. Contrarily, the new treatment of the capillary and gravity forces assumes that the imbibition of water in matrices can occur only in a portion of the oil zone invaded by water. This results in a substantial initial water production.

CONCLUSIONS

1. A three-dimensional, three-phase numerical simulator is developed for modeling naturally fractured reservoirs. The simulator incorporates the rigorous treatment of the gravity and capillary forces for a fracture/matrix media into a general purpose double porosity model designed for field applications.
2. Gravity and capillary forces in a matrix-fracture interaction are calculated considering the performance of a single matrix block, the number of matrix blocks in the grid block and the water or gas levels inside a grid block. The same approach (using the properties of a single matrix block and combining it for the whole grid block) was used for the derivation of the matrix-fracture transmissibilities and geometrical factor.
3. A special initial equilibration method has been developed for the description of the discontinuous water saturation in the matrix rock.
4. The conditions were discussed when the matrix to matrix flow can be neglected and when the performance of the double porosity system is approaching a single porosity case.
5. The example problem demonstrates a significant effect of the new treatment of the gravity and capillary forces on the predicted performance of fractured reservoirs.

Table 1 - Reservoir Properties for the Example Problem

Reservoir grid: 6 * 6 * 3

Grid dimensions: $\Delta x = \Delta y = 220$ ft. for all grid blocks, $\Delta z_1 = 250$ ft., $\Delta z_2 = 300$ ft., $\Delta z_3 = 700$ ft.

Matrix block size: $L_x = L_y = L_z = 10$ ft. for all grid blocks

Matrix properties:

	ϕ_m (%)	k_m (md)	S_{om}	S_{wm}
Layer 1	10.0	1.0	0.55	0.45
Layer 2	10.0	1.0	0.422	0.578
Layer 3	7.5	1.0	0.0	1.0

Fracture properties:

	ϕ_f (%)	$k_{xf}=k_{yf}$ (md)	k_{zf} (md)	S_{of}	S_{wf}
Layer 1	1.5	2000	500	1.0	0.0
Layer 2	1.5	1000	500	1.0	0.0
Layer 3	1.5	100	50	0.0	1.0

Fracture rock compressibility: $C_{rf} = C_{rm} = 0.000004$ 1/psi

Fracture relative permeabilities are nearly a straight line.

Fracture capillary pressure was equal to zero.

Martix relative permeabilities end points:

$S_{or} = 0.20$, $S_{wc} = 0.30$, $S_{gc} = 0.02$

Matrix capillary pressure for oil-water system:

$P_{cm} = 4.0$ psia for $S_w = 0.3$

$P_{cm} = 2.95$ psia for $S_w = 0.5$

$P_{cm} = 0.85$ psia for $S_w = 0.7$

Undersaturated oil properties:

- slope of B_o is -0.00001 RB/STB/psi,
- slope of μ_o is -0.0004 cp/psi

Saturated oil properties:

- at p = 5014 psia:
  $B_o = 1.381$ RB/STB, $\mu_o = 1.11$ cp, $R_{so} = 760$ SCF/STB
- at the bubble point pressure 2464 psia:
  $B_o = 1.233$ RB/STB, $\mu_o = 1.27$ cp, $R_{so} = 403$ SCF/STB
- at p = 14.7 psia:
  $B_o = 1.0$ RB/STB, $\mu_o = 5.52$ cp, $R_{so} = 0$ SCF/STB

Table 1 - Reservoir Properties for the Example Problem (cont'd.)

Water properties:

- at p = 5014 psia: B_w = 0.984 RB/STB, μ_w = 0.3 cp
- at p = 14.7 psia: B_w = 1.0 RB/STB, μ_w = 0.3 cp

Gas properties:

- at p = 5014 psia: B_g = 0.0039 RCF/SCF, μ_g = 0.0262 cp
- at p = 14.7 psia: B_g = 0.1556 RCF/SCF, μ_g = 0.0138 cp

Stock tank fluid densities:

ρ_o = 55 lb/cf, ρ_w = 62.43 lb/cf, ρ_g = 0.0514 lb/cf

Initial pressure is 3485 psia at the depth of 8250 ft.

The producing wells are completed in the upper two layers. Wells operate in the pressure specified mode with limiting bottom-hole flowing pressure of 1000 psia for layer 1 and 1236 psia for layer 2. Model productivity indexes are 0.72 for layer 1 and 0.68 for layer 2.

NOMENCLATURE

A = area, sq. ft. (m^2);
  = coefficient submatrix of a set of simultaneous equations
B = formation volume factor, RB/STB (res m^3/stock tank m^3);
  = right-hand side vector for set of simultaneous equations
C = compressibilities, psi^{-1} (kPa^{-1})
CG = capillary and gravity term in the flow equations, SCF/D (std m^3/D)
D = depth of grid block, ft. (m)
E = main diagonal coefficient submatrix
k = permeability, md (mm^2)
k_r = relative permeability, fraction
L = length of matrix block, ft. (m)
n = number of normal sets of fractures, n = 1,2,3
N = number of matrix blocks in one grid block
p = fluid pressure, psi (kPa)
P_c = capillary pressure, psi (kPa)
q = production rate, STB/D (stock tank m^3/D)
R_s = solution gas/oil ratio, SCF/STB (std m^3/std m^3)

S = saturation, fraction
t = time
T = fluid transmissibility, STB/D-psi (std m^3/D-kPa)
V = volume, STB (stock tank m^3)
x,y,z = Cartesian coordinates
Z = depth to water or gas level in a grid block, ft (m)
γ = specific weight, psi/ft (kPa/m)
δ = time step difference,
$\delta = \delta x = x_{n+1} - x_n$
Δ = finite-difference operator
Δt = time increment, $t_{n+1} - t_n$
$\Delta(T\Delta p) = \Delta_x(T_x \Delta_x p) + \Delta_y(T_y \Delta_y p) + \Delta_z(T_z \Delta_z p)$
$\Delta_x(T_x \Delta_x p) = T_{i+\frac{1}{2}}(p_{i+1} - p_i) - T_{i-\frac{1}{2}}(p_i - p_{i-1})$
Δx = grid block dimension in x direction, ft.
Δy = grid block dimension in y direction, ft.
Δz = grid block dimension in z direction, ft.
μ = viscosity, cp (Pa x s)
ρ = density, lb/cu. ft. (kg/m^3)
σ = matrix shape factor, 1/sq. ft. (1/m^2)
$\emptyset$ = porosity, fraction

SUBSCRIPTS

c = capillary effect
e = effective
f = fracture
g = gas
G = gravity effect
CG = combine effect of gravity and capillary pressure
m = matrix
mf = matrix/fracture flow
o = oil
t = time or total
w = water
x,y,z = x, y and z directional flow properties
α = phase (oil, water, gas)

ACKNOWLEDGMENTS

I thank the management of Texaco Inc. for permission to publish this paper.

REFERENCES

1. Barenblatt, G., Zheltov, Y., and Kochina, I. (1960). J. Applied Math. and Mechanics, 24, No. 5, 1286
2. Warren, J. and Root, P. (1963). Soc. Pet. Eng. J., No. 9, 245
3. Kazemi, H. (1969). Soc. Pet. Eng. J., No. 9. 451
4. Yamamoto, R., Padgett, J., Ford, W., and Boubeguera, A (1971). Soc. Pet. Eng. J., No. 6, 113
5. Saidi, A. (1975). Proceedings 9th World Petroleum Congress, Tokyo, 175
6. Reiss, L., Bossie-Codreanu, D., Lefebvre du Pray, E. (1973). Paper SPE 4343 presented at the Second Annual European meeting of AIME held in London, England
7. Rossen, R. (1977).Soc. Pet. Eng. J., No. 6, 201
8. De Swaan, A. (1978). Soc. Pet. Eng. J., No. 4, 117
9. Bossie-Codreanu, D., Bia, P., Sabathier, J. (1982). SPE paper 10977 presented at the 57th Annual Fall SPE Meeting, New Orleans, LA
10. Kazemi, H., Merrill, L., Jr., Porterfield, K., and Zeman, P. (1976). Soc. Pet. Eng. J., No. 12, 317
11. Gillman, J., Kazemi, H. (1982). SPE paper 10511 presented at 6th Symposium on Reservoir Simulation, New Orleans, LA
12. Thomas, L., Dixon, T., Pierson, R. (1980). SPE paper 9305 presented at SPE-AIME 55th Annual Fall Meeting, Dallas, Texas
13. Blaskovich, F., Cain, G., Sonier, F., Waldren, D., Webb, S. (1983). SPE paper 11480 presented at the Middle East Technical Conference of the SPE, Manama, Bahrain
14. Saidi, A. (1983). SPE paper 12270 presented at the Seventh SPE Symposium on Reservoir Simulation, San Francisco, CA
15. Reiss, L. (1980). "The Reservoir Engineering Aspects of Fractured Formation," Gulf Publishing Co., Houston
16. Van Golf-Racht, T. (1982). "Fundamentals of Fractured Reservoir Engineering," Elsevier Scientific Publishing Co., Amsterdam-Oxford-New York

MIXING DURING SINGLE-PHASE FLOW IN RESERVOIR ROCKS: MODELS, EFFECTS OF PORE STRUCTURE AND INTERPRETATION OF EXPERIMENTS[1]

Robert E. Bretz
Robert M. Specter[2]
Franklin M. Orr, Jr.[3]

New Mexico Petroleum Recovery Research Center
New Mexico Institute of Mining and Technology
Socorro, New Mexico

ABSTRACT

This paper examines the relationships between pore structures observed in thin sections and stable, miscible displacement behavior in reservoir core samples. We report results of miscible displacement experiments to characterize reservoir core samples. The displacements were performed with fluids of matched density and viscosity in seven reservoir and outcrop core samples, three sandstones and four carbonates. In addition, we examine the representation of one-dimensional miscible displacements in terms of three models: the convection-

---

[1]Support for the work described in this paper was provided by the U.S. Department of Energy, The New Mexico Energy Research and Development Institute, the Abu Dhabi National Reservoir Research Foundation, the Amoco Production Co., ARCO Oil and Gas, Cities Service, Conoco, Gulf, Japan National Oil, Marathon, Mobil, Petro-Canada Resources, Shell, Societie Nationale Elf Aquitaine, Sohio, Sun and Tenneco. That support is gratefully acknowledged.

[2]Present address: Department of Finance, College of Business Administration, Arizona State University, Tempe, AZ.

[3]Present address: Department of Petroleum Engineering, Stanford University, Stanford, CA.

dispersion equation, the Coats-Smith model, and a model based on a representation of diffusion in and out of porous spheres around which fluid flows. We compare calculated and measured effluent compositions and report parameters which produce the best matches between experimental and calculated values.

Examination of thin sections taken from the core samples indicates that the structures observed are qualitatively consistent with displacement behavior in the core samples. We report pore size distributions measured from the thin sections, and we describe a simple technique for characterizing whether the pore space is connected in such a way as to generate preferential flow paths. We argue that early solvent breakthrough and a long period of tailing require the existence of a relatively wide pore size distribution and pore connections which produce preferential flow paths. We detect the existence of such pore connections by measuring the average diameter of pores surrounding a reference pore. Preferential flow paths exist if, on average, small pores are connected to other small pores, while large ones are connected to other large ones. Relationships between observations of thin sections and the parameters determined by matching miscible displacements are discussed. Finally, problems with scaling results of laboratory displacements to reservoir-scale processes are considered.

I. INTRODUCTION

Enhanced oil recovery (EOR) processes such as CO_2 flooding, condensing and vaporizing gas drives and surfactant floods depend on the mixing of injected fluids with those in place in the reservoir to generate in situ fluid compositions which lead to high local displacement efficiency. In CO_2 flooding, for example, extraction of hydrocarbons from the oil by dense CO_2 can create mixtures which can displace oil without forming two phases. Because local displacement efficiency depends on the composition path followed by the fluids in such processes, the mixing that occurs as the fluids flow in the reservoir rocks must also affect process performance. The combined effects of phase behavior and flow can be represented mathematically if the mixing is simple. For example, Helfferich (1981) examined the composition paths of two-phase, three-component flows under the assumption that mixing results only from convection of whatever phases are present. In one-dimensional displacements, that assumption is equivalent to a statement that fluids in a plane transverse to the flow direction are completely mixed (and at chemical equilibrium), but that fluids in adjacent planes do not mix by dispersion. In any real porous medium, however, dispersive mixing will occur. That mixing alters composition paths and hence changes local displacement efficiency, as was shown by Gardner, Orr and Patel (1981). They found that high dispersion leads to lower oil recovery in CO_2 floods. Gardner and Ypma (1982) found that viscous instability has an analogous effect in CO_2 floods because mixing of CO_2 within a viscous finger with uncontacted oil in adjacent unswept regions reduces to some extent the beneficial effects of phase behavior. Dai and Orr (1984) reached a similar conclusion for CO_2 floods in a porous medium which is divided into flowing and stagnant fractions. The exchange of fluid by mass transfer between the flowing and stagnant fractions causes composition paths to fall deeper into the two-phase region, so that local displacement efficiency is lower than for a homogeneous porous medium. Thus, if the performance of EOR processes that depend on compositional behavior is to be predicted quantitatively, the mixing that occurs as fluids flow in real pore structures must be examined and characterized.

Mixing during flow in a reservoir rock must be strongly influenced by the structure and properties of the pore network. Reservoir rock pore structures, of course, are heterogeneous on scales which range from the pore level to interwell distances. In the discussion that follows we attempt to relate observations of heterogeneities that are visible on the scale of a thin section (20 x 40 mm) to the mixing behavior of stable, single-phase displacements in laboratory test cores. There are obvious limitations to this approach since there are heterogeneities at larger scales that are not considered at all. It is, however, a first step toward relating flow and mixing behavior to the geological structures which occur in reservoir rocks.

To examine mixing in reservoir rock samples, we present results for stable, miscible displacements in seven reservoir and outcrop core samples, three sandstone and four carbonates. We interpret the results in terms of three linear differential models: the convection-dispersion equation (Aronofsky & Heller, 1957), the Coats-Smith model (Coats & Smith, 1964), and a porous sphere model (Rao et al., 1980). We compare Coats-Smith model parameters, particularly the flowing fraction, to observations and measurements of geometrical parameters from thin sections taken from the same rock samples used to make test cores, and we discuss the relationship between the lumped parameter approach used in the Coats-Smith model and the more detailed diffusion calculations of the porous sphere model. Finally, we examine the uncertainties that arise from translating the results of small-scale displacements to larger flow systems.

II. MATHEMATICAL MODELS

In a typical miscible displacement experiment, a fluid containing tracer is injected into a test core, and the effluent concentration of the tracer is monitored. The shape of the effluent concentration curve plotted against time can be used to characterize the nature of the fluid-rock behavior. In any displacement, the effluent concentration curve will not be the sharp step change that was injected, but will be smeared. This smearing is due to the action of molecular diffusion, mixing caused by non-uniform flow velocities within individual pores, a distribution of mean velocities through different flow paths and distribution of the lengths of flow paths through the porous medium. These mechanisms are all related in some way to the transport properties (diffusivity and viscosity) of the fluids and the geometry of the rock pore space.

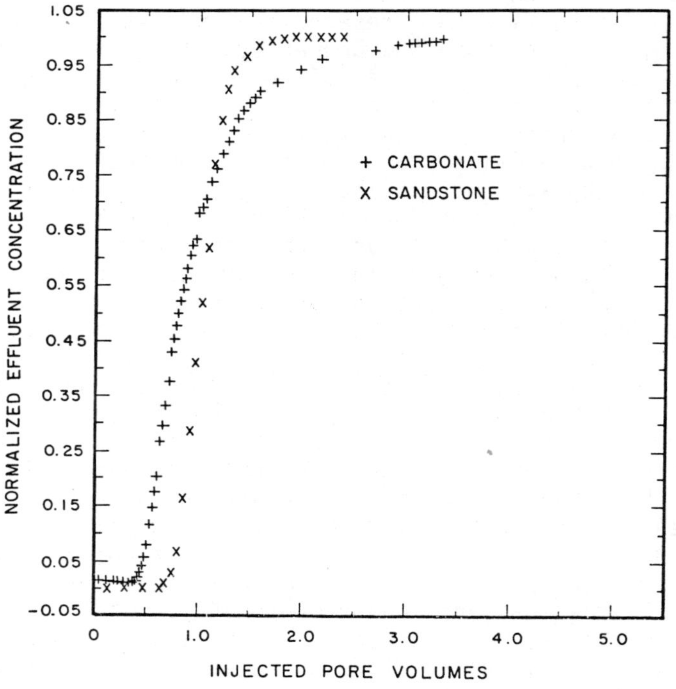

Figure 1. Step input displacement data demonstrating the differences between homogeneous (sandstone) and heterogeneous (carbonate) fluid-rock systems.

In an homogeneous flow system, the concentration of tracer in the effluent will reach 50% of the inlet value when approximately one pore volume of fluid has been injected, and the curve will be nearly symmetrical about the 50% concentration point. In an heterogeneous system, the 50% concentration point will appear in the effluent fluids well before one pore volume has been injected. This early arrival indicates that some portions of the pore volume are bypassed by the flowing stream or that some portions of the pore space have flowing velocities very much slower than others, that is, that preferential flow paths exist. In such displacements, the effluent composition curve is not symmetric, with a long tail produced after the 50% concentration point appears. Fig. 1 gives examples of the two types of rock-fluid systems.

The displacement results illustrated in Fig. 1 can be represented mathematically by one of several models. Here we consider three: the convection-dispersion equation, the Coats-Smith model and a model of flow in an assemblage of porous spheres. For all three models, we examine stable, one-dimensional flow of one phase. We also assume that there is no volume change on mixing and that the effects of viscous instability and gravity segregation are absent. Those conditions were closely approximated in the experiments performed. In this section, we summarize the differential equations and boundary conditions used to model the displacement experiments described below.

A. Convection-Dispersion Equation

If the porous medium is uniform, then the flow of one phase with dispersion is described well by the convection-dispersion equation (Aronofsky & Heller, 1957), which in dimensionless form is

$$\frac{\partial C}{\partial \tau} + \frac{\partial C}{\partial \xi} - \frac{1}{Pe} \frac{\partial^2 C}{\partial \xi^2} = 0 \qquad 0 \leq \xi \leq 1 . \tag{1}$$

The symbols used are defined in the Nomenclature below. The parameter in Eq. (1) is the Peclet number, $Pe = UL/D_L$, the ratio of the rate of mass transport due to convection to that due to dispersion.

B. Coats-Smith Model

If the porous medium can be represented as divided into flowing (f) and stagnant (1-f) fractions, the resulting mass balance equations are (Coats & Smith, 1964)

$$f \frac{\partial C}{\partial \tau} + \frac{\partial C}{\partial \xi} - \frac{1}{Pe} \frac{\partial^2 C}{\partial \xi^2} + (1-f) \frac{\partial C^*}{\partial \tau} = 0 \quad 0 \leq \xi \leq 1 \tag{2a}$$

$$(1-f) \frac{\partial C^*}{\partial \tau} = a(C-C^*) . \tag{2b}$$

The equations depend on three parameters; the Peclet number ($Pe = UL/D_L$), the flowing fraction (f), and the Damköhler number ($a = K_m L/U$). Since there are no assumptions about the pore geometry of stagnant and flow volumes, all three parameters are determined by fitting the model to experimental data. If there were truly stagnant and flowing regions in the porous medium, the flowing fraction, f, would be constant for displacements at different velocities in the same porous medium. However, both Pe and a must be determined for each velocity because the dispersion and mass transfer coefficients are not constant if velocity changes.

A basic assumption of the Coats-Smith model is that the two fluid fractions, the stagnant and flowing volumes, are well stirred. This assumption allows description of the diffusive mass transfer between the two fluid fractions by an overall mass transfer coefficient which lumps the molecular diffusion coefficient and the resistance to molecular diffusion into a single parameter. The exact relationship between the molecular diffusion coefficient and the resistance to diffusion, which must somehow include both the area across which diffusion occurs as well as the distance through which it must occur, is not known.

C. Porous Sphere Model

The porous sphere model is based on the assumption that the porous medium is an aggregation of uniform porous spheres, or at least is a matrix with porous spheres uniformly dispersed throughout. Further, the size of the pores within the spheres is much smaller than the interstitial voids between the spheres or in the matrix. Hence when fluid flows through such a porous medium, the velocity through the matrix pores is much greater than that through the sphere pores. Therefore, we assume that mass is transferred between the sphere interiors and the matrix porosity by molecular diffusion. The convection-dispersion equation (1) describes mixing in the matrix porosity.

For model purposes, we assume that the spheres are of uniform size and have constant porosity and tortuosity. In practice this uniformity could be replaced by average values if the average values do not change for samples of the porous medium taken at different places. This last restriction on the uniformity of average properties is a requirement for scaling experimental results from one length to another.

Similar models have been proposed by Rao et al. (1980) and Passioura (1971). In dimensionless form, the model is

$$\frac{\partial C_m}{\partial \tau} + \frac{\partial C_m}{\partial \xi} - \frac{1}{Pe_m}\frac{\partial^2 C_m}{\partial \xi^2} + \frac{\theta_s}{\theta_m}\frac{\partial \bar{C}_s}{\partial \tau} \tag{3a}$$

$$\frac{\partial \bar{C}_s}{\partial \tau} = 3\int_0^1 \frac{\partial C_s}{\partial \tau}\eta^2 \, d\eta \tag{3b}$$

In spherical coordinates, the rate of transfer to the sphere is given by

$$\frac{\partial C_s}{\partial \tau} = \frac{1}{Pe_s}\left[\frac{\partial^2 C_s}{\partial \eta^2} + \frac{2}{\eta}\frac{\partial C_s}{\partial \eta}\right] \quad 0 \leq \eta \leq 1 \tag{3c}$$

The porous sphere model also depends on three parameters, Pe_m, Pe_s, and θ_s/θ_m. The parameters $Pe_m = U_m L/D_L$ and θ_s/θ_m correspond to the Coats-Smith parameters Pe and f, while $Pe_s = b^2 U_m/D_e L$ is similar to a, the Damköhler number. Pe_m is a Peclet number based on convection and dispersion in the pore space surrounding the spheres, while Pe_s is a ratio of characteristic times for diffusion in the sphere to that for flow in the surrounding pore space. D_e is an effective diffusion coefficient within the sphere, taken to be $D_e = D\phi/\gamma$. The molecular diffusion coefficient, D, can be measured as can core length, L, and average interstitial velocity, U_m. This leaves ϕ, the sphere porosity, γ, the sphere tortuosity, and b, the sphere radius to be determined. These unknown constant parameters may be combined into a single parameter, either $b^2\gamma/\phi$ or b^2/D_e, which is constant in all displacements regardless of the displacement velocity or core length.

D. Initial and Boundary Conditions

All three models are first order in time and second order in space so that one initial condition and two boundary conditions must be specified. The initial condition is

$$C(\xi, 0) = 0 \quad 0 \leq \xi \leq 1 \tag{4a}$$

For boundary conditions, we follow Brigham's (1974) reasoning and use

B.C. I

$$C(0, \tau) = C_0 - \frac{1}{Pe}\frac{\partial C}{\partial \xi}\bigg|_{\xi = 0} \qquad \tau > 0 \tag{4b}$$

$$C_0 = 1 \qquad 0 \leq \tau \leq \beta$$

$$C_0 = 0 \qquad \tau > \beta$$

B.C. II

$$C(\xi \rightarrow \infty, \tau) = 0 \tag{4c}$$

$$C' = C(1, \tau) - \frac{1}{Pe}\frac{\partial C}{\partial \xi}\bigg|_{\xi = 1} \tag{4d}$$

The implementation of the exit boundary condition in the finite difference formulation is discussed in the Appendix. More thorough discussions on choices of boundary conditions are given by Parker (1984) and Bretz and Orr (1985). Since both the Coats-Smith and the porous sphere models are extensions of the convection-dispersion equation, the same initial and boundary conditions apply.

All of the models were solved for specific sets of parameters with a fully explicit finite difference representation. The effects of numerical dispersion were accounted for in determining the overall level of dispersion. Details of the solution technique for the convection-dispersion equation and the Coats-Smith model are given by Orr and Taber (1984). Solutions of the porous sphere model were obtained in a similar manner with the concentration distribution in the spheres, Eq. (3c), solved using a forward time, backward space algorithm. The sphere grid mesh and time step sizes were chosen small enough to minimize truncation errors.

III. DISPLACEMENT APPARATUS AND PROCEDURE

Fig. 2 is a schematic of the apparatus used for miscible displacements. The configuration shown is for a pulse input displacement, but step input displacements were also performed. In a typical experiment, fluid is pumped through a sample valve into a test core. A sample loop of known volume is loaded with a miscible fluid with matched viscosity and density containing a tracer. To begin a tracer displacement, the sample valve is switched so that flow is redirected through the sample loop and then into the test core. Fluids leaving the test core pass

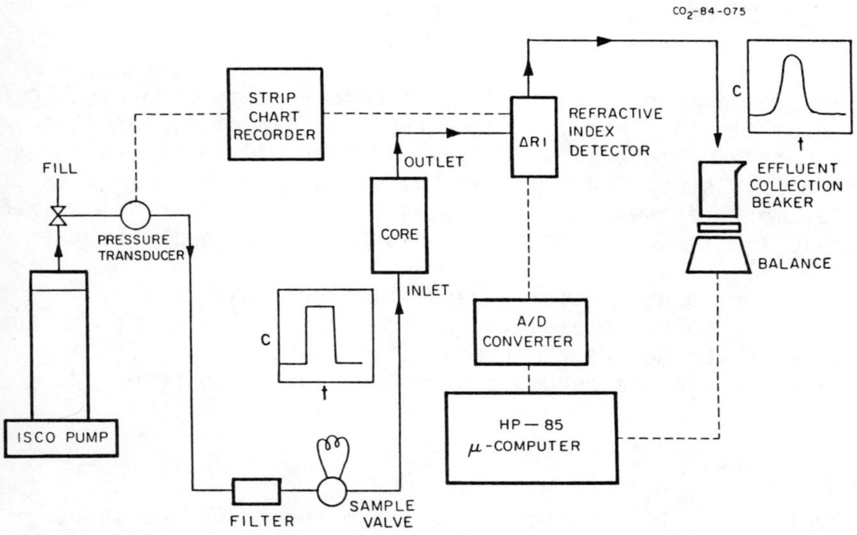

Figure 2. Schematic of miscible displacement apparatus.

through a differential refractometer, where the concentration of tracer is monitored continuously, and into a reservoir on an electronic balance which monitors the displacement rate. Effluent tracer and rate data are collected automatically using a microcomputer.

In most of the displacements described here, the cores were saturated with brine (2.0 wt % NaCl, 1.0 wt % $CaCl_2$, 0.01 wt % NaN_3, saturated with $CaSO_4$) to which 0.4% sucrose was added. The brine in the injected pulse was the same except that the sucrose concentration was 0.52 wt %. In the concentration range used, refractometer output was linear. In addition, some experiments were performed with a nitrate brine (1 wt. % $NaNO_3$, 1 wt % KNO_3, 1 wt. % $Ca(NO_3)_2$, 0.01 wt % NaN_3, saturated with $CaSO_4$) with sucrose as a tracer, with normal decane containing normal undecane as a tracer, and with ethylbenzene and ethylbutyrate. Additional details of those experiments and associated analytical techniques are given by Orr and Taber (1984).

Care was taken to minimize the dead volumes present in tubing, fittings, end caps on the cores, and the composition detection cells. Core pore volumes were normally about 10 cm^3 in cores about 25 cm in length, while total dead volume was usually less than 0.3 cm^3, though some earlier data (also reported here) were obtained for larger cores with larger dead volume.

IV. RESULTS OF DISPLACEMENT EXPERIMENTS

For comparison we separate test core results according to the broad mineral description of the core material, sandstone or carbonate. Properties of the sandstone and carbonate cores used in this study are reported in Tables I and II. Results of displacements at several velocities in each core are summarized in Table III for the sandstone cores and in Table IV for the carbonates.

The experimental results were all fit to either the convection-dispersion equation or the Coats-Smith model and the least-error parameters reported. Limited results of fitting the same data to the porous sphere model are also reported.

A. Sandstone Cores

Typical results of displacements in the three sandstones, Berea, Frannie and Rock Creek, are shown in Figs. 3-5. Values of Coats-Smith parameters (f, a, Pe, D_L and K_m), the flow velocity U, and the pulse size in pore volumes, β, are shown on the plots as well as in Table III. Also shown are the standard error and the overall mass balance, V_o/V_{in}, defined as the volume of pulse fluid produced divided by the volume of pulse fluid injected. Those data give some indication of the quality of the data and the history match. Effluent concentration plots are given by Orr and Taber (1984) for all of the displacements listed in Tables III and IV.

Differences between the sandstone displacements were small. In all the displacements the 50% concentration appeared at about one pore volume injected, and there was little tailing. Differences in peak height were due primarily to differences in dimensionless pulse volume which result when the same injection loop is used with cores having different pore volumes. Dispersion coefficients determined either by fitting results to the Coats-Smith model or to the convection-dispersion equation are plotted against displacement velocity in Fig. 6. Clearly, the dispersion coefficients are approximately linear with the average interstitial velocity, as has been observed by others (Perkins & Johnston, 1963; Spence & Watkins, 1980; Salter & Mohanty, 1982). The two straight lines shown have slopes of 1.0 and 1.14. Because solutions to the convection-dispersion equation depend only on the Peclet number, UL/D_L, approximately linear variation of the dispersion coefficient with velocity produces Peclet numbers, and hence solutions which are nearly independent of velocity.

Table I Summary of Sandstone Core Properties

Core	Description	Rock Dimensions		Pore Volume cm^3	Porosity %	Air Perm. md	Exp't. Dead Volume cm^3
		Diameter cm	Length cm				
Berea	Berea Sanstone Outcrop	3.81	15.05	31.7	18.4	180	2.16[1] 0.185[2]
B-1	Berea Sandstone Outcrop (Experiments 7-9)	3.81	13.9	28.9	18.2	180	0.40
B-1/2	Berea Sandstone Outcrop (Experiments 10-12)	1.27	12.07	3.8	25	1000	0.53
B-5/8	Berea Sandstone Outcrop (Experiments 13-15)	1.59	22.1	9.25	21.1	500	0.33
F-2	Frannie Reservoir Sandstone (Experiments 16-18)	3.81	14.92	34.17	20.1	380	1.1
R-1	Rock Creek Reservoir Sandstone (Experiments 19-21)	1.59	21.8	10.55	24.5	--	0.365

[1]Dead volume experiment 1.

[2]Dead volume experiments 2-6.

Table II Summary of Carbonate Core Properties

Core	Description	Rock Dimensions Diameter cm	Rock Dimensions Length cm	Pore Volume cm^3	Porosity %	Air Perm. md	Exp't. Dead Volume cm^3
H-1	San Andres Dolomite, Seminole	3.81	16.2	33.7	18.2	40	0.85[1]
WW-2	San Andres Dolomite, Willard Unit, Wasson Field (Experiments 27-32)	1.27	13.7	3.70	21.3	90	0.575
M-1	San Andres Dolomite, Maljamar Field (Experiments 33-35)	1.59	31.5	5.67	9.1	6	0.358
SAO	San Andres Dolomite, Outcrop (Experiments 36-38)	3.81	66.7	235.16	17	124	1.262

[1] Experimental dead volume for experiments 24-26. Dead volumes for experiments 22-23 are not available.

Table III Summary of Stable Miscible Displacement Results in Sandstone Cores

Exp't. No.	Core	Average Interstitial Velocity cm/sec	Pulse Size, Pore Volumes	Flowing Fraction	Peclet No. $\frac{UL}{D_L}$	Dispersion Coefficient cm^2/sec	Damköhler No. $\frac{K_m L}{U}$	Mass Transfer Coefficient sec^{-1}	Mass Balance Closure $\frac{V_o}{V_{in}}$	Remarks
1	Berea	3.6×10^{-3}	--	1.0	98	5.6×10^{-4}	0	0	0.963	Step input, ethylbenzene/ethylbutyrate, GC composition determination, manual sample collection.
2	Berea	3.8×10^{-3}	--	1.0	100	5.8×10^{-4}	0	0	1.00	" " "
3	Berea	8.7×10^{-3}	--	1.0	102	1.3×10^{-3}	0	0	0.996	" " "
4	Berea	4.8×10^{-3}	--	1.0	87	8.4×10^{-4}	0	0	0.972	" " "
5	Berea	2.0×10^{-3}	--	1.0	148	2.0×10^{-4}	0	0	0.986	" " "
6	Berea	1.3×10^{-3}	--	1.0	155	1.2×10^{-4}	0	0	0.983	" " "
7	B-1	2.6×10^{-3}	--	1.0	90	4.8×10^{-4}	0	0	--	Step input, ethylbenzene/ethylbutyrate, GC composition determination, auto-sample collection.
8	B-1	4.3×10^{-3}	0.5	1.0	79	7.4×10^{-4}	0	0	--	Pulse input, 2%/3% NaCl brine, on-line conductivity cell composition determination.
9	B-1	4.3×10^{-3}	0.5	1.0	78	7.5×10^{-4}	0	0	--	Pulse input, C_{10}/C_{11}, on-line refractive index composition determination.

Table III (continued)

Exp't. No.	Core	Average Interstitial Velocity cm/sec	Pulse Size, Pore Volumes	Flowing Fraction	Peclet No. $\frac{UL}{D_L}$	Dispersion Coefficient cm^2/sec	Damköhler No. $\frac{K_m L}{U}$	Mass Transfer Coefficient sec^{-1}	Mass Balance Closure $\frac{V_o}{V_{in}}$	Remarks
10	B-1/2	3.9×10^{-2}	0.55	1.0	62	7.5×10^{-3}	0	0	1.002	Pulse input, C_{10}/C_{11}, on-line refractive index composition determination.
11	B-1/2	3.9×10^{-3}	0.55	1.0	84	5.6×10^{-4}	0	0	0.976	" " "
12	B-1/2	3.9×10^{-2}	0.55	1.0	76	6.2×10^{-3}	0	0	1.029	Pulse input, 2%/3% NaCl brine, on-line conductivity composition determination.
13	B-5/8	3.5×10^{-4}	0.57	1.0	732	1.1×10^{-5}	0	0	0.988	Pulse input, 0.4%/0.52% sucrose tracer, on-line refractive index composition determination.
14	B-5/8	3.5×10^{-3}	0.57	1.0	230	3.4×10^{-4}	0	0	0.969	" " "
15	B-5/8	3.5×10^{-2}	0.57	1.0	83	9.05×10^{-3}	0	0	1.011	" " "
16	F-2	3.5×10^{-4}	0.42	1.0	165	3.2×10^{-5}	0	0	1.068	" " "
17	F-2	3.5×10^{-3}	0.42	1.0	140	3.7×10^{-4}	0	0	1.10	" " "
18	F-2	3.5×10^{-2}	0.42	0.974	129	2.8×10^{-3}	$<10^{-4}$	$<10^{-7}$	1.12	" " "
19	R-1	4.6×10^{-4}	0.503	0.998	180	5.5×10^{-5}	0.159	3.3×10^{-6}	0.968	" " "
20	R-1	3.5×10^{-3}	0.503	0.982	236	3.2×10^{-4}	0.076	1.2×10^{-3}	0.985	" " "
21	R-1	3.5×10^{-2}	0.503	0.936	88	8.7×10^{-3}	0.693	1.1×10^{-3}	0.950	" " "

Table IV Summary of Miscible Displacement Results in Carbonate Cores

Exp't. No.	Core	Average Interstitial Velocity cm/sec	Pulse Size, Pore Volumes	Flowing Fraction	Peclet No. $\frac{UL}{D_L}$	Dispersion Coefficient cm^2/sec	Damköhler No. $\frac{K_m L}{U}$	Mass Transfer Coefficient sec^{-1}	Mass Balance Closure $\frac{V_o}{V_{in}}$	Remarks
22	H-1	4.4×10^{-3}	0.42	0.687	14.1	5.1×10^{-3}	0.192	5.2×10^{-5}	--	Pulse input, ethylbenzene/ethylbutyrate, automatic sample collection, GC composition determination.
23	H-1	3.5×10^{-4}	0.42	0.714	13	4.4×10^{-4}	0.155	3.3×10^{-6}	--	" " "
24	H-1	2.6×10^{-2}	0.42	0.658	8.5	5.0×10^{-2}	0.226	3.6×10^{-4}	1.062	Pulse input, 0.4%/0.52% sucrose tracer in Cl^- brine, on-line refractive index composition determination.
25	H-1	2.6×10^{-3}	0.42	0.698	9.9	4.3×10^{-3}	0.119	1.9×10^{-5}	0.939	" " "
26	H-1	2.6×10^{-4}	0.42	0.657	18.1	2.3×10^{-4}	$<10^{-4}$	$<10^{-7}$	0.965	" " "
27	WW-2	2.8×10^{-2}	0.51	0.946	15.4	2.5×10^{-2}	$<10^{-4}$	$<10^{-7}$	1.026	" " "
28	WW-2	2.8×10^{-3}	0.51	0.949	17.6	2.2×10^{-3}	$<10^{-4}$	$<10^{-7}$	1.014	" " "
29	WW-2	2.8×10^{-4}	0.51	0.946	24.9	1.5×10^{-4}	$<10^{-4}$	$<10^{-7}$	0.946	" " "
30	WW-2	2.8×10^{-2}	0.51	0.884	26.9	1.4×10^{-2}	$<10^{-4}$	$<10^{-7}$	1.063	Pulse input, C_{10}/C_{11}, on-line reflective index composition determination.
31	WW-2	2.8×10^{-3}	0.51	0.897	27.8	1.4×10^{-3}	$<10^{-4}$	$<10^{-7}$	1.024	" " "
32	WW-2	2.8×10^{-4}	0.51	0.880	35.1	1.1×10^{-4}	$<10^{-4}$	$<10^{-7}$	1.051	" " "

Table IV (continued)

Exp't. No.	Core	Average Interstitial Velocity cm/sec	Pulse Size, Pore Volumes	Flowing Fraction	Peclet No. $\frac{UL}{D_L}$	Dispersion Coefficient cm^2/sec	Damköhler No. $\frac{K_m L}{U}$	Mass Transfer Coefficient sec^{-1}	Mass Balance Closure $\frac{V_o}{V_{in}}$	Remarks
33	M-1	4.6×10^{-4}	0.34	0.697	7.0	2.0×10^{-3}	$<10^{-4}$	$<10^{-7}$	1.06	Pulse input, 0.4%/0.52% sucrose tracer in Cl^--brine, on-line refractive index composition determination.
34	M-1	2.5×10^{-3}	0.34	0.621	6.5	1.2×10^{-2}	0.117	9.2×10^{-6}	0.947	" " "
35	M-1	4.6×10^{-3}	0.34	0.561	7.7	1.9×10^{-2}	0.176	2.6×10^{-5}	1.05	" " "
36	SAO	1.6×10^{-3}	--	0.809	21.9	4.8×10^{-3}	0.375	8.9×10^{-6}	0.980	Step input, ethylbenzene/ethylbutyrate, manual sample collection, GC composition determination.
37	SAO	2.5×10^{-3}	--	0.771	21.7	7.8×10^{-3}	0.395	1.5×10^{-5}	0.954	" " "
38	SAO	6.3×10^{-3}	--	0.766	19.6	2.1×10^{-2}	0.402	3.8×10^{-4}	0.953	" " "

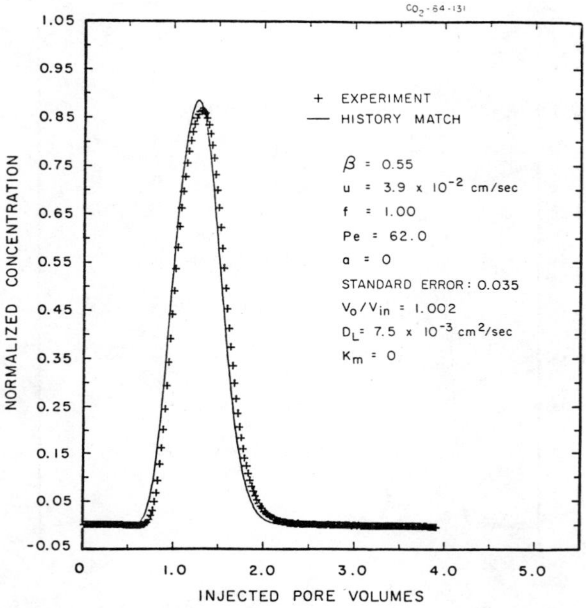

Figure 3. Effluent composition history for displacement no. 10 in a Berea sandstone core (B-1/2).

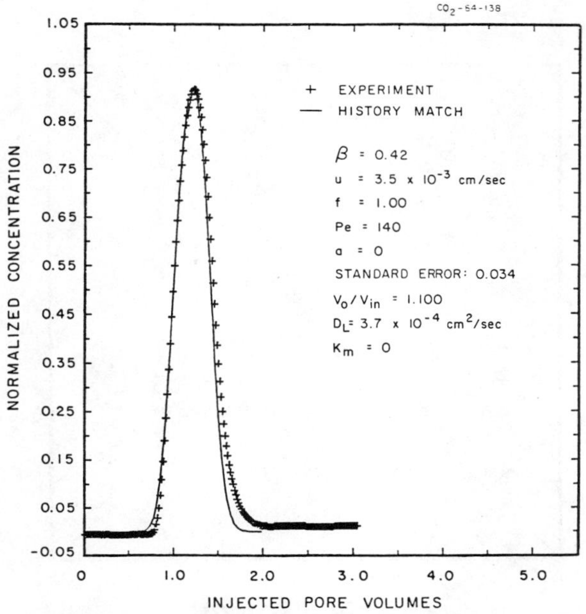

Figure 4. Effluent composition history for displacement no. 17 in a Frannie sandstone core (F-2).

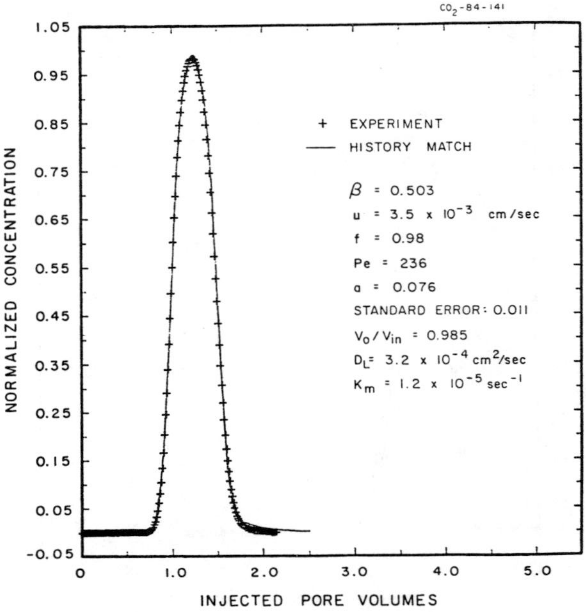

Figure 5. Effluent composition history for displacement no. 20 in a Rock Creek sandstone core (R-1).

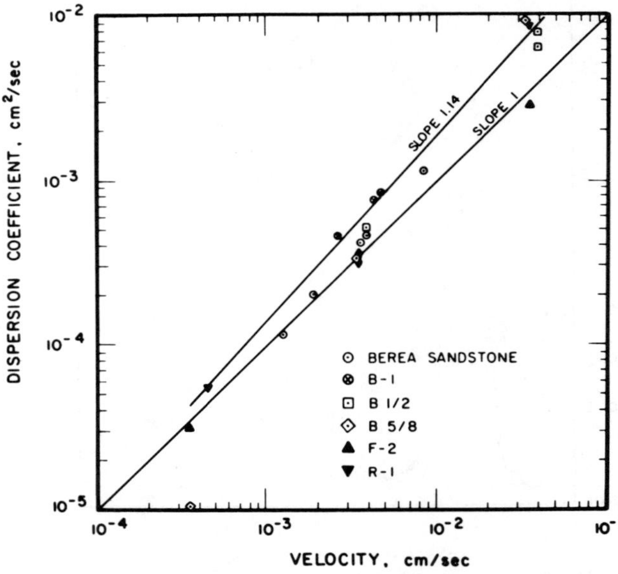

Figure 6. Dispersion coefficients for sandstone cores.

Dispersion coefficients obtained here are consistent with values reported for sandstones by other investigators. Fig. 7 compares dispersion coefficients reported by Baker (1977), Spence and Watkins (1980) and Batycky, Maini, and Fisher (1982) with the values reported in Table III. Though there is some range of values at any velocity, it is clear that for the limited number of sandstones studied, differences in dispersive mixing were small.

The behavior of the displacements in the sandstone cores was consistent with the idea that the pore space was relatively uniform. The fact that the 50% concentration appeared at about one pore volume injected suggests that no preferential flow paths were present. To examine the sensitivity of the Coats-Smith model, parameter fits were performed for several of the sandstone displacements, even though they were well modeled by the Peclet number alone. Fig. 8 compares Peclet numbers from the two models. Clearly, there was very little difference between the values obtained by the two methods.

Flowing fractions determined with the Coats-Smith model are also given in Table III. All are near one. In most cases, the mass transfer coefficients were very low, though, of course, they are only weakly determined if the flowing fraction is near one. It seems likely that the variations in flowing fraction obtained for the sandstones are a reflection of the precision of the measurement. Small errors in the pump rate, for instance, could lead to errors in the arrival time of the injected pulse. In the current version of the equipment, pump rates are checked directly by weight, but in earlier displacements pump rates alone were used to determine the time scale. Because the flowing fraction is sensitive to the arrival times of the leading and trailing edges of the pulse, small errors in measurements of those times would translate directly into errors in the flowing fraction. Material balance errors, which arise from analytical error or error in the time scale, have a similar effect, because the fitting scheme adjusts the arrival time of the calculated pulse to minimize the error in fitting the leading and trailing edges of pulse largely by adjusting the flowing fraction. The variations in the flowing fraction are of the same magnitude as typical material balance errors, as is indicated in Table III. Thus, most of the flowing fractions obtained do not appear to be sufficiently different from one to justify any conclusions about trends with changes in velocity. The one possible exception is the Rock Creek sandstone. Observations of thin sections suggest that a broader pore size distribution and nonuniform distribution of clay minerals might lead to preferential flow paths, as is discussed in Section V below.

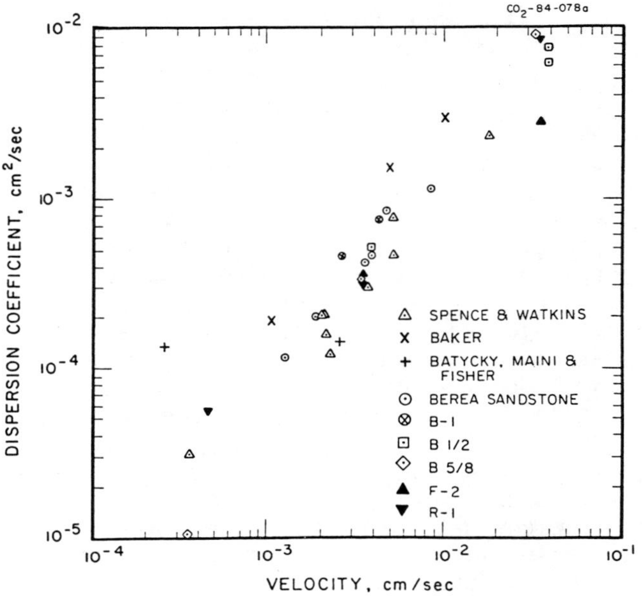

Figure 7. Comparison of dispersion coefficients for sandstone cores.

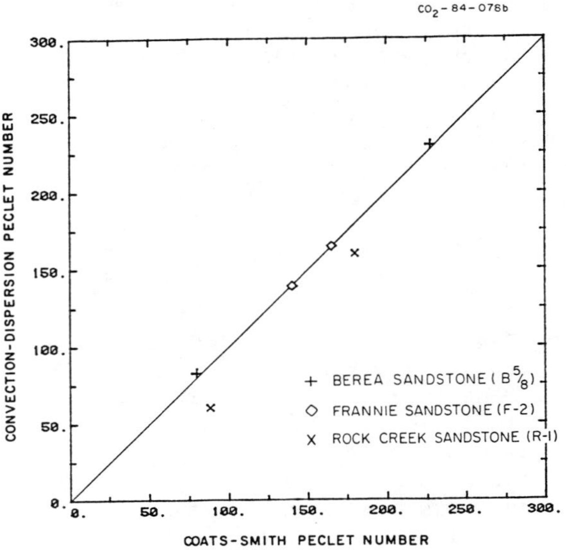

Figure 8. Comparison of Peclet numbers estimated using the Coats-Smith model and the convection-dispersion equation for several sandstone core samples.

B. Carbonate Cores

Table IV summarizes results of single-phase displacements in carbonate core samples. Displacements were performed in cores from three fields, Seminole, Wasson and Maljamar, in which CO_2 floods are now underway. Properties of those cores are summarized in Table II. All of those cores were from the San Andres formation. Also reported are displacement results for a San Andres outcrop core taken from the Algerita escarpment in northeastern New Mexico.

Figs. 9-12 give typical effluent composition histories for the four carbonate samples. Similar plots for all of the displacements are reported by Orr and Taber (1984). Three of the four carbonate cores showed significantly earlier breakthrough of injected fluid and more tailing than did the sandstones, as is shown in Figs. 9-11 for displacements in the Seminole, Maljamar and San Andres outcrop cores. The fourth core, from the Willard unit of the Wasson field, showed later breakthrough and a much more symmetric effluent concentration curve (Fig. 12). Coats-Smith parameters determined for the four samples reflect the breakthrough pattern. As Fig. 13 indicates, the Wasson core showed values of f near one, while values obtained for the other cores were significantly lower: around 0.75-0.80 for the other San Andres cores. Flowing fractions determined for the carbonate samples showed, at most, only mild dependence on velocity. Only the San Andres outcrop and Maljamar samples showed any dependence. For both samples, flowing fractions decreased slightly as the velocity increased,and pore size distributions were quite broad (see Section VI below). Three phenomena could explain the effects observed. It is possible that there is incomplete mixing by diffusion in the largest pores at higher velocities; or there is some flow in the portion of the pore space represented as stagnant; or that this velocity dependence is an artifact of time averaging of the mass transfer coefficient (see Section VI below).

Coats-Smith parameters for the Wasson core (WW-2) were determined in two sets of experiments, one with C_{10}/C_{11} mixtures and the other using sucrose as a tracer. In both cases, the Coats-Smith parameter fits produced flowing fractions less than one despite the fact that mass transfer coefficients were negligibly small. If the effects of mass transfer were small, then the flowing fraction should have been one. In addition, the flowing fractions determined in the two sets of experiments were different (0.95 and 0.89). Several possible explanations for the low values of the flowing fractions were investigated, though no clear explanation for that behavior was found. A detailed discussion of potential explanations is given by Orr and Taber (1984).

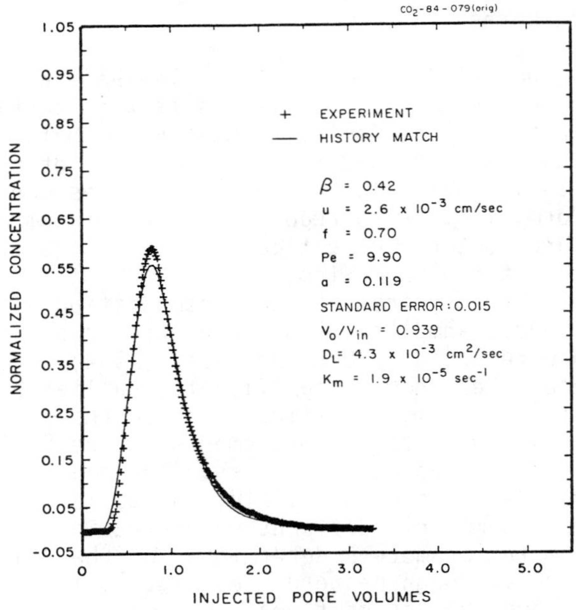

Figure 9. Effluent composition history for displacement no. 25 in a Seminole San Andres core (H-1).

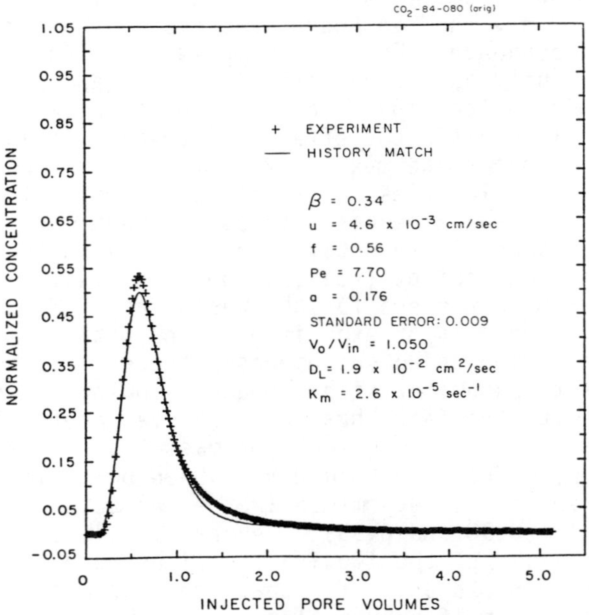

Figure 10. Effluent composition history for displacement no. 35 in a Maljamar core (M-1).

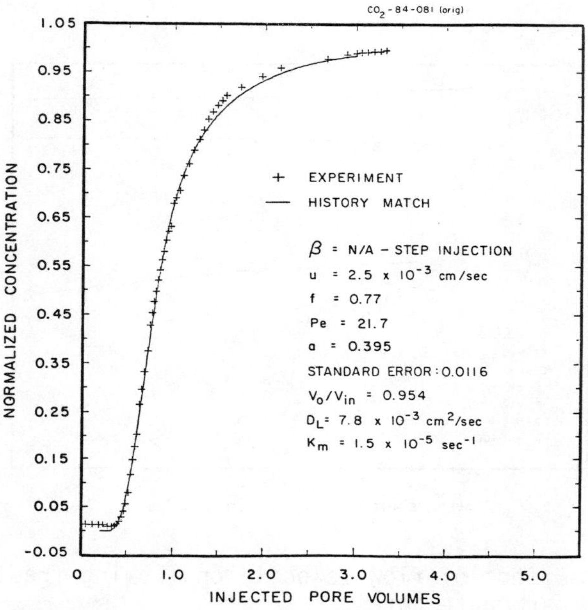

Figure 11. Effluent composition history for displacement no. 37 in a San Andres outcrop core (SAO).

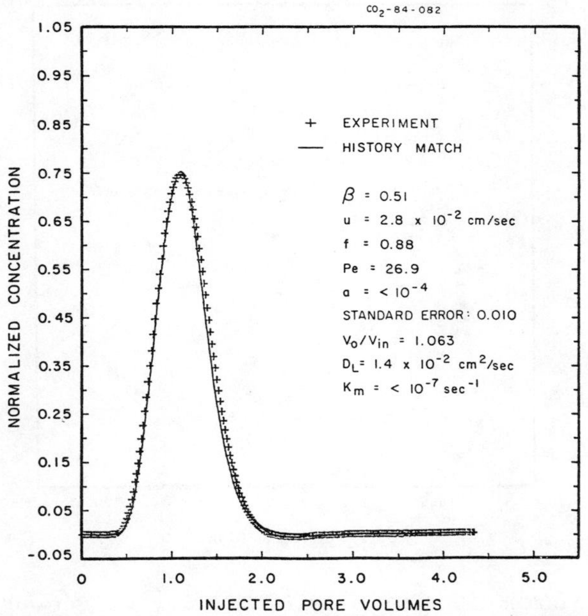

Figure 12. Effluent composition history for displacement no. 30 in a Wasson core (WW-2).

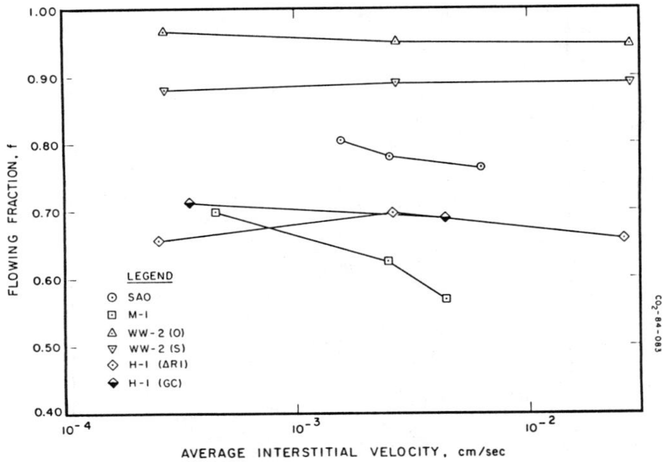

Figure 13. Effect of flow velocity on flowing fraction in carbonate cores.

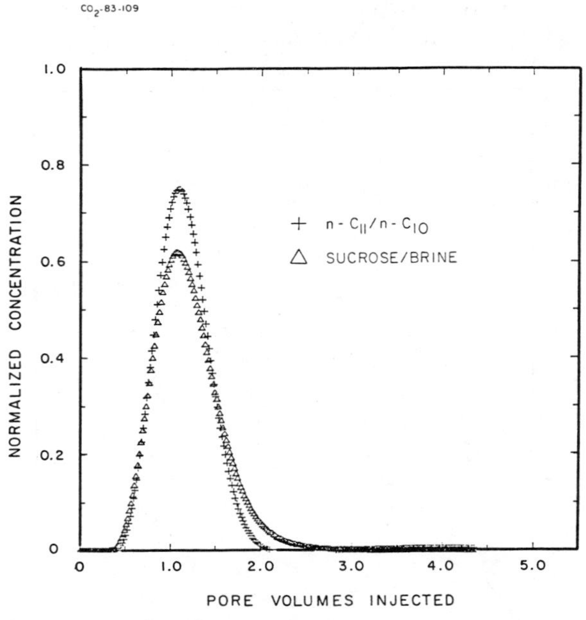

Figure 14. Comparison of displacements (runs 27 and 30) using different tracers in a Wasson San Andres core (WW-2).

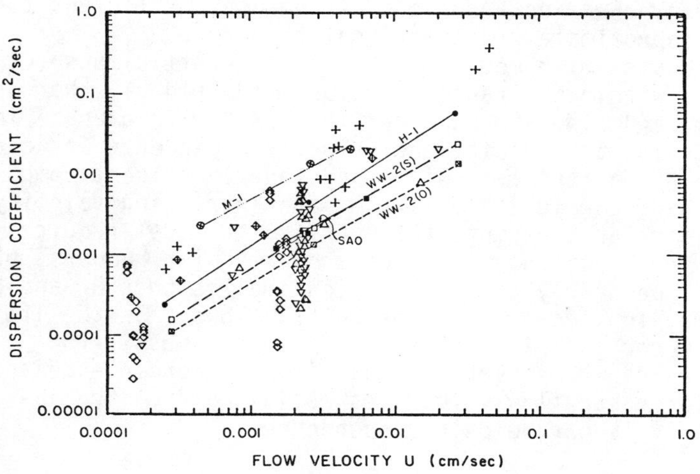

Figure 15. Dispersion coefficients for carbonate core samples. Sources of data; + Batycky, Maini, and Fisher (1982); △ Spence and Watkins (1980), carbonate A; ▽ Spence and Watkins (1980), carbonate B; ◇ Yellig and Baker (1980); ⊕ Baker (1977); □ Core WW-2 (sucrose tracer); ⊠ Core WW-2 (C_{10}/C_{11}); ⊗ Core M-1; • Core H-1; ▣ Core SAO.

Despite the uncertainty concerning the flowing fraction, all the displacements in core WW-2 could be modeled accurately by accounting for the effects of dispersion alone. Dispersion effects did not appear to be the same in the sequences of displacements with different fluids. Fig. 14 compares displacement results for the two fluid systems. The dispersion with the sucrose/brine system is greater than with C_{10}/C_{11} system. The differences in dispersion can be adequately explained by differences in the molecular diffusivities of the two tracers. These differences are also discussed in more detail by Orr and Taber (1984).

Dispersion coefficients for all of the carbonate cores are compared with those reported by other investigators in Fig. 15. Even with the uncertainty concerning the different values of dispersion coefficients obtained with different fluids, the values reported here fall well within the range of values reported elsewhere. Values obtained for the carbonates were significantly larger than those of sandstones. Dispersion coefficients increased consistently with velocity, and the dependence on velocity was similar to that observed for sandstones. For instance, the dispersion coefficient for core H-1 was approximately proportional to $U^{1.16}$.

The Coats-Smith model mass transfer coefficients (Fig. 16) also fell within the range of values reported by other investigators, and also showed clear increases with average interstitial velocity. Again, the velocity dependence was close to linear. As was the case with sandstones, effluent composition curves changed relatively little with increasing velocity. To produce similar behavior, the Coats-Smith model requires that the flowing fraction and Peclet and Damköhler numbers, UL/D_L and K_mL/U be nearly constant, which implies that D_L and K_m are approximately linear in the velocity. Thus, for the limited number of rock samples tested to date, at least, flow velocities can be chosen for experimental convenience and values for other flow velocities estimated from the approximately linear velocity dependence.

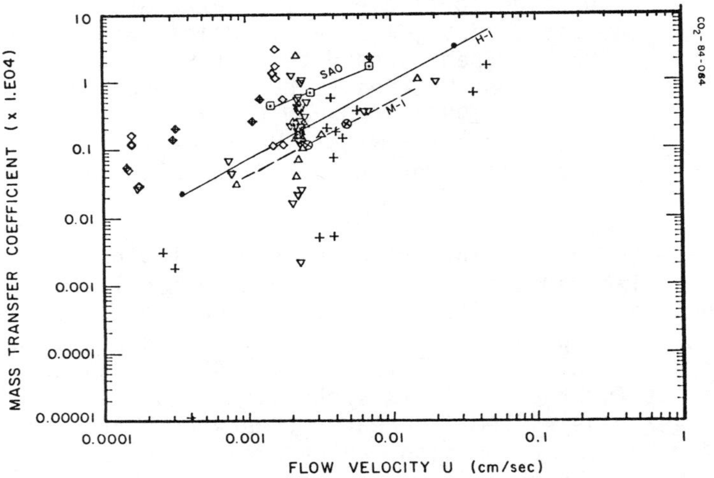

Figure 16. Dependence of mass transfer coefficients on flow velocity for carbonate core samples. Sources of data: + Batycky, Maini, and Fisher (1982); △ Spence and Watkins (1980), carbonate A; ▽ Spence and Watkins (1980), carbonate B; ◇ Yellig and Baker (1980); ⊕ Baker (1977); • Core H-1; ⊗ Core M-1; ▣ Core SAO.

C. Porous Sphere Model: Interpretation of Displacement Results

Examples of solutions to the porous sphere model for fixed values of Pe_m and θ_s/θ_m and three values of Pe_s are shown in Fig. 17a. Also shown in Fig. 17b are displacement results for the San Andres outcrop core. Displacements for that core were chosen for interpretation with the porous sphere model because spherical particles containing pores much smaller than those between spheres were observed in thin sections of that rock sample (see Section V below). The value of $\theta_s/\theta_m = 0.38$ was measured directly on photomicrographs of the thin sections. The value of $Pe_m = 30$ was that obtained from Coats-Smith model fits. Then values of Pe_s were adjusted to yield a best fit. As can be seen from Fig. 18, the resulting fit for $Pe_s = 10$ is slightly better than that obtained from fitting the Coats-Smith model to the displacement results.

To examine the effects of scaling in the porous sphere model, solutions were calculated for the displacements at other velocities with b^2/D_e held constant. The resulting predictions are compared with displacement results in Figs. 18a and b. At the lower velocity (Fig. 18a), the calculated solution agrees well with the experimental results. At the higher velocity, the agreement is less satisfactory. The Coats-Smith model also did a poor job of fitting the displacement data. Whether that behavior is due to the fact that fluid in the spheres is not truly stagnant, to problems with the experimental data, or to some other source is not known. In any case, the porous sphere model using a value of θ_s/θ_m obtained from thin sections, values of Pe_m from Coats-Smith model fits and values of Pe_s obtained by fitting one displacement and then calculating values of Pe_s from the velocity for the other two displacements match the results of displacement experiments at least as well as the best fits of the Coats-Smith model.

V. ANALYSIS OF THIN SECTIONS

In this discussion we attempt to identify features of the pore space that lead to high or low levels of dispersion and to flowing fractions less than one. In traditional petrophysical analysis of rock pore systems, properties such as porosity, permeability, pore size and shape, and average coordination number and aspect ratio are determined. It is not obvious, however, how these properties relate to the sort of averaged mixing observed in the core flood experiments described above. Individually, none of these properties seems likely to control

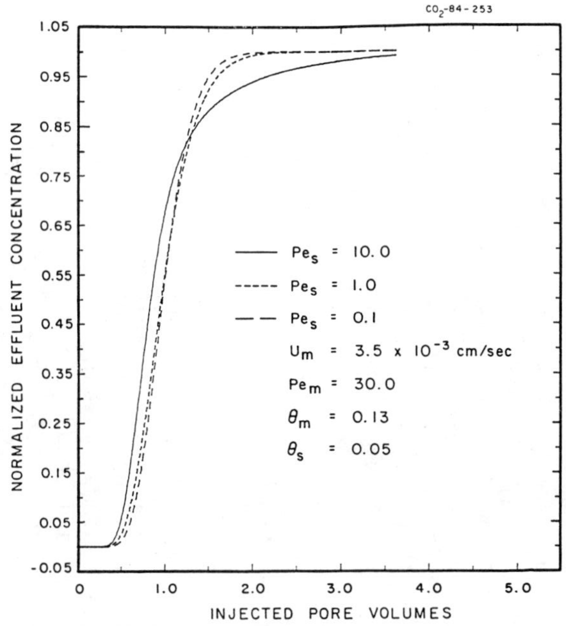

Figure 17a. Effect of variations in Pe_s on computed porous sphere model concentration histories. The Pe_m was obtained from a Coats-Smith fit to data from core SAO displacement no. 37. The θ_m and θ_s were measured from photomicrographs of a thin section of SAO material.

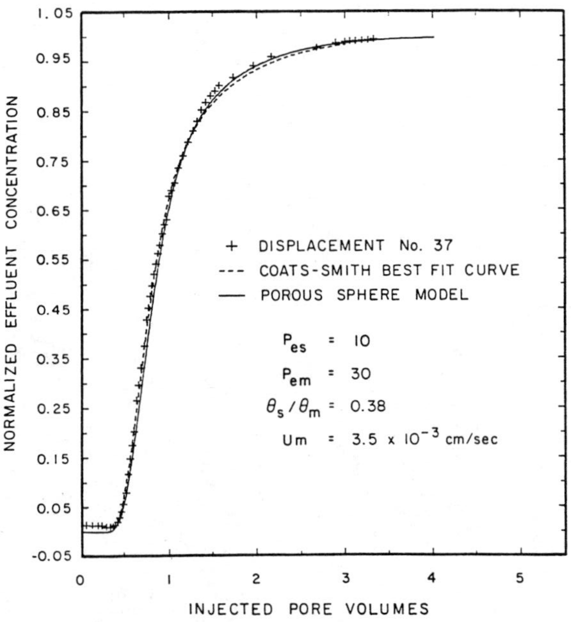

Figure 17b. Comparison of displacement data and corresponding best fit curve using the Coats-Smith model with the curve generated using the diffusion coefficient model for Pe_s = 10.0, Pe_m from the Coats-Smith fit and θ_s/θ_m from measurements of a photomicrograph from test core SAO.

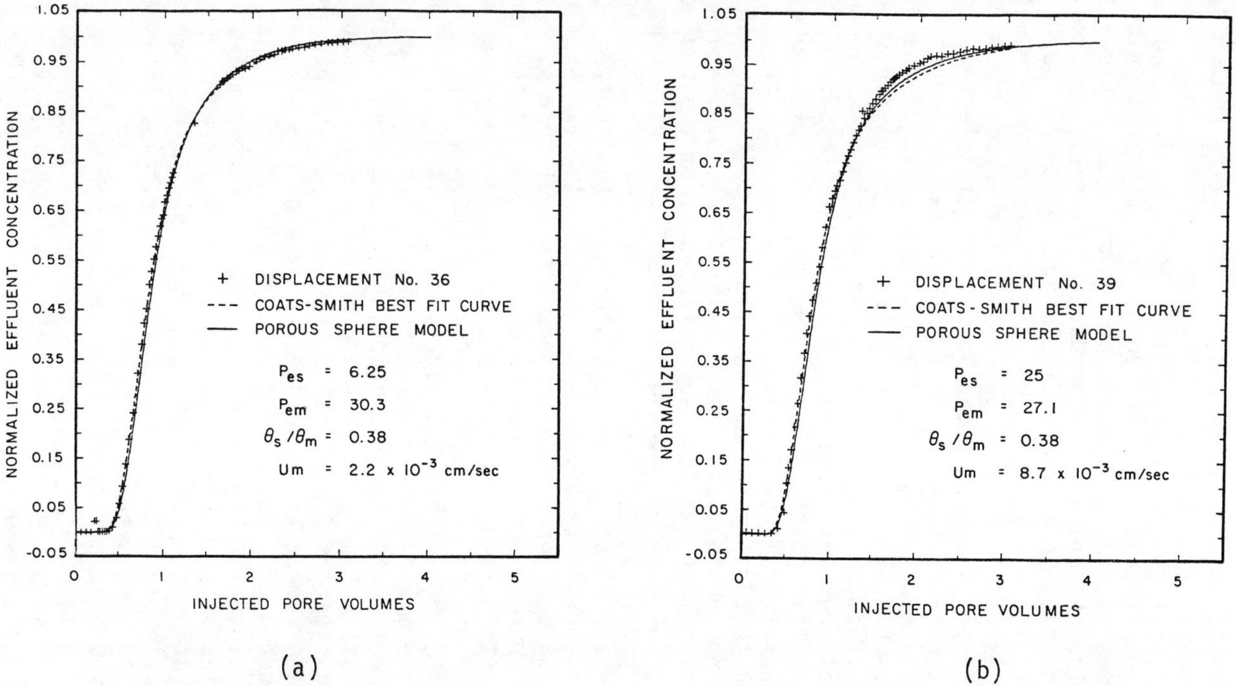

Figure 18. Comparison of displacement data and corresponding best fit curves generated using the Coats-Smith model with the curves generated using the porous sphere model for values of Pe_s predicted from the fit value of Figure 17b, but accounting for the differences in displacement velocity. Values of Pe_m are obtained from the Coats-Smith fit, and θ_s/θ_m are the same as those in Figure 17.

mixing behavior. It is easy to imagine, for instance, rock samples which have approximately the same porosity and permeability but exhibit totally different mixing behavior. It seems unlikely that pore size alone, for the pore sizes typical of many reservoir rocks, strongly influences mixing. For example, two geometrically similar pore structures differing only in the average size of the pores can be imagined. It seems likely that such samples would produce similar effluent composition histories in miscible displacements as long as the displacement rate was low enough that diffusion could equalize concentrations within typical pores. In displacements in unconsolidated sphere packs, the dispersion coefficient is found to be independent of sphere size for beads larger than about 30 mesh. Smaller beads show slightly higher dispersion coefficients. Such behavior is usually attributed to difficulties in obtaining uniformly sized particles and uniform packings when the particles are small (see Perkins & Johnston, 1963 for references on this point). Pore shape must also play some role in the mixing of miscible fluids, but if pores are, on average, small enough that diffusion can equalize compositions within them, it is not obvious that rectangular pores, say, would produce displacement results very different from those of circular pores. Studies of unconsolidated packs of particles other than spheres suggest that larger dispersion coefficients are obtained for nonspherical particles than for spheres, but the effect is not large (Perkins & Johnston, 1963). Aspect ratio, the ratio of pore body to pore throat diameters, and coordination number, the number of other pores connected to a given pore, strongly affect displacement performance when capillary forces are important (Wardlaw & Cassan, 1978; Wardlaw, 1980; Chatzis et al., 1983), but it is not clear how they influence mixing in miscible displacements.

One property which clearly does influence mixing is the pore size distribution. Perkins and Johnston report that several investigators found that dispersion coefficients for sphere packs increase as the width of the particle size distribution increases. Spence and Watkins (1980) found that flowing fractions less than one occurred for rock samples with wide pore size distributions. It seems likely, however, that a wide distribution of pore sizes alone does not guarantee a flowing fraction less than one. For instance, Fig. 19 shows four idealized pore systems with the same pore size distribution which would have very different mixing behavior. In Fig. 19a, the large and small pores are randomly distributed. Flow in such a pore system would be characterized by a flowing fraction of one and a dispersion coefficient higher than that of a totally uniform set of pores, but still relatively low. In Fig. 19b, pores with the same distribution of sizes are distributed in clusters of small and large pores, but the

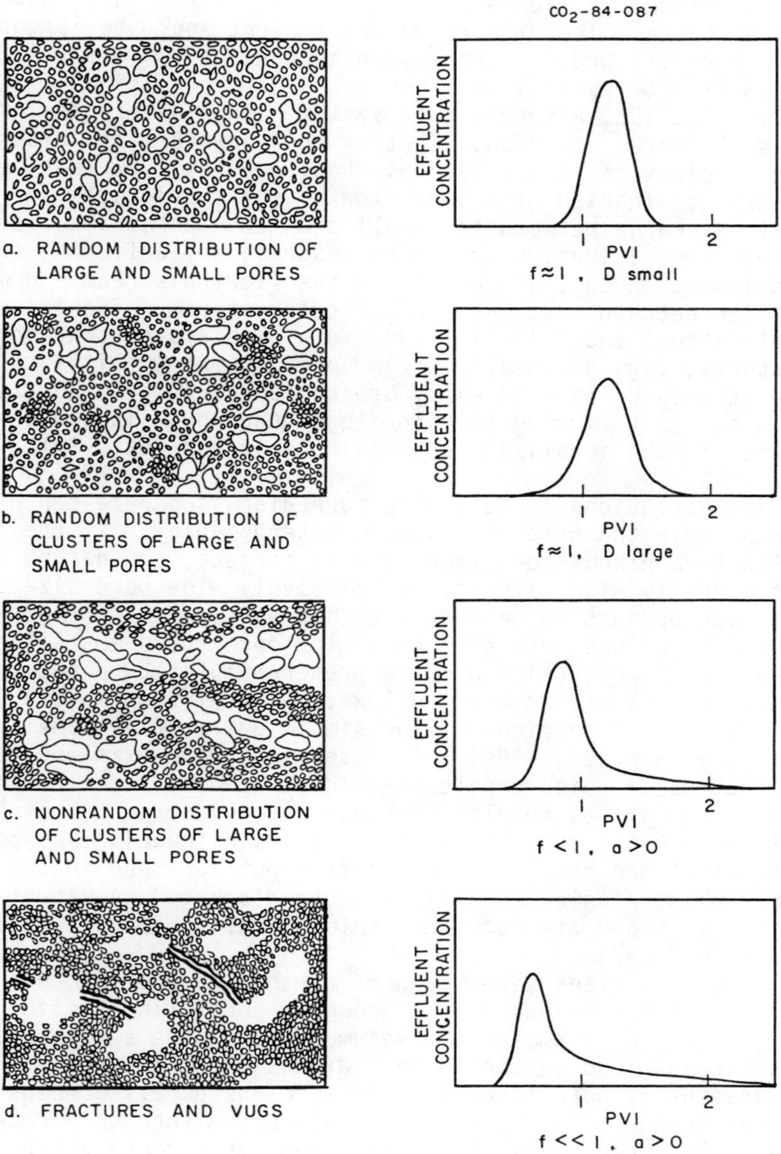

Figure 19. Idealized pore structures.

clusters themselves are randomly distributed so that there are no preferential flow paths for large samples compared to the average length over which clusters correlate. Such a pore system would still have a flowing fraction of one because the local variations in flow would average out over a large enough flow distance. The dispersion coefficient would be larger, reflecting the broader transition zone which would result from the local flow variations.

In Fig. 19c, a second pore system with clusters of large and small pores is shown. In this one, however, the distribution of clusters is not random. Preferential flow paths exist, and the combination of slower flow in the small pores and diffusive exchange between the small and large pores leads to effluent compositions which can be described by a flowing fraction less than one, dispersion in the flowing stream and mass transfer between a stagnant and flowing stream. The last idealization, shown in Fig. 19d, is a pore system containing fractures, vugs and small matrix pores. Again, this sort of pore structure leads to early breakthrough of injected fluid and hence to a flowing fraction less than one. It comes closest to the idealization envisioned in the Coats-Smith model.

The variations in pore structure discussed here can all be lumped under the name microscopic heterogeneity. If the simple qualitative discussion given here is correct, the nature of the heterogeneity also matters. A relatively wide pore size distribution appears to be necessary but not sufficient to produce flowing fractions less than one. The pore space must also be connected in ways which produce preferential flow paths.

In the discussion that follows, we attempt to relate observations of thin sections to the simple qualitative models presented (Specter, 1984). Thin sections were examined for the seven cores in which single-phase displacements were performed. Table V summarizes results of those observations. Also reported in that table are values of the flowing fraction, dispersion coefficient and mass transfer coefficient obtained in the single-phase displacement experiments discussed previously. The values shown are for the displacements at about 10 ft/day (3.5×10^{-3} cm/sec).

Pore body sizes were measured as the diameter of the smallest circumscribed circle around a pore. This definition exaggerates the diameter and volume of the pore somewhat, because it is based on the longest dimension of a pore. Measurement of pore sizes in thin sections partly compensates for the exaggeration because a random slice through a rock shows only a few pores which have been cut through their longest dimension. For the qualitative discussion here, any consistent definition is probably adequate. Approximately 300 randomly selected pores were measured for each thin section. The resulting pore size distributions are reported as the percent of total porosity contained in pores of a given size. Porosity was estimated based on the area of circles of the pore diameters divided by the total area.

To describe the connected pore space, we use a simple measurement which compares the average size of pores near a reference pore to the size of the reference pore. We reason that if pores of similar size are close to each other, then they are likely to be connected. If like-sized pores are connected, especially if the large pores are connected to other large pores, then preferential flow paths are also likely to exist. Conversely, if the average size of nearest neighbor pores are not correlated to the size of reference pores, the connections between pores are random with respect to size and thus are not likely to generate preferential flow paths.

The procedure for measuring the spatial distribution of pore sizes begins by dropping a brad onto a photomicrograph of a thin section. The point of the brad will come to rest on a pore which is designated the reference pore. The size of the reference pore is measured, and a semicircle with a diameter three times the mean diameter of all pores, taken from the pore size distribution previously determined, is centered on the reference pore and oriented with respect to the direction of fluid flow. The number and sizes of all pores within the semicircle are determined. From these counts, the mean size of nearest neighboring pores is determined. Fig. 20a illustrates this procedure.

Figs. 20b and c illustrate the types of results one might expect from plots of mean size of neighboring pores vs. reference pore size. If the mean size of the neighboring pores is not correlated with the size of the reference pores, then the plot will reflect this randomness (Fig. 20b). We argue that preferential flow paths are not likely in such pore structures. If, however, like-sized pores are grouped together, then a correlation will be reflected in the plots (Fig. 20c), which is an indication that preferential flow paths are likely.

A. Sandstones

Figs. 21-27 include photomicrographs, pore size distribution plots and spatial distribution of pore size plots for all seven cores. Fig. 21a is a photomicrograph of a Berea sandstone outcrop sample, core B-1, and Fig. 21b is the pore size distribution. Pore sizes in this relatively uniform sandstone are distributed in a single, narrow mode. Pores

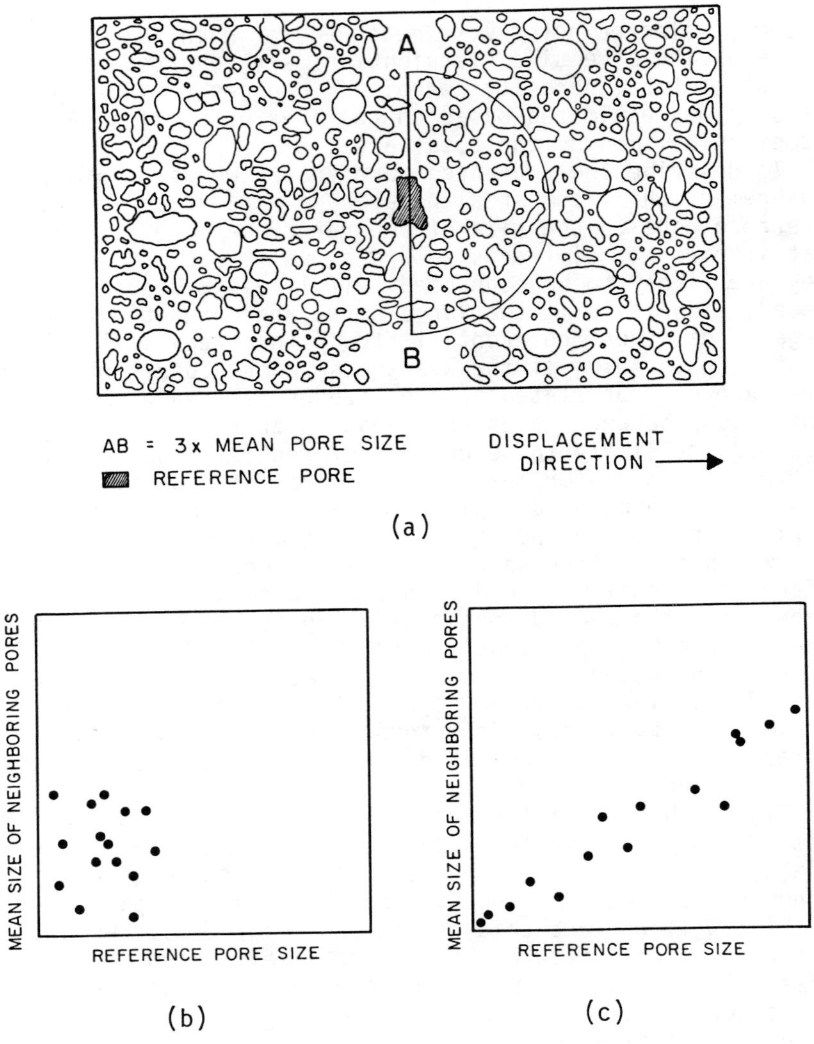

Figure 20. Determination and interpretation of the spatial distribution of pore sizes. (a) schematic of the procedure; (b) random distribution of pore connections; (c) like-sized pore connections.

smaller than 80 μm are mostly intergranular pore throats, elongated passages which connect larger and more irregular pore bodies. Small (10-25 μm) crystals of aluminosilicates (clay minerals) line some of the pore walls, and patches of dolomite cement are sparsely distributed around some of the sand grains. Both precipitates appear to be randomly distributed and present in amounts small enough that they offer no significant restriction to flow.

Fig. 21c shows the spatial distribution for the Berea outcrop sample. There appears to be little correlation between the mean size of nearest neighbor pores and a reference pore size indicating random connectedness and no preferential flow paths.

The Frannie sandstone (Fig. 22) is very similar to the Berea material in both pore shape and genetic type. The principal differences between them are the smaller grain sizes, and therefore, pore sizes, and the lack of clay mineral precipitates in the Frannie. The spatial distribution for the Frannie sandstone (Fig. 22c) is also similar to the Berea sample. There is little correlation between the mean size of nearest neighbor pores and a reference pore size indicating random connectedness and no preferential flow paths.

In the displacements for the Berea and Frannie sandstones (see Figs. 3 and 4), symmetric effluent concentration histories were obtained, and the displacements were modeled reasonably well by a flowing fraction of one and a dispersion coefficient. Thus, the pore structures in the Berea and Frannie sandstones appear to be qualitatively similar to that idealized in Fig. 19a.

In contrast, the Rock Creek sandstone (Fig. 23) has a substantial matrix (~20%) of clay minerals and micaceous laths which are distributed as coherent masses or "pods" around quartz grains. These pods appear to be nonuniformly distributed at the thin section scale. The clay minerals within these permeability barriers are tightly woven together and may limit access of flowing fluids to a portion of the pore space. The pore size distribution is also broader, additional evidence of microscopic heterogeneity in the Rock Creek sample. Fig. 23c shows correlation between the mean size of neighboring pores and reference pore size. As shown in Table V (and in Table II), some indication of flowing fractions less than one was found in the displacement experiments at the higher velocities. Such behavior may have been due to relatively less efficient diffusional mixing between pores containing clays at the higher velocities. Thus, it appears that the more complex pore structure of the Rock Creek sandstone sample could account for the observation of flowing fractions slightly less than one.

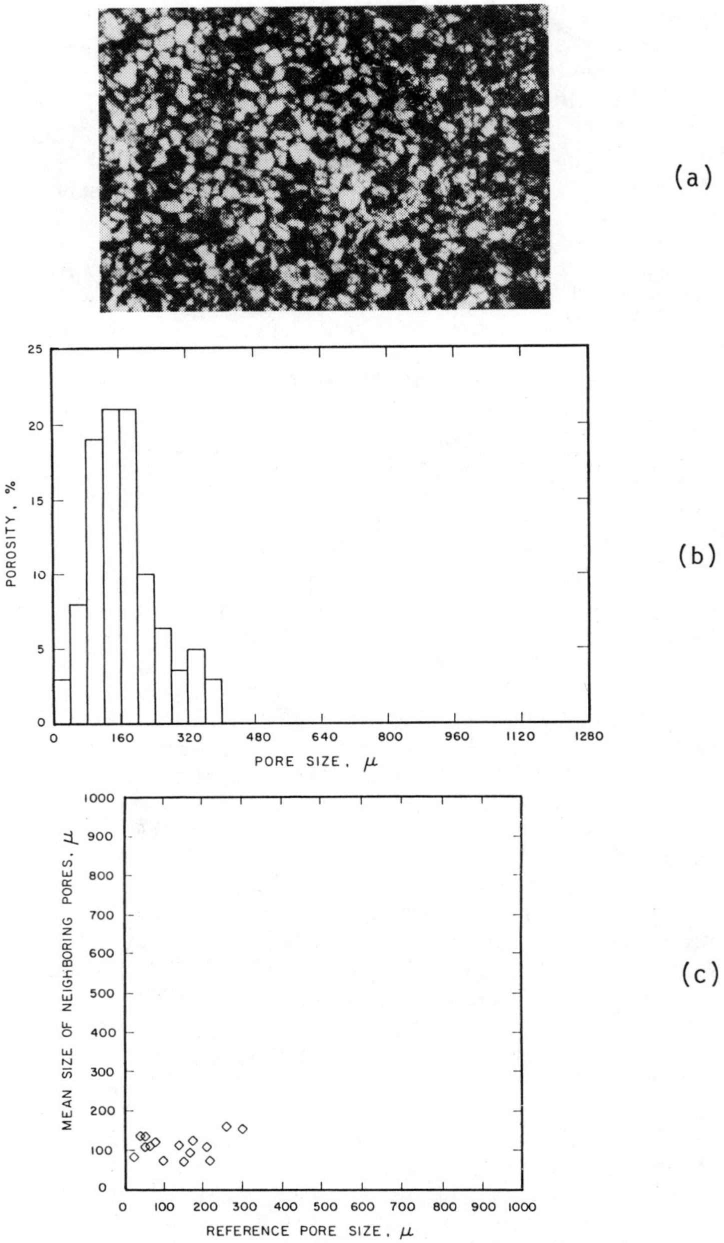

Figure 21. Berea sandstone, core B-1. (a) photomicrograph of a thin section, porosity is black; (b) distribution of pore body sizes; (c) spatial distribution of pores.

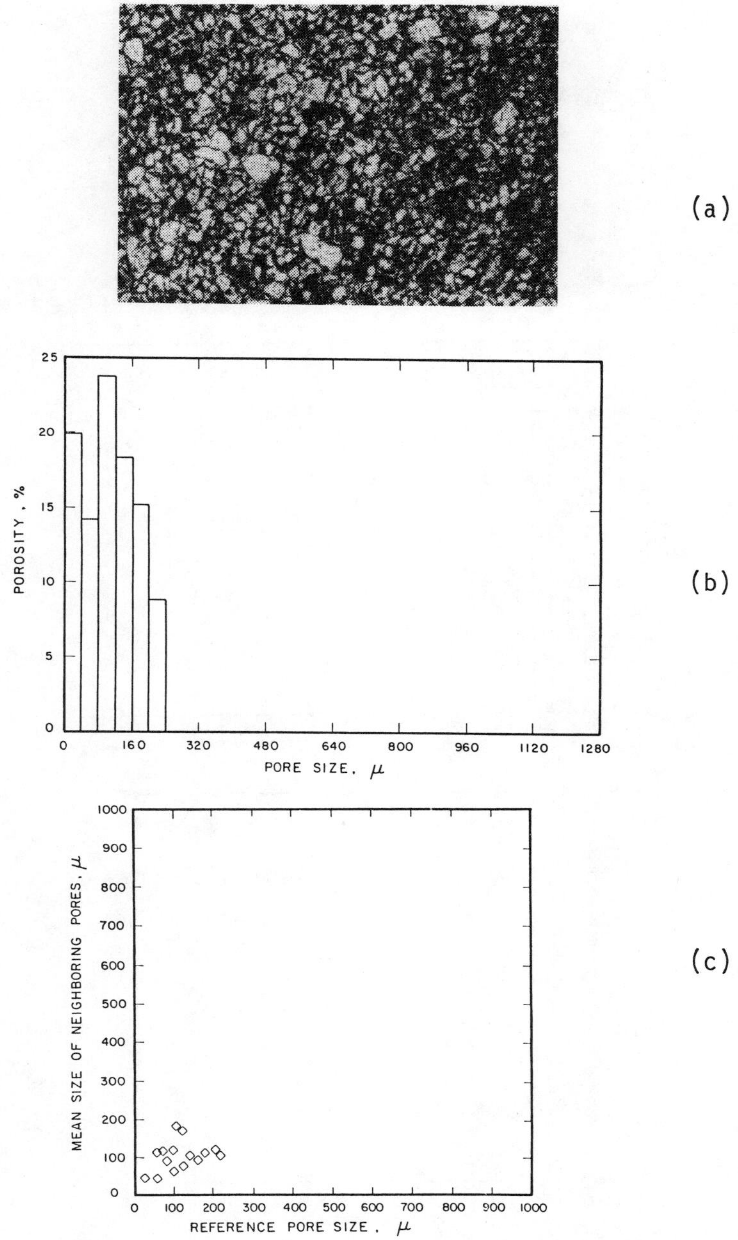

Figure 22. Frannie sandstone. (a) photomicrograph of a thin section, porosity is black; (b) distribution of pore body sizes; (c) spatial distribution of pores.

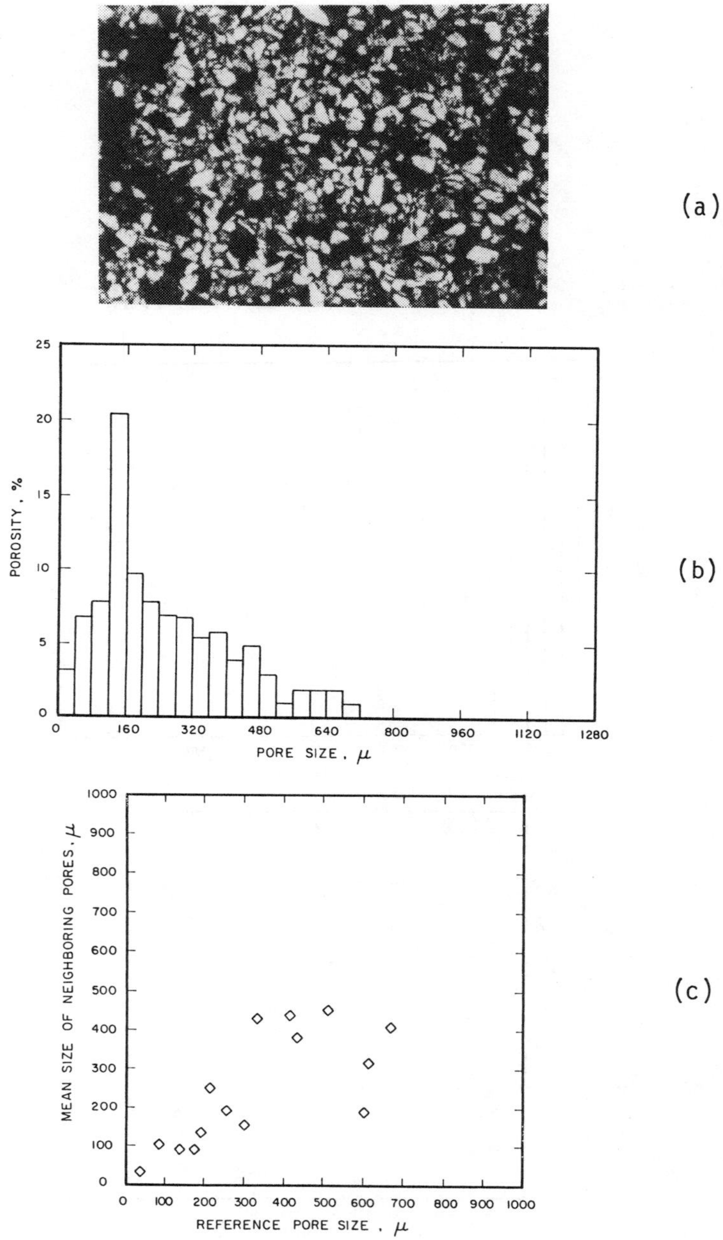

Figure 23. Rock Creek sandstone, core R-1. (a) photomicrograph of a thin section, porosity is black; (b) distribution of pore body sizes; (c) spatial distribution of pores.

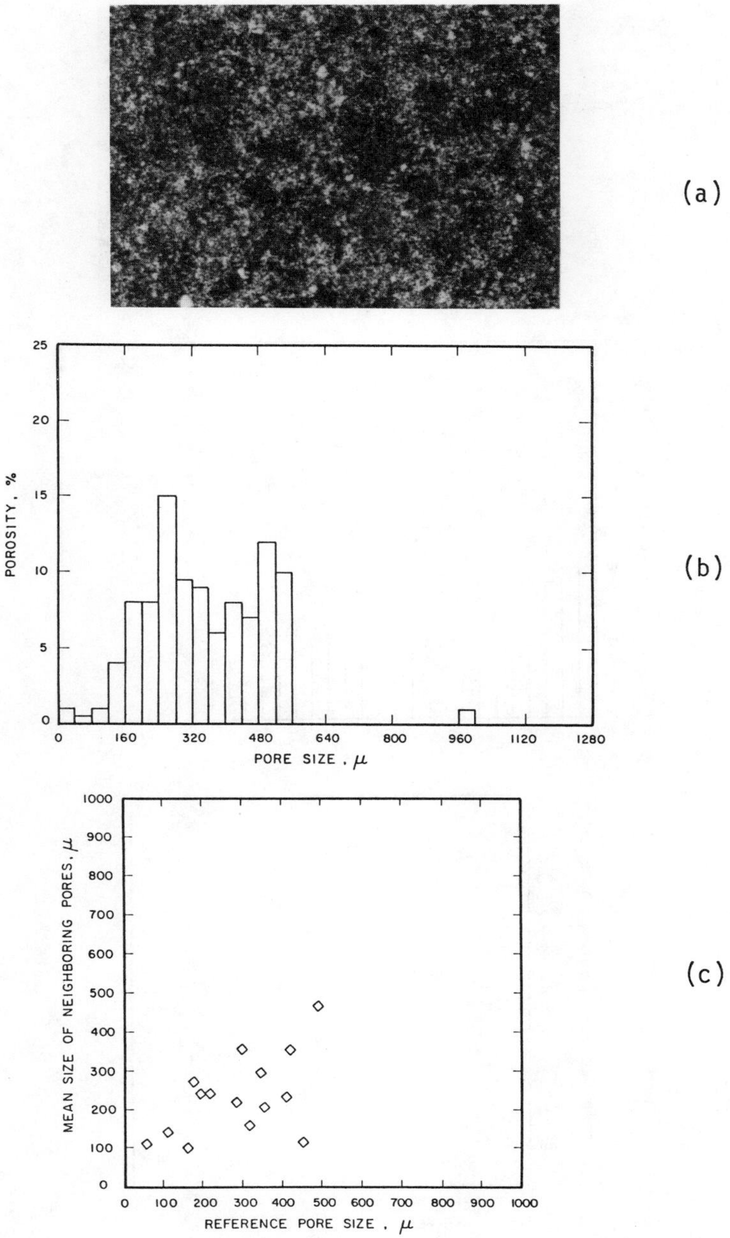

Figure 24. Wasson San Andres, core WW-2. (a) photomicrograph of a thin section, porosity is black; (b) distribution of pore body sizes; (c) spatial distribution of pores.

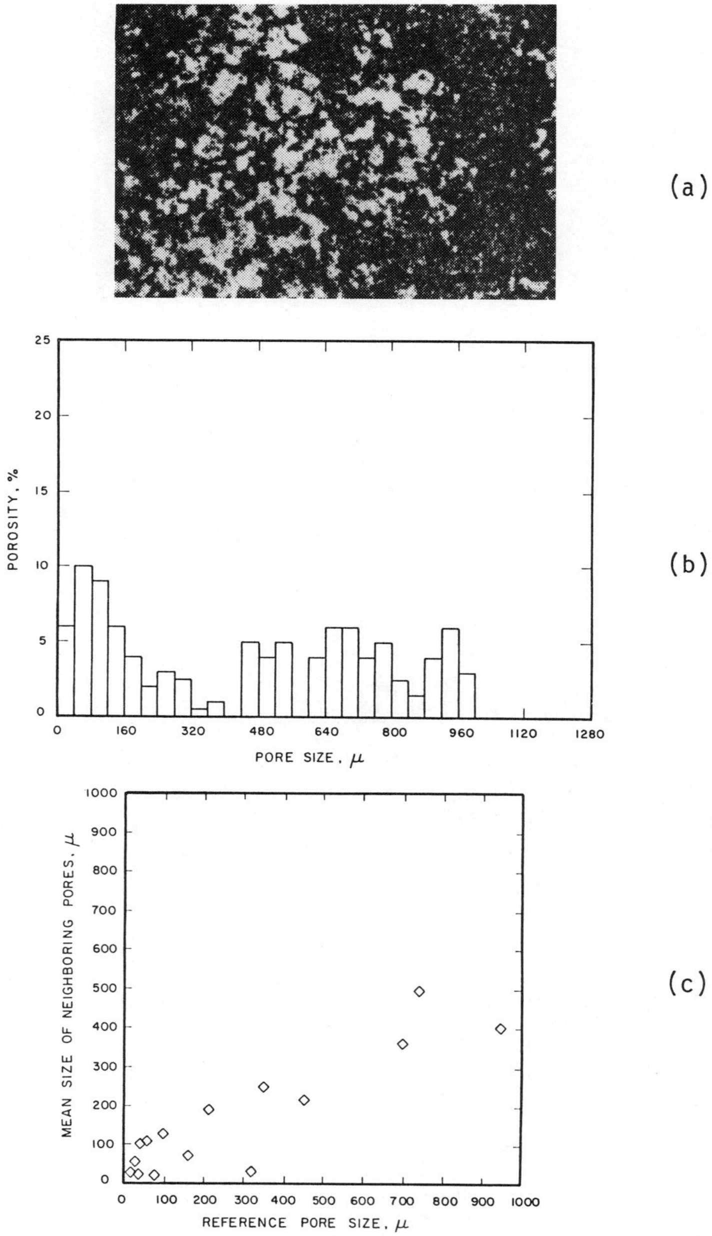

Figure 25. Seminole San Andres, core H-1. (a) photomicrograph of a thin section, porosity is black; (b) distribution of pore body sizes; (c) spatial distribution of pores.

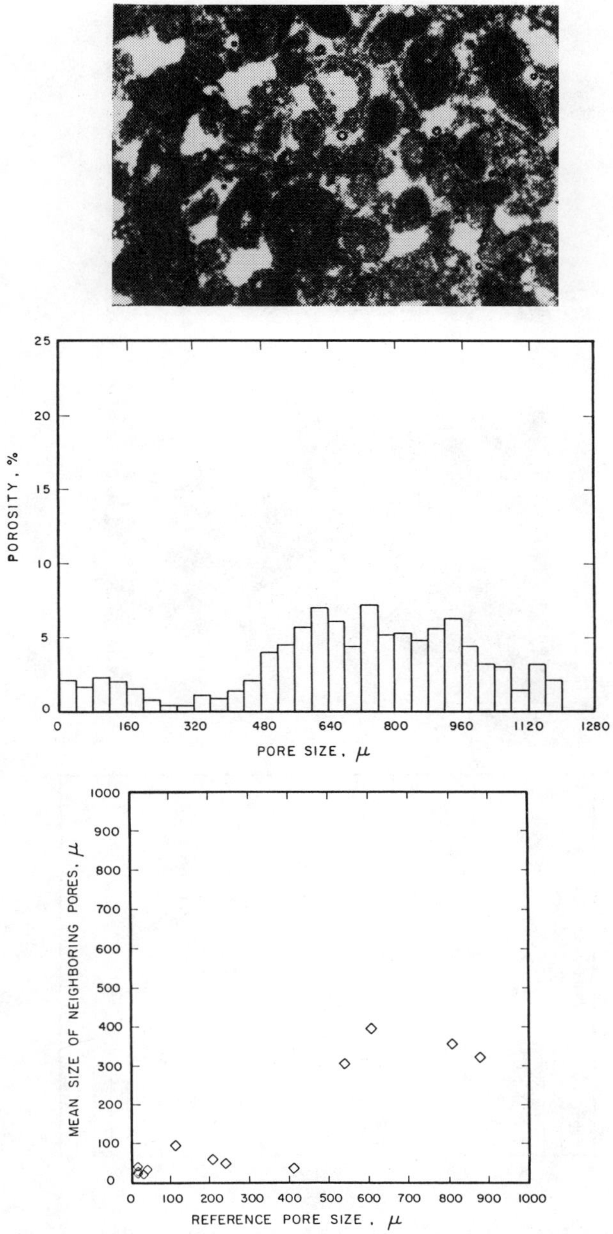

(a)

(b)

(c)

Figure 26. San Andres outcrop, core SAO. (a) photomicrograph of a thin section, porosity is white; (b) distribution of pore body sizes; (c) spatial distribution of pores.

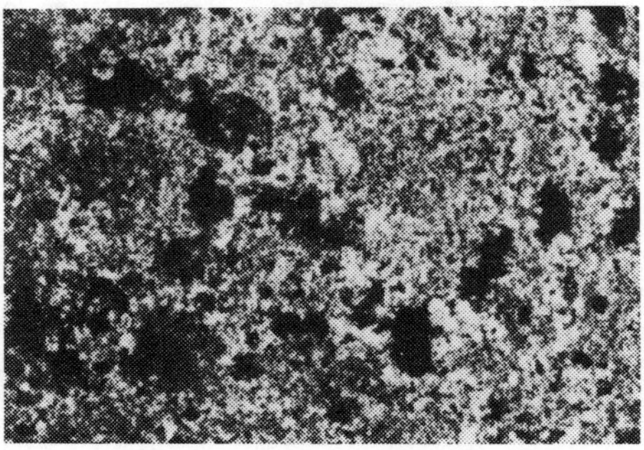

(a)

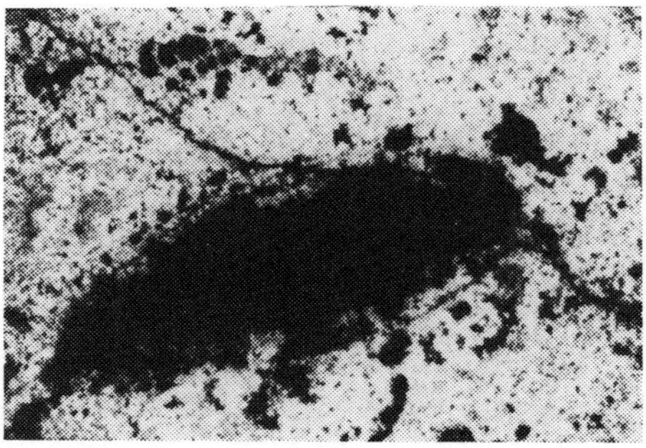

(b)

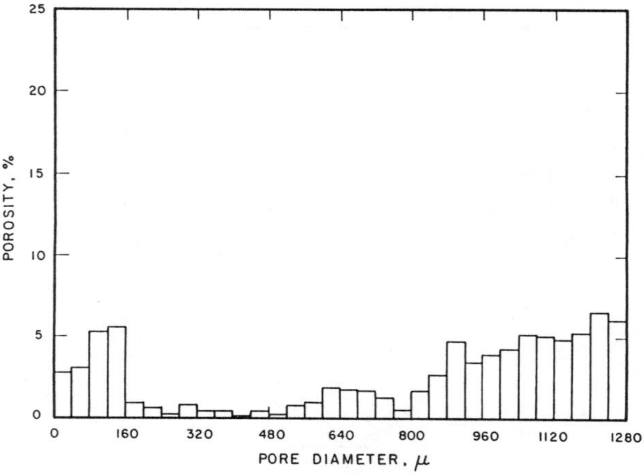

(c)

Figure 27. Maljamar San Andres, core M-1. (a) photomicrograph of a thin section, porosity is black; (b) photomicrograph of a thin section showing vugs and fractures, porosity is black; (c) distribution of pore body sizes for the lower 84% of porosity;

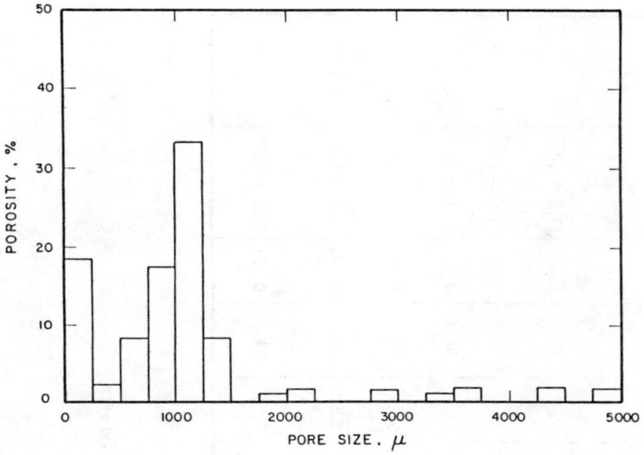

(d)

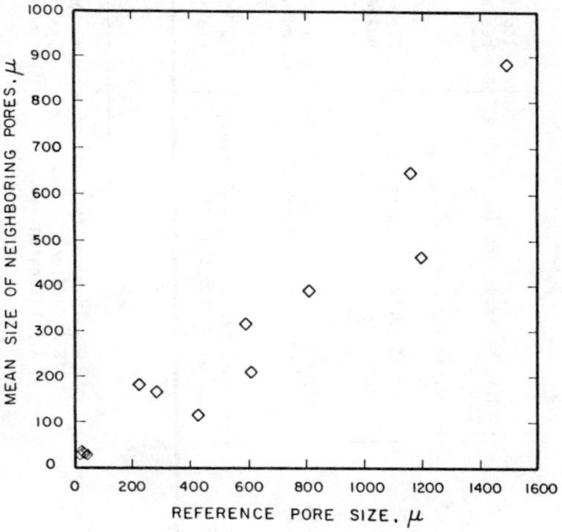

(e)

Figure 27. (continued) (d) distribution of pore body sizes; (e) spatial distribution of pores.

Table V Description of Thin Sections

Sample	Pore Size Distribution		Non-Random Permeability Barriers	Preferential Flow Paths	Pore Type	Exp't. No.	f	D_L cm^2/sec	K_m sec^{-1}
	# Modes	Mean ±Std. Dev. (μm)							
Berea SS (B-1)	One	169.44 ± 83.24	None	None	G	9	1.0	7.5×10^{-4}	0
Frannie SS (F-2)	One	108.48 ± 62.98	None	None	G	17	1.0	3.7×10^{-4}	$<10^{-7}$
Rock Creek SS (R-1)	One	259.80 ±165.62	Clay mineral pods (matrix)	Flow through type G pores around clay pods	G	20	0.982	3.2×10^{-4}	1.2×10^{-5}
Wasson DM (WW-2)	One	343.23 ±129.42	Minor anhydrite plugging	None	SC	31	0.897	1.4×10^{-3}	$<10^{-7}$
Seminole DM (H-1)	Three	129.55 ± 90.95 500.00 ± 33.81 786.19 ±117.19	Anhydrite plugging	Flow through vugs inuirectly connected by type-C pores	SC,V	22	0.687	5.1×10^{-3}	5.2×10^{-5}
San Andres FM (SAO) DM	Two	110.74 ± 66.60 769.84 ±200.81	Type-C pores arranged in spherical clusters	Flow through well-coordinated type-SG and type-V pores	C,SC,V	37	0.771	7.8×10^{-3}	1.5×10^{-5}
Maljamar DM (M-1)	Many	Widest distribution of sizes	Preponderance of type-C pores, anhydrite plugging	Flow through fracture/stylolite-connected vugs	C,V,F,M	34	0.621	1.2×10^{-2}	9.2×10^{-6}

Key to Pore Types:

C = Intercrystalline <25μm
F = Fracture/Stylolite
G = Intergranular
M = Moldic
S = Solution Enhanced (Prefix)
V = Vugular

B. Carbonates

The carbonate samples examined showed much greater variation than did the sandstones. Fig. 24 shows the thin section photograph, pore size distribution and spatial distribution for the Wasson core. The pore size distribution for the Wasson material was not greatly different from the Rock Creek material, as the average pore sizes and standard deviations given in Table V indicate. In fact, the pore size distribution was also similar to that of the Berea core. Pores in the Wasson material appeared to be more nearly spherical than they did in the Berea sample, probably because the dolomitization and subsequent dissolution of carbonate material during diagenesis transformed the original carbonate materials of the sediments. No clays were observed, but some anhydrite was present, particularly in pores with diameters greater than 200 μm. The anhydrite did not fill the pores, but appeared in thin sections as small, patchy areas. Figure 24c indicates a weak, if any, correlation between the mean size of neighboring pores and a reference pore size. The apparent uniformity of the pore structure of the Wasson material was reflected in the displacements in that core (see Fig. 12). All were described well by a dispersion coefficient only, even though flowing fractions less than one were obtained. In any case, both displacement data and thin sections indicate that from the standpoint of mixing, the Wasson core material behaved much more like the sandstones than like the other carbonate samples.

Pore size distributions, spatial distributions and a photograph for the Seminole core material are shown in Fig. 25. Fig. 26 gives the same data for the San Andres outcrop sample. Both samples showed much broader distributions of pore sizes than the sandstones or Wasson material. The spatial distributions obtained for both samples indicate some correlation between the mean size of neighboring pores and a reference pore size, therefore indicating the existence of preferential flow paths. In the San Andres outcrop sample, there was considerable interparticle porosity due to extensive dissolution of cements and grains. Also present in the San Andres outcrop material were ooid grains within which small intraparticle pores were observed. The pore space within the ooid grains was surrounded by low permeability, fine crystalline dolomite. Presumably, flow into and out of such pore space was much slower than in the well connected interparticle pore space. Thus, fluid in the ooid grains would exchange with fluid in the preferential flow path by slow flow and diffusion. The San Andres outcrop sample provided the initial impetus for an investigation of the porous sphere model. Thus, that sample appears to be similar to the idealization of Fig. 19: a wide pore size distribution with pore connections that generate preferential flow paths. It showed a flowing fraction of much less than one, a large dispersion coefficient and a significant mass transfer coefficient in displacement tests (Table V and Fig. 11).

The Seminole core material showed similar displacement behavior, but the details of the pore structure observed in thin sections were significantly different (Figs. 25 and 9). Extensive recrystallization of the carbonate material apparently obliterated the original ooids, leaving a dolomite fabric containing many small pores and some medium size and large pores. A significant fraction (31%) of the pore volume was in pores with diameters less than 160 μm. The larger pores appeared to be connected to each other or located with only small zones of smaller pores between them. Figure 25c shows a definite correlation between the mean size of neighboring pores and a reference pore size, especially the larger pores. Thus, it appears that preferential flow paths could have been present. Flow in the small matrix porosity must have been much slower. Some anhydrite was observed to block some pores of intermediate size and may have further restricted access of flowing fluid to the matrix in some portions of the pore space. Thus, the Seminole material also appears to fit the idealization of Fig. 19c, but with pore geometry different from that of the San Andres outcrop sample.

Samples of Maljamar dolomite (core M-1) exhibited the widest pore size distribution and the most definite correlation between mean size of neighboring pores and reference pore size of the cores examined in this study (Fig. 27). The rock is composed of tightly woven euhedral to anhedral dolomite crystals. Seventeen percent of the pore space consists of intercrystalline voids with diameters less than 25 m (Fig. 27a). Large vugs directly connected by fractures and stylolites dwarf the surrounding intercrystalline micropores (Fig. 27b). Most of the porosity was in very large vugs (Fig. 27c and d). The fact that at least some of the vugs were connected by fractures must have generated preferential flow paths, and exchange of fluid between those flow paths and the fine matrix porosity must have caused the long tail observed in the displacement experiments (Fig. 10). The Maljamar core showed the earliest breakthrough and hence the lowest flowing fraction of any of the cores tested. Thus, this sample is similar to the idealization of Fig. 19. It seems clear from the observations of thin sections of the Maljamar core material that the displacement behavior is consistent with features of the pore structure present in the thin section.

Throughout the discussion here, two aspects of microscopic heterogeneity have been considered. The first is a distribution of pore sizes. If the distribution of pore sizes is broad, early breakthrough may occur, but only if the pore space is also connected in ways that generate preferential flow paths. Microscopic fractures, solution pores through fine matrix porosity and no doubt other pore structures, can generate such flow paths. Qualitative observations of thin sections appear to be consistent with displacement behavior in stable, single-phase miscible displacements. In multiphase displacements, the distribution of phases within the pore structure must also influence mixing. Wetting behavior will control the distribution of phases between small and large pores and hence must also affect mixing. Thus, the simple analysis presented here is only a first step, but a necessary one, toward understanding mixing in the complex flows which occur in actual CO_2 floods in reservoir rocks.

VI. DISPLACEMENTS AT LARGER SCALE

The comparison of observations of thin sections with displacement behavior indicates that heterogeneities of pore structure observed at the scale of a thin section have a recognizable impact on displacements at the laboratory core scale. The question of what effect such features have in larger scale flows is important in the assessment of the effect of the interaction of phase behavior and heterogeneity on local displacement efficiency at reservoir scale (Dai & Orr, 1984). The conventional argument (Coats & Smith, 1964; Stalkup, 1983) is that the effects of mass transfer into a stagnant volume are minimal at field scale. That argument is based on examination of the Peclet and Damköhler numbers, $Pe = uL/D_L$ and $a = K_mL/U$. A key assumption is that the dispersion and mass transfer coefficients do not change as the length scale increases. If so, both parameters become large. It can be shown (Baker, 1977) that for large L, a miscible displacement behaves as if there is no stagnant fraction, but the effective dispersion coefficient, D_k is given by

$$D_k = D_L + \frac{U^2(1-f)^2}{K_m}. \tag{5}$$

Thus, the level of dispersion would appear to be larger than would be observed if there were no stagnant fraction. For example, with typical values of D_L, f and K_m for one of the carbonates, L = 142 m (20 acre spacing), and a velocity of 3.53 x

10^{-4}cm/sec (1 ft/day), the effective dispersion coefficient and Peclet number would be 3.7 x 10^{-3} cm^2/sec and 1360, respectively. Thus, even if the effects of mass transfer to a stagnant fraction were included, dispersion would have a smaller effect on the growth of a transition zone at field scale than at laboratory core scale. Similar arguments can be made for the porous sphere model. If this argument is correct, then the interaction of dispersion with phase behavior probably has a relatively small effect at field scale.

In many reservoir rocks, however, heterogeneities with characteristic length scales larger than those sampled by a laboratory core will appear as the scale of the flow system increases. For example, layers, fractures and areal variations in porosity and permeability will be observed in field displacements, but such variations will not be present in small laboratory core samples. In some reservoirs, at least, those heterogeneities will be distributed in ways that lead to preferential flow paths. That is, there will be portions of the pore space in which the flow will be significantly faster than that in other portions. Presumably the low viscosity of injection fluids such as CO_2 will further accentuate the contrast in velocities.

In such situations, it will be the rate of exchange of material between the preferential flow paths (flowing fraction) and the remainder of the pore space (stagnant fraction) that determines how local displacement efficiency is affected in EOR processes such as a CO_2 flood. If the rate of interchange is very high, then the pore space acts as if it is completely mixed, and the flow can be modeled in terms of dispersion as discussed above. If the rate of interchange has some intermediate value, then interactions with phase behavior will lead to higher local residual oil saturations in the swept zone (Dai & Orr, 1984). If the rate is very low, then that interaction would have only a small effect, but a smaller fraction of the pore space would be swept. In field-scale flows, the exchange of material will be the result of slow, but finite flow in the so-called stagnant fraction, diffusion, transverse dispersion, viscous crossflow (Zapata & Lake, 1981) capillary pressure and wetting effects (Campbell & Orr, 1983), and so on.

Unfortunately, models that allow direct estimation of the lumped parameters of the Coats-Smith model for field-scale displacements are not available. However, the porous sphere model can be used to give an indication of how Coats-Smith parameters for larger scale flows might behave. The Coats-Smith model represents the exchange of material between the flowing and stagnant fractions in terms of an overall mass transfer coefficient, while the porous sphere model accounts for the distribution of material as it diffuses within a sphere. The two models should be equivalent when diffusion is fast enough that the sphere can be treated as well mixed (Cussler, 1984). When concentration within the sphere varies significantly with position, then the more detailed porous sphere model should give a more accurate description of the overall mixing behavior.

Rao et al. (1980) compared the two types of models. They synthesized a porous medium of porous ceramic beads dispersed in a packing of glass beads of the same size. Prior to displacement experiments, the effective diffusion coefficient, D_e, of the ceramic beads was measured by independent experiments. In addition, they found that an overall mass transfer coefficient could be calculated, based on the geometry and effective diffusion coefficient, but that the value of the overall mass transfer coefficient varied during the time of the displacement. Therefore, they found a time-averaged value to compare with displacement data. Rao et al. concluded that the overall mass transfer coefficient-type model is adequate for describing laboratory displacement data.

What Rao et al. did not address was the problem of scaling from laboratory experiments to larger scale displacements. The porous sphere model can be used, however, to estimate Coats-Smith parameters that might result from displacements in porous media with larger scale heterogeneities. To make such estimates we consider displacements in which the flow length and sphere radius are larger. We calculate effluent composition data with the porous sphere model and then fit that data to the Coats-Smith model to determine parameters consistent with larger scale heterogeneities and displacement lengths.

Results of calculated displacements at two length scales are shown in Fig. 28a and b. The small scale, about 20 m, is consistent with displacement lengths that might be encountered in a single well tracer test. The larger scale, about 140 m, is appropriate to a displacement in a field with 20 acre spacing. At both scales, we consider the effect of spheres with radius 33 cm. It is unlikely, of course, that actual spheres would be distributed in the way that the model is formulated, but the model calculations do give a qualitative indication of the effects of heterogeneities with similar characteristic lengths. In all of the calculations, θ_s/θ_m was taken to be about .4 (θ_m = .7), a value consistent with the fraction of preferential flow paths in smaller scale displacements. Peclet numbers in the matrix were estimated using $D_L = \alpha U$ with α chosen as 3 cm (0.1 ft) for the single well case and 61 cm (2.0 ft) for the 20 acre case. That range of values is reported by Stalkup (1983) to be applicable to reservoirs. In both cases the displacement velocity was $U = 3.53 \times 10^{-4}$ cm/sec (1 ft/day).

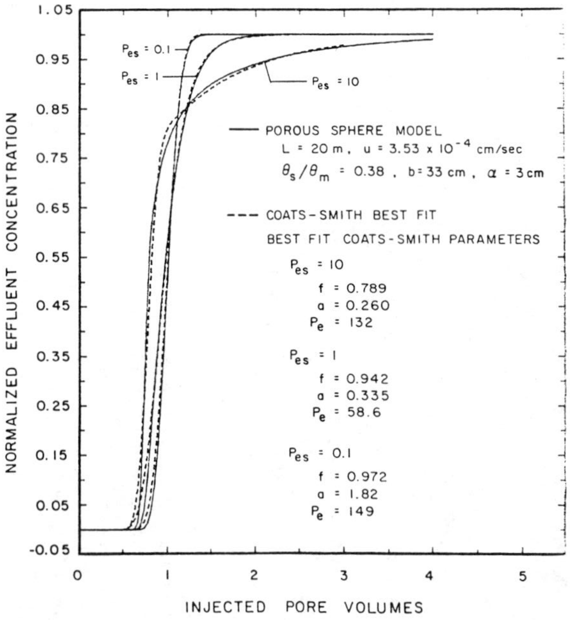

(a)

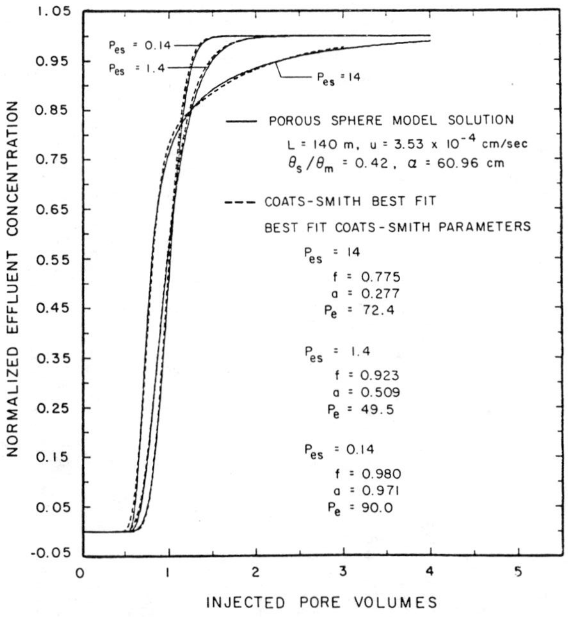

(b)

Figure 28. Concentration histories for hypothetical displacements predicted by the porous sphere model and fit by the Coats-Smith model. (a) L = 20 m corresponding to a typical distance of a single well tracer test; (b) L = 140 m corresponding to the distance between wells for 20 acre spacing.

A range of values of D_e was used to estimate spherical Peclet numbers, Pe_s, for each case. For the single well case, the base value of D_e was chosen to be 2×10^{-5} cm^2/sec, a typical value of a transverse dispersion coefficient at that velocity. The resulting value of Pe_s was 1. Values of Pe_s an order of magnitude larger and smaller were also used. The value of Pe_s used is a measure of the rate of interchange between the flowing and stagnant fractions, but it is not clear what value is appropriate. The values used appear to be reasonable estimates, but are not justified by measurements. For the larger scale case, a range of Pe_s from 0.014 to 14 was used.

In both cases, the presence of larger scale heterogeneities, as represented by the porous sphere model, produced dimensionless Coats-Smith parameters not very different from the values appropriate to laboratory scale displacements. In all the cases considered, the Coats-Smith model matched the porous sphere model quite well. For example, for the single well scale (Fig. 28a) with $Pe_s = 1$, the Coats-Smith flowing fraction was $f = 0.942$, the Damköhler number, $a = 0.335$ and the Peclet number $Pe = 58.6$. Thus, the Peclet number in the Coats-Smith model was an order of magnitude smaller than that estimated for the matrix in the porous sphere model. That result is consistent with observations that dispersion coefficients obtained in single well tracer tests are significantly larger than those measured at laboratory scale.

When the rate of transfer is reduced in the porous sphere model ($Pe_s = 10$), the resulting Coats-Smith flowing fraction and dimensionless mass transfer (Damköhler) numbers are smaller, and the Peclet number is larger. Slower mass transfer leads to earlier breakthrough and a longer tail in both models. Faster mass transfer ($Pe_s = 0.1$) leads to effluent composition curves that can be represented as the result of dispersion alone, though with a dispersion coefficient a factor of three larger than that of the matrix in the porous sphere model.

Similar results were obtained for the calculated displacements for 20 acre spacing. Results for $Pe_s = 0.14$, 1.4 and 14 are given in Fig. 28b. A solution for $Pe_s = 0.014$ was calculated but was indistinguishable from that for $Pe_s = 0.14$ and hence is not shown in Fig. 28b. As in the single well case, assignment of 30% of the pore space to regions of slow flow with a characteristic length of 33 cm leads to estimates of low flowing fractions and a significant mass transfer coefficient in the Coats-Smith model.

The estimated Coats-Smith parameters obtained here suggest that the interaction of phase behavior with mixing behavior could have a significant effect on local displacement efficiency in some reservoir settings. For example, experimental results presented by Spence and Watkins (1980) and model calculations given by Dai and Orr (1984) indicate that for a CO_2 flood in a core with f = 0.71, a = 0.47 and Pe = 20, the residual oil saturation was 11.4%, about three times higher than was observed for a more uniform core with f = .97, a = 0.12 and Pe = 83. The calculations presented here suggest that similar values of the Coats-Smith parameters might be reasonable for some reservoir rocks, even though the scale of the displacement is larger than in the laboratory, because the scale of heterogeneities present is also larger. Hence, a significant interaction of phase behavior with heterogeneity (and no doubt with the accompanying viscous instability) appears possible.

VII. SUMMARY AND CONCLUSIONS

In this paper, we have shown that heterogeneities observed in thin sections account for the behavior of miscible displacements in laboratory cores. A simple technique for detection of preferential flow paths was demonstrated. It is based on the argument that such paths will exist if, on average, large pores are surrounded by other large pores while small ones are close to other small ones. Examination of thin sections for carbonate and sandstone core samples indicated that early breakthrough and significant tailing were observed for those samples with wide pore size distributions and an indication of the existence of preferential flow paths.

Calculations performed with the convection-dispersion equation, the Coats-Smith model and a porous sphere model indicate that for displacements at a fixed scale, a lumped parameter model, such as the Coats-Smith model, fits displacement behavior reasonably well. A more detailed model, such as the porous sphere model, requires that fewer parameters be determined by fitting experimental data to the model if geometric parameters can be estimated from observations of thin sections or by other means. Parameters which describe the transfer of material between the porous sphere and the matrix still require such estimation. A difficulty with the Coats-Smith model is that it is not clear how appropriate parameter values change with the scale of the flow system. The porous sphere model does, however, allow estimation of the qualitative impact of heterogeneities with larger characteristic lengths on displacements at field scale. Estimates of Coats-Smith parameters made using the porous sphere model for larger-scale displacements suggest that for some types of heterogeneities Coats-Smith parameters will have magnitudes similar to those observed in laboratory core displacements. Thus, interactions of phase behavior and heterogeneity, demonstrated previously to have significant impact on residual oil saturations to CO_2 floods in laboratory cores, may also have significant effects at field scale.

NOMENCLATURE

a - Damköhler number, K_mL/U, which relates a first order mass transfer rate constant to a characteristic time for convection

b - outside radius of a porous sphere, cm

C - normalized tracer concentration

C* - normalized tracer concentration in well stirred stagnant volume

C' - normalized tracer concentration in fluid exiting the core

C_m - normalized tracer concentration for fluids flowing in the matrix outside of spheres

C_s - normalized tracer concentration for fluids within the porous sphere pore space

$\bar{C}_s$ - average normalized tracer concentration for fluid in the porous sphere

D - molecular diffusion coefficient, cm^2/sec

D_L - longitudinal dispersion coefficient, cm^2/sec

D_e - effective diffusion coefficient, $D\phi/\gamma$, cm^2/sec

D_k - effective dispersion coefficient, cm^2/sec

f - flowing fraction, the fraction of the pore volume through which convection occurs

K_m - overall mass transfer coefficient, sec^{-1}

L - length of the porous medium, cm

Pe - macroscopic Peclet number, UL/D_L

Pe_m - macroscopic Peclet number based on the velocity of fluids in the matrix, U_mL/D_L

Pe_s - Peclet number based on the size of the porous spheres, but scaled for the length of the porous media, b^2U_m/D_eL

r - radius of the porous spheres, cm

t - time, sec

U - the average interstitial velocity based on the total pore space, both flowing and non-flowing fractions, cm/sec

U_m - average interstitial velocity based on assumption that all convection occurs in the matrix pore space outside of the porous spheres

x - length in the direction of flow, cm

α - mixing coefficient, cm

β - width of pulse in dimensionless time on pore volumes

η - dimensionless sphere radius, r/b

γ - tortuosity

ϕ - porosity of the porous spheres

ξ - dimensionless length, x/L

τ - dimensionless time in pore volumes, either tU/L or tU_m/L

θ_m, θ_s - the fraction of the total (bulk) volume of the porous medium occupied by the matrix and sphere porosity, respectively

REFERENCES

Aronofsky, J.S. and Heller, J.P. (1957), "A Diffusion Model to Explain Mixing of Flowing Miscible Fluids in Porous Media," Trans., AIME 210, 345-349.

Baker, L.E. (1977), "Effects of Dispersion and Dead-End Pore Volume in Miscible Flooding," Soc. Pet. Eng. J. 17, 219-227.

Batycky, J.P., Maini, B.B. and Fisher, D.B. (1982), "Simulation of Miscible Displacement in Full Diameter Carbonate Cores," Soc. Pet. Eng. J. 22, 647-658.

Bretz, R.E. and Orr, F.M., Jr. (1985), "Comparison of Analytical and Explicit Finite Difference Solutions of the Convection-Dispersion Equation," PRRC Report 85-6, New Mexico Petroleum Recovery Research Center.

Brigham, W.E. (1974), "Mixing Equations in Short Laboratory Cores," Soc. Pet. Eng. J. 14, 91-99; Trans., AIME, 257.

Campbell, B.T. and Orr, F.M., Jr. (1983), "Flow Visualization for CO_2-Crude Oil Displacements," paper SPE 11958 presented at the 58th Annual Technical Conference and Exhibition, San Francisco, Oct. 5-8; accepted for publication J. Pet. Tech., Oct. 1985.

Chatzis, I., Morrow, N.R. and Lim H.T. (1983), "Magnitude and Detailed Structure of Residual Oil," Soc. Pet. Eng. J. 23, 311-336.

Coats, K.H. and Smith, B.D. (1964), "Dead-End Pore Volume and Dispersion in Porous Media," Soc. Pet. Eng. J. 4, 73-84

Cussler, E.L. (1984), Diffusion Mass Transfer in Fluid Systems, Cambridge University Press, Cambridge, UK, 1-11.

Dai, K.K. and Orr, F.M., Jr. (1984), "Prediction of CO_2 Flood Performance: Interaction of Phase Behavior with Microscopic Pore Structure Heterogeneity," paper SPE 13115 presented at the 59th Annual Technical Conference and Exhibition, Houston, Sept. 16-19.

Gardner, J.W. and Ypma, J.G.J. (1982), "An Investigation of Phase Behavior-Macroscopic Bypassing in CO_2 Flooding," paper SPE 10686 presented at SPE/DOE Third Joint Symposium on Enhanced Oil Recovery, Tulsa, April 4-7.

Gardner, J.W., Orr, F.M., Jr. and Patel, P.D. (1981), "The Effect of Phase Behavior on CO_2 Flood Displacement Efficiency," J. Pet. Tech. 33, 2067-2081.

Helfferich, F.G. (1981), "Theory of Multicomponent, Multi-Phase Displacement in Porous Media," Soc. Pet. Eng. J. 21, 51-62.

King, M.J. (1985), personal communication, Sohio Petroleum Co., Cleveland, OH.

Orr, F.M., Jr. and Taber, J.J. (1984), "Displacement of Oil by Carbon Dioxide," Final Report to U.S. Department of Energy, Report No. DOE/BC/10221-13, March.

Parker, J.C. (1984), "Analysis of Solute Transport in Column Tracer Studies," Soil Sci. Soc. Am. J. 48, 719-724.

Passioura, J.B. (1971), "Hydrodynamic Dispersion in Aggregated Media 1. Theory," Soil Sci. 111, 339-344.

Perkins, T.K. and Johnston, O.C. (1963), "A Review of Diffusion and Dispersion in Porous Media," Soc. Pet. Eng. J. 3, 70-75.

Rao, P.S.C., Rolston, D.E., Jessup, R.E. and Davidson, J.M., (1980), "Solute Transport in Aggregated Porous Media: Theoretical and Experimental Evaluation," Soil Sci. Soc. Am. J. 44, 1139-1146.

Salter, S.J. and Mohanty, K.K. (1981), "Multiphase Flow in Porous Media: I. Macroscopic Observations and Modelling," paper SPE 11017 presented at 57th Annual Fall Technical Conference and Exhibition, New Orleans, September 26-29.

Specter, R.M. (1984), "Effects of Pore Structure on Mixing in Stable, Single-Phase Miscible Displacements," M.Sc. Thesis, New Mexico Institute of Mining and Technology, Socorro, NM, July.

Spence, A.P. and Watkins, R.W. (1980), "The Effect of Microscopic Core Heterogeneity on Miscible Flood Residual Oil Saturation," paper SPE 9229 presented at 55th Annual Fall Meeting, Dallas, September 21-24.

Stalkup, F.I., Jr. (1983), Miscible Displacement, Monograph Series, SPE, New York, 8.

Wardlaw, N.C. (1980), "The Effects of Pore Structure on Displacement Efficiency in Reservoir Rocks and in Glass Micromodels," paper SPE 8843 presented at First SPE/DOE Symposium on Enhanced Oil Recovery, Tulsa, April 20-23.

Wardlaw, N.C. and Cassan, J.P. (1978), "Estimation of Recovery Efficiency by Visual Observation of Pore Systems in Reservoir Rocks," Can. Pet. Geol. Bull. 26, 572.

Yellig, W.F. and Baker, L.E. (1980), "Factors Affecting Miscible Flooding Dispersion Coefficients," presented at the 31st Annual Technical Meeting of the Petroleum Society of CIM in Calgary, May 25-28.

Zapata, V.J. and Lake, L.W. (1981), "A Theoretical Analysis of Viscous Crossflow," paper SPE 10111 presented at the 56th Annual Fall Technical Conference and Exhibition, San Antonio, Oct. 5-7.

APPENDIX

EXIT BOUNDARY CONDITION AND FINITE DIFFERENCE SOLUTIONS TO THE CONVECTION-DISPERSION EQUATION.

The exit boundary condition, BCII,

$$C(\xi \to \infty, \tau) = 0 \tag{4c}$$

$$C^* = C(1, \tau) - \frac{1}{Pe}\frac{\partial C}{\partial \xi}\bigg|_{\xi = 1} \tag{4d}$$

for flow in a semi-infinite porous medium is commonly used for analytical solutions to the convection-dispersion equation (1). In such application, the problem is solved for a semi-infinite domain, and the flux at a given length is evaluated from that solution as the sum of convection and dispersion contributions (Eq. 4d). Problems arise in numerical solutions using this boundary condition because the numerical solutions are obtained for a finite domain (King, 1985). For instance, we used a forward time, backward first space, centered second space finite difference form of the material balance equation

$$\frac{C_i^{k+1} - C_i^k}{\Delta\tau} + \frac{C_i^k - C_{i-1}^k}{\Delta\xi} - \frac{1}{Pe\Delta\xi}\left[\frac{C_{i+1}^k - C_i^k}{\Delta\xi} - \frac{C_i^k - C_{i-1}^k}{\Delta\xi}\right] = 0 \tag{A1}$$

where $i = 1, 2, \ldots, n$. The space index is i and the time level is indicated by k. The inlet grid block corresponds to $i = 1$ and the outlet block to $i = n$. As noted above, the exit boundary condition at $C = \infty$ (4c) cannot be imposed with this form. Instead we used a linear extrapolation to estimate the value of $\partial C/\Delta\xi$ at the outlet. In other words the composition of fluid leaving the last grid block was assumed to be the result of convection and dispersion (Eq. 4d), with the dispersion contribution represented as

$$\frac{1}{Pe}\left[\frac{C_n^k - C_{n-1}^k}{\Delta\xi}\right] \tag{A2}$$

The effect of this extrapolation is to impose an exit boundary condition of the form

$$\frac{\partial^2 C}{\partial \xi 2}\bigg|_{\xi = 1} = 0 \tag{A3}$$

Thus, for the last grid block, the dispersive fluxes at the upstream-downstream outlet block faces are the same.

WORKSHOP SESSION REPORT

Larry W. Lake

Department of Petroleum Engineering
University of Texas at Austin
Austin, Texas

Herbert B. Carroll, Jr.

NIPER
P.O. Box 2128
Bartlesville, Oklahoma

I. MOTIVATION

The idea of a workshop germinated at the very first meeting of the steering committee. It was motivated by two desires:

1) Do something different from the norm. Modern researchers are deluged with many technical meetings, which, though they may be quite different in content, are often identical in the methods of information exchange. Workshops are fairly common in non-technical meetings, and we wanted to see if this method would work in technical meetings, also.

2) Involve more people in technical discussions. In an average technical conference, less than 5 percent of the attendees contribute to the question and answer periods even though session chairmen literally plead for participation. Moreover, the proceedings of a conference make far more of an impression to someone who feels he has contributed in some way.

II. PURPOSE

The main purpose of this workshop is to stimulate all the attendees to participate in the workshop. A second purpose is to expose the state of the art in reservoir characterization through discussion of several common issues. The final purpose is to direct many of the new ideas being presented in the workshop to these same issues.

III. PROCEDURE

The attendees were divided into four roughly equal groups, each pre-selected for a certain emphasis. Group one was about 60 percent engineers; group two, 60 percent geologists; group three, 60 percent stochastic modelers; and group four about 60 percent engineers.

Each group was given two questions, one of which, the first, was the same for all groups. The other question was tailored to the individual group to provoke discussion and even controversy. The groups were to prepare an oral response to the questions that was to be a consensus, but, nevertheless, brief. Each group was led by a member of the steering committee and one of the keynote speakers. Upon completion of this task, an appointed leader of the group was to report back the findings to the general assembly. The following summarizes the results of the committee reports. This paper concludes with a summary of the comments on the discussion session based on attendee evaluations.

IV. ALL GROUP QUESTION

"One of the main difficulties with reservoir characterization as a scientific discipline is the inability to validate procedures. There are no rigorous analytical solutions or models, and, even if there were, few of us would be able to interpret the results of an experiment unambiguously.

"Imagine that you have unlimited time and resources. Outline a field experiment that could be done so that you are sure that your characterization procedure did or did not work. Absoluteness is difficult to come by outside of ecclesiastical circles, so you should be as sure of the results of your experiment as you would be from a relatively simple core flood. You are not limited to any process or reservoir type, although these could be specified in your experiment.

"Specifically, you should say what the process is, what things you would measure beforehand, what data you would collect afterward, and how you would interpret it."

A. Group One Response

Since no one characterization procedure is valid, we emphasize the value of redundance in the procedure. Specific steps in the validation would be:

1) Fully define the use of the characterization. This means that the process, design and implementation should be selected with the validation goal in mind.
2) Employ several of the standard characterization techniques beforehand. A sufficient amount of data should be collected to develop a reservoir model.
3) Plan on interactive verification. Two things are possible here: Use one or more pilot tests before the large test, and deliberately sample redundant data.
4) Anticipate in-process verification. Sample procedures that could be done here include tracer tests, monitor wells (gamma-ray emitters), injection tests, production histories and pressure transient tests.
5) Budget for post-mortem verification.

B. Group Two Response

The group agreed that the best process to attempt a validation experiment would be a carbon dioxide flood. The procedure would begin by selecting a geologically easy reservoir. From this reservoir would be extracted a data set that is specifically designed to aid the description. The data set should have a degree of redundancy so that information may be checked both before and after the process is completed. This would provide one measure of validation: checking the consistency of geological modeling. Engineering validation would come from performing flow experiments after project completion.

C. Group Three Response

Validation would be from developing a model for multiphase fluid displacements in a pattern, near-surface formation. The group would validate the model by a fluid displacement experiment and physical sampling.

The data required for this would be extensive coring, logging and pressure testing from the candidate formation, coupled with statistical characterization from analog outcrops. Prior to the test the group suggests constructing predictions of the displacement behavior using a complete set and subsets of the available data. The prediction could also be used to predict the results of detailed sampling. The data to be matched consist of the passage of tracers and the change of saturation at intermediate wells. These data should be matched over as much of the pattern as possible. The final step would be to physically sample (including digging up) the formation to verify the description of variability.

D. Group Four Response

This group said that given "infinite" time and resources they would mine the reservoir. However, with a little less than this available they propose the following:

1) Select a reservoir with a realistic degree of heterogeneity. That the heterogeneity would be realistic would come from studying outcrops or shallow excavations (mine faces, road cuts, for example) which are analogous to the candidate reservoir.
2) Build a prototype reservoir model which includes the desired depositional features. Simulate the prototype with a fine grid simulator.
3) Sample the candidate reservoir and compute averages.
4) Check the data collected in step 3 and the derived statistics against the prototype. This will provide a degree of geological validation.
5) Run the simulator, with practical resolution, on a flow problem.
6) Check the simulator prediction against an actual flow experiment in the prototype. Repeat the simulation with progressively increased or decreased degrees of heterogeneity until a match is achieved.

V. GROUP ONE QUESTION

"When I first started working I was assigned to study a field in which there was an enhanced oil recovery project going on. Buried in the mountain of data that was given to me was several pages of "routine" core analysis (permeability, porosity, densities and saturations at one-foot intervals).

Inasmuch as the youngest of these data were two decades old, there were no physical cores available.

"I am sure you've had a similar experience. It seems there should be something to do with this data, but it's not exactly clear what.

"How can this data be useful in reservoir characterization? Discuss exactly what can be learned, how it can be calculated, and how it is to be applied. Do not consider the accuracy of the measurements to be at issue; however, you can suggest what additional information you could take now to make this data more useful."

Response: The group agreed that old data could be of great value. However, its utility could be greatly enhanced with other information--logging and drilling data, a geologic model, and even production history--available. Without this additional information, the data can largely be treated in a statistical fashion: correlations (for example, permeability-porosity), averages, measures of heterogeneity (Dykstra-Parsons coefficient), and analysis of structure. The group also remarked that old data can be used as a basis for requesting additional data.

VI. GROUP TWO QUESTION

"It's been said that the number of interpretations of a reservoir is greater than or equal to the number of geologists making the interpretation. Is this really true? Is geological interpretation subjective or is this all the grumbing of the engineering staff.

"Decide within your group whether the above statement has any accuracy. Give specific cases where the interpretation was observer-dependent and the resulting change(s) this caused. Give as many factors as you can think of that would cause an interpretation to be ambiguous and enumerate how these uncertainties impact on subsequent strategies. Give additional information which might resolve the ambiguities."

Response: The group agreed that there is a degree of inconsistency in geological interpretations. However, this is usually prompted by one of two reasons: 1) Uncertainty about whether the characterization purpose is primarily for geology or whether it is for engineering, and 2) the fragmentary nature of the available data, particularly in the early stages of a project.

The group suggested that a means to test the reliability of geologic models would be to use different teams, each with identical data sets, to make the same description. Each team would be given new data to check their procedures at realistic intervals, but the measure of consistency would be to compare the models produced whan all data is in.

VII. GROUP THREE QUESTION

"Several of the papers in this conference have dealt with stochastic parameter assignments and descriptions. This is a relatively new idea in the petroleum literature which has been dominated by deterministic assignments.

"What are the relative merits of stochastic and deterministic assignments? Discuss the advantages and disadvantages of each, how both would confront an extremely complex mechanistic process description (for example, in situ combustion or micellar/polymer flooding), and how the predictions of each can and should be used in planning. Most particularly, discuss how geology can be used in both areas. Be as specific as possible."

Response: Stochastic parameter assignment in simulation has the following advantages:

1) It allows the structure of regionalized variables to be accounted for.
2) It is the best current application for geostatistics.
3) It is ideally suited for planning and screening because it will predict a range of possible outcomes and associate a probability to each. This will be helpful in risk analysis.
Its disadvantages are that under the current state of the art it is not advanced enough for more than parameter assignments in simple miscible floods.

Tying stochastic modeling to geology remains one of the challenges of current usage, both for statisticians and geologists. The best way to relate the two would be to develop a catalog of statistical characteristics for generic depositional environments. This would amount to a catalog of depositional and, perhaps, diagenetic features.

VIII. GROUP FOUR QUESTION

"Interpreting data from a single well is difficult enough, but infering reservoir characteristics between wells with any degree of confidence is well nigh impossible.

"Basically, there are four ways to develop interwell properties: geology, seismic tests, well tests and tracers.

Discuss, through specific examples, how each of the above has been used in this connection. Is there a general procedure for any of these that will guarantee a successful description regardless of the reservoir type? Many of the measurements are aggregates, that is, they sample a comparatively large volume of the reservoir. Discuss how these measurements may be used to reconcile a large number of point measurements. If two techniques give different results, how will you determine which is correct? If the tracer data indicate a definitive description, do we really need the others? (If it looks like a duck, can we treat it as a duck even if it's really not a duck?)"

Response: The group stated that high angle or horizontal drilling is probably the solution for interwell properties. However, at present they offer the following suggestions:

1) Use well tests in comparison with core permeabilities averaged in various ways to derive a basic configuration of the reservoir between wells. This procedure will indicate a degree of homogeneity and layering which may be adequate for many purposes.
2) To reduce the gap in scale between plug data and well tests, one should design cylindrical models based on core, log and geological/stochastic/outcrop analog data and test these against well test results.
3) Compare the reservoir with analog prototypes of similar facies. For this to be most effective, one should have a catalogue of quantitative properties of the expected heterogeneities organized and classified for that facies.
4) Seismic data can be useful in selected cases. Consider using seismic, three-dimensional and variable offset seismic profiles.
5) Tracer tests are of great value because they reflect the actual flow processes relevant to oil production. However, one should relate the results of a proper geological configuration (fractures or high permeability streaks).
6) Early recognition of production problems or anomolies is very important because it allows for better data collection and planning.

The group also discussed several topics related to these issues. One should investigate the accuracy of various data to judge implications of possible errors on the results. The mass of data collected should be interpreted by an interdisciplinary team which has available to it an efficient means of data presentation. Finally, some indication of the importance (sensitivity) of interwell assignments is desireable.

IX. SUMMARY

The comments extracted from the conference evaluation forms generally reflected favorably upon the discussion session, particularly in meeting the stated goal of increasing attendee participation. Most felt that the session was less informative than the ordinary proceedings but was still useful, nevertheless.

Conference evaluations pointed out two general problems. The size of the discussion sessions (about 30 per group) was too large. This made it more likely that each group would be dominated by one or two strong personalities, a situation not unlike the general meeting. Also, the questions themselves were too open-ended, particularly given the time available for their completion. This was exclusively the result of our inexperience in such matters; we actually feared that the questions would be trivial or too easy.

Finally, the reader should appreciate that group responses necessarily lack some of the directness of individual responses. This, plus the shortness of time, accounts for what seems to us to be a certain vagueness and repetition in some of the responses. After all, a camel is a horse designed by a committee. We do feel, however, that the discussion session proceedings brought forth several interesting ideas: data redundancy, statistical/geological depositional catalogues, and the idea of risk analysis from stochastic assignments.

Index

R